Abbreviations of the Chemical Elements

Ac	actinium	H	hydrogen	Pr	praseodymium
Ag	silver	He	helium	Pt	platinum
Al	aluminum	Hf	hafnium	Pu	plutonium
Am	americium	Hg	mercury	Ra	radium
Ar	argon	Ho	holmium	Rb	rubidium
As	arsenic	In	indium	Re	rhenium
At	astatine	Ir	iridium	Rh	rhodium
Au	gold	K	potassium	Rn	radon
B	boron	Kr	krypton	Ru	ruthenium
Ba	barium	La	lanthanum	S	sulfur
Be	beryllium	Li	lithium	Sb	antimony
Bi	bismuth	Lw	lawrencium	Sc	scandium
Bk	berkelium	Lu	lutetium	Se	selenium
C	carbon	Md	mendelevium	Si	silicon
Ca	calcium	Mg	magnesium	Sm	samarium
Cd	cadmium	Mn	manganese	Sn	tin
Ce	cerium	Mo	molybdenum	Sr	strontium
Cm	curium	N	nitrogen	Ta	tantalum
Co	cobalt	Na	sodium	Tb	terbium
Cr	chromium	Nb	niobium	Tc	technetium
Cs	cesium	Nd	neodymium	Te	tellurium
Cu	copper	Ne	neon	Th	thorium
Dy	dysprosium	Ni	nickel	Ti	titanium
Er	erbium	No	nobelium	Tl	thallium
Es	einsteinium	Np	neptunium	Tm	thulium
Eu	europium	O	oxygen	U	uranium
F	fluorine	Os	osmium	V	vanadium
Fe	iron	P	phosphorus	W	tungsten
Fm	fermium	Pa	protactinium	Xe	xenon
Fr	francium	Pb	lead	Y	yttrium
Ga	gallium	Pd	palladium	Yb	ytterbium
Gd	gadolinium	Pm	promethium	Zn	zinc
Ge	germanium	Po	polonium	Zr	zirconium

STRUCTURE–PROPERTY RELATIONS IN NONFERROUS METALS

STRUCTURE–PROPERTY RELATIONS IN NONFERROUS METALS

Alan M. Russell
Department of Materials Science and Engineering
Iowa State University

Kok Loong Lee
Corus Group

A JOHN WILEY & SONS, INC., PUBLICATION

Copyright © 2005 by John Wiley & Sons, Inc. All rights reserved.

Published by John Wiley & Sons, Inc., Hoboken, New Jersey.
Published simultaneously in Canada.

No part of this publication may be reproduced, stored in a retrieval system, or transmitted in any form or by any means, electronic, mechanical, photocopying, recording, scanning, or otherwise, except as permitted under Section 107 or 108 of the 1976 United States Copyright Act, without either the prior written permission of the Publisher, or authorization through payment of the appropriate per-copy fee to the Copyright Clearance Center, Inc., 222 Rosewood Drive, Danvers, MA 01923, 978-750-8400, fax 978-646-8600, or on the web at www.copyright.com. Requests to the Publisher for permission should be addressed to the Permissions Department, John Wiley & Sons, Inc., 111 River Street, Hoboken, NJ 07030, (201) 748-6011, fax (201) 748-6008.

Limit of Liability/Disclaimer of Warranty: While the publisher and author have used their best efforts in preparing this book, they make no representations or warranties with respect to the accuracy or completeness of the contents of this book and specifically disclaim any implied warranties of merchantability or fitness for a particular purpose. No warranty may be created or extended by sales representatives or written sales materials. The advice and strategies contained herein may not be suitable for your situation. You should consult with a professional where appropriate. Neither the publisher nor author shall be liable for any loss of profit or any other commercial damages, including but not limited to special, incidental, consequential, or other damages.

For general information on our other products and services please contact our Customer Care Department within the U.S. at 877-762-2974, outside the U.S. at 317-572-3993 or fax 317-572-4002.

Wiley also publishes its books in a variety of electronic formats. Some content that appears in print, however, may not be available in electronic format.

Library of Congress Cataloging-in-Publication Data:

Russell, Alan M., 1950–
   Structure–property relations in nonferrous metals / Alan M. Russell, Kok Loong Lee.
     p. cm.
   Includes bibliographical references and index.
   ISBN-13 978-0-471-64952-6 (cloth)
   ISBN-10 0-471-64952-X (cloth)
   1. Nonferrous metals—Textbooks. I. Lee, Kok Loong, 1976– II. Title.

TA479.3.R84 2005
620.1'8—dc22
                                                                                         2004054807

Printed in the United States of America.

10 9 8 7 6 5 4 3 2 1

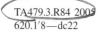

To Laurel, Jeffrey, Kristen, and Mark

CONTENTS

Preface xiii

PART ONE

1 Crystal and Electronic Structure of Metals 1

    1.1 Introduction 1
    1.2 Crystal Structures of the Metallic Elements 1
    1.3 Exceptions to the Rule of the Metallic Bond 4
    1.4 Effects of High Pressure on Crystal Structure 8
    1.5 Effect of Electronic Structure on Crystal Structure 9
    1.6 Periodic Trends in Material Properties 13

2 Defects and Their Effects on Materials Properties 18

    2.1 Introduction 18
    2.2 Point Defects 18
    2.3 Line Defects (Dislocations) 20
    2.4 Planar Defects 24
    2.5 Volume Defects 27

3 Strengthening Mechanisms 28

    3.1 Introduction 28
    3.2 Grain Boundary Strengthening 28
    3.3 Strain Hardening 32
    3.4 Solid-Solution Hardening 34
    3.5 Precipitation Hardening (or Age Hardening) 35

4 Dislocations 38

    4.1 Introduction 38
    4.2 Forces on Dislocations 38
    4.3 Forces Between Dislocations 40
    4.4 Multiplication of Dislocations 42
    4.5 Partial Dislocations 44
    4.6 Slip Systems in Various Crystals 48
    4.7 Strain Hardening of Single Crystals 52
    4.8 Thermally Activated Dislocation Motion 55

4.9 Interactions of Solute Atoms with Dislocations 58
4.10 Dislocation Pile-ups 59

5 Fracture and Fatigue — 61

5.1 Introduction 61
5.2 Fundamentals of Fracture 61
5.3 Metal Fatigue 69

6 Strain Rate Effects and Creep — 76

6.1 Introduction 76
6.2 Yield Point Phenomenon and Strain Aging 76
6.3 Ultrarapid Strain Phenomena 76
6.4 Creep 77
6.5 Deformation Mechanism Maps 82
6.6 Superplasticity 83

7 Deviations from Classic Crystallinity — 85

7.1 Introduction 85
7.2 Nanocrystalline Metals 85
7.3 Amorphous Metals 90
7.4 Quasicrystalline Metals 96
7.5 Radiation Damage in Metals 98

8 Processing Methods — 102

8.1 Introduction 102
8.2 Casting 102
8.3 Powder Metallurgy 109
8.4 Forming and Shaping 112
8.5 Material Removal 118
8.6 Joining 122
8.7 Surface Modification 127

9 Composites — 130

9.1 Introduction 130
9.2 Composite Materials 130
9.3 Metal Matrix Composites 130
9.4 Manufacturing MMCs 132
9.5 Mechanical Properties and Strengthening Mechanisms in MMCs 134
9.6 Internal Stresses 137
9.7 Stress Relaxation 139
9.8 High-Temperature Behavior of MMCs 142

PART TWO

10 Li, Na, K, Rb, Cs, and Fr — 146

  10.1 Overview   146
  10.2 History, Properties, and Applications   146
  10.3 Sources   154
  10.4 Structure–Property Relations   154

11 Be, Mg, Ca, Sr, Ba, and Ra — 156

  11.1 Overview   156
  11.2 History and Properties   156
  11.3 Beryllium   156
  11.4 Magnesium   166
  11.5 Heavier Alkaline Metals   177

12 Ti, Zr, and Hf — 179

  12.1 Overview   179
  12.2 Titanium   179
  12.3 Zirconium   197
  12.4 Hafnium   204

13 V, Nb, and Ta — 205

  13.1 Overview   205
  13.2 History and Properties   205
  13.3 Vanadium   205
  13.4 Niobium   210
  13.5 Tantalum   216

14 Cr, Mo, and W — 221

  14.1 Overview   221
  14.2 Chromium   221
  14.3 Molybdenum   227
  14.4 Tungsten   233

15 Mn, Tc, and Re — 243

  15.1 Overview   243
  15.2 History and Properties   243
  15.3 Manganese   244
  15.4 Technetium   252
  15.5 Rhenium   253

16 Co and Ni — 259

16.1 Overview 259
16.2 Cobalt 259
16.3 Nickel 270

17 Ru, Rh, Pd, Os, Ir, and Pt — 290

17.1 Overview 290
17.2 History, Properties, and Applications 290
17.3 Toxicity 298
17.4 Sources 299
17.5 Structure–Property Relations 299

18 Cu, Ag, and Au — 302

18.1 Overview 302
18.2 Copper 302
18.3 Silver 321
18.4 Gold 329

19 Zn, Cd, and Hg — 337

19.1 Overview 337
19.2 Zinc 337
19.3 Cadmium 349
19.4 Mercury 353

20 Al, Ga, In, and Tl — 358

20.1 Overview 358
20.2 Aluminum 358
20.3 Gallium 387
20.4 Indium 389
20.5 Thallium 390

21 Si, Ge, Sn, and Pb — 392

21.1 Overview 392
21.2 Silicon 392
21.3 Germanium 399
21.4 Tin 402
21.5 Lead 410

22 As, Sb, Bi, and Po — 419

22.1 Overview 419
22.2 Arsenic 419
22.3 Antimony 423
22.4 Bismuth 427
22.5 Polonium 430

23 Sc, Y, La, Ce, Pr, Nd, Pm, Sm, Eu, Gd, Tb, Dy, Ho, Er, Tm, Yb, and Lu — 433

23.1 Overview   433
23.2 History   433
23.3 Physical Properties   435
23.4 Applications   440
23.5 Sources   446
23.6 Structure–Property Relations   446

24 Ac, Th, Pa, U, Np, Pu, Am, Cm, Bk, Cf, Es, Fm, Md, No, and Lr — 451

24.1 Overview   451
24.2 History and Properties   451
24.3 Thorium   454
24.4 Uranium   456
24.5 Plutonium   464
24.6 Less Common Actinide Metals   470

25 Intermetallic Compounds — 474

25.1 Overview   474
25.2 Bonding and General Properties   474
25.3 Mechanical Properties   475
25.4 Oxidation Resistance   475
25.5 Nonstructural Uses of Intermetallics   477
25.6 Stoichiometric Intermetallics   478
25.7 Nonstoichiometric Intermetallics   483
25.8 Intermetallics with Third-Element Additions   485
25.9 Environmental Embrittlement   487

Index — 489

ADDITIONAL INFORMATION

References, Appendixes, Problem Sets, and Metal Production Figures are available at *ftp://ftp.wiley.com/public/sci_tech_med/nonferrous*

PREFACE

We have written this book as a textbook suitable for junior/senior-level undergraduate materials science students, but it is also intended to serve as a reference for practicing engineers. The book describes the relationships between the atomic structure, crystal structure, and microstructure of metals and their physical behavior (e.g., strength, ductility, electrical conductivity, corrosion, etc.). In Part One we present basic principles of the atomic and crystal structure, defects, and processing of metals. In Part Two we describe the properties and uses of all the metallic elements except iron. We are grateful for the reviews and helpful suggestions of Karl Gschneidner, Jr., L. Scott Chumbley, Alan Renken, Bruce Cook, Fran Laabs, Joel Harringa, and Timothy Ellis, who graciously contributed their time to improving both parts of the book.

The periodic table contains 83 metallic elements. We have observed that most metals texts focus primarily on iron, the dominant commercial metal. The 82 nonferrous metals are often lumped into one or two chapters and treated as "aluminum, copper, plus scattered comments about all those other metals." Many excellent books exist that are devoted entirely to one metal, but the reader seeking a broader, more comprehensive understanding of all nonferrous metals has heretofore found no single volume that addresses the entire spectrum of nonferrous metal properties and engineering applications. With this book we have striven to meet that need in a readily comprehensible format that emphasizes the needs of the engineer. References, appendixes, problem sets, and information about leading metal-producing nations and their annual production tonnages are available at *ftp://ftp.wiley.com/public/sci_tech_med/nonferrous*.

We have found that students learn most efficiently when specific, real-world examples can be given to support theory and general principles. The university climate biases authors to focus on graduate education, and the result is a wealth of "scholarly" books containing lengthy mathematical derivations and extended narratives on modeling. These books are inherently ill-suited to convey the excitement of engineering practice to undergraduate students. Even at their best, such books often do little to explain *why* the student should care about these fundamental principles. The result is that talented students may lose motivation and opt out of materials science for other careers. This book devotes a substantial portion of its content to examples of how the fundamental concepts and equations apply to real engineering practice. It shares with the reader the odd, surprising, and exciting ways that the structures of metals affect their behavior in modern engineering systems.

Two special features of the book are (1) short examples of the properties and uses of specific metals, and (2) features on structure–property relations that expand on a particular concept, explaining how it applies to the metal in question. These often describe little-known, unusual, or exotic features of the metallic elements, and it is our hope that they will maintain the interest of the student and motivate him or her to persevere through the sometimes arduous but unfailingly rewarding study of metals.

ALAN RUSSELL
K. L. LEE

April 2004

1 Crystal and Electronic Structure of Metals

> To do exactly as your neighbors do is the only sensible rule.
> —Emily Post, *Etiquette*, 1922

1.1 INTRODUCTION

One of the most remarkable qualities of metals is their precise, repeating atomic packing sequences. Periodic crystalline order is the equilibrium structure of all solid metals, and heroic measures in alloying or ultrarapid cooling (Sec. 7.3) are required to suppress it. Metals crystal structures are a dominant factor in determining their mechanical properties, and crystal structures play an important role in magnetic, electrical, and thermal properties as well. Thus, an understanding of the crystal structures of metals should precede any attempt to understand their properties.

1.2 CRYSTAL STRUCTURES OF THE METALLIC ELEMENTS

If atoms are assumed to behave as impenetrable solid spheres, many different atom-packing arrangements are geometrically possible. If the atoms in the model are permitted to deviate from hard spheres to assume ellipsoidal shapes that may encroach slightly on their neighboring atoms outer electron shells, the number of possible atom arrangements becomes infinite. Yet, despite the large number of possible crystal structures, more than 80% of all metallic elements (Fig. 1.1) have one of the four simple crystal structures shown in Fig. 1.2.

Three of the structures shown in Fig. 1.2 [all except body-centered cubic (BCC)] achieve the densest possible packing for equal-sized spheres (i.e., the volume inside the atom spheres occupies 74% of total crystal volume); and the BCC structure is nearly as efficient at filling space, with its 68% packing efficiency. When one considers the curve of bonding energy versus interatomic distance for two atoms (Fig. 1.3), it is clear that the bonding energy is greatest at interatomic spacings that correspond to atoms in the hard-sphere model touching each other. Note in Fig. 1.3 how the repulsive energy is essentially zero until the atoms are almost contiguous, then increases very steeply (typically about as r^{-8}!) as the outermost electron orbitals begin to overlap. These two factors combine to produce the greatest bonding energy when the atoms are closely packed; gaps between neighboring atoms are energetically disfavored. In addition, total metallic bonding energy is increased when each atom has the greatest possible number of nearest-neighbor atoms.

The hexagonal close-packed (HCP), face-centered cubic (FCC), and α-La hexagonal structures are nearly identical, differing only in the stacking sequence of their densest-packed planes

---

Structure–Property Relations in Nonferrous Metals, by Alan M. Russell and Kok Loong Lee
Copyright © 2005 John Wiley & Sons, Inc.

2 CRYSTAL AND ELECTRONIC STRUCTURE OF METALS

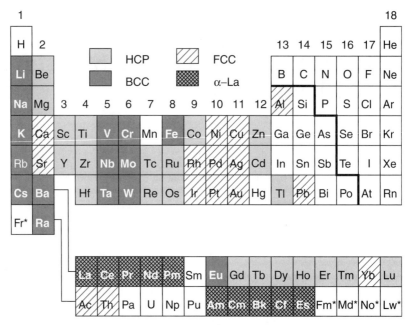

Fig. 1.1 Crystal structures of the metallic elements at 20°C and 1 atm pressure. (The asterisk denotes unknown crystal structure.)

(labeled *A*, *B*, and *C* in Fig. 1.2). Each atom in these three structures is surrounded by 12 nearest-neighbor atoms, the maximum number possible for equal-sized spheres. Since a BCC crystal is not a densest-packed structure, it might seem odd that it "outcompetes" the most densely packed structures to become the equilibrium structure of 15 metallic elements at room temperature. This is attributable to two effects. First, although each atom has only eight nearest neighbors in a BCC structure, the six second-nearest-neighbor atoms are closer in the BCC structure than in the FCC structure. Calculations indicate that BCC's second-nearest-neighbor bonds make a significant contribution to the total bonding energy of BCC metals. In addition, the greater entropy of the "looser" BCC structure gives it a stability advantage over the more tightly assembled FCC structure at high temperature. The BCC structure has unusually low resistance to vibration in the $\langle 110 \rangle$ direction, making lattice vibration in that mode especially easy. In the Gibbs free energy expression, ΔG is related to enthalpy (ΔH), absolute temperature (T), and entropy (ΔS) by

$$\Delta G = \Delta H - T \Delta S$$

The easier lattice vibration (greater ΔS factor) in BCC metals makes ΔG lower for BCC vis-à-vis FCC at higher temperatures. For this reason many metals have closest-packing structures at low temperature that transform to BCC at higher temperatures (Fig. 1.4). Half of all metallic elements possess one of the densest-packed structures at low temperature and transform to the slightly less densely packed BCC structure at higher temperature. The close-packed → BCC transformation usually occurs well above room temperature, but two of the low-melting alkali metals, Li and Na, are BCC at room temperature, with a transformation to a closest-packed structure (α-Sm, hR3, a variation of the α-La structure) at cryogenic temperatures.

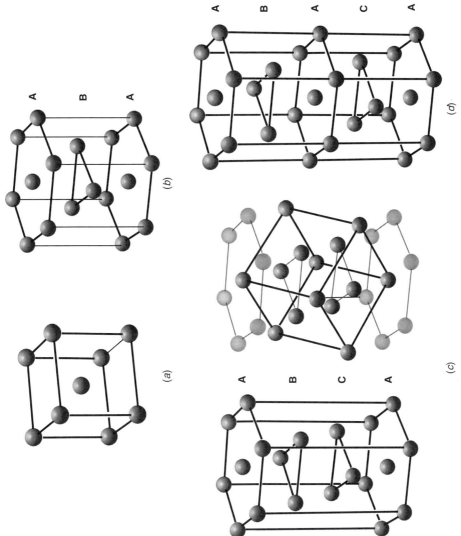

Fig. 1.2 Atomic packing sequences of (a) body-centered cubic; (b) hexagonal close-packed; (c) face-centered cubic (note that the face-centered cubic structure can be visualized in two seemingly different, but geometrically identical, depictions); (d) α-La hexagonal. Atoms are shown disproportionately small for clarity.

4 CRYSTAL AND ELECTRONIC STRUCTURE OF METALS

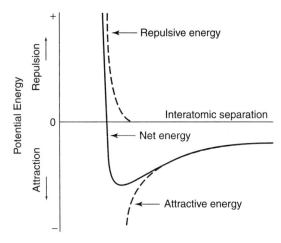

Fig. 1.3 Bonding energy versus interatomic spacing for two atoms. The solid curve is the sum of the attractive and repulsive energies. Although this plot represents the behavior of only two atoms, nearest-neighbor atoms in crystals behave similarly. The equilibrium interatomic spacing occurs at the minimum of the net bond energy curve.

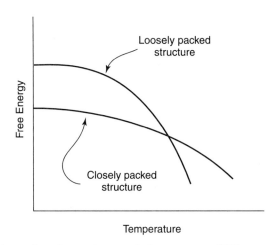

Fig. 1.4 Many metals transform from a closest-packed structure to a BCC structure at high temperature. The greater entropy (ΔS term) in the loosely packed BCC structure lowers ΔG more rapidly in BCC structures than in close-packed structures.

1.3 EXCEPTIONS TO THE RULE OF THE METALLIC BOND

Since stronger bonding occurs when metal atoms are in densely packed structures (HCP, FCC, α-La, or BCC), the key question is not why so few crystal structures predominate when so many are possible, but rather, why all metals don't crystallize in a closely packed structure? The answer to the latter question involves an intriguing shared bonding arrangement in some metals that is partially metallic and partially covalent.

Before embarking on a case-by-case narrative of metals that do not crystallize into HCP, FCC, α-La, or BCC structures, it is helpful to recall the nature of the metallic bond. Unlike ionic and covalent bonds, which have a specific electron exchange among a small number of atoms and a certain directional character, the metallic bond is a delocalized release of the bonding electrons into the crystal. For this reason, metallic bonds are not strongly directional and can exist only when substantial numbers of atoms are present. For example, vaporized Au forms diatomic Au_2 molecules, but these molecules are covalent, not metallic. When many Au_2 molecules condense into a solid crystal, the bonding becomes metallic.

The 20% of metallic elements that possess less densely packed crystal structures at 20°C deviate from the norm because their bonding is partly covalent. As Fig. 1.1 shows, the metals in the lower left region of the periodic table all possess one of the densely packed crystal structures typical of nearly pure metallic bonding. However, covalent tendencies appear as one moves closer to the nonmetals on the periodic table. For example, Be lies next to the nonmetal B. Be is an HCP metal, but its c/a ratio is unusually low at 1.57 (1.633 is ideal for close-packed spheres). This distortion of the Be atom and the unusually high elastic modulus of Be (296 GPa) result from a covalent component in its bonding (Sec. 11.3). Contributions from covalency are also seen in Zn and Cd, which are both HCP metals with c/a ratios greater than 1.85. This lowers their packing density to about 65%, considerably less than the 74% of the ideal HCP structure (Sec. 19.2.2.2).

Of course, some metals do more than show distortions in a densely packed structure. As one moves rightward across the periodic table, progressively greater numbers of metals have "odd" structures (i.e., not HCP, FCC, BCC, or α-La). Most metals bordering the nonmetals possess more complex structures with lower packing density because covalent effects play a large role in determining their crystal structures; the directionality of covalent bonding dictates fewer nearest neighbors than exist in densely packed crystals.

Mn is the first odd crystal structure encountered as one moves left to right across the main section of the periodic table. All Mn's neighboring elements in the periodic table possess simple, conventional crystal structures, but Mn has extraordinarily complex structures. These are attributed to magnetic effects (Sec. 15.3.1). The room-temperature phase, α-Mn (cI58) (Sidebar 1.1) is stable below 727°C. From 727 to 1100°C, the equilibrium β-Mn structure is also quite complex (cP20). At higher temperatures, lattice vibrations nullify magnetic effects and Mn transforms to "normal" FCC and BCC phases.

Aside from Mn, all the remaining transition elements display normal metal crystal structures except Hg. Hg is a liquid at room temperature, but when it finally solidifies at −39°C, it forms a rhombohedral crystal (hR1) with 61% packing efficiency. Hg is weakly bonded because it is difficult to ionize the filled subshells of Hg (Xe core $+ 4f^{14}5d^{10}6s^2$). Hg's ionization energy (1007 kJ/mol) is the highest of all metals, and those of its congener elements Zn and Cd are only slightly lower. Since the d shells are filled completely in these group 12 metals, only s and p electrons are available for bonding, making them substantially weaker and lower melting than the neighboring group 11 elements. Zn and Cd hybridize to form $4s^1 + 4p^1$ and $5s^1 + 5p^1$ outer electron structures, which provide two electrons for bonding. Hg does this less well (Sec. 19.4.2.1), making Hg the lowest melting of all the metallic elements.

The lanthanide metals display a fascinating array of polymorphic phases below room temperature and at high pressure, but all lanthanides possess densely packed crystal structures at 20°C. The α-Sm structure is a stacking variation of the α-La structure, but it is a closest-packed structure. The room-temperature crystal structures of the actinide metals, however, are not as simple. Ac and Th are FCC, and the higher Z-number actinides (Am to Es) all have a densest-packed α-La structure at room temperature. However, the elements Pa through Pu have $5f$ electron participation in bonding, which produces low-symmetry structures at 20°C: tetragonal (tI2) Pa, orthorhombic U (oC4) and Np (oP8), and monoclinic Pu. Even in Pu's simpler high-temperature FCC structures (δ-Pu and δ'-Pu), the coefficient of thermal expansion is negative, a clear manifestation of a covalent component to the bonding (Sec. 24.5.3.1).

For metals near the nonmetals on the right side of the periodic table, electronegativities are high, covalency becomes a major part of the bonding, and dense-packed crystal structures are seen only in Al (FCC), Pb (FCC), and Tl (HCP). Ga (oC8) possesses a weakly bonded structure (Fig. 1.5) that provides each Ga atom with one nearest neighbor (0.244 nm away) and three

SIDEBAR 1.1: PEARSON SYMBOL NOTATION

In an effort to catalog the thousands of possible crystal structures observed in elements and compounds, crystallographers have developed the Pearson notation for crystal structures. This simple combination of two letters and a number provides a description of the structure according to the following precepts:

- The first letter (lowercase) identifies the crystal family:

    a = triclinic

    m = monoclinic

    o = orthorhombic

    t = tetragonal

    h = hexagonal, trigonal (rhombohedral)

    c = cubic

- The second letter (uppercase) identifies the Bravais lattice type:

    P = primitive

    I = body-centered*

    F = all cell faces contain a face-centered atom

    C = side or base faces are face-centered

    R = rhombohedral

- The number indicates the total number of atoms per unit cell.

Examples: Cu is a face-centered cubic metal; its Pearson notation is cF4 (cubic, Face-centered on all cell faces, 4 atoms per unit cell). Mg is a hexagonal close-packed metal; its Pearson notation is hP2 (hexagonal, Primitive, 2 atoms per unit cell).

Although the Pearson notation does not always define a unique crystal structure (e.g., somewhat different structures can all be mC2), it provides a concise description of the general nature of the crystal. Table 1.1 correlates the Pearson notation with the Bravais lattices.

*English speakers may wonder why the letter "I" is used for body-centered atoms rather than "B." The "I" stands for *innenzentrierte*, German for "produced from the inside center."

additional pairs of next-nearest neighbors that are 0.270, 0.273, and 0.279 nm away. This odd structure somewhat resembles covalently bonded Ga_2 "molecules" within a metal lattice. When Ga melts (30°C), it loses this unusual structure and contracts 3.4% (Sec. 20.3.2). In's structure is body-centered tetragonal (tI2), although its distortion from the BCC structure is small.

The lighter group 14 elements (Si and Ge) crystallize in the diamond structure (cF8). In their ground state (s^2p^2), these elements have only two p electrons in an unfilled subshell, but they hybridize to a s^1p^3 configuration that provides four electrons from partially filled subshells, all of which participate in forming covalent bonds with the four nearest-neighbor atoms (Fig. 1.6). The energy required (95 kcal/mol) to promote one electron from a Si atom's outer s subshell to the outer p subshell is more than offset by the stronger bonding afforded by four electrons rather

TABLE 1.1 Pearson Symbols for the 14 Bravais Lattices

Pearson Symbol	Bravais Lattice
aP	Triclinic
mP	Simple monoclinic
mC	Base-centered monoclinic
oP	Simple orthorhombic
oC	Base-centered orthorhombic
oF	Face-centered orthorhombic
oI	Body-centered orthorhombic
tP	Simple tetragonal
tI	Body-centered tetragonal
hP	Hexagonal
hR	Rhombohedral
cP	Simple cubic
cF	Face-centered cubic
cI	Body-centered cubic

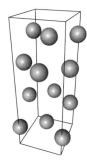

Fig. 1.5 oC8 structure of Ga. This layered structure provides one nearest neighbor for each atom and three pairs of next-nearest neighbors. Ga's packing efficiency is low (39%), and its density actually increases upon melting. Atoms are shown disproportionately small for clarity.

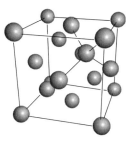

Fig. 1.6 cF8 diamond cubic structure of C (diamond), Si, Ge, and α-Sn. Covalent bonding predominates over metallic bonding in these tetravalent elements. Each atom has only four nearest neighbors, yielding a low packing efficiency (34%). The nearest neighbors of one atom are indicated by lines to the neighboring atoms. Atoms are shown disproportionately small for clarity.

than two. Si and Ge are classic metalloids, possessing a crystal structure determined by their predominantly covalent bonding but displaying somewhat metal-like optical and conductivity behavior (Sec. 21.2.2). At room temperature, β-Sn has a low density (53% packing efficiency) tI4 structure with $5s^25p^2$ electronic structure displaying both covalent and metallic bonding. At

8 CRYSTAL AND ELECTRONIC STRUCTURE OF METALS

13°C, β-Sn transforms to the even less densely packed α-Sn cF8 diamond cubic structure with s^1p^3 hybridized covalent bonding.

The group 15 metals As, Sb, and Bi all share an hR2 structure that allows each trivalent atom to bond covalently with three nearest neighbors in a sort of "puckered layer" geometry. The interatomic spacing within an As layer is 0.252 nm, and the spacing between layers is 0.312 nm. These elements still retain metal-like electrical conductivity, although the crystal structure is clearly dictated by the covalent bonding component (Fig. 22.1). Finally, Po, the only group 16 element with predominantly metallic character, has a simple cubic (cP1) structure with six nearest neighbors for each atom and a packing efficiency of 52% (Sec. 22.5.2).

Physical properties important to the engineer are strongly influenced by crystal structure. Perhaps the most important property related to crystal structure is the metal's ductility (or lack thereof). Densely packed structures usually allow free dislocation motion on one or more slip planes, permitting the metal to deform plastically without fracturing. Ductility is vital for easy formability and for fracture toughness, two properties that give metals a great advantage over ceramic materials for many engineering uses. Metals with an HCP, FCC, BCC, or α-La crystal structure are usually ductile at room temperature; most metals with less densely packed structures have little or no ductility until they approach their melting temperatures.

1.4 EFFECTS OF HIGH PRESSURE ON CRYSTAL STRUCTURE

The electrons of a metal form a "band structure" of electron energies. As Fig. 1.7a shows, the half-filled 3s band in Na provides the electrons for the metallic bond. Only a tiny amount of extra energy is needed to promote electrons in the 3s band to higher, previously unoccupied energy levels in the band. This allows Na to carry electric current easily, and it contributes to thermal conductivity as well. In Mg the structure is somewhat different (Fig. 1.7b), but the resulting physical properties are similar; the filled 3s band overlaps the empty 3p band, and the energy needed to promote a 3s electron into the empty 3p band is negligibly small, conferring classic metallic behavior on Mg. For the transition metals Sc through Ni, a partially filled 3d band overlaps the 4s band. This provides fairly easy access for electron promotion to unoccupied states, although complex interactions between these two bands reduce conductivity somewhat. The exceptionally high electrical conductivity of group 11 metals (Cu, Ag, Au) results from a half-filled s outer subshell and the lack of interaction with the filled d band (Sec. 18.1). Similar behaviors occur in all the other metals of the periodic table, although elements near the

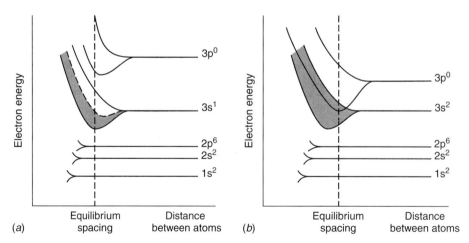

Fig. 1.7 Simplified electronic band structure depictions for crystals of (a) Na and (b) Mg.

nonmetals, such as Si and Ge, have a small gap between the filled lower-energy electron band and the empty higher-energy electron band. This sharply reduces electrical conductivity, because only a small fraction of the electrons can acquire enough energy from thermal excitation to jump the bandgap (Sec. 21.2.2). For most nonmetals, the bandgap is so large that thermal excitation cannot promote appreciable numbers of electrons; these materials lack mobile electrons and are electrical insulators.

Note in Fig. 1.7 that the $3s$ and $3p$ subshells form bands, but the noninteracting inner orbitals ($1s^2 2s^2 2p^6$) have discrete energy levels (rather than bands) at the equilibrium spacing between neighboring atoms. Of course, single atoms of Na and Mg vapor would have no bands; all the electron levels would be single values like the $1s$, $2s$, and $2p$ levels shown in Fig. 1.7. The Pauli exclusion principle requires that only two electrons may have the same energy in the crystal, so enormous numbers of extremely closely spaced but different energy levels exist, forming the energy bands.

In the Na case, the $3s$ band is only half full, so electrical conductivity is high. In the Mg case, conductivity is also high because the filled $3s$ band overlaps the empty $3p$ band. By applying pressure to a metal, the dashed vertical line marking the equilibrium spacing of atoms in Fig. 1.7b effectively moves leftward. This changes the overlaps and gaps between bands, altering the number and types of electrons available for bonding. Under high pressure, this effect can convert semiconducting metalloids into ordinary metals, and vice versa.

Si provides an example of how a metalloid element with lower packing density can be changed into simpler, more densely packed metallic structures at high pressure. At normal pressures, Si has predominantly covalent bonding, a coordination number of 4, and rather low electrical conductivity. However, increasing pressure drives Si through a series of transformations (Appendix A, *ftp://ftp.wiley.com/public/sci_tech_med/nonferrous*) that progressively increase the coordination number first to 6 (tI4), then 8 (hP1), and finally to 12 (cF4, FCC). The electrical conductivity increases by several orders of magnitude as pressure rises, and the final result is a "normal" FCC metal whose properties are much different from those of diamond cubic (cF8) Si. It should be noted that the pressures necessary to achieve these transformations are exceptionally high. One atmosphere is 0.1 MPa. Si's transformation to tI4 occurs at 12 GPa (120,000 atm), and the transformation to cF4 requires 78 GPa (780,000 atm). For comparison, the water pressure at the greatest ocean depth (>11,000 m) is "only" 0.1 GPa (1150 atm).

High pressure does not always cause transformation to simpler crystal structures. In some metals (e.g., Rb, Bi) pressures in the range 15 to 30 GPa actually convert comparatively simple crystal structures into phases with hundreds of atoms per unit cell. Intriguingly, some materials will retain nonequilibrium, high-pressure crystal structures after the applied pressure is removed. A familiar example of this behavior occurs in the commercial process of transforming graphite to diamond in high-pressure cells; the resulting diamond is metastable, but it persists indefinitely at ambient temperature and pressure. Research efforts in high-pressure materials science are producing many new materials, some of which survive the decompression from high pressure to ambient pressure, yielding ultrahard materials (e.g., cubic BN and c-BC_2N), optoelectronic semiconductors, and magnetic materials. The "workhorse" research apparatus for studying high-pressure phase transformations is the diamond-anvil cell, but most units can compress only milligrams of material. However, other technologies are capable of producing pressures almost as high as the diamond-anvil cell on quantities large enough to be technologically useful (in the range of cubic centimeters). High-pressure phases that can survive the decompression to ambient pressure promise to greatly expand the material choices available to design engineers in the decades to come (Sidebar 1.2).

1.5 EFFECT OF ELECTRONIC STRUCTURE ON CRYSTAL STRUCTURE

The *metallic bond* is often described as a gas or cloud of electrons adrift in the lattice of their parent ions. This simple description is useful in providing an easily visualized model consistent

> **SIDEBAR 1.2: METALLIC HYDROGEN AND JUPITER'S CORE**
>
> In 1935 theorists predicted that H would transform to a metal at ultrahigh pressures. Some calculations even suggest that once formed, metallic H might be metastable at 1 atm of pressure and that metallic H might be an electrical superconductor at room temperature. Unfortunately, until recently no laboratory system existed to test this prediction. The definitive test for transformation to a metallic state is a large increase in electrical conductivity, but diamond-anvil cells cannot test conductivity while applying such high pressures.
>
> In 1996 a team of scientists at Lawrence Livermore Laboratory subjected liquid H to impact from a metal plate moving 7 km/s. During a 200-ns period after initial impact, the electrical conductivity of the H increased by a factor of 10^4 as the pressure in the H exceeded 140 GPa (1.4 million atm). This was the first direct evidence of successful metallization of H on Earth. Although the experiment was not designed to produce persistent samples of metallic H, it is the first step toward developing production processes. If metastable metallic H could be produced, it might be an excellent ultralightweight structural material. The engineering benefits of a room-temperature superconductor are obvious. If it is possible to trigger a sudden transformation of metallic H to normal H, a powerful explosive or potential rocket fuel might result.
>
> Metallic hydrogen production remains a daunting materials engineering challenge. However, astronomers have strong evidence that the interiors of the massive gas planets Jupiter and Saturn are mostly metallic H (Fig. 1.8). Jupiter's intense magnetic field matches theoretical predictions of highly conductive H beneath the planet's outer atmosphere. Recent discoveries of large numbers of massive planets orbiting other stars suggest that metallic H planets may be quite common throughout the universe. If so, metallic H may be the most exotic terrestrial metal but the most common metal in the universe as a whole.

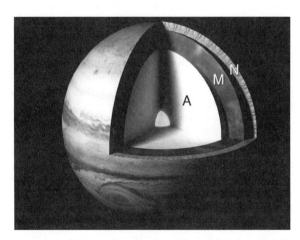

Fig. 1.8 Pictorial section view of the interior of Jupiter. Most of the planet is believed to be metallic H containing smaller amounts of other elements. The regions are labeled as N (normal molecular hydrogen), M (metallic molecular hydrogen), and A (metallic atomic hydrogen). Mixed rock and ice comprise the small central core.

with ordinary metals' high electrical conductivity and nondirectional bonding. However, it does not provide a quantitatively accurate description of metallic heat capacity and other physical properties. Throughout most of the twentieth century, physicists and materials scientists worked to refine the models of electron behavior in metals to describe their crystal structures and mechanical, electrical, and magnetic properties more accurately.

In principle it should be possible to use the Schrödinger equation from quantum mechanics to calculate the wave function of a free electron gas in a metallic crystal. This can be done if it is assumed that the electrons are not affected by the atoms in the metal lattice. Since the Pauli exclusion principle requires that no more than two electrons may have the same energy within the crystal, a population distribution of electrons occupying extremely closely spaced (but separate) energy levels results, as plotted in Fig. 1.9. The curve is a plot of the density-of-states function:

$$n(E) = \left(\frac{V}{2\pi^2}\right) E^{1/2}$$

where $n(E)$ is the density of states, V the crystal volume, and E the electron kinetic energy. At 0 K a sharp cutoff of occupied states occurs at the Fermi energy, E_F; above 0 K some electrons in states just below E_F will be thermally excited to occupy formerly empty states above E_F. This "rounds off" the sharp step at E_F, but at room temperature, the effect is rather small, and the overall appearance of the plot in Fig. 1.9 changes little.

The plot in Fig. 1.9 is based on the simplifying assumption that the ion cores of the atoms in the metal lattice do not affect the electron cloud. Not surprisingly, this assumption introduces error into the prediction of the electron behavior within a metal. Since the first Schrödinger equation calculations for Na metal (a particularly simple case) were published in 1933, a continuous progression of improvements and refinements in these computations has occurred. It is much more difficult to perform these computations for polyvalent metals and transition metals because multiple interactions must be accounted for, which greatly increases the complexity of the computation. Of course, most of the periodic table is comprised of polyvalent metals and transition metals. Indeed, this proved to be one of the great intellectual challenges of the twentieth century, and the process is still a work in progress. Even the best computations are still only estimates of electron behavior in metals, but the accuracy of the estimates has become quite impressive in recent years.

Figure 1.10 shows the result of density-of-states computations for two nontransition metals (i.e., all bonding electrons are s and p type, no d electrons). Al provides an example of a metal that conforms closely to the free-electron-model density of states in Fig. 1.9. By contrast, Be deviates sharply from that case and is nearly a semiconductor because its density-of-states plot approaches zero near the Fermi energy. A gap in the plot at E_F would greatly decrease electrical

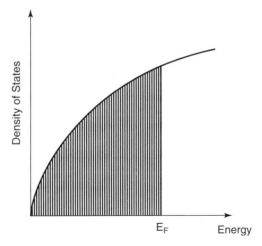

Fig. 1.9 Density of states (i.e., the energy distribution) of an idealized free electron cloud in a metal crystal at 0 K (solid line). E_F is the Fermi energy, the highest occupied electron energy level in the crystal at 0 K.

12 CRYSTAL AND ELECTRONIC STRUCTURE OF METALS

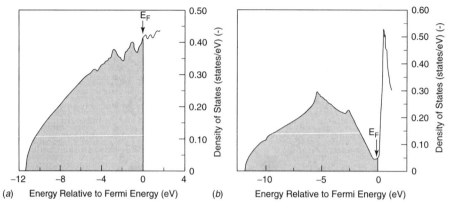

Fig. 1.10 (a) The density-of-states function calculated for Al closely matches the idealized nearly free electron model of Figure 1.9. (b) By contrast, calculations for Be show substantial deviation from the idealized DOS. Be nearly has a bandgap at E_F, which would make it a semiconductor. (Redrawn from Moruzzi et al., 1978, pp. 41 and 53.)

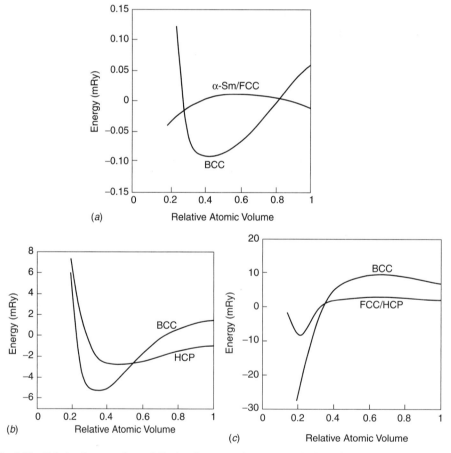

Fig. 1.11 Relative free energies at 0 K of various crystal structures calculated for (a) Na, (b) Mg, and (c) Al. Relative atomic volume (actual atomic volume/equilibrium atomic volume) is 1 at atmospheric pressure. The mRy is an energy unit equal to 0.0136 eV. (Redrawn from Moriarity and McMahon, 1982.)

conductivity, since thermal excitation would be required to promote electrons across the gap into empty states, where they could move in response to a voltage gradient in the crystal.

Calculations can also be made to predict crystal structure transformations in metals as a function of applied pressure. The results of such computations are shown in Fig. 1.11 for Na, Mg, and Al. The lower curve is the structure predicted to be stable at that relative atomic volume (i.e., pressure). At relative atomic volume = 1 (the right edge of each plot), we see that the computations correctly predict the crystal structures for the case of no applied pressure at 0 K (i.e., α-Sm for Na, HCP for Mg, and FCC for Al). Only 1 GPa of pressure is needed to reach the transformation point (compression to about 0.8 of original atomic volume), where Na's α-Sm structure shifts to BCC. In the Mg case, the transformation from HCP to BCC is predicted to occur at about 0.6, which corresponds to a pressure of 57 GPa. In Al, a transformation from FCC to BCC is predicted for about 0.35, which occurs at 130 GPa.

Predictions of transition metals' crystal structures are difficult due to the role of the d electrons in bonding, the large number of interactions that must be considered, and the magnetic effects in the first row of the transition metals. Figure 1.12 shows the results from computations performed without accounting for magnetic effects for the $3d$ transition metals. Note that the equilibrium crystal structures predicted at ambient pressure ($\Omega/\Omega_0 = 1.0$) are correct for the nonmagnetic metals Sc, Ti, V, and Cu. The predictions for Cr and Ni are also correct, even though they have magnetic behavior that should be considered for best accuracy. The prediction for Mn is HCP (like its congeners, Tc and Re); the complex α-Mn structure (cI58) was not among the cases calculated. The prediction for Co is FCC, although the energy calculated for the true equilibrium phase (HCP) is nearly as low. The calculations for Fe predict a closest-packed structure; Fe's BCC structure at room temperature is a consequence of its ferromagnetism.

Similar computations generate some fascinating predictions that await experimental confirmation. For example, the group 10 metals Ni, Pd, and Pt are predicted to have their sp band move through the Fermi energy as pressure increases, eventually creating a bandgap that would make these metals semiconducting at sufficiently high pressure. This is a difficult prediction to verify experimentally in diamond-anvil cells, which are poorly suited to resistivity measurements.

1.6 PERIODIC TRENDS IN MATERIAL PROPERTIES

The periodicity of atoms electronic structures powerfully influences properties such as density (Fig. 1.13), electronegativity (Fig. 1.14), melting point, elastic modulus, compressibility, and coefficient of thermal expansion. In the alkali metals of group 1, the single outer s electron is the only participant in bonding, resulting in weak, low-melting crystal structures. The comparatively large spacings between the atoms give the alkali metals the lowest densities and the lowest electronegativities of all metallic elements. The alkaline earth metals (group 2) hybridize to provide one s and one p outer bonding electron. This additional bonding electron makes them substantially denser and more electronegative than the alkali metals, but these are still among the lightest and most reactive of metals (Table 11.1).

As progressively larger numbers of d electrons become available to participate in bonding, the transition metals of groups 3 to 8 display ever stronger bonding, and the shorter interatomic spacings that result from this stronger bonding raise densities. Maximum density is reached in the middle of each row of transition metals. Densities begin to decline before the d subshell approaches its maximum of 10 electrons, because it is the number of *unpaired* d electrons, not total d electrons, that determines how many can participate in bonding. In W, the ground-state electron structure is (Xe core) $+ 4f^{14}5d^46s^2$; however, the additional energy needed to promote a $6s$ electron to the $6p$ subshell is only 8 kcal/mol, so each W atom contributes

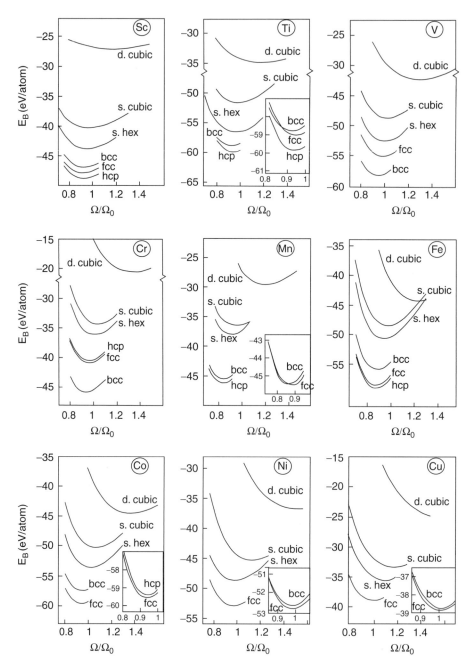

Fig. 1.12 Relative bonding energies of 3d transition metals at 0 K calculated for various crystal structures, not accounting for magnetic effects. The Ω/Ω_0 ratio is actual atomic volume/equilibrium atomic volume ($\Omega/\Omega_0 = 1$ at ambient pressure). (Redrawn from Paxton et al., 1992.)

six electrons for bonding rather than four. As a result, W is the highest-melting and one of the densest metallic elements. As the d subshell fills in groups 9 to 11, the total number of unpaired electrons available for bonding decreases, and the metals become lower melting and less dense.

PERIODIC TRENDS IN MATERIAL PROPERTIES   **15**

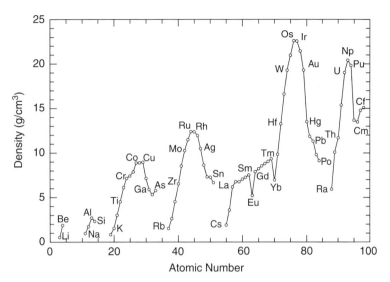

Fig. 1.13 Densities of the metallic elements in their equilibrium room-temperature crystal structures.

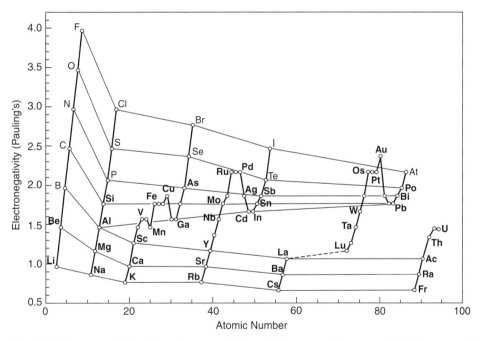

Fig. 1.14 Electronegativities of the elements, excluding the inert gases. Electronegativity is the power of an atom in a molecule to attract electrons to itself.

For the metals in groups 13 to 16, the increasing covalent component in the bonding favors less densely packed crystal structures that have lower densities. Even though the total atomic mass is greater in these metals, the densities are generally lower. Electronegativity, however, reaches a minimum at or near groups 12 or 13 since the d subshell is now filled and the p subshell is only beginning to fill.

SIDEBAR 1.3: HIGH-DENSITY AMMUNITION

In military ammunition, the density of the projectile is a key design parameter. The higher the projectile density, the more effectively it can penetrate heavy armor plate. The highest-density materials (Fig. 1.13) are Os, Ir, Pt, and Re, all of which are expensive. However, there are a few other materials (e.g., W, U) that are much less costly and have densities almost as high. Among the latter group, U offers an appealing combination of high density, reasonable cost, and high fracture toughness.

Most U is used for nuclear reactor fuel rods and for atomic bombs. For both of those applications, the U is first "enriched" to increase its content of fissionable isotope ^{235}U from the rather low 0.7% present in natural U (Sec. 24.4.2.1). For reactor fuel, a ^{235}U content of 3% is typical; atomic bombs use U containing 90% or more ^{235}U. The enrichment process produces large quantities of U that has been depleted of its ^{235}U content, leaving mostly ^{238}U, which is not readily fissionable, as a by-product. Thousands of tons of depleted U have been produced, and there is relatively little market for it.

In the 1980s, the U.S. Defense Department began using depleted U alloys to produce *long-rod kinetic energy penetrator rounds* able to penetrate heavily armored vehicles such as tanks. Figure 1.15 shows a cross-sectional view of a 120-mm projectile that can be fired from a tank cannon. These projectiles and others like them were used with devastating effectiveness in the 1991 and 2003 wars in Iraq. Their ability to penetrate thick armor from great range allows them to destroy the target vehicle and kill its occupants while the target is quite distant. The dense depleted-U alloy central rod (about 19 g/cm^3) penetrates armor more effectively than lighter materials such as steel (7.8 g/cm^3) or Pb (11.3 g/cm^3).

Fig. 1.15 A 120-mm-diameter armor-penetrating cannon shell. The long central rod is made of a depleted-U alloy; its high density makes it an effective armor-penetrating projectile. (Courtesy of the U.S. Department of Defense.)

It might seem that the group 11 metals should have electronegativities fairly close to those of the group 1 metals because each has a filled subshell plus one s electron (p^6s^1 for the group 1 metals and $d^{10}s^1$ for the group 11 metals). However, as Fig. 1.14 shows, this is not the case. The group 1 metals have low electronegativities, whereas the group 11 metals' electronegativities are among the highest of all metals. This occurs because a filled d subshell is much less effective than a filled p subshell at screening the nuclear charge from the outer s electron. Consequently, Au, Ag, and Cu are dense, oxidation-resistant metals (Sec. 18.1), whereas the alkali metals are quite the opposite. Au won't oxidize even when pure O_2 is bubbled through the molten metal, whereas Cs is so reactive that it erupts in flames when exposed to room-temperature air.

The densities of most of the lanthanides display a steady increase as the atomic number increases. This "lanthanide contraction" occurs because the $4f$ electron orbitals are not spherical shells but are highly directional. Thus, the nuclear charge is imperfectly screened by these inner electron orbitals, and the outermost electrons are more strongly attracted to the nucleus as the $4f$ subshell fills. Consequently, lanthanide element atoms (with two exceptions, Sec. 23.3) become smaller as their atomic number increases, and their densities increase since the atoms are both heavier and more closely spaced.

The lanthanide contraction gives congener metals in the $4d$ row (Y to Cd) and $5d$ row (La to Hg) nearly equal atomic radii. For example, the atomic radius of Mo is 0.136 nm, and its congener element W has a radius of 0.137 nm. The substantial increase in atom size that occurs as one moves down group 6 from Cr (0.125 nm) to Mo (0.136 nm) is nearly absent upon moving down another row to W. Similar behavior occurs in other $5d$ transition metals, producing much higher densities in the $5d$ row of transition metals (Sidebar 1.3).

ADDITIONAL INFORMATION

References, Appendixes, Problem Sets, and Metal Production Figures are available at
ftp://ftp.wiley.com/public/sci_tech_med/nonferrous

2 Defects and Their Effects on Materials Properties

> Crystals are like people, it is the defects in them which tend to make them interesting!
>
> —Colin Humphreys, *Introduction to Analytical Electron Microscopy*

2.1 INTRODUCTION

An incontrovertible law of nature states: "Nothing is perfect." This law applies both to people and to crystalline solids. The perfect crystals described in Chapter 1 are built up by repeated translation of the basic unit cell along three crystallographic axes. In practice, however, a crystal with a perfectly regular arrangement of atoms cannot exist; imperfections, irregularities, and defects are present to some extent in all crystals. For example, some of a crystal's unit cells may have too few atoms, whereas others may have an extra atom or an impurity atom. Defects in crystals are important in daily life. The price one pays for a diamond of a given size is determined primarily by the number and type of its crystal defects, and it is generation, accumulation, and interaction of defects in different parts of a car that define its mileage when it arrives at the junkyard.

Although the word *defect* usually carries a negative connotation, crystal defects often improve the properties and performance of materials. Commonly used processing methods such as alloying and heat treating are done deliberately to increase concentrations of defects that strengthen metals. Precipitation-hardened Al 7075 alloy (Sec. 20.2.4.6), for example, has a yield strength hundreds of times greater than that of high-purity single-crystal Al. In fact, the principal task of the metallurgical engineer could be described concisely as crystal defect engineering. All defects and imperfections can conveniently be grouped into four categories, based on their geometry: point defects, line defects (dislocations), planar defects, and volume defects.

2.2 POINT DEFECTS

A *point defect* is an irregularity in the lattice associated with a missing atom (vacancy), an extra atom (interstitial), or an impurity atom. For all temperatures above 0 K, there is a thermodynamically stable concentration of vacancies and interstitial atoms. Introducing a point defect into a crystal increases its internal energy vis-à-vis a perfect crystal. The number of defects at equilibrium at a certain temperature can be determined as

$$N_d = N \exp\left(\frac{-E_d}{kT}\right) \quad (2.1)$$

where N_d is the number of defects present, N the total number of atomic sites present, E_d the activation energy necessary to form the defect, k the Boltzmann constant, and T the absolute temperature.

Structure–Property Relations in Nonferrous Metals, by Alan M. Russell and Kok Loong Lee
Copyright © 2005 John Wiley & Sons, Inc.

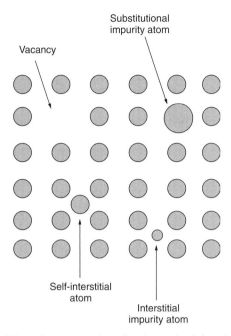

Fig. 2.1 Schematic representation of various point defects in a crystal.

A *vacancy* occurs when an atom is missing from its place in the lattice (Fig. 2.1). Vacancies are present in all bulk metals. By successive jumps of atoms, just like playing Chinese checkers, it is possible for a vacancy to move in the lattice structure, thereby assisting diffusion of atoms through the lattice. Familiar processes such as annealing, sintering, surface hardening, oxidation, and creep all involve, to varying degrees, the transport of atoms through the lattice with the help of vacancies. Vacancies can result from imperfect packing during the original crystallization, or they may arise from atoms' thermal vibrations. Vacancy concentrations increase as temperature increases, because as thermal energy increases, each atom's probability of jumping out of its lowest-energy position also increases. The number of vacancies at equilibrium at each temperature can be calculated from equation (2.1), in which E_d is the energy necessary for an atom to migrate from a regular site in the crystal to the surface or to a grain boundary. When a solid is heated, a new, higher equilibrium number of vacancies exists, generally first near the crystal surfaces and then in regions near dislocations and grain boundaries (Sec. 2.4.1), which provide sites to which the vacating atoms can move after leaving their normal lattice sites. These vacancies spread slowly throughout the crystal from the surfaces into the bulk. On cooling, the vacancy concentration is lowered by diffusion of vacancies to dislocations or grain boundaries, which act as sinks. The rate at which vacancies move from point to point in the lattice decreases exponentially with decreasing temperature, and consequently, in many metals it is possible to retain the high-temperature concentration at room temperature by rapidly quenching from the equilibrating temperature. On quenching from a temperature near the melting point, most of the vacancies have too little time to diffuse to sinks and are said to be "frozen in." This gives a significantly greater ("nonequilibrium") concentration of vacancies in quenched samples than that predicted by equation (2.1).

Equation (2.1) predicts that just below the melting point there is about one vacancy for every 10^4 lattice sites in a Cu crystal ($E_d = 0.9$ eV/atom in Cu). The same calculation performed for Cu at 20°C predicts one vacancy per 10^{15} lattice sites, but in reality it is nearly impossible to achieve such a low vacancy density, due to the sluggishness of diffusion near room temperature and the constant creation of new vacancies (Sec. 7.5) from cosmic and other radiation sources in the environment. Although concentrations on the order of 10^{-4} and smaller may seem inconsequentially small, vacancies control the rate of atom diffusion in metals, making them a key operator in a host of solid-state processes.

Atoms occupying lattice positions that are unoccupied in the perfect crystal are called *interstitial defects* (Fig. 2.1). An interstitial defect's energy of formation is considerably higher than a vacancy's energy of formation, so by equation (2.1) the equilibrium density of interstitials is several orders of magnitude lower than that of vacancies. The interstitial atom can be the same species as the lattice atoms or a foreign atom. Unless the interstitial atom is much smaller than the rest of the atoms in the crystal, it will push the surrounding atoms farther apart (Fig. 13.5) and distort the lattice planes. When the atom occupies a nearby interstitial position, leaving a vacancy at the original lattice site, it is known as a *Frenkel defect*. Interstitial atoms may be produced by severe local distortion during plastic deformation or by irradiating crystals with high-energy particles (Sec. 7.5).

No metal is 100% pure; some foreign atoms are always present in the crystal. Undesired elements that harm the properties of a metal are regarded as impurities. Elements added deliberately to improve properties are called alloying elements. Foreign atoms may reside in the matrix of a metal as either substitutional or interstitial atoms (i.e., foreign atoms either occupy lattice sites from which the regular atoms are missing or occupy positions between the atoms of the ideal crystal) (Fig. 2.1). Impurity defects occur in covalent, metallic, and ionic solids and strongly influence solid-state processes such as diffusion, dislocation motion, phase transformation, and electrical conductivity.

Point defects typically strengthen a metal and lower its ductility by impeding the motion of dislocations (Sec. 2.3). Point defects also increase electrical resistivity because the defects interact with moving conduction electrons, reducing their mean free path through the metal. Since metals have high conduction-electron concentrations (about one electron per atom), any trapping of electrons by defects causes only a small drop in the conduction-electron concentration. Vacancies have the greatest effect on resistivity when they are dispersed individually throughout the lattice; when vacancies cluster together, a metal's electrical resistivity decreases. If a metal's vacancies are all in clusters containing about 100 vacancies, the vacancy-induced increase in resistivity is only half of what it would be with the same number of vacancies distributed individually throughout the lattice. Vacancy clusters in quenched or irradiated samples can contain up to several thousand vacancies (Figs. 7.12 and 7.13). In general, the larger the number of vacancies per cluster, the smaller the effect of each vacancy on increasing a metal's resistivity. Vacancies are of interest to designers of microprocessors because they cause current noise (i.e., rapid and seemingly random fluctuations in current) in the thin metal films used in such devices. The noise results from changes in the electrical resistance caused in part by the creation and annihilation of vacancies. This effect is determined by the lifetimes of the vacancies and is therefore temperature dependent.

Computer simulations have shown that interstitials raise the bulk modulus of W, whereas vacancies lower it. Measurements of the elastic modulus of Al in different crystalline directions over the temperature range 30 to 660°C show that the elastic modulus decreases nonlinearly with temperature. One reason for this nonlinear decrease is the nonlinear change in the number of equilibrium vacancies with temperature.

2.3 LINE DEFECTS (DISLOCATIONS)

The principal one-dimensional crystal defect is the *dislocation*. In a perfect crystal, the atoms lie in planes in the lattice. However, if half a plane of atoms is missing, a *line defect* exists along the bottom edge of the half-plane that remains. Line defects are called dislocations because atoms are dislocated from their positions in a perfect crystal. The presence of dislocations explains two key observations about the plastic deformation of crystalline metals: (1) The stress required to deform a crystal plastically is much less than the stress calculated for plastic flow in a defect-free crystal; and (2) metals work-harden. After a metal has been plastically deformed, it requires greater stress to deform further. Although the dislocation concept was first proposed in 1934, it was not until 1947 that the existence of dislocations was verified experimentally. It took another 10 years before electron microscopy techniques were advanced enough to show dislocations moving through a material. Being able to see dislocations as they move through a structure gives materials scientists a fascinating insight into the mechanisms of plastic deformation.

There are two fundamental types of dislocations: the pure edge and the pure screw dislocation (Fig. 2.2). *Edge dislocations* consist of an extra half-plane of atoms in the crystal structure (Fig. 2.2*a*). The imperfection may extend in a straight line all the way through the crystal, or it may follow an irregular path. *Screw dislocations* (Fig. 2.2*b*) occur when one end of a crystal undergoes a shear stress and has moved at least one interplanar distance, whereas the other end of the crystal has not moved from its initial position. The name comes from the fact that the crystal planes around the dislocation line are converted into a helix by the dislocation. Although the geometry of a screw dislocation is a little more difficult to visualize than that of an edge dislocation, many of its properties are similar. The linear distortion that either type of dislocation represents produces a stress field in the surroundings that strains the crystal. The larger part of this dislocation strain energy is pure elastic strain in the lattice surrounding the dislocation. The remaining, nonelastic strain energy is attributed to the dislocation's center, known as the *core*. The exact size of the core is difficult to measure, but the diameter is typically estimated to be on the order of one to four atomic distances. The strain energy per unit length is about 50% greater for edge dislocations than for screw dislocations.

The Burgers circuit and Burgers vector are used to characterize dislocations. A *Burgers circuit* is any atom-to-atom path, taken in a crystal with dislocations, which forms a closed loop around the dislocation line. Figure 2.3*a* shows a Burgers circuit taken around an edge dislocation in a crystal. When the same Burgers circuit is taken through this portion of the crystal after the dislocation has moved away, the circuit now fails to close itself; the vector required to complete the circuit is the *Burgers vector* for the dislocation (Fig. 2.3*b*). For an edge dislocation, the Burgers vector is perpendicular to the dislocation line. A similar analysis for the screw dislocation case (Fig. 2.4) yields a Burgers vector parallel to the screw dislocation. In general, dislocations are usually combinations of edge and screw. As dislocations are hardly ever entirely straight, different parts of the dislocation will have different degrees of edge and screw character. When a dislocation bifurcates, the sum of the Burgers vectors at a node equals zero (Fig. 2.5). Therefore, dislocations can never end abruptly inside a crystal, they are either present as loops or they terminate at grain boundaries or at the crystal's free surface.

Dislocations have two basic types of movement, glide and climb. *Glide* is dislocation movement in the plane containing both the dislocation line and the Burgers vector. During each glide

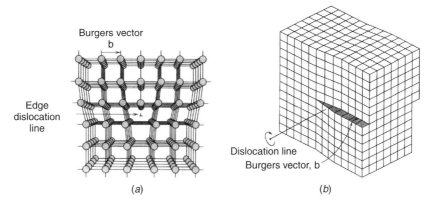

Fig. 2.2 Atom positions near (*a*) an edge dislocation and (*b*) a screw dislocation. (From Callister, 2003, pp. 74 and 75; with permission.)

22   DEFECTS AND THEIR EFFECTS ON MATERIALS PROPERTIES

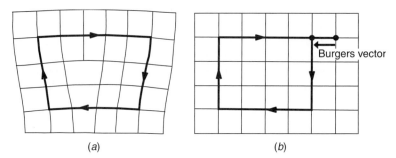

Fig. 2.3 Burgers circuit around (*a*) an edge dislocation and (*b*) a perfect crystal. (From Hull and Bacon, 1984, p. 19; with permission.)

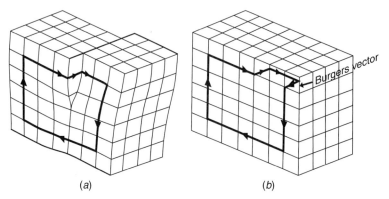

Fig. 2.4 Burgers circuit around (*a*) a screw dislocation and (*b*) a perfect crystal. (From Hull and Bacon, 1984, p. 20; with permission.)

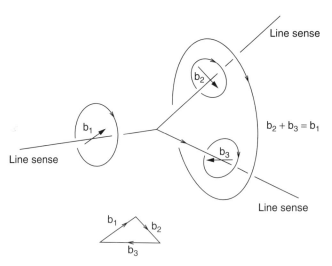

Fig. 2.5 Conservation of the Burgers vector at a dislocation node. (From Hull and Bacon, 1984, p. 22; with permission.)

LINE DEFECTS (DISLOCATIONS)

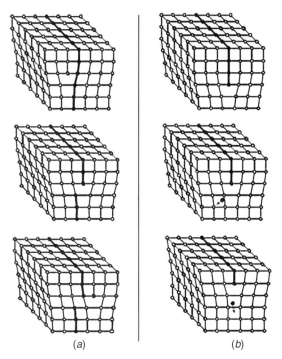

(a)   (b)

Fig. 2.6 Glide and climb of a dislocation: (*a*) glide displacing the upper grain with respect to the lower part, and (*b*) seen sequentially from top to bottom, positive climb occurs by emission of interstitial atoms.

step a single row of atoms changes closest-neighbor atoms, and the passage of the dislocation clearly displaces the upper part of the grain with respect to the lower (Fig. 2.6*a*). Simultaneous glide of many identical dislocations under an applied stress, known as *slip*, is the typical mechanism of plastic deformation in crystalline materials. The second type of dislocation movement, *climb*, is coupled directly with emission or absorption of vacancies. For an edge dislocation, climb can be shown as a receding of the extra half-plane (Fig. 2.6*b*). This movement results in surplus atoms in the dislocation core, a condition that can be resolved by the absorption of vacancies. If vacancies are not available in large quantities, climb cannot continue, and thus the rate of climb is coupled to the rate of diffusion. Although glide can occur at all temperatures, climb is practically nonexistent unless the temperature is about $0.4T_{m.pt.}$ or higher.

Almost all crystalline metals contain some dislocations; concentrations of about 10^5 mm of dislocation line length per cubic millimeter of metal (i.e., 10^5 mm^{-2}) are typical in cast metals. These dislocations were introduced during solidification or as a result of small deformations induced by thermal contraction of the metal during cooling. Dislocation interactions within a metal are the primary means by which metals are deformed. Without dislocations, movements of atoms relative to each other would be much more difficult, and plastic deformation would be restricted to such mechanisms as twinning, kink band formation, and diffusion. About 5% of deformation energy is retained internally (primarily as dislocations) when a metal is deformed plastically. When metals deform by dislocation motion, the more barriers the dislocations meet, the stronger the metal. The density of dislocations in a metal (number per unit volume) increases continuously during plastic deformation, reaching values as high as 10^{10} mm^{-2} in heavily deformed metal. Deformation by dislocation motion is one of the features of metals that make them such useful engineering materials. The metallic bond is nondirectional, so strains to the crystal can be accommodated by dislocation motion. Most metals can stand considerable plastic deformation before failing. It is remarkable that these line defects may reduce the strength of a material by a factor of 10,000, even though only about 50 atoms in a billion lie along a line of defect!

Several different techniques exist to observe dislocations: surface methods, decoration methods, transmission electron microscopy (TEM), x-ray diffraction, and field ion microscopy. Of these, TEM is the most powerful method available today for observing dislocations. This method uses a very thin sample (0.2 to 3 μm) which is placed into a vacuum chamber and bombarded with a focused, high-energy electron beam. The strain in the lattice surrounding a dislocation diffracts the high-energy electrons, bending them into paths away from the microscope's optical axis; thus, the metal adjacent to a dislocation appears dark, making the dislocation visible as a dark line (Fig. 18.18).

2.4 PLANAR DEFECTS

Planar (interfacial) defects occur wherever the crystalline structure of the material is discontinuous across a plane. Some examples of planar defects are free surfaces, grain boundaries, twins, and phase boundaries.

2.4.1 Grain Boundary

Most metals are composed of many small crystallites called *grains*. The interfaces between these grains are grain boundaries which are one or two atom spacings thick. A *grain boundary* is a general planar defect that separates regions of different crystalline orientation (i.e., grains) within a polycrystalline solid (Fig. 2.7). This commonly occurs when two crystals begin growing separately and then meet. Most atoms at the boundaries are located in highly strained and distorted positions, and their free energy is higher than that of the atoms in the regular, undisturbed part of the crystal lattice. Impurities often segregate at grain boundaries because this high surface energy lowers their defect energies. Since the width of grain boundaries (*width* being defined as the disordered region at the junction of the two crystallites) is less than two atomic diameters, grain boundaries in fine-grained polycrystalline metal with an average grain size of 1 μm constitute only 0.03% of the volume of the metal. This volume fraction increases sharply, however, to 50% at a grain size of 2 nm.

The two main types of grain boundaries are pure tilt boundaries and pure twist boundaries (Fig. 2.8). *Pure tilt boundaries* occur when the axis of rotation is parallel to the plane of the grain boundary; *pure twist boundaries* occur when the axis of the rotation is perpendicular to

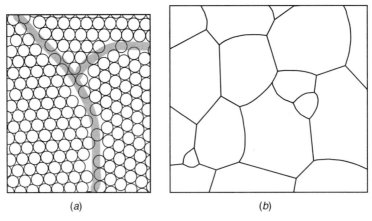

Fig. 2.7 Grain boundaries viewed on (*a*) the atomic scale and (*b*) the micrometer scale.

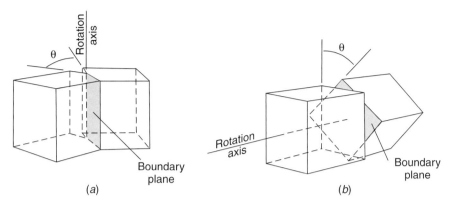

Fig. 2.8 (*a*) Tilt grain boundary; (*b*) twist grain boundary. (Redrawn from Porter and Easterling, 2001, p. 117.)

the plane of the grain boundary. Tilt boundaries in crystals are comprised of arrays of parallel edge dislocations. Twist boundaries are comprised of crossing arrays of screw dislocations.

In polycrystalline samples the individual grains usually have a random crystallographic orientation with respect to one another, and the grain structure is referred to as being *randomly oriented*. The importance of the orientation change across a grain boundary to the process of slip has been shown by experiments on bicrystal materials, where it was found that as the orientation difference between the two crystals decreases, the direction and plane of slip being relatively the same in both crystals, the mechanical properties change progressively from polycrystalline behavior to single-crystal behavior. In some instances, however, deformation or solidification effects align most grains such that they have the same orientation to within a few degrees. In this instance the metal is said to have a *preferred orientation* or *texture* (Secs. 3.3 and 18.4.5).

Grain boundaries influence the mechanical (Secs. 3.2 and 7.2) and electrical properties of metals. In many metal-processing operations, metal is cold-worked and recrystallized to reduce grain size; this generally raises its strength compared to as-cast metal, without reducing ductility. At temperatures where the grain boundary diffusion rate is low, small grain size enhances strength (Sec. 18.2.6.2). However, when the grain boundary diffusion rate is high, the material can exhibit extraordinarily large elongation under load (superplastic behavior, Sec. 19.2.6) or can exhibit high creep rates under moderate or low conditions of loading (Sec. 6.4). For certain applications, such as the combustion zone turbine blades of a jet engine (Sec. 16.3.6.2) or semiconducting electronic devices (Sec. 21.2.6), grain boundaries are eliminated to improve creep performance or optimize electronic performance.

2.4.2 Twinning and Kink Band Formation

Mechanical twinning is a coordinated movement of a large number of atoms that deforms a portion of a crystal by an abrupt shearing motion (Fig. 2.9). Mechanical twinning causes serrations in stress–strain curves and "tin cry," audible clicking sounds caused by the formation of twins. Mechanical twinning is especially common in Sn, Cd, Zn, and Mg, but it occurs in many other metals as well (Sec. 12.3.5.1 and Fig. 12.22). Twinned regions are often bounded by parallel or nearly parallel sides (Fig. 2.10) which correspond with planes of low indices known as the *twin habit* or *twinning plane*. Etching behavior and x-ray diffraction techniques indicate that the twinned region differs considerably in orientation from the grain in which it formed. Shear stress along the twin plane causes atoms to move a distance that is proportional to the distance from the twin plane. Atom motion with respect to the atom's nearest neighbors is less than one atomic spacing. Twinning can also occur during recrystallization (Sec. 16.3.6.1), but in this case the twin results from a "stacking error" in the recrystallizing metal, not from stress.

26   DEFECTS AND THEIR EFFECTS ON MATERIALS PROPERTIES

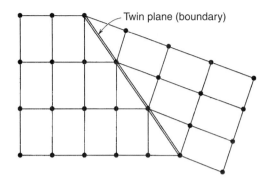

Fig. 2.9   Depiction of atom positions near a twin boundary. (From Callister, 2003, p. 79; with permission.)

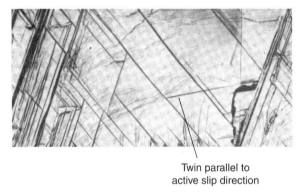

Fig. 2.10   Micrograph showing twins at a Zn crystal surface fractured in a tensile test. (From Honeycombe, 1982, p. 205; with permission; originally from Deruytere and Greenough, 1956.)

Mechanical twinning subdivides the grains and therefore increases the number of barriers to slip and the work-hardening rate. It also contributes to plastic deformation due to twinning shear, which not only changes the crystal's shape but also reorients the twinned portion such that slip may begin afresh on planes that were not previously well oriented for slip (Sec. 24.4.3.1).

In BCC and HCP metals, mechanical twinning is common at low temperatures and high strain rates because the shear stresses needed to move dislocations are high under these conditions (i.e., dislocation motion becomes so much more difficult that twinning becomes the lower energy response to applied stress). In BCC metals, twinning normally occurs prior to macroyielding, and in many cases it is prevented by significant plastic deformation. In FCC metals, which have a much lower strain rate sensitivity but higher work-hardening ability, twinning often occurs after considerable plastic deformation, which raises the corresponding stress level. Twinning is an essential and unavoidable mechanism in polycrystalline samples, but in single crystals, the orientation, the stress level, and the temperature determine whether twinning will occur.

Kink band formation is a less common deformation mechanism somewhat similar to twinning. Kink band formation is most often observed during compressive loading, when the direction of applied stress is nearly perpendicular to the principal slip plane [e.g., compressive loading nearly parallel to the c axis in Zn and Cd, whose primary slip plane is (0001)]. In this mechanism, the metal shears by formation of dislocation arrays that produce a buckling motion along the slip plane direction that shears the metal several degrees away from its previous position. Kink band formation differs from twin formation in that the atom positions do not possess mirror plane symmetry after the shearing displacement.

2.4.3 Phase Boundaries

A *phase* is a homogeneous, physically distinct, and mechanically distinguishable part of the material with a given chemical composition and structure. Phases may be interstitial or substitutional solid solutions, ordered alloys, amorphous regions, or even pure elements. A crystalline phase in the solid state may be either polycrystalline or exist as a single crystal. Phase boundaries are similar to grain boundaries in their effects on mechanical and electrical properties, but they can be more varied in size and shape (Secs. 20.2.4.4 and 12.2.3). Deliberate introduction of phase boundaries is one of the most effective ways to strengthen metals (Sec. 20.2.4.4).

2.4.4 Stacking Faults

Closest-packed planes in FCC, HCP, and α-La crystals are arranged in stacking sequences (Fig. 1.2*b*–*d*). If one denotes a closest-packed plane as *A*, the layer above can be placed in either of two positions (denoted *B* or *C*), the layer above that can also be placed in either of two positions, and so on. For perfect HCP crystals, the stacking sequence is *ABABABA....* For perfect FCC crystals, the stacking sequence is *ABCABCABCA....* In α-La, the stacking sequence is *ABACABACA....* An occasional deviation from these stacking sequences is called a *stacking fault*.

The total energy of a crystal with the atomic planes stacked in a regular manner is lower than that of the same crystal with a stacking fault. The energy difference is known as the *stacking-fault energy*. If planes of atoms glide past one another in some direction, the stacking-fault energy increases up to a maximum, which corresponds to an unstable configuration in which two identical planes are adjacent (i.e., the stacking sequence *ABCCAB*). As the atomic planes slide farther, the stacking-fault energy is lowered until it reaches a local minimum, where the lattice is stable but not in a perfect bulk configuration. This stable configuration is called an *intrinsic stacking fault*. An example of an intrinsic stacking fault in an FCC metal would be the stacking sequence *ABCBCABCABC...* (i.e., removal of one plane from the ordinary stacking sequence, Fig. 4.13). Stacking faults can be thought of as one plane of HCP stacking within an FCC lattice, or as one plane of FCC stacking within an HCP lattice. In FCC metals, slip in the ⟨112⟩ direction by partial dislocations (Sec. 4.5) is common because the energy of the unstable configuration is lowest in that direction. *Extrinsic stacking faults* also occur, in which a change in sequence occurs by introduction of an extra layer.

Stacking-fault energy plays a part in determining deformation textures in FCC and HCP metals. The rolling texture of high-purity Cu, for example, does not conform to any simple texture and can best be described by the irrational texture {146}⟨211⟩. Stacking faults also influence the plastic deformation of metals. Metals with wide stacking faults (due to low stacking-fault energy) strain-harden more rapidly and twin more easily during annealing than do metals with narrow stacking faults caused by high stacking-fault energy (Secs. 18.2.6.1 and 11.3.6).

2.5 VOLUME DEFECTS

Metals frequently have some volume that contains no solid material; such defects include voids, gas bubbles, and cracks. These defects are similar in some ways to a free surface, but the volume defect is mostly or entirely surrounded by metal on all sides. Volume defects are typically deleterious to properties, but there are instances (e.g., the void volume in foamed metals) in which the defects improve certain properties, such as the bending moment/weight ratio.

ADDITIONAL INFORMATION

References, Appendixes, Problem Sets, and Metal Production Figures are available at
ftp://ftp.wiley.com/public/sci_tech_med/nonferrous

3 Strengthening Mechanisms

She did make defect perfection.

—William Shakespeare, *Antony and Cleopatra*

3.1 INTRODUCTION

Since dislocation motion is the primary deformation mechanism in metals (Sec. 2.3), circumstances that prevent or impede dislocation motion strengthen metals. Two general approaches have been developed to strengthen metals. One is to produce metal with no dislocations at all; the other is to change a metal's microstructure to make dislocation motion difficult.

Whiskers of single-crystal metal can be made that contain no dislocations. These whiskers have exceptionally high strength, but they are small, difficult to produce, and quite expensive. Although whiskers can be added to composites as a reinforcing phase, it is generally impractical to strengthen metals by making them dislocation-free because this would prevent all use of common fabrication and shaping processes. Polycrystalline metals, although weaker than whiskers, are used in nearly all engineering applications (see, however, Secs. 7.3, 16.3.6.2, and 21.2.6). Polycrystalline metals contain dislocations and generate more dislocations under stress, so their strengthening strategies focus on imposing barriers to dislocation motion such as grain boundaries, phase boundaries, other dislocations (strain hardening), and solid-solution atoms. These methods can increase strength by as much as two or even three orders of magnitude over the strength of high-purity single-crystal metal.

3.2 GRAIN BOUNDARY STRENGTHENING

Metals normally contain huge numbers of randomly oriented grains, or crystals, separated by grain boundaries. Whereas a single crystal has a free surface and can deform by dislocation glide on a single active slip system, grains in a polycrystal with differing orientations of their lattice and slip systems are forced to conform to the overall strain. In general, for a given plastic strain the dislocation density will be higher in a polycrystal than in a single crystal due to the presence of geometrically necessary dislocations (Fig. 3.1) that result from nonuniform strain in the polycrystal. A polycrystal is deformed by disassembling it into its constituent grains and allowing each to slip according to Schmid's law (Sec. 4.6), thereby introducing statistical dislocations (not shown in Fig. 3.1). When the crystals are subsequently reassembled, they no longer "fit" together (Fig. 3.1*b*). In some areas adjacent grains have moved apart, leaving a gap, whereas in other areas grain overlap occurs. To shift metal from the overlapped regions to the gaps, geometrical dislocations, in the form of prismatic loops, tilt boundaries, and so on, may be imposed on the structure (Fig. 3.1*c*) until the grains again fit together as shown in Fig. 3.1*d*. The number of dislocations required to put the polycrystal back together should be roughly proportional to the strain times the grain size (i.e., the displacement) times a "geometrical constant."

Structure–Property Relations in Nonferrous Metals, by Alan M. Russell and Kok Loong Lee
Copyright © 2005 John Wiley & Sons, Inc.

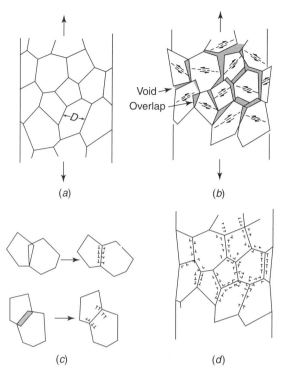

Fig. 3.1 If each grain of a polycrystal shown in (*a*) deforms in a uniform manner, overlap and voids appear (*b*). These can be corrected by introducing geometrically necessary dislocations, as shown in (*c*) and (*d*). Statistical dislocations are not shown in this figure. (From Ashby, 1970; with permission.)

Figure 3.2 compares the stress–strain behavior of single crystal and polycrystalline Al tensile specimens. Curves 1 to 4 were measured from polycrystalline specimens with various grain sizes, and these display generally greater strength than curves 5 to 7, which were measured from single-crystal specimens with various orientations. Although the general form of curves 1 to 4 is similar, the coarser-grained samples work-hardened substantially less than the finer-grained samples. Crystal 6 is typical of single crystals deforming on several slip systems, and its strength falls somewhat short of the coarsest polycrystalline sample. The soft orientation of crystal 7 has a much lower work-hardening capacity than that of any of the polycrystalline samples because it slips predominantly on one slip plane and dislocation interactions are minimal. Curiously, crystal 5 (oriented with its tensile axis nearly parallel to the [$\bar{1}$11] direction) hardens even more rapidly than polycrystalline samples, probably because this particular orientation is especially favorable for the formation of Lomer–Cottrell dislocation locks (Sec. 4.5).

The higher strength of fine-grained polycrystalline specimens (curves 1 to 4 in Fig. 3.2) was first quantified by Hall (1951) and Petch (1953) in the now familiar Hall–Petch relationship:

$$\sigma_y = \sigma_0 + k_y d^{-1/2} \qquad (3.1)$$

where σ_0 is the friction stress, d the average grain size, and k_y the stress intensity for plastic yielding across polycrystalline grain boundaries. This behavior (Fig. 18.19) is typical also of other types of boundaries, such as second-phase particles, mechanical twins, and martensite plates. In general, more closely spaced barriers to dislocations produce greater strength. A complete and fundamental understanding of the mechanisms behind the Hall–Petch equation still eludes materials scientists. Two types of theories have been proposed to explain Hall–Petch

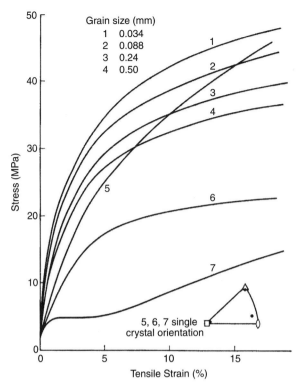

Fig. 3.2 Effect of grain size on stress–strain curves of pure Al. (From Honeycombe, 1982, p. 238; with permission.)

strengthening in metals. One invokes dislocation pile-ups, and the other could be called a grain boundary source model.

3.2.1 Dislocation Pile-up Model

Petch attributed the grain size dependence of yield stress to dislocation pile-ups at grain boundaries. He assumed that yielding begins first in a single grain when one dislocation source (i.e., a Frank–Read source, Sec. 4.4) becomes active and emits dislocation loops. As shown in Fig. 3.3, the dislocation loops move to the grain boundary, where they are stopped. When a sufficient number of dislocations pile up in grain 1, slip is finally induced in grain 2. The dislocation pile-up acts to concentrate stress, and if this stress concentration is sufficiently large, a source in a neighboring grain will be activated to emit dislocation loops. These new loops in turn will produce a stress concentration in yet another grain and initiate plastic deformation there. Plastic yield thus spreads from one grain to the next.

Consider a dislocation source at the center of a grain sending out dislocations to pile up at the grain boundary (Fig. 3.3). It can be shown from dislocation theory that the number of dislocations in the pile-up before yielding is given by

$$n = \frac{\alpha \pi \tau_s d}{4 G b} \tag{3.2}$$

where α is a constant near unity, τ_s the average resolved shear stress in the slip plane, G the metal's shear modulus, and b the Burgers vector length. The stress at the grain boundary at the

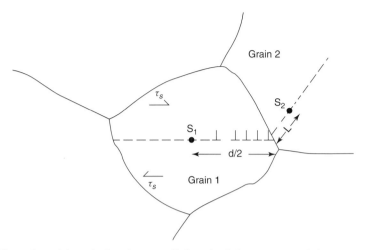

Fig. 3.3 Pile-up formed in grain 1 under an applied resolved shear stress τ_s. S_2 is a source in grain 2. The trace of the preferred slip plane in each grain is marked by a dashed line.

head of the pile-up, $\tau_i = n\tau_s$ (where τ_i is the shear stress induced in grain 2, n the number of dislocations piled up at the grain boundary in grain 1, and τ_s the shear stress in grain 1 where slip begins); thus,

$$\tau_i = \frac{\alpha \pi \tau_s^2 d}{4\,Gb} \qquad (3.3)$$

However, if the applied stress τ_s is opposed by a "frictional" stress τ_0, the effective value of τ_s will be reduced:

$$\tau_s^{\text{effective}} = \tau_s - \tau_0 \qquad (3.4)$$

Thus, equation (3.3) can be rewritten as

$$\tau_i = \frac{\alpha \pi (\tau_s - \tau_0)^2 d}{4\,Gb} \qquad (3.5)$$

The frictional stress may be due to the Peierls stress (Sec. 4.3), fine precipitates, impurity atoms, temperature, and/or strain. The frictional stress reduces the stress at the grain boundary by reducing the number of dislocations in a pile-up.

From equation (3.5) we obtain

$$\tau_s = \tau_0 + \sqrt{\frac{4\,Gb\tau_i}{\alpha \pi}}\, d^{-1/2} \qquad (3.6)$$

Note that

$$\sigma = M\tau \qquad (3.7)$$

where M is an orientation factor (also called the *Taylor factor*), which is equal to about 3. Thus, equation (3.6) can be rewritten as

$$\sigma = \sigma_0 + \sqrt{\frac{4\,MGb\tau_i}{\alpha \pi}}\, d^{-1/2} \qquad (3.8)$$

$$\sigma = \sigma_0 + k d^{-1/2} \qquad (3.9)$$

which is in agreement with equation (3.1).

3.2.2 Grain Boundary Source Model

Although Petch's model seems plausible, there is inadequate experimental evidence that dislocation pile-ups actually occur in pure metals. This led to an alternative explanation for the grain size dependence of yield stress. In this alternative model, there is no requirement for dislocation pile-ups at grain boundaries. Rather, the boundary may act as a source of dislocations, and the capacity to emit dislocations is dependent on the character of the grain boundary.

If L is the total length of dislocation emitted per unit grain boundary area at yielding, the dislocation density is

$$\rho = \frac{1}{2} L \frac{S}{V} \tag{3.10}$$

where S is the grain surface area, V the grain volume, and the $\frac{1}{2}$ factor accounts for two grains sharing one boundary. Assuming the grain to be spherical, we get

$$\frac{S}{V} = \frac{4\pi(d/2)^2}{\frac{4}{3}\pi(d/2)^3} = \frac{6}{d} \tag{3.11}$$

Substituting this into equation (3.10), we obtain

$$\rho = \frac{3L}{d} \tag{3.12}$$

From the work-hardening relationship,

$$\tau = \tau_0 + \alpha G b \rho^{1/2} \tag{3.13}$$

we can use equation (3.12) to obtain

$$\tau = \tau_0 + \alpha G b \sqrt{3L}\, d^{-1/2} \tag{3.14}$$

or

$$\sigma = \sigma_0 + \alpha' G b \sqrt{3L}\, d^{-1/2} \tag{3.15}$$

which is also in agreement with equation (3.1). If a suitable physical understanding of L as a parameter could be developed, this grain boundary source concept could prove an important model for the Hall–Petch effect. There is experimental evidence that grain boundary dislocation sources are important [and in nanoscale grains, dominant (Sec. 7.2)] in yielding.

3.3 STRAIN HARDENING

The phenomenon of strain hardening, in which yield stress rises with increasing plastic deformation, is an important industrial process to harden pure metals or alloys, particularly those that do not respond to heat treatment. However, the toughness and ductility of the metal decrease when it is strain hardened. Figure 3.4 compares the yield strength of a metal in three different conditions: annealed, after straining beyond the yield stress in a normal tensile test; and after wire swaging. It can be seen that the yield stress of the swaged wire has been raised well above the tensile strength of the annealed metal because compressive radial stresses from the die prevent the onset of necking. However, the increase in strength comes with a corresponding decrease in ductility. Besides this, other physical properties are also affected. There is a slight decrease in density (about a few tenths of a percent), an appreciable decrease in electrical conductivity

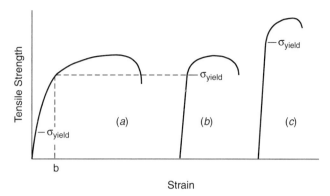

Fig. 3.4 Strengthening by strain hardening (*a*) annealed metal; (*b*) after straining plastically in tension to point b; (*c*) after wire swaging.

due to an increased number of scattering centers, and a small increase in the coefficient of thermal expansion.

All theories of strain hardening depend on the assumption that cold work causes an enormous increase in the dislocation density and a corresponding decrease in the ease with which these dislocations can glide. After a deformation of only a few percent, dislocation densities increase from an initial value of about 10^5 mm^{-2} to as high as 10^8 mm^{-2}. As these dislocations move, they become piled up at grain boundaries and other lattice defects, or because the dislocations act on many different slip planes, they intersect each other, producing numerous jogs (see Sidebar 4.1), and generally become tangled. This makes it more difficult for other dislocations to glide through the metal. Generally, metals with a cubic structure (which have several slip planes) show a higher rate of strain hardening, whereas those HCP metals that slip only on the basal plane (e.g., Zn) show a lower rate of strain hardening. Increasing temperature also lowers the rate of strain hardening.

During cold work, the grains of polycrystalline metal become elongated in the direction of maximum strain, and the crystallographic planes and directions of easiest slip are typically seen to tilt to positions that permit the lowest energy deformation. As a result, the statistical distribution of these grains orientations is no longer random but becomes "clustered" around directions that promote easier deformation. This preferred orientation is frequently referred to as *texture*. It is usually described in terms of ideal orientation such as [*uvw*] for the fiber texture produced by swaging or drawing, and (*hkl*) [*uvw*] for the sheet textures produced by rolling, for which the (*hkl*) planes are oriented parallel to the plane of the sheet with the [*uvw*] direction along the rolling direction. The properties of textured metals become anisotropic; that is, they vary with direction. Sheet metal with a texture is not suitable for press work because it does not deform uniformly. On the other hand, textures in soft ferromagnetic metals for transformer laminations can greatly enhance the magnetic properties because the energy losses are minimized by orienting the grains in the direction of easy magnetization (Sec. 16.2.2.4).

Most of the work of plastic deformation is converted to heat, which raises the temperature of the metal while it is being deformed. However, a small but significant fraction of the work is stored in the metal in the form of lattice defects, mainly dislocations. Grain boundaries can also be considered as a dense arrangement of tangled dislocations and contribute significantly to the stored strain energy. A distorted crystal containing a high density of lattice defects is thermally unstable. Heating such a deformed crystal above a certain temperature initiates a number of processes that reduce the stored energy and return the crystal to its initial degree of perfection. One of these is the formation and growth of new, strain-free grains in the matrix by a process known as *recrystallization*. The nuclei of these new grains are quite small and normally form in regions of severe or complex distortion, for example, near grain boundaries. The new grains have

34 STRENGTHENING MECHANISMS

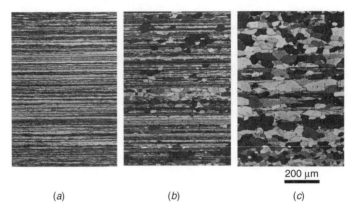

Fig. 3.5 Changes in microstructure of cold-worked 70–30 brass with annealing: (*a*) cold-worked 40%; (*b*) 440°C, 15 min; (*c*) 575°C, 15 min (150×). (From ASM, 1972, p. 243; with permission.)

high-angle boundaries separating them from the strained matrix, and their orientations change significantly from that of the grains they consume. After recrystallization, the metal's strength and hardness are reduced and ductility increases. The rate of recrystallization is determined by the annealing temperature, amount of cold work, and purity of the metal.

During recrystallization, boundaries must be created between the strained and unstrained grains, these boundaries have surface energy. For recrystallization to proceed, the energy released by removing stored defects must be greater than the energy required to create the new boundaries. A minimum temperature, which is a function of the amount of cold work done on this metal, is required before recrystallization can begin. Typically, the recrystallization temperature is ∼$0.4T_{m.pt.}$, where $T_{m.pt.}$ is the metal's absolute melting temperature. Recrystallization begins at regions with the greatest defect energy (e.g., regions where dislocations have accumulated at grain boundaries). Increasing the degree of cold work increases the number of such high-energy regions, resulting in a greater number of recrystallized grains. The final grain size is generally determined by the ratio between the rate of nucleation and the rate of grain growth of the new grains. The higher the ratio, the smaller the recrystallized grain size. Since the new grains do not all nucleate at the same time, there is a range of grain sizes when recrystallization is completed.

Metals held at a high temperature after recrystallization is completed display an increasing average grain size known as *grain growth*. As the mean grain size increases, the total grain surface area for a given volume of metal decreases. Accordingly, the total surface energy is reduced, which lowers the energy level of the metal. This energy difference provides the driving force for grain growth whereby the larger grains continue to grow by the absorption of smaller grains. Thus, a cold-worked microstructure recrystallizes to a fine-grained structure, and this, in turn, becomes a coarser-grained microstructure as grain growth proceeds (Fig. 3.5). For the majority of engineering applications, the most favorable properties are associated with the finest possible grain size, but any strain-hardening effect is lost to recrystallization and grain growth if the metal is annealed or used at elevated temperatures. Deformation, recrystallization, and grain growth can all occur in rapid succession in metals deformed at high temperature ($T > 0.4T_{m.pt.}$ or $0.5T_{m.pt.}$), where recrystallization and grain growth follow each plastic deformation step almost immediately.

3.4 SOLID-SOLUTION HARDENING

Foreign atoms dissolved in a metal's lattice increase the metal's yield strength. There are two types of solid solutions. If the solute atoms and matrix (solvent) atoms are roughly the same size, the solute atoms occupy lattice sites in the crystal normally occupied by solvent atoms; this is a *substitutional solid solution*. If the solute atoms are much smaller than the matrix atoms, they occupy interstitial positions in the matrix lattice; this is an *interstitial solid solution*. Solute atoms interact with dislocations attempting to move through the crystal. The impurity's effect on strength depends on the atoms' size difference and the amount of impurity present. A solute atom smaller than the matrix atom will create an approximately spherical tensile field around itself that attracts the compressive regions around mobile dislocations. In contrast, larger solute atoms attract the tensile region of nearby dislocations. On average, dislocations will maneuver so as to lower their strain energies by associating with the nonuniform strain fields around impurities. This association impedes dislocation motion, which inhibits plastic flow and increases the yield stress. Solid-solution hardening is only a moderately effective strengthening mechanism in commercial metals, largely because the solubility is usually rather low for impurity atoms that produce the greatest strengthening effect (i.e., those that have the greatest size difference with the solvent atoms).

The classical model for explaining solid-solution hardening attributes the increase in yield strength to a combined effect arising from the differences in size and shear modulus between solute and matrix atoms, respectively. An interaction force between a single solute and a dislocation can be calculated by taking the dislocation type and the nature of the interaction (size and/or modulus effect) into consideration. For example, the interaction force between a solute and a dislocation of general character is given by Fleischer's relation:

$$F = Gb^2 x R^3 (\varepsilon_G - \alpha \varepsilon_b) \frac{1}{3\pi r^4} \tag{3.16}$$

where G is the shear modulus, b the Burgers vector length, x the distance between the solute and the dislocation along the slip plane, R the atomic radius, r the nearest interaction distance, and ε_G and ε_b the strains caused by the differences in modulus and size, respectively; α varies depending on whether the interaction is with a dislocation of edge or screw character.

3.5 PRECIPITATION HARDENING (OR AGE HARDENING)

A fourth method of strengthening metals is that of producing a fine distribution of hard particles which impede dislocation motion in the matrix. The increase in yield stress depends principally on the strength, structure, spacing, size, degree of coherency, orientation, shape, and distribution of the precipitate particles. The increase in alloy strength is most often caused by the presence of precipitated particles, and in that case the strengthening effect is called *precipitation hardening*.

The fundamental requirement of a precipitation-hardening alloy system is that the solid solubility decrease with decreasing temperature, as shown in Fig. 20.13. Here an Al–Cu alloy exists as a single-phase solid solution at high temperature and as a two-phase material at room temperature. To precipitate a fine dispersion, the alloy is quenched from the high-temperature single-phase condition, and the cooling is so rapid that formation of the second phase is suppressed temporarily. This is known as the *solution-treated condition*, and such metal is comparatively soft and ductile. Formation of the second phase as a fine dispersion throughout the α matrix occurs either spontaneously over time at room temperature or by heating to a temperature below the single-phase region. Known as *aging treatment*, this increases hardness and strength but reduces ductility (Sec. 20.2.4.4).

When a dislocation is moved by an applied stress on the matrix slip planes, it encounters second-phase particles (precipitate particles) and interacts with them. Depending on the nature of the precipitate and on its crystallographic relationship with the matrix, different interactions may occur. The particles may be either impenetrable or penetrable by dislocations. In the former case, a dislocation is forced by the applied stress to bow around the particle and bypass it, leaving a dislocation loop (often called an *Orowan loop*) around the particle. In the latter case, the particle is sheared by the dislocation as the latter moves through the crystal (Fig. 20.16). This

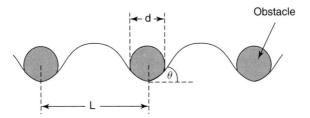

Fig. 3.6 Interaction of a dislocation with a row of obstacles.

can occur only under the special condition that the precipitate is coherent with the matrix (i.e., has alignment of the crystal planes in both phases). In general, small (coherent) particles may or may not be penetrable, and large particles are incoherent and thus impenetrable.

Both the Orowan and shearing mechanisms can be visualized in terms of a single dislocation interacting with a coplanar array of particles of diameter d and a center-to-center distance L (Fig. 3.6). One can develop an expression for the critical resolved shear stress needed for the dislocation to overcome this row of equally spaced obstacles in the slip plane. The dislocation line is forced against the row by the applied stress and bends between the particles with a bending angle θ. Angle θ depends on the interparticle spacing L, the particle diameter d, the increase in the applied shear stress $\Delta\tau$ due to the interaction with the particles, the Burgers vector length b, and the line tension Γ (which acts to straighten bowed dislocations) according to the equation

$$\Delta\tau b(L-d) = 2\Gamma \sin\theta \tag{3.17}$$

The right side of this equation describes the pinning force exerted by each particle on the dislocation.

$$f_{\text{pin}} = 2\Gamma \sin\theta \tag{3.18}$$

There exists a maximum force, $f_{\max}$, which the particle can sustain. This value depends on the distance of the slip plane with respect to the particle center. If the bending angle becomes 90° before $f_{\max}$ is reached, the dislocation bypasses the particle by looping around it (the Orowan mechanism). However, if the maximum force is reached before the bending angle becomes 90°, the dislocation shears through the particle. Assuming that this is the beginning of plastic deformation, the following equations show the increase in the critical resolved shear stress, $\Delta\tau_0$ for this action at 0 K:

Orowan mechanism:

$$\Delta\tau_0 = \frac{2\Gamma}{b(L-D)} \quad \text{for } f_{\max} \geq 2\Gamma \tag{3.19}$$

Cutting mechanism:

$$\Delta\tau_0 = \frac{f_{\max}}{b(L-D)} \quad \text{for } f_{\max} < 2\Gamma \tag{3.20}$$

If the precipitates are small and coherent, they act on dislocations much as solute atoms do. In this case, the dislocation interaction with the obstacle occurs only along a short part of the dislocation's total length. The obstacles are then known as *point obstacles* or *localized obstacles*. The obstacles considered so far have been arranged in a regular manner. In real metals, the obstacles are distributed nonuniformly; the average obstacle spacing then becomes dependent on the stress, and equation (3.20) becomes

$$f_{\max} = \tau_c bL \tag{3.21}$$

The statistical problem involving dislocation–obstacle interactions is given by the Friedel relation:

$$L = \left(\frac{2\Gamma b}{\tau_c c}\right)^{1/3} \quad (3.22)$$

where L is the average spacing between obstacles and c is the volume fraction of obstacles. Combining equations (3.21) and (3.22) gives

$$\tau_c = \frac{f_{\max}^{3/2} c^{1/2}}{b^2 (2\Gamma)^{1/2}} \quad (3.23)$$

This equation describes the contribution of a random array of localized obstacles to the flow stress at 0 K. At finite temperatures, this theory has to be combined with the theory of thermal activation (Sec. 4.8). Although great progress has been made in characterizing dislocation–obstacle interactions, a comprehensive model for localized and long-range interactions above 0 K remains a "work in progress."

Two, three, or even all four of these strengthening methods are often present in the same engineering alloy, and combinations of these methods can produce metals possessing both high tensile strength and ductility (Sidebar 14.3). For example, cold-drawn steel piano wire uses all four strengthening mechanisms, and it can have a yield strength of 4800 MPa parallel to the wire axis. A wire of this metal only 2.5 mm in diameter can hold the weight of a large automobile. Still stronger alloys, such as heavily cold-rolled Mo–Re, have achieved ultimate strengths of 8000 to 9000 MPa.

ADDITIONAL INFORMATION

References, Appendixes, Problem Sets, and Metal Production Figures are available at
ftp://ftp.wiley.com/public/sci_tech_med/nonferrous

4 Dislocations

Perfection has one grave defect: it is apt to be dull.
—William Somerset Maugham, *The Summing Up*

4.1 INTRODUCTION

Dislocations affect all the physical properties of crystalline materials to some degree, and they alter mechanical properties profoundly. Dislocations are line defects that can change a crystal's shape permanently without fracturing it. A crystalline solid without mobile dislocations is a less versatile material because it can deform plastically only by twinning, kink band formation, or diffusion. Dislocations greatly improve a crystal's ductility, but they also weaken the metal. The concept of dislocation was introduced in Chapter 2; in this chapter we discuss dislocation theory in more detail.

4.2 FORCES ON DISLOCATIONS

When sufficient force is applied to a ductile, crystalline material, its dislocations move and produce slip. Figure 4.1 shows a dislocation line moving in the direction of its Burgers vector under the influence of a shear stress τ. When an element of the dislocation line of Burgers vector **b** moves forward a distance ds, the crystal planes above and below the slip plane are displaced relative to one another by **b**. The average shear displacement of the crystal surface produced by glide of segment dl is

$$\frac{ds\,dl}{A}b \tag{4.1}$$

where A is the entire slip plane area. The applied force on this area is τA, so that the work done (work = force × distance) is

$$dW = \tau A \frac{ds\,dl}{A} b \tag{4.2}$$

The force on a dislocation is defined as force F per unit length of dislocation line. $F = dW/ds$, and knowing that this is a force per unit length dl gives

$$F = \frac{dW}{ds\,dl} = \frac{dW}{dA} = \tau b \tag{4.3}$$

The stress is the shear stress in the slip plane resolved along **b**, and the force F acts normal to the dislocation line at each point along its length.

A dislocation is a line defect that causes elastic strain energy in the surrounding lattice, so it possesses strain energy per unit length. As a result, it will minimize its energy if it can reduce

---

Structure–Property Relations in Nonferrous Metals, by Alan M. Russell and Kok Loong Lee
Copyright © 2005 John Wiley & Sons, Inc.

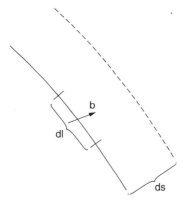

Fig. 4.1 Force acting on a dislocation line. (Redrawn from Hull and Bacon, 1984, p. 82.)

its length, which means that it possesses a line tension analogous to the surface tension of a soap bubble or liquid. The dislocations seen in crystals are seldom straight, but dislocations are observed to "smooth out irregularities" along their length while moving under stress. The line tension of a dislocation per unit length is given by

$$\Gamma = \alpha G b^2 \qquad (4.4)$$

where α is a numerical constant, typically between 0.5 and 1.0, G the shear modulus, and b the Burgers vector of the dislocation.

Suppose that a shear stress is acting on a dislocation to curve it (Fig. 4.2). The line tension will produce a force that acts to straighten the line (to reduce the total energy of the line), and the line will remain curved if there is a shear stress producing a force on the dislocation line in the opposite sense. The shear stress τ_0 maintains a radius of curvature R that can be calculated as follows. The angle subtended by an element of arc dl is $d\theta = dl/R$, assumed to be $\ll 1$. The outward force on the dislocation line (along OA) is $\tau_0 b\, dl$, and the opposing inward force (along OA) due to the line tension Γ at the ends of the element is $2\Gamma \sin(d\theta/2)$, which simplifies to $\Gamma\, d\theta$ for small values of $d\theta$. For the dislocation line to remain in the curved position,

$$\Gamma\, d\theta = \tau_0 b\, dl$$

or

$$\tau_0 = \frac{\Gamma}{bR} \qquad (4.5)$$

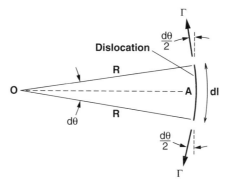

Fig. 4.2 Forces on a curved dislocation line. (Redrawn from Hull and Bacon, 1984, p. 82.)

40 DISLOCATIONS

Substituting for Γ from equation (4.4), the shear stress required to bend a dislocation to a radius R can be expressed as

$$\tau_0 = \frac{\alpha G b}{R} \qquad (4.6)$$

This equation provides an adequate approximation for most dislocations. However, it assumes that edge, screw, and mixed dislocation segments have the same energy per unit length (they do not), and it also assumes that the curved dislocation is the arc of a circle (true only if the material's Poisson's ratio is zero). Nevertheless, these two imperfect assumptions do not seriously compromise the validity of equation (4.6).

4.3 FORCES BETWEEN DISLOCATIONS

The forces acting on dislocations depend on two factors: (1) the Peierls stress, which is the intrinsic resistance to dislocation movement through a lattice, and (2) interactions with other dislocations. Consider two parallel edge dislocations lying on the same slip plane (Fig. 4.3). They can have either the same sign (i.e., have parallel Burgers vectors) (Fig. 4.3a) or opposite sign (i.e., have antiparallel Burgers vectors) (Fig. 4.3b). When the dislocations are widely separated, the total elastic strain energy per unit length of the dislocations in both parts (a and b) will be

$$\alpha G b^2 + \alpha G b^2 \qquad (4.7)$$

When the two dislocations in Fig. 4.3a are close together, the arrangement can be approximated by a single dislocation with a Burgers vector having magnitude 2**b**. For this situation, the elastic strain energy will be given by

$$\alpha G (2b)^2 \qquad (4.8)$$

which is twice the energy of widely separated dislocations. Therefore, the dislocations tend to repel each other to reduce their total elastic strain energy. When two parallel edge dislocations

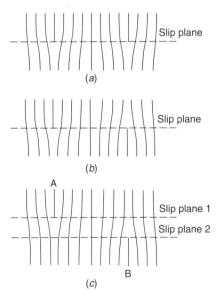

Fig. 4.3 Arrangement of edge dislocations: (*a*) two like edge dislocations on the same slip plane; (*b*) Two unlike edge dislocations on the same slip plane; (*c*) Two unlike edge dislocations on slip planes separated by a few atomic spacings. (Redrawn from Hull and Bacon, 1984, p. 86.)

of *opposite* sign lie close together on the same slip plane, the effective magnitude of their combined Burgers vectors is zero, and the corresponding long-range elastic energy is also zero. The negative and positive edge dislocations (Fig. 4.3b) will attract each other and annihilate to reduce their total elastic energy. A more complicated situation arises (Fig. 4.3c) when the two edge dislocations do not lie on the same slip plane. That case is discussed below.

Consider the two dislocations shown in Fig. 4.4. Each dislocation is assumed to lie parallel to the z axis, and the slip planes for both are parallel to xz. The force exerted on edge dislocation II located at (x, y) by edge dislocation I located at the origin has two components, F_x and F_y. These x- and y-direction force components (force/unit length of dislocation line) can be described by

$$F_x = \sigma_{xy} b \tag{4.9}$$

$$F_y = \sigma_{xx} b \tag{4.10}$$

where σ_{xy} and σ_{xx} are the stresses caused by dislocation I evaluated at the position (x, y) of II. The stress field of an edge dislocation is given by

$$\sigma_{xy} = Dx \frac{x^2 - y^2}{(x^2 + y^2)^2}$$

and

$$\sigma_{xx} = -Dy \frac{3x^2 + y^2}{(x^2 + y^2)^2} \tag{4.11}$$

where $D = Gb/2\pi(1 - v)$. Substitution from equation (4.11) gives

$$F_x = \frac{Gb^2}{2\pi(1 - v)} \frac{x(x^2 - y^2)}{(x^2 + y^2)^2} \quad \text{and} \quad F_y = \frac{-Gb^2}{2\pi(1 - v)} \frac{y(3x^2 + y^2)}{(x^2 + y^2)^2} \tag{4.12}$$

Since edge dislocations are confined primarily to their original slip plane, the force component along the x direction (F_x) is more important in determining dislocation behavior. Figure 4.5 is a plot of the variation of force between edge dislocations F_x, against distance x, where x is expressed in units of y. The solid curve represents two edge dislocations of the same sign, and the dashed curve represents two edge dislocations of opposite sign. The glide force F_x will be repulsive between two dislocations of the same sign on the same glide plane and on different

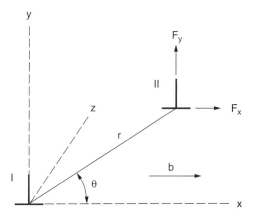

Fig. 4.4 Interaction between two edge dislocations.

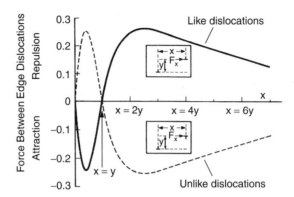

Fig. 4.5 Force between parallel edge dislocations with parallel Burgers vectors from equation (4.12). Solid curve represents two edge dislocations of the same sign; dashed curve represents two edge dislocations of opposite sign. (Adapted from Hull and Bacon, 1984, p. 86; with permission.)

glide planes as long as $x > y$. But if $x < y$, two dislocations of the same sign are attracted into alignment directly above one another.

4.4 MULTIPLICATION OF DISLOCATIONS

There are too few dislocations normally present in an unstressed crystal to produce large plastic strains. If every dislocation present in a typical annealed crystal moves all the way to the nearest free surface or barrier, plastic strains of only a few percent are possible. To achieve the large plastic strains that are actually observed in ductile crystals, many new dislocations must be produced, and they must be produced on a fairly small number of slip planes to account for the slip bands seen on the crystal's surfaces.

The best-known mechanism of dislocation generation is the Frank–Read source (Fig. 4.6), a dislocation segment pinned between two node points (A and B). The node points can be point defects, or they can be points at which the dislocation intersects dislocations lying in other planes. As discussed in Sec. 4.2, the dislocation possesses a line tension, tending to shorten its length (Fig. 4.6a). Under the action of an applied shear stress, the dislocation will bow out, decreasing its radius of curvature until it reaches an equilibrium position in which the line tension balances the force due to the applied shear stress. As applied stress rises, the dislocation loop adopts a semicircular pattern (Fig. 4.6b), at which point the loop becomes unstable and expands as shown in parts (c) and (d). It is important to note that as the loop expands, every point on the dislocation loop still has the same Burgers vector as the original dashed straight dislocation line shown in Fig. 4.6a. Between stages (c) and (d), the two parts of the loop below AB (which now have screw character and antiparallel Burgers vectors) meet and annihilate each other to form a closed loop that expands into the slip plane, leaving behind a new dislocation line between A and B. An almost unlimited amount of slip can thus be generated on this plane by the repeated creation of dislocation loops. This sequence is considered as the typical dislocation multiplication mechanism. Figure 4.7 shows a Frank–Read source; the dislocation is held at each end by other parts of the dislocation network. Because these other dislocations are not in the plane of the loops, the microscope's electron diffraction contrast is unfavorable for them, and they are not visible on the micrograph. The dislocation lines in Fig. 4.7 are not circular, but rather, align parallel to specific directions in which they possess a minimum energy; this shows the strong anisotropy of Si.

Figure 4.8 shows another mechanism of multiplication that is similar to that of a Frank–Read source. Part AB of a dislocation line in the lower slip plane of screw orientation has cross-slipped

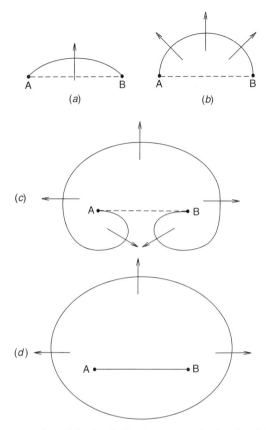

Fig. 4.6 Schematic representation of the Frank–Read source mechanism for dislocation multiplication. The arrows represent the direction of motion of the dislocation line, not Burgers vectors.

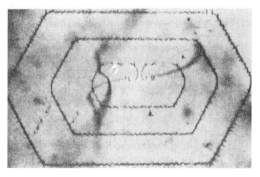

Fig. 4.7 Transmission electron micrograph image of a Frank–Read source in Si decorated with Cu. (From Dash, 1956; with permission.)

(Sec. 4.5) on plane *ABCD* to the plane above and has subsequently bowed itself out in this plane to form loop *CEFD*. This loop may in turn cross-slip and become a source to generate a loop on a still higher plane. In this way, a single dislocation loop is able to expand, multiply, and spread to nearby parallel slip planes to create a wide slip band. This process of multiplication by cross-slip has been seen in metals with high stacking-fault energy, such as BCC metals, and

44 DISLOCATIONS

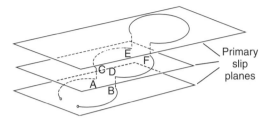

Fig. 4.8 Dislocation multiplication by cross-slip. (Redrawn from Smallman, 1970, p. 268.)

in ionic crystals such as LiF and NaCl. Multiple cross-slip is a more effective process than a Frank–Read source, since it can produce more rapid dislocation multiplication.

4.5 PARTIAL DISLOCATIONS

Movement of *perfect* dislocations through a distance of one Burgers vector places atoms in the same configuration they had before movement. The atom positions before and after the dislocation passed by are identical. If, however, a different sort of dislocation movement causes a new atomic configuration, the dislocation is said to be *imperfect* or *partial*. In many crystals, a perfect dislocation will split into two or more imperfect dislocations if in doing so, the energy of the crystal is reduced.

FCC metals slip most easily on {111} planes in ⟨110⟩ directions. But there is more than one vector sum that can accomplish the {111}⟨110⟩ slip displacement, and this leads to dislocation reactions that provide alternative combinations of dislocations that cause the same net plastic deformation. Figure 4.9 shows the close-packed array of atoms on the (111) planes in an FCC crystal. The lattice vector $\mathbf{b}_1$ in the FCC crystal ($\frac{1}{2}a\langle 110\rangle$) joins a cube corner atom to a neighboring face-centered atom and defines the FCC slip direction observed. Recalling that

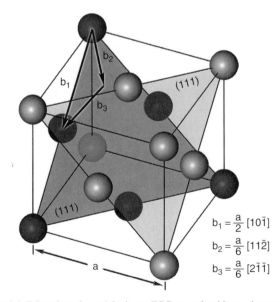

Fig. 4.9 Slip by partial dislocations b_2 and b_3 in an FCC crystal achieves the same deformation as that produced by perfect dislocation b_1.

{111} planes in FCC metals have an *ABCABCABC* ... stacking sequence (Sec. 2.4.4), glide on a (111) plane requires that atoms that sit in *B* positions above the *A* plane must move. The *B* atoms will move most easily initially along the vector marked $\mathbf{b}_2$, followed by the motion labeled $\mathbf{b}_3$. Consequently, to produce a macroscopic slip movement along $[10\bar{1}]$, the atoms might be expected to take the zigzag path marked by $\mathbf{b}_{\bar{2}}$ and $\mathbf{b}_{\bar{3}}$. The perfect dislocation with Burgers vector $\mathbf{b}_1$ therefore splits into two dislocations, $\mathbf{b}_2$ and $\mathbf{b}_3$, according to the reaction

$$b_1 \rightarrow b_2 + b_3$$

or

$$\frac{a}{2}[10\bar{1}] \rightarrow \frac{a}{6}[2\bar{1}\bar{1}] + \frac{a}{6}[11\bar{2}] \tag{4.13}$$

Note that the dislocation reaction is (1) algebraically correct, since the sum of the Burgers vector components of the two partial dislocations are equal to the components of the Burgers vector of the unit dislocation as shown by

$$\tfrac{1}{2}a[10\bar{1}] \rightarrow \tfrac{1}{6}a[2+1], \tfrac{1}{6}a[\bar{1}+1], \tfrac{1}{6}a[\bar{1}+\bar{2}] \tag{4.14}$$

and (2) energetically favorable, since the sum of strain energy values for the pair of half dislocations is less than the strain energy value of the single unit dislocation, where the initial dislocation energy is proportional to $b_1^2 = a^2/2$ and the energy of the resulting partials is lower, $b_2^2 + b_3^2 = a^2/3$. The two partial dislocations, known as *Shockley partials*, are coupled together by a stacking fault because the atoms between the two partials have been moved from the normal *ABCABC* stacking sequence. The complete *extended dislocation*, as the configuration of two partials and their stacking fault is called, glides as a single entity in the slip plane.

The total energy E_T of the extended dislocation depends on the width of the stacking-fault ribbon w according to the equation

$$E_T = E_1 + E_2 + E_{12} + \gamma w \tag{4.15}$$

where E_1 and E_2 are the energies of the individual partials, E_{12} the interaction energy between them, and γw the energy of a unit length of stacking fault with width w. Two opposing factors determine the width of the stacking fault ribbon between the two partial dislocations. The elastic strain energy (E_{12}) produces a repulsive force between the two partial dislocations, tending to move them far apart. However, the energy of the stacking fault, E_T, is minimized when the partial dislocations are as close to one another as possible. The final equilibrium separation of the two dislocations (i.e., the stacking fault width) is the value that minimizes the sum of these two terms:

$$w = \frac{Ga^2}{24\pi\gamma} \tag{4.16}$$

where G is the shear modulus, a the lattice parameter, and γ the energy per unit area of stacking fault.

Figure 4.10a shows sets of extended dislocations lying in parallel slip planes. The stacking fault ribbon between two partials appears as a fringe pattern of bold parallel black lines. The individual partials are not always clearly visible, and their positions are drawn pictorially in Fig. 4.10b. An estimate of γ can be made by examining extended dislocations in the TEM. The stacking fault energies of some familiar FCC metals range from about 20 mJ/m^2 for Ag, 40 mJ/m^2 for Cu, 45 for Au, 150 for Al, to 250 mJ/m^2 for Ni. The width of the stacking-fault ribbon affects the phenomenon of cross-slip, in which a dislocation changes from one slip plane to another intersecting plane. Cross-slip can occur only if the slip plane contains both the Burgers vector and the line of the dislocation (i.e., the dislocation must have a pure screw orientation). If the dislocation is extended, the partials must be brought together to form an unextended

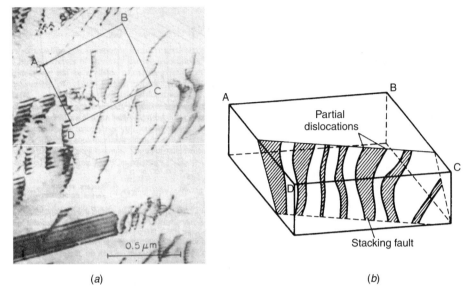

Fig. 4.10 (*a*) Transmission electron micrograph of extended dislocations in Cu–7% Al alloy. (*b*) Schematic arrangement of dislocations in the inset in part (*a*). Note that the stacking faults intersect both the upper and lower surfaces of the thin foil specimen photographed in the TEM. (From Howie, 1961; with permission.)

(perfect) dislocation before the dislocation can move onto the cross-slip plane (Fig. 4.11). The constriction necessary for cross-slip is aided by thermal activation; consequently, the tendency to cross-slip increases with increasing temperature. The constriction is also more difficult to form for widely separated partials (i.e., having low stacking-fault energy). This explains why cross-slip is common in Al (Fig. 4.12), which has a very narrow stacking-fault ribbon, but is rare in Cu or Au, which have wider stacking-fault ribbons.

In the preceding discussion, the Burgers vector of the partial lies in the fault plane, and hence the dislocation can glide and is called *glissile*. However, when the vector does not lie in the fault plane, it cannot glide in the normal fashion and hence is called *sessile*. A sessile dislocation cannot move readily except by the diffusion of atoms or vacancies to or from the fault (i.e., by climb, Sec. 2.3 and Fig. 2.6). Sessile dislocations are natural obstacles to glissile dislocations and thus participate in phenomena where dislocations move with increasing difficulty

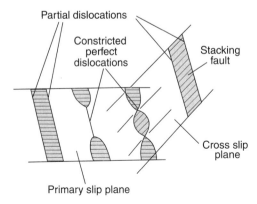

Fig. 4.11 A dissociated screw dislocation must first collapse to a perfect dislocation before it can cross-slip.

PARTIAL DISLOCATIONS   47

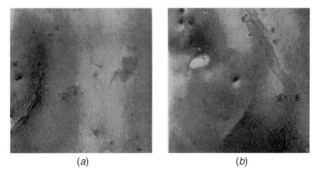

Fig. 4.12 Transmission electron micrograph of slip and cross-slip of dislocation B in Al foil as shown by comparison of (a) and (b). (From William et al., 1969, p. 296; with permission.)

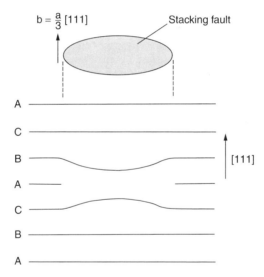

Fig. 4.13 Frank sessile dislocation is a stacking fault with Burgers vector $a/3[111]$. (Adapted from Smallman, 1970, p. 220.)

(e.g., work hardening). Such a dislocation can be produced by removing a portion of a close-packed plane (Fig. 4.13). When vacancies aggregate on the central A plane, the adjoining parts of the neighboring B and C planes would collapse to fit in close-packed formation. The Burgers vector of the dislocation line bounding the collapsed sheet is perpendicular to the plane with $b = a/3[111]$. Such a dislocation encloses an area of stacking fault that could move only by following a "cylindrical path" (i.e., moving perpendicular to its plane), with the dislocation producing a circular depression when it finally intersects the crystal's surface. It is difficult to apply a stress to the crystal that will induce this motion, so for all practical purposes, the sessile loop moves only by diffusion. A second type of Frank sessile dislocation loop can be obtained by insertion of an extra part plane between two normal planes. The Frank dislocation loop for both cases contains a stacking fault. For the loop formed from vacancies by removing, say, an A layer of atoms, the stacking sequence changes from the normal $ABCABCA\ldots$ to $ABCBCABCA\ldots$. Inserting a layer of atoms makes the sequence $ABCBABCA\ldots$. These *intrinsic* and *extrinsic stacking faults* are not necessarily confined to small regions bounded by dislocation loops, but can extend across the entire crystal (Secs. 16.2.2.1 and 16.3.6.1).

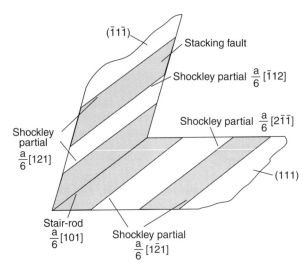

Fig. 4.14 Sessile dislocation: stair-rod dislocation as part of a Lomer–Cottrell barrier. (Adapted from Smallman, 1970, p. 220.)

A sessile dislocation can also be formed between extended dislocations on intersecting {111} planes (Fig. 4.14). The combination of the leading partial dislocation lying in the (111) plane with that which lies in the intersecting plane forms another partial dislocation, at the junction of two stacking-fault ribbons, by the reaction

$$\frac{a}{6}[121] + \frac{a}{6}[1\bar{2}1] = \frac{a}{6}[101] \tag{4.17}$$

The new partial with vector a/6[101] is often referred to as a *stair-rod dislocation* (Fig. 4.14), since it lies along the line of intersection of two stacking faults. It is a low-energy dislocation with $|\mathbf{b}| = \mathbf{a}\sqrt{2}/6$, and since its Burgers vector does not lie in either of the {111} planes but in the (010) plane, it is sessile. It exerts a repulsive force on the two remaining Shockley partials, and these three partial dislocations form a stable sessile arrangement. It acts as a strong barrier to the glide of further dislocations and is known as a *Lomer–Cottrell barrier*. Such barriers have been observed by electron microscope in metals with low stacking-fault energies, and they play an important role in the work hardening of FCC metals.

4.6 SLIP SYSTEMS IN VARIOUS CRYSTALS

The most important deformation mechanism in metallic crystals, *crystallographic slip*, involves the movement of dislocations in certain planes and directions in each crystal. The combination of slip plane and slip direction constitutes a slip system (e.g., {110}⟨111⟩). The greater the number of such sets of slip planes with different orientations in the crystal, the greater the number of directions on each slip plane in which slip can occur, and the easier it is to deform that metal. Crystals slip most readily on their more densely packed planes (e.g., {111} is preferred in FCC over {210}) and along relatively close-packed directions. There are, of course, exceptions to this tendency, when, for example, the process involves the close approach of dissimilar atoms in intermetallic compounds (Sec. 25.6.1). Table 4.1 shows the more common slip planes and slip directions for a number of metals at relatively low temperatures.

In HCP metals, slip usually occurs along ⟨11$\bar{2}$0⟩ directions, since these have the most atoms per unit length, but the active slip plane depends on the value of the axial ratio, c/a. For ideal

TABLE 4.1 Predominant Slip Planes and Slip Directions of Various Metals

Structure	Metal	Slip Plane	Slip Direction
FCC	Al	{111}	$\langle \bar{1}01 \rangle$
	Cu	{111}	$\langle \bar{1}01 \rangle$
	Ag	{111}	$\langle \bar{1}01 \rangle$
	Au	{111}	$\langle \bar{1}01 \rangle$
	Ni	{111}	$\langle \bar{1}01 \rangle$
BCC	α-Fe	{110}{211}{321}	$\langle \bar{1}11 \rangle$
	Mo	{211}	$\langle \bar{1}11 \rangle$
	W	{211}	$\langle \bar{1}11 \rangle$
	K	{321}	$\langle \bar{1}11 \rangle$
	Na	{211}{321}	$\langle \bar{1}11 \rangle$
	Nb	{110}	$\langle \bar{1}11 \rangle$
HCP	Mg	{0001}{10$\bar{1}$1}	$\langle 2\bar{1}\bar{1}0 \rangle$
	Cd	{0001}	$\langle 2\bar{1}\bar{1}0 \rangle$
	Zn	{0001}	$\langle 2\bar{1}\bar{1}0 \rangle$
	Be	{0001}	$\langle 2\bar{1}\bar{1}0 \rangle$
	Re	{10$\bar{1}$0}	$\langle 2\bar{1}\bar{1}0 \rangle$
	Ti	{10$\bar{1}$0}{0001}{10$\bar{1}$1}	$\langle 2\bar{1}\bar{1}0 \rangle \langle 11\bar{2}0 \rangle \langle 11\bar{2}2 \rangle$
	Zr	{10$\bar{1}$0}{0001}{10$\bar{1}$1}	$\langle 2\bar{1}\bar{1}0 \rangle \langle 11\bar{2}0 \rangle \langle 11\bar{2}2 \rangle$
	Hf	{10$\bar{1}$0}{0001}{10$\bar{1}$1}	$\langle 2\bar{1}\bar{1}0 \rangle \langle 11\bar{2}0 \rangle \langle 11\bar{2}2 \rangle$

packing of spheres in the HCP configuration, the axial ratio is 1.633. However, this ideal ratio does not occur in any HCP metal. Cd and Zn have axial ratios considerably higher than the ideal (1.886 and 1.856, respectively), so their planes of greatest atomic density are the (0001) basal planes, and slip occurs primarily on that plane (see, however, Sec. 19.3.6). When the axial ratio is smaller than 1.633, other slip planes may become active. For example, in Zr the axial ratio is 1.589, and slip occurs on {10$\bar{1}$0} prism planes at room temperature and also on pyramidal planes at high temperatures.

The dominant slip system in FCC metals has a Burgers vector of $a/2 \langle 110 \rangle$ on {111} planes. The {111} planes are the most densely packed planes and also the planes on which stacking faults form. In addition, $a/2 \langle 110 \rangle$ is the smallest possible perfect-dislocation Burgers vector. However, slip has been observed at high temperature in the $a/2 \langle 110 \rangle$ direction in Al on {110} planes, as has slip on the {100} and {211} planes. Dislocations with a $\langle 100 \rangle$ Burgers vector have been observed in FCC metals, but they have never been observed to slip. This finding is consistent with the prediction that the large Burgers vector $\langle 100 \rangle$ dislocations should have a higher Peierls stress than the $a/2 \langle 110 \rangle$ type.

In BCC metals, slip almost always occurs in the $\langle 111 \rangle$ directions. The shortest lattice vector (i.e., the Burgers vector of the perfect slip dislocation) is of the type $a/2 \langle 111 \rangle$. One exception to this is $\langle 100 \rangle$ slip in Nb single crystals with $\langle 111 \rangle$ tensile axes. This uncommon behavior results from elastic anisotropy. BCC metals' slip planes are {110}, {112}, and {123}. The preference of slip planes in these metals is often influenced by temperature, and a preference is normally shown for {112} below $T_{m.pt.}/4$, {110} from $T_{m.pt.}/4$ to $T_{m.pt.}/2$, and {123} at high temperatures. Thus, there are 48 possible BCC slip systems, but since the planes are not so close packed as in FCC structures, higher shearing stresses are usually required to produce slip. The multiplicity of possible slip systems is sometimes reflected in the wavy appearance of slip bands in many BCC metals. This occurs because screw dislocations can easily move from one type of plane to another by cross-slip, causing irregular, wavy slip bands.

The availability of active slip systems is crucial in determining whether a polycrystalline metal can deform with extensive ductility or fracture with little or no ductile flow. Von Mises predicted that polycrystalline metals would display extensive ductility if they possessed what he termed five "independent" slip systems. The difficulty in achieving polycrystalline ductility is illustrated by Fig. 3.1, which depicts the various shape accommodations that must occur as each grain changes shape under the applied stress. The von Mises criterion's five independent slip systems do not count a slip system such as the $\{111\}\langle110\rangle$ of the FCC lattice as being one slip system, but rather, count the various geometric options in three-dimensional space that the grains can use to change their shape. By von Mises's precept, the $\{111\}\langle110\rangle$ of the FCC lattice provides the necessary five independent slip systems. An example of von Mises's counting of independent slip systems is given in Sec. 11.3.2.2 for the case of polycrystalline Be metal. Many metals with complex crystal structures (e.g., α-Mn, Sec. 15.3.1) as well as most intermetallic compounds (Sec. 25.6.1) fail to satisfy von Mises's criterion, and these metals are brittle. Some metals possess the five independent slip systems required by von Mises's precept but fail to display extensive polycrystalline ductility for other reasons (e.g., Cr, Sec. 14.2.2.2).

As stated above, *shear* stress drives dislocation slip on specific crystallographic planes. Often, the *tensile* stress applied to a test specimen is known, but it would be more useful to know what portion of that tensile stress is actually resolved into a *shear* stress to move dislocations on slip planes. If one knows the orientation of a tensile force vector with respect to the slip plane and slip direction, it is possible to calculate the shear component of that vector that is resolved onto the slip plane in the slip direction. Consider the three cylindrical single-crystal tensile test specimens depicted in Fig. 4.15. The single crystal in Fig. 4.15a is oriented with its slip plane almost perpendicular to the tensile force F. The portion of the tensile force that acts to produce a shear force (S_1) on the slip plane is fairly small because the angle between the slip plane and F is nearly 90°. In Fig. 4.15b the slip plane is more steeply inclined, so the magnitude of resolved shear force S_2 is greater than S_1. In Fig. 4.15c, the slip plane is steeper yet, so the resolved shear force S_3 is greater still. It is clear that the orientation shown in Fig. 4.15c produces the largest shear force of the three cases shown; however, dislocations move in response to shear *stress*, not shear *force*. The shear stress (τ) is the shear force divided by the area (A)

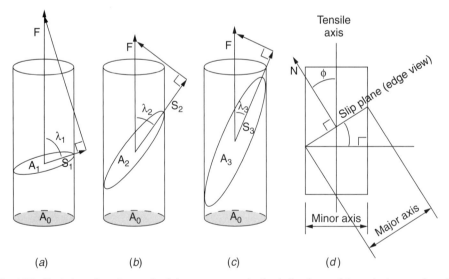

Fig. 4.15 Depiction of varying resolved shear stresses on inclined slip planes of three single-crystal tensile test specimens with different orientations.

of the slip plane:

$$\tau = \frac{S}{A} \qquad (4.18)$$

In Fig. 4.15c, the area of the slip plane, A_3, is larger than A_2 in Fig. 4.15b, which in turn is larger than A_1. Thus, it is not immediately obvious which of the three orientations depicted in Fig. 4.15 produces the greatest shear stress.

A simple geometric relationship exists between F and S such that $F(\cos \lambda) = S$. The slip planes in these specimens are all elliptical, and the area of an ellipse is given by

$$A = \frac{\pi}{4}(\text{major axis})(\text{minor axis}) \qquad (4.19)$$

In addition, the major axis of the ellipse is related to the minor axis by

$$\text{major axis} = \frac{\text{minor axis}}{\cos \phi} \qquad (4.20)$$

so the expression for the area of the ellipse can be written as

$$A = \frac{\pi}{4} \frac{\text{minor axis}}{\cos \phi}(\text{minor axis}) \qquad (4.21)$$

where ϕ is defined as the angle between the tensile axis and a vector normal to the slip plane (Fig. 4.15d). Since the minor axis of the ellipse is just the diameter of the cylindrical tensile specimen, the area of the ellipse is related to the circular cross-sectional area (A_0) of the cylinder as

$$A = \frac{A_0}{\cos \phi} \qquad (4.22)$$

because A_0 is just $\pi/4(\text{minor axis})^2$, the area of a circle. Finally, the expression for shear stress in equation (4.18) can now be rewritten as

$$\tau = \frac{S}{A} = \frac{F \cos \lambda}{A_0/\cos \phi} = \frac{F}{A_0}(\cos \lambda)(\cos \phi) \qquad (4.23)$$

Since F/A_0 is the definition of tensile stress, this quotient can be replaced by σ. Thus, the relation between tensile stress (which is usually easily determined) and shear stress on a given slip plane in the slip direction (which is less easily determined) is defined by the relation (Schmid's law)

$$\tau = \sigma \cos \lambda \cos \phi \qquad (4.24)$$

Schmid's law allows the shear stress driving dislocation motion to be calculated from the tensile stress if the orientations between the slip plane, the slip direction, and the tensile stress axis are known. It turns out that the optimal orientation to maximize shear stress occurs when $\lambda = \phi = 45°$; in that orientation, $\tau = 0.5\sigma$. For all other orientations, $\tau < 0.5\sigma$. If either λ or ϕ is 90°, τ will be zero no matter how large σ may be. Although Fig. 4.15 has been drawn to show the slip direction as the line on the slip plane with the greatest slope, the slip direction in general can be in any position on the slip plane [i.e., $(\lambda + \phi)$ will not necessarily be 90°, as they appear to be in Fig. 4.15].

Schmid's law can be used to determine the minimum shear stress needed to initiate dislocation motion on the slip plane. This value, called the *critical resolved shear stress* (τ_{CRSS}), varies with the slip system, metal composition, and temperature. For some situations τ can be quite small (e.g., 0.08 MPa for basal slip in Zn at 300 K); in other cases it can be much greater (e.g., 5000 MPa for $\{11\bar{2}2\}\langle 11\bar{2}3\rangle$ slip in Be near 0 K).

4.7 STRAIN HARDENING OF SINGLE CRYSTALS

In ductile metals dislocation density increases greatly during plastic deformation, causing work hardening (strain hardening). Strain hardening often doubles flow stresses in single crystals of ductile metals, and in polycrystalline metals a quadrupling of flow stress is possible (Sec. 15.5.2). Strain hardening is caused by dislocation interactions that impede dislocation motion. Strain-hardening behavior is more complex in cubic metals than in HCP metals because of the variety of slip systems available. The resolved shear stress–resolved shear strain curve for deformation of FCC single crystals can be partitioned into different stages (Fig. 4.16). Stage I, the region of easy glide, immediately follows the yield point and displays very low and constant strain hardening (typically, 10^{-5} to $10^{-4}G$ at 20°C, where G is the shear modulus). During easy glide, dislocations move long distances without encountering barriers. The small amount of strain hardening that does occur during stage I results primarily from the elastic interaction of dislocations of opposite sign approaching each other on neighboring planes. The strain-hardening rate is insensitive to temperature because there are few dislocation intersections (Sidebar 4.1) and therefore little need for thermal activation.

The ends of the single-crystal tensile specimen are not free to move sideways as the tensile testing machine elongates the specimen vertically, so the shearing action of stage I slip eventually rotates the slip planes into a position where the first planes to slip no longer have the most advantageous Schmid factor. When this occurs, one or more other planes begin to slip. Stage II, the linear hardening region, shows a rapid increase in work hardening (about $4 \times 10^{-3}G$), and the new, steeper slope of the curve is approximately independent of applied stress, temperature, orientation, or purity. Since two or more slip systems are now active, dislocation intersections become much more common, and tangles start to develop. These ultimately result in the formation of a dislocation cell structure consisting of regions almost free of dislocations enclosed by areas with high dislocation density (Fig. 18.18a). Stage III is a region of decreasing strain-hardening rate. In this stage the flow stress becomes sensitive to temperature and strain rate. At sufficiently high stress in this region, dislocations held up in stage II are able to move by cross-slip to bypass an obstacle in the original slip plane, making further deformation easier. Stage III slip is sensitive to temperature; raising the temperature allows easier slip, which suggests that the intersection of forests of dislocations is the main strain-hardening mechanism.

Figure 4.16 illustrates the general behavior of FCC metals, but certain deviations from a three-stage flow curve are possible. For example, at 77 K, Al has no clearly developed stage II slip, and stage III starts before stage II becomes at all predominant. Cu behaves similarly at room temperature. This difference between Al and Cu is attributed to the difference in stacking-fault energies, which affects the width of extended dislocations. Cu has a lower stacking-fault energy (40 mJ/m^2) than Al (150 mJ/m^2), so dislocations in Cu will have wider stacking faults separating its partial dislocations, making cross-slip more difficult (Sec. 4.5). Each stage depends in a characteristic way on the purity, crystallographic orientation, temperature, size, shape, and surface condition of the specimen.

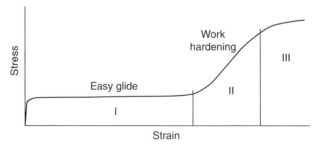

Fig. 4.16 Stress–strain behavior for a single FCC crystal that is oriented favorably for plastic flow.

SIDEBAR 4.1: DISLOCATION INTERSECTIONS

Fully annealed metal usually has a low dislocation density (about 10^5 mm of dislocation line length per mm^3 of metal). If the metal is plastically deformed, Frank–Read and cross-slip dislocation sources (Figs. 4.6-8) raise dislocation densities as high as 10^{10} mm^{-2}. Such high densities make intersections of dislocations a frequent occurrence. The effects of such intersections on the dislocations range from nil to mild to severe. The governing precept for predicting the effect of dislocation intersections is: When two dislocations intersect, a jog forms in each dislocation that is equal in direction and length to the Burgers vector of the other dislocation.

In Fig. 4.17a edge dislocation XY approaches edge dislocation AB. Their Burgers vectors are perpendicular. After intersection (Fig. 4.17b), dislocation AB has acquired a *jog* (PP') with Burgers vector **b**$_2$. Dislocation XY could be said to have a jog also, but since that jog lies parallel to dislocation line XY, it is actually nonexistent. As a consequence of this intersection, XY experiences no perceptible change, and AB has a jog of pure edge character. AB's energy is increased by the jog, but the jog can move when AB moves, so the dislocation with its new jog can still glide through the lattice. Figure 4.17c and d illustrate the outcome from the intersection of two edge dislocations with parallel Burgers vectors; both dislocations acquire a screw-character jog called a *kink*, but kinks do not prevent the dislocations from moving. In general, edge dislocations survive intersections with their mobility nearly unimpaired. They may acquire jogs or kinks that raise their energy somewhat, but they can still glide through the crystal.

Screw dislocations, however, suffer more serious consequences from intersections. In Fig. 4.17e and f, screw dislocation AB cuts through screw dislocation XY. In Fig. 4.17g and h, edge and screw dislocations intersect. All three of these screw dislocations acquire edge-character jogs from the intersections. An edge-character jog can glide only in the plane containing the jog and its Burgers vector (this plane is crosshatched in Fig. 4.17i); therefore, each of these three jogs is free to glide up and down the length of the screw dislocation. However, the normal glide motion of these screw dislocations would force the jogs to climb, which can occur only by generating a vacancy (or an interstitial if the screw dislocation moves in the opposite direction) for every atomic space the screw dislocation travels. Producing vacancies is energetically "costly" (about 1 eV per vacancy), and producing interstitials requires even more energy. Thus, the mobility of screw dislocations containing edge-character jogs is seriously impaired because jogs must climb nonconservatively through the lattice to accommodate this motion. Larger shear stresses are required to move jogged screw dislocations since a string of vacancies or interstitials must be produced as they move. Alternatively, they can move their jogs by a thermally activated process of diffusion to provide the atom motions needed to permit their jogs to climb as the screw dislocation glides (Sec. 6.4.3.1). Thus, jogged screw dislocations move much more readily as the temperature rises, and this is a major contributor to the loss of strength and to the creep of metals at high temperature.

Slip lines on the crystal's surface have different arrangements in each stage of strain hardening (Fig. 4.18). They identify the crystallographic nature of slip and also indicate the heterogeneity and dynamics of slip. Dislocation structures in Cu have been particularly well studied and provide a useful example. For Cu, stage I slip lines on the surface are long, straight, and uniformly spaced, and each slip line is small, corresponding to passage of only a few dislocations. In stage II the dislocations are stopped by elastic interaction when they pass too close to an existing tangled region with high dislocation density. The long-range internal stresses due to the dislocations piling up behind are partially relieved by secondary slip, which transforms the discrete pile-up

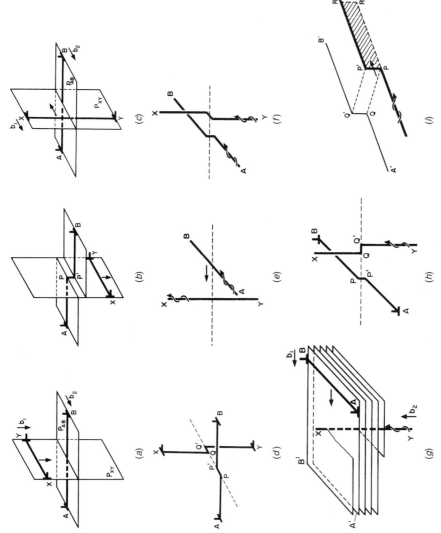

Fig. 4.17 Effects of dislocation intersections: (*a–d*) edge–edge intersections, (*e* and *f*) screw–screw intersections, (*g* and *h*) edge–screw intersections, (*i*) nonconservative motion of and edge-character jog in a screw dislocation. (From Hull and Bacon, 1984, pp. 142–145; with permission.)

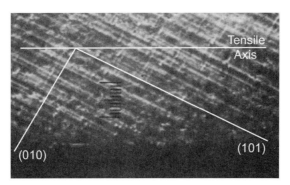

Fig. 4.18 Optical micrograph of stage II slip lines on a strained single crystal of CuY intermetallic compound. Lines from slip on two different slip planes are visible on the polished specimen surface. The fine graduations on the scale bar are 40 μm apart. (From Gschneidner, 2003, p. 587; with permission.)

into a region of high dislocation density. These regions of high dislocation density act as new obstacles to dislocation glide. As deformation proceeds, the number of obstacles increases, the distance of dislocation glide decreases and therefore the slip lines become shorter and become branched, with varied slip heights. The onset of stage III is accompanied by cross-slip; the slip lines are broad, deep, and consist of segments joined by cross-slip traces (Fig. 4.18).

The dislocation structure that develops during deformation of FCC polycrystalline metals follows the same general precepts as in single crystals. The understanding of polycrystalline hardening on the basis of single-crystal behavior requires that two additional microstructural elements be considered: the average grain size and the distribution of grain orientations (i.e., the texture). For pure FCC metals, texture has a greater influence than grain size on the dislocation structures.

Strain hardening allows metals to be given an extra increment of yield strength by cold work (Sec. 3.3), it provides the driving force for recovery and recrystallization (Sec. 3.3), and it is the operating mechanism in fatigue (Sec. 5.3). Strain hardening also adds a vitally important safety factor to structural components that often prevents or delays catastrophic failure. Consider a part of approximately uniform cross section that is loaded in tension to the point where plastic deformation just begins. Due to small variations in section or stress concentrations, the first deformation will be localized. Local elongation will decrease the cross-sectional area of that local region and increase stress in the deformed region. If strain hardening did not occur, the first local plastic deformation would lead to immediate failure. Strain hardening strengthens the first deformed regions enough to counteract the accompanying decrease in section, so failure does not occur until the load is increased significantly beyond the magnitude where the first local deformation begins. The deformation is distributed uniformly over the length of the member until necking finally begins at a stress substantially higher than the yield stress.

4.8 THERMALLY ACTIVATED DISLOCATION MOTION

Most rate-dependent behavior in metals is attributable to thermally activated processes; these, in turn, depend on the stress, strain rate, deformation, and temperature. The rate at which deformation proceeds is described by an *Arrhenius rate equation*:

$$\dot{\varepsilon} = \dot{\varepsilon}_0 \left(-\frac{\Delta Q}{kT} \right) \tag{4.25}$$

where $\dot{\varepsilon}$ is the tensile strain rate, $\dot{\varepsilon}_0$ a material constant, ΔQ the change in activation energy, k is Boltzmann's constant, and T the temperature. Dislocations moving through a crystal may

56 DISLOCATIONS

encounter two types of barriers: (1) long-range barriers 10 atom diameters or larger in size, and (2) short-range barriers smaller than 10 atom diameters. Barriers larger than 10 atom diameters are called *athermal barriers*, τ_μ, since they produce long-range stress fields within the crystal that are not affected by temperature (apart from the variation of the shear modulus G with temperature). Two common athermal barriers are large second-phase particles and other dislocations on parallel slip planes. Barriers smaller than 10 atom diameters are called *thermal barriers*, τ^*, because they can be overcome by thermal activation. It is these thermal barriers that govern the dynamic aspects of plastic flow. Examples of thermal barriers are the stress fields of coherent precipitates and solute atoms. Dislocation climb and cross-slip allow dislocations to overcome thermal barriers, and both of these processes are temperature dependent.

Figure 4.19 illustrates the combined effects of short- and long-range internal stress fields encountered by a dislocation moving through a crystal. Short-range stress fields are superimposed on a long-range stress of wavelength λ. The total internal stress can be described by the equation

$$\tau_i = \tau_\mu + \tau^* \tag{4.26}$$

Negative values of τ_i assist dislocation motion, and positive values oppose it. The rate-controlling process will consist of overcoming the strongest short-range obstacle situated near the top of the opposing (+) long-range stress field. Plastic flow occurs at 0 K (where thermal fluctuations do not exist) only when the applied stress τ equals or exceeds τ_0. However, above 0 K, thermal fluctuations aid the applied stress, and plastic flow can occur when $\tau < \tau_0$. The energy that thermal fluctuations must supply is approximated by the small crosshatched area in Fig. 4.19. In general, more energy is needed to surmount a long-range obstacle than a short-range obstacle. As the temperature increases, τ can decrease because more aid is received from thermal fluctuations, until, in the limit, $\tau = \tau_\mu$, which is the top of the opposing long-range stress field. Further increase in temperature will not allow further decrease in applied stress because the energy barrier (which now consists of the area under the long-range stress against the distance curve of Fig. 4.19) is too large for thermal fluctuations to make a significant contribution.

Figure 4.20 represents a type of short-range barrier where the activation energy is not a function of τ^* [e.g., the intersection of dislocations and the movement of jogs (Sidebar 4.1)]. When a stress τ is applied ($\tau > \tau_\mu$), the dislocation moves up the force barrier to position F,

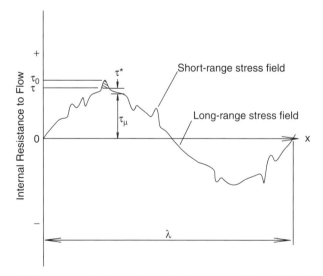

Fig. 4.19 Schematic diagram showing internal stress fields encountered by a dislocation moving through the crystal lattice. (Redrawn from Conrad, 1964.)

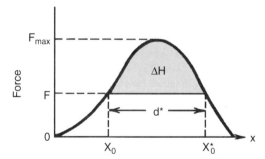

Fig. 4.20 Depiction of a thermal activation barrier. (Redrawn from Dieter, 1988, p. 312.)

given by $F = \tau^* b \ell^*$, where **b** is the Burgers vector and ℓ^* is the length of the dislocation segment involved in the thermal fluctuation. Additional energy, ΔH, must be supplied by thermal fluctuations to overcome the barrier. The work done by the applied stress during thermal activation is

$$W = \tau^* b \ell^* (x_0^* - x_0) = \tau^* b \ell^* d^* \quad (4.27)$$

The term ΔH is defined by the area under the force–distance curve from x_0 to x_0^* (designated ΔH^*) minus the work done by the applied stress during the thermal activation:

$$\Delta H = \int_{x_0}^{x_0^*} [F(x) - \tau^* b \ell^*] dx = \Delta H^* - v^* \tau^* \quad (4.28)$$

where ΔH^* is the activation energy for zero applied stress and v^* is the *activation volume*, the average volume of dislocation structure involved in the deformation process. From Fig. 4.20, the activation volume is given by $v^* = l^* \mathbf{b} d^*$ for a process in which l^* does not change with stress, $l^* \mathbf{b}$ is the atomic area involved in the flow process and d^* the distance the atoms move during this process. Activation volume provides useful information on the deformation mechanism because it has a fixed value and stress dependence for a given atomistic process.

The equation of activation volume, v^*, is given by

$$v^* = -\left(\frac{\partial \Delta Q}{\partial \tau}\right)_{T,m} \quad (4.29)$$

where ΔQ is the change in activation energy and the subscripts T and m indicate that both temperature and the dislocation microstructure are constant. Equation (4.29) is used to determine the mechanism controlling the thermally activated deformation process by comparing values of v^* and ΔQ with values calculated from specific dislocation models. The v^* and ΔQ values can be determined through different mechanical tests, which require a change in related experimental parameters. For example, v^* can be measured by the stress change that results from strain rate jumps during constant strain rate tests or stress jumps during creep experiments. Since both activation parameters are defined for constant dislocation microstructures, uncertainty exists about possible microstructure changes that may accompany such jumps in strain rate or stress. It is generally assumed to be acceptable, since the plastic strains taking place during the transient tests are small, so changes in the dislocation microstructure (if any) should not greatly alter the value of v^* determined. However, some studies have shown that structural changes during these transient tests can give activation parameter values that differ considerably from the actual thermodynamic values, so the term *apparent activation parameter* is often used.

4.9 INTERACTIONS OF SOLUTE ATOMS WITH DISLOCATIONS

Isolated solute atoms (especially atoms much smaller or larger than the solvent atoms) elastically distort the lattice just as dislocations do; this causes elastic interaction between solute atoms and dislocations. Specifically, the elastic stress fields that surround misfitting solute atoms interact with the cores of edge dislocations. The motion of pure screw dislocations is unimpeded in the presence of such solutes, since screw dislocations have no hydrostatic stress field. The misfit field surrounding a substitutional solute atom is expressed by a misfit parameter δ, which is defined as

$$\delta = \frac{1}{a}\frac{da}{dc} = \frac{d\ln a}{dc} \tag{4.30}$$

where a is the lattice parameter of the alloy and c is the concentration of solute atoms.

Consider a substitutional solute atom that changes the volume of the lattice by ΔV and produces a spherically symmetrical distortion in the surrounding lattice. An edge dislocation produces a compressional field above the slip plane ($0 < \theta < \pi$) and a dilational field below it ($\pi < \theta < 2\pi$). The interaction force between a dislocation at the origin and the solute atom at polar coordinate (r, θ) in the x-direction (i.e., the slip direction) is

$$F^P = \frac{Gb\,\Delta V}{\pi z^2}\frac{2(x/z)}{[1+(x/z)^2]^2} \tag{4.31}$$

where the parameters x and z are as defined in Fig. 4.21. The maximum in this function occurs at $x = z/\sqrt{3}$. A solute atom near the slip plane hinders the dislocation motion (i.e., F^P increases with decreasing z). As $z \to 0$, however, linear elasticity theory becomes inadequate, and the minimum value of z is therefore set equal to $b/\sqrt{6}$, which is one-half the separation of the {111} planes in the FCC lattice [i.e., $b/\sqrt{6} = (a/2)\sqrt{3}$]. Also, if the volume change is expressed as

$$\Delta V = 3\Omega\delta \tag{4.32}$$

where Ω is the atomic volume of the base metal, we obtain for the maximum interaction force in the FCC lattice

$$F_m^P \propto Gb^2\Omega\delta \tag{4.33}$$

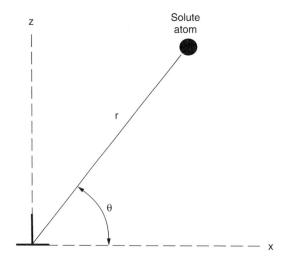

Fig. 4.21 Solute atom at a distance r from an edge dislocation.

A different sort of interaction between solute atoms and dislocations can occur if the solute atoms locally alter the modulus of the solvent matrix. If an atom of the matrix is replaced by an impurity atom of the same size but with different elastic constants, there is no lattice distortion from the solute atom, but an interaction between a dislocation and the solute atom occurs from the work the dislocation must do to move near an elastically harder impurity atom. Therefore, there is a repulsion if the impurity has a higher shear modulus (harder) than the matrix and an attraction if the solute atom has a lower shear modulus (softer) than the matrix. The change in energy, ΔU_G, due to this interaction is given by

$$\Delta U_G = \frac{G \varepsilon'_G b^2 R^3}{6 \pi r^2} \quad (4.34)$$

where ε'_G is the modulus difference between solvent and solute and is defined thus:

$$\varepsilon'_G = \frac{\varepsilon_G}{1 - \varepsilon_G/2} \quad \text{and} \quad \varepsilon_G = \frac{1}{G} \frac{dG}{dc} \quad (4.35)$$

where c is the atomic concentration of solute, G the shear modulus, R the atom radius of solute, and r the distance between solute atom and dislocation.

Electrical interaction can also occur between solute atoms and dislocations because some of the charge associated with solute atoms of different valence remains localized around the solute atom. The solute atoms develop into charge centers and interact with dislocations that have electrical dipoles. The resulting interaction energy is fairly small compared to the elastic and modulus interaction and isn't significant unless matrix and solute valence difference is large.

As discussed in Sec. 4.5, it is energetically favorable for a slip dislocation in a close-packed lattice to dissociate into a pair of partial dislocations separated by a stacking fault. When this occurs, the solid solubility in the fault region is often different from that in the surrounding matrix. This can cause a heterogeneous distribution of solute atoms around a dislocation that can be considered a chemical interaction with the dislocation (often called *Suzuki locking*).

4.10 DISLOCATION PILE-UPS

Dislocations generated from a Frank–Read source will tend to pile up against any obstacle present on the slip plane (Fig. 4.22). These obstacles can be grain boundaries, second phases, or sessile dislocations. The later dislocations "push" on the leading ones so the dislocations near the obstacle will be much more closely spaced than the last ones originating from the source. This causes a stress concentration on the leading dislocation in the pile-up. When many dislocations are restricted in the pile-up, the stress on the dislocation at the head of a pile-up can initiate yielding on the other side of the barrier or nucleate a crack at the barrier. Dislocation pile-up against a barrier exerts a back stress τ_b on the dislocation source which will continue to produce dislocations until the magnitude of the back stress equals that of the applied stress less the stress τ_a required to activate the source, such that

$$\tau_b = \tau - \tau_a \quad (4.36)$$

where τ is the applied stress.

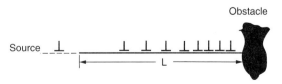

Fig. 4.22 Schematic illustration of dislocation pile-up at an obstacle.

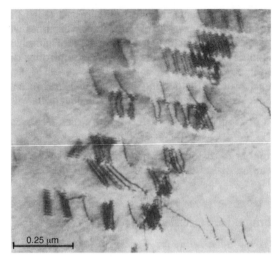

Fig. 4.23 Dislocation pile-ups in stainless steel. (From Honeycombe, 1982, p. 46; with permission.)

The dislocations in the pile-up will be tightly packed together near the head of the array (Fig. 4.23) because the applied shear stress forces all the dislocations from the source toward the obstacle. Eshelby et al. (1951) studied the distribution of dislocations of like sign in a pile-up along a single slip plane and found that the number of dislocations that can be piled into a distance L along the slip plane between the source and the obstacle is given by

$$n = \frac{k \pi \tau L}{Gb} \tag{4.37}$$

where n is the number of dislocations in the pile-up and $k = 1$ for screw and $1 - \nu$, where ν is Poisson's ratio, for edge dislocations. The number of dislocations is directly proportional to the stress and inversely proportional to the shear modulus and the Burgers vector length. Other factors that may influence this are the type of barrier, the material, the orientation relationship between the slip plane and the structural features at the barrier, and temperature.

ADDITIONAL INFORMATION

References, Appendixes, Problem Sets, and Metal Production Figures are available at
ftp://ftp.wiley.com/public/sci_tech_med/nonferrous

5 Fracture and Fatigue

A major portion of the upper crown skin and structure of section 43 separated in flight causing an explosive decompression of the cabin.... [T]he flight attendant was immediately swept out of the cabin through a hole in the left side of the fuselage.

—National Transportation Safety Board Accident Report for
Aloha Airlines Flight 243

5.1 INTRODUCTION

A materials *failure* is simply an unacceptable deformation or fracture. When a material is subjected to an applied force, it can fail by fracture or buckling. In their less severe manifestations, failures delay delivery of products and services, causing inconvenience and lost profits. In more critical components, failures can cause injury, illness, death, and huge financial losses. In many materials, a certain amount of deformation—elastic or plastic—is tolerable and in some cases, desirable (e.g., elastic deformation in fishing poles and plastic deformation of automobile "crush-zone" structures). Therefore, any classification of failure is arbitrary and normally involves subjective judgment. In this chapter we present the fundamental concepts involved with metal fracture.

5.2 FUNDAMENTALS OF FRACTURE

Fracture is the separation of a body under stress into two or more pieces as the end result of deformation. The nature of fracture differs with materials and is influenced by the nature of the applied stress, geometrical features of the sample, temperature, and strain rate. Fracture is a two-step process; it starts with crack initiation and requires crack propagation to reach failure. Crack propagation, in turn, can be divided into two categories: brittle fracture and ductile fracture. *Brittle fracture* in metals occurs by rapid crack propagation after little or no plastic deformation. Such cracks typically occur perpendicular to the direction of maximum applied tensile stress. By contrast, *ductile fracture* is preceded by significant plastic deformation, and ductile fracture has slower crack propagation resulting from the formation and coalescence of voids. Ductile fracture surfaces show extensive gross deformation. Although no failure is welcome, ductile fracture is less likely to be catastrophic, because it requires more energy (Sidebar 5.1), which sometimes delays fracture long enough to mitigate damage and provide warning that fracture is imminent.

5.2.1 Brittle Fracture: Griffith Theory

The first explanation of the discrepancy between the theoretical fracture strength of a flawless material and the actual fracture strength was given by Griffith. His theory was based on the assumptions that the material was completely brittle and that fracture occurs by growth of preexisting sharp flaws. He showed that growth of a crack is encouraged by release of elastic

---

Structure–Property Relations in Nonferrous Metals, by Alan M. Russell and Kok Loong Lee
Copyright © 2005 John Wiley & Sons, Inc.

62 FRACTURE AND FATIGUE

SIDEBAR 5.1: USING PLASTIC DEFORMATION FOR ENERGY ABSORPTION

In a high-speed frontal collision, the steering column of an automobile can thrust forcefully into the driver's chest, breaking ribs, crushing internal organs, and in extreme cases, even impaling the driver. Safety engineers use seat belts, shoulder harnesses, and air bags to reduce this hazard, but a less visible safety feature is a "crush column" inside the steering column. This part is designed to deform plastically under load to reduce the severity of injury to the driver. Collapsing the crush column absorbs energy that would otherwise be delivered to the driver's body, a classic example of designing for plastic deformation.

---

strain energy around the crack. Relieving this energy lowers the overall energy of the material. If this were the only factor involved, any stressed object containing the slightest flaw would fracture immediately under load. Obviously, real-world objects do not do this. Brittle objects with large flaws can bear a load without fracturing because energy must be added to the object to form the new surfaces created by crack growth. Griffith showed that an object's total energy reaches a maximum when the rate of stress field energy release from crack propagation is exactly balanced by the rate of surface energy input required to form new crack surfaces. Griffith's work produced the now-famous expression

$$\sigma_f = \left(\frac{2E\gamma}{\pi a}\right)^{1/2} \tag{5.1}$$

where σ_f is the fracture strength, E is Young's modulus of elasticity, γ the surface energy, and a the crack length. This expression predicts that brittle materials with strong bonds (high E) and high surface energies (high γ) have better fracture resistance. It also makes the reasonable prediction that small cracks are less likely than large ones to cause fractures.

Griffith performed tests on glass to determine that $\gamma = 1000$ ergs/cm^2. (More recent experiments indicate that the true γ value for glass is about 7000 ergs/cm^2 in dry air or vacuum. Atmospheric humidity sharply lowers glass's surface energy.) Griffith's equation is valid only for purely elastic materials in quasistatic situations; that is, slow-moving cracks in perfectly brittle materials where the kinetic energy of separation of the fractured parts and the variation of γ with the speed of crack propagation do not come into play. Many ostensibly brittle materials actually experience a small amount of plastic deformation at the crack tip during fracture. This small degree of plasticity can be accommodated in Griffith's theory by adding a plastic work term, W, to the surface energy 2γ. (Total surface energy is 2γ because an advancing crack is forming two new surfaces, each having an intrinsic energy γ.) Irwin and Orowan modified the total energy term to

$$2\gamma + W = \zeta_c \tag{5.2}$$

In real materials, W is typically hundreds of J/m^2 and γ is typically only 1 or 2 J/m^2. Thus, γ is negligibly small in comparison to W. So Griffith's equation becomes

$$\sigma_f = \left(\frac{E\zeta_c}{\pi a}\right)^{1/2} \tag{5.3}$$

This modification extends the equation's usefulness to many more materials.

Griffith's original theory was postulated for an infinitely large, very thick plate to avoid accounting for edge effects and thickness effects. Of course, real objects have finite size, so the next advance in fracture mechanics was Irwin's introduction of a stress intensity factor K. The tensile and shear stresses near a crack tip all contain a common factor $K = \sigma(\pi\alpha a)^{1/2}$, where σ is the overall applied tensile stress, α a geometrical factor, and a the crack length. The K factor determines the overall stress magnitude near the crack tip, which Irwin called the *stress intensity factor*, with dimensions of stress·(length)$^{1/2}$ (e.g., MPa·m$^{1/2}$). The theory states that the crack will propagate if K reaches a critical value K_c, which is the material's fracture toughness. To account for plasticity effects at the crack tip, a term αr_p is added to the actual crack length a to create an effective crack length $a_{\text{eff}} = a + \alpha r_p$. In this expression, r_p is the crack tip plastic zone radius and α is a constant determined experimentally to be about 1. The American Society for Testing and Materials (ASTM) defines three loading modes in a cracked body (Fig. 5.1). Mode I is a crack-opening mode, and this mode is of greatest concern in designing engineering systems. The critical value of stress intensity factor, K_C, varies with the dimensions of the object containing the crack in two ways:

1. Free surfaces near the crack tip modify the stress intensity factor from the case of a semi-infinite plate. Calibration curves are published for various crack–plate geometries to produce accurate "K calibrations" for different combinations of plate and crack dimensions.

2. The plate thickness (measured parallel to the crack tip) affects the amount of plastic relaxation possible at the crack tip. Typical end effects on a crack propagating through a plate are shown in Fig. 5.2. In a thin sheet, too little material is present to support stresses parallel to the crack tip, so all stresses lie within the plane of the sheet. Thus, for thin plates, a plane stress situation prevails. This type of loading causes shear plastic flow through the sheet thickness, and it thins the sheet at the tip of the crack, producing visible "dimples" on both sheet surfaces. The opposite extreme of the *plane stress* situation is the *plane strain* case, which occurs in thick plate. In a thick plate, deformation of the crack tip is constrained, preventing strains from developing through the plate's thickness, so strain occurs only within the plane of the plate. In this case the crack tip stresses are triaxial. (They are biaxial in the plane stress situation.) There is only limited plasticity at the crack tip in the plane strain situation, and fracture toughness is lower in this case because crack tip deformation is the largest energy-absorbing phenomenon in crack propagation.

This variation in fracture toughness with specimen thickness is shown in Fig. 5.3, where the critical stress intensity approaches the plane strain value, K_{IC} (pronounced "kay-one-see"), asymptotically as the plate is made thicker. In the strict mathematical sense, even very thick plates do not achieve a perfect plane strain situation. However, ASTM has defined that plane strain occurs when the specimen thickness, b_0, is at least about 40 times the plastic zone radius [i.e., $b_0 \geq 2.5(K_{\text{IC}}/\sigma_y)^2$]. In such cases the crack tip plastic zone is long, and the free surface effects at its ends are such a small portion of its length that they can be disregarded. K_{IC} is an inherent property of a material, similar to elastic modulus or resistivity, and values of K_{IC} are tabulated for many materials. In the same fashion, plane strain fracture energy ζ is designated ζ_{IC}, and K_{IIC}, K_{IIIC}, ζ_{IIC}, and ζ_{IIIC} exist for modes II and III.

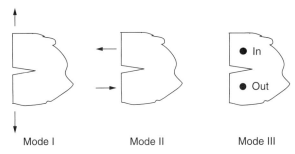

Fig. 5.1 Three modes of crack displacement. Mode I is a tensile or crack-opening action, which is often the greatest concern in engineering systems, mode II is a shearing (sliding) action, and mode III is the sort of action used to tear the edge of a piece of paper.

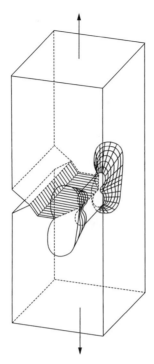

Fig. 5.2 End effects on a crack propagating through a plate. (Redrawn from Totten and MacKenzie, 2003.)

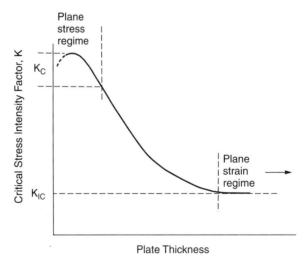

Fig. 5.3 Schematic depiction of the variation of critical stress intensity with plate thickness.

For a semi-infinite plate, ζ and K are related. For plane stress

$$K_I^2 = E\zeta \qquad (5.4)$$

and for plane strain

$$K_I^2 = \frac{E\zeta}{1-\nu^2} \qquad (5.5)$$

where ν is Poisson's ratio. This relationship is derived from the expressions for the stress intensity factor and for crack surface displacement and is valid for all geometries. Similarly, the energy release rate ζ of a cracked specimen can be expressed in terms of the applied load P and elastic compliance C by

$$\zeta = \frac{P^2}{2}\frac{dC}{da} \tag{5.6}$$

where a is crack length. The stress intensity factor for plane stress, K_I, can be expressed as

$$K_I^2 = E\zeta = \frac{1}{2}P^2 E\frac{dC}{da} \tag{5.7}$$

Ductile materials (e.g., metals and polymers) that deform extensively before fracture need other fracture parameters. A crack tip in a ductile material opens a certain critical distance before fast crack propagation occurs. This distance is called the *crack opening displacement* (COD) or δ, and δ can be related to ζ by the relationship

$$\zeta = M\sigma_y\delta \tag{5.8}$$

where M is a factor that converts a material's uniaxial yield stress into the value needed for triaxially constrained conditions. Experiments have shown that $M = 2.1$.

In the ASTM-recommended fracture toughness testing method (ASTM E 399-90), a machined notch is cut into the specimen, and this is extended into a sharp crack by fatiguing the specimen. The overall scale of the specimen is dependent on the properties of the particular material. Testing is done until fast fracture occurs, and posttest data analysis produces a critical load for fast fracture propagation. This critical load, together with the specimen dimensions, allows a provisional plane strain fracture toughness value, K_Q, to be calculated. The K_Q value can then be used to determine whether the specimen dimensions had been sufficient to ensure plane strain fracture conditions. Thus, the test may have to be repeated with a larger specimen to assure that true plane strain fracture has occurred. Due to the considerable labor and machining costs associated with this testing protocol, ASTM K_{IC} fracture toughness testing is done less frequently than other, simpler mechanical property tests. The Charpy and Izod fracture tests can be performed to measure fracture energy by breaking the specimen with a heavy swinging pendulum, but these tests cannot determine K_{IC} values.

5.2.2 Brittle Fracture

Brittle fracture occurs when the metal does not plastically deform before it breaks. In plastic deformation the metal's planes of atoms slide over one another, like the action of playing cards fanned across a tabletop. In brittle fracture, the planes pull apart completely, similar to "cutting" a deck of cards into two separate stacks. Brittle fracture occurs most often in metals that are extremely hard, particularly at low temperatures. Brittle fracture is important because sudden failure from a brittle fracture can have catastrophic consequences in ships' hulls, pressure vessels, bridges, and gas transmission lines.

Brittle fracture can occur by cleavage along crystal planes, or it may be intergranular (i.e., along grain boundaries). Metals that fracture by cleavage as single crystals often fail in a manner similar to that of polycrystals at low temperature. Under normal circumstances FCC metals do not fail by cleavage fracture. Extensive plastic deformation will occur before the cleavage stress is reached unless the strain rate is extremely high, as in an explosion. Cleavage fracture is more common in BCC metals, particularly when interstitial impurity atoms are present. A number of HCP metals (e.g., Mg, Zn, and Be) undergo brittle fracture along the basal plane at low temperature. It is not possible to use a single criterion for the fracture plane (e.g., closest-packed plane or that of lowest surface energy), but metals of the same crystal structure generally behave similarly. Three main factors contribute to a brittle-cleavage type of fracture: (1) low temperature, (2) triaxial stress state, and (3) high strain rate. However, all three factors need not be present in brittle fracture. Cleavage fracture, primarily a low-temperature phenomenon, disappears at high deformation temperatures. There is thus a transition from ductile to brittle behavior; this normally occurs over a narrow temperature range called the *transition temperature*. For pure metals, the transition temperature varies with purity and grain size. Unlike tensile forces, compressive forces may be transmitted across existing cracks without stress concentrations. As a result, brittle materials are invariably stronger in compression than in tension. Consequently, engineers using brittle materials (e.g., concrete) normally exploit their high compressive strength while striving to minimize or eliminate tensile loading.

Brittle tensile fracture surfaces have a bright, granular appearance and exhibit little or no necking. They are generally perpendicular to the direction of maximum tensile stress. Zn single crystals display ideal cleavage fracture surfaces. When cleaved at 77 K, Zn fracture surfaces are remarkably smooth, with little evidence of plastic deformation. However, Zn does show fine steps or "river markings" running in the crack propagation direction (Fig. 5.4). These markings occur because the cleavage surface is not all on the same plane but is stepped; the crack has been deflected at the steps to other parallel planes by screw dislocations that emerge at 90° to the fracture surface. As the crack passes the dislocation, it moves on two different levels.

5.2.3 Hydrogen Embrittlement

Hydrogen embrittlement results from H diffusion from the environment into the metal (Sidebar 5.2). In a hydrogen environment, materials can fail at loads much lower than H-free material can sustain. Due to its small atomic radius, H diffuses rapidly in most metals at room temperature. Metals can absorb H during production, processing, and/or service. If the H pickup occurs in production or processing, the component is embrittled before it goes into service and may fail immediately in use. Alternatively, if H enters the metal from its service environment, the metal is initially able to bear the design load but is progressively embrittled as its H content rises in service, causing a delayed-onset failure. H severely degrades plasticity in metals and alloys, causing H-induced cracking. Whether the metal actually fails depends on several variables, such

Fig. 5.4 Zn crystal partly cleaved, then fully fractured, in liquid N_2. River markings on fracture surface are shown by arrows. (From Gilman, 1955; with permission of AIME.)

> **SIDEBAR 5.2: HYDRIDE–DEHYDRIDE PROCESSING FOR POWDER METALLURGY**
>
> Although H embrittlement is almost always undesirable, there are circumstances where it is used to advantage in processing operations. It is fairly easy to produce fine powders of brittle materials (e.g., Al_2O_3, As, and Si_3N_4) simply by milling large pieces in a tumbling or oscillating vial with hard balls. This fractures the particles repeatedly, comminuting them rapidly to fine powder. However, this approach is largely ineffective in comminuting large chunks of ductile metals, because they deform and cold-weld to one another, forming plastic masses inside the milling vial that may be heavily deformed and cracked, but are not reduced to powder. However, ductile metals can be comminuted if they are first embrittled by diffusing H interstitial impurity atoms into the metal. After milling, the metal powder can be made ductile again by heating (either as powder or as a sintered piece) in a vacuum to a temperature where H solubility is much lower (Fig. 13.4). This causes the H atoms to diffuse to the surface and be pumped away by the vacuum pump. This is sometimes referred to as the "hydride–dehydride process."

as stress intensity, environment, chemical composition, and microstructure. H embrittlement problems are most frequent in ferrous metals, because they are the most widely used metals, but nonferrous metals are also susceptible. H can enter metals from use of H-containing fuels in heat-treating furnaces or welding torches (e.g., acetylene gas) (Sec. 18.2.3.1), electrolytic or corrosion reactions that generate H on the metal's surface (Sidebar 12.4), or water vapor present in the air (Sec. 25.6.2).

H embrittlement failures normally occur by brittle cleavage or quasicleavage fracture with little plastic deformation. Strong materials are usually more susceptible to H embrittlement. This can be explained by the decohesion theory of embrittlement, which states that dissolved H migrates into a triaxially stressed area and embrittles the lattice by lowering the cohesive strength between metal atoms. During stress cracking, maximum stress is found at the localized region in front of the crack tip. On the crack surface, H_2 gas will physisorb near the crack tip. The H_2 then separates into atomic H via chemiadsorption, and the atomic H preferentially diffuses to the area of localized maximum stress in front of the crack tip. At this point the H is believed to weaken the metallic bonds and could coalesce to form minute voids in front of the advancing crack tip, leading to brittle fracture. In some metals, H accumulates preferentially at grain boundaries, causing intergranular fracture.

Two prevention strategies can be pursued to eliminate H-induced failures: (1) minimizing the metal's H content both before and after it is placed in service, and (2) lowering the internal stresses imposed on the metal. Embrittlement can be reversed with treatment that effectively rids the metal of H, usually by heating the metal. However, a majority of the work against H damage focuses on preventing H from entering the metal. The solutions are to improve processing techniques to avoid trapping H within the metal and to match the type of metal according to the conditions that the final product is expected to encounter in operation. A comprehensive analysis of possible H embrittlement risks should be part of the early design planning for all engineered systems for both safety and economic reasons.

5.2.4 Ductile Fracture

Ductile fracture is accompanied by appreciable plastic deformation prior to and during crack propagation. In extremely pure single crystals and polycrystals that are virtually free of second-phase particles, it is possible for tensile plastic deformation to continue until the sample necks to a point or line (i.e., tensile test ductility of near 100% reduction in area). Such failures are a geometric consequence of slip deformation but are seldom seen in ordinary polycrystalline metals. Ductile fracture normally occurs in several stages: (1) the sample begins necking, and small cavities form in the necked region; (2) the cavities coalesce and form a small crack in the center of the sample which expands perpendicular to the applied stress outward toward the sample surface; and (3) the crack direction changes to an angle of 45° to the tensile axis as it approaches the sample surface. These processes produce the *cup and cone fracture* surfaces shown in Fig. 5.5. One of the hallmarks of the plastic deformation on such fracture surfaces is the irregular and fibrous appearance of the central interior region. The initial cavities normally form at inclusions where the gliding dislocations "pile up" and produce sufficient stress to form a void or microcrack.

The fibrous or cup region of ductile fracture surfaces usually show a continuous pattern of dimples or shallow depressions. These vary from 0.5 to 20 µm in diameter and are the result of the linkage of the cavities formed in the necked region (Fig. 5.6). Void growth (due to dislocation movements and slip displacements) leading to fracture will be much more rapid in the necked portion of a tensile sample following instability than during stable deformation, since the stress system changes in the neck from uniaxial tension to plane strain tension. The voids, which are the main source of ductile fracture, are nucleated heterogeneously at places where compatibility of deformation is difficult. The preferred sites for void formation are inclusions,

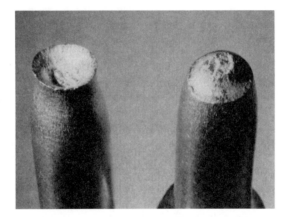

Fig. 5.5 Ductile cup and cone fracture in Al. (From Callister, 2003; with permission.)

Fig. 5.6 Scanning electron micrograph image of the fibrous inner region of the fracture surface of a Cu tensile test specimen. The depressions on the metal surface are artifacts of the voids that formed prior to fracture. (Courtesy of F. C. Laabs, Ames Laboratory, U.S. Department of Energy.)

fine oxide particles, or second-phase particles. In extremely pure metals, voids can form at grain boundary triple points. The overall mode of failure is sensitive to the method of testing, the resulting stress distribution in the neck, and work-hardening features of the metal.

Since brittle fracture is a more dangerous failure mode than ductile fracture, various mathematical predictors have been developed to indicate whether ductile or brittle fracture is more likely to occur for a given metal (Appendix D, *ftp://ftp.wiley.com/public/sci_tech_med/nonferrous*; Secs. 17.5; and 14.4.2.2). Ductility is important to providing a "fail-safe" design that avoids a catastrophic (unpredictable) fracture (Sidebar 11.1). When failure involves plastic deformation, work hardening can increase the strength of a component, preventing immediate fracture and giving visible warning that failure is near. Besides this, any plastic flow at the tip of a crack will blunt the crack, and as a result the stress may not be sufficient to continue crack propagation. Thus, ductility prevents many failures and minimizes losses from some failures that do occur. Ductility is dependent on the following factors:

1. *Hydrostatic stress*. Ductility increases as hydrostatic stress increases (Sec. 14.4.2.2). Compressive ultimate strengths increase with increasing superimposed hydrostatic stress in both nonmetals (e.g., marble and sandstone) and metals. This is one reason why compressive metal-forming processes are more widely used.

2. *Strain rate*. Materials tend to be more brittle at higher strain rates because diffusion-dependent processes (e.g., nonconservative dislocation motions such as climb or motion of jogged screw dislocations) have less time to act at high strain rates. At extremely high strain rates where stress waves occur, spallation may result from reflections of compressive waves at free surfaces.

3. *Temperature*. Increasing the temperature has effects similar to those of decreasing the strain rate. As temperature increases, metals become more ductile because additional slip systems are active at higher temperatures, and all thermally activated dislocation movements are accelerated. Also, at higher temperatures, recrystallization can dynamically relieve cold work as deformation progresses, avoiding the loss of ductility that deformation causes below the recrystallization temperature, altering the mode of failure.

5.3 METAL FATIGUE

Fatigue is defined as failure under a repeated or varying load that is never large enough to cause failure in a single application. It has been estimated that 80 to 90% of metal failures in practice arise from fatigue. Overworked metal will most often result in fatigue. Fatigue is a prime concern for engineers who wish to ensure the longevity of a product under stress, as well as for metallurgists trying not to overwork the metal during production. Overworking may be the result of overforming during processing or the result of excessive use by the customer. The term *fatigue* was introduced in 1839, when Poncelet first described metallic structures as being "tired" or worn out. Wohler conducted some of the first systematic fatigue experiments in 1860 of railroad axles, and he created the concept of the *fatigue endurance limit*. Although fatigue has been recognized for more than 150 years, fatigue failures still occur with unpleasant frequency (Sidebar 5.3).

To address design engineers' needs to predict a material's ability to withstand cyclic loading, cyclic fatigue data have been produced and presented in the form of families of curves. Normally known as S/N (ε/N) curves, they plot applied stress (S) versus cycles of loading (N) or strain (ε) versus cycles to failure (N). S/N curves are normally used when stress is maintained at a constant level regardless of possible changes in specimen condition, whereas the ε/N curves are produced using test machines that control the total strain of the specimen. S (usually known as $\Delta\sigma$, the stress range) is the applied stress and ε is the strain experienced due to the applied stress. Similarly, ε is simply written in the place of the strain range, $\Delta\varepsilon$. Generally, stress is controlled for testing high-cycle fatigue ($N > 10^5$ cycles), and strain is controlled for testing

SIDEBAR 5.3: CATASTROPHIC FATIGUE IN AIRCRAFT

In 1988, Aloha Airlines' flight 243 suffered a catastrophic fatigue failure as it flew from Hilo Airport to Honolulu at 7300 m. This aircraft was used primarily for short flights between the islands in the state of Hawaii and had been subjected to tens of thousands of cabin pressurization and depressurization cycles. As passengers boarded the plane for Flight 243, one passenger noticed a long crack in the Al fuselage skin, but she did not mention this to anyone before the accident. The crew managed to land the plane after the fuselage skin failed (Fig. 5.7), but one flight attendant died when she was swept out of the aircraft, and several passengers were injured.

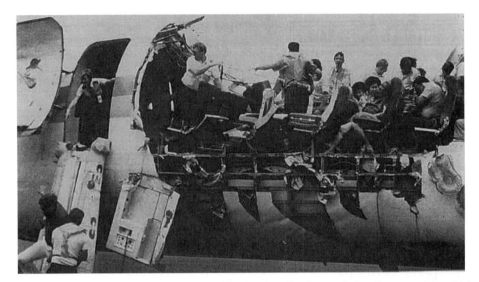

Fig. 5.7 Example of fatigue failure. In 1988, this aircraft suffered an explosive decompression at high altitude, killing one crew member and injuring several passengers. The pilot made a successful emergency landing. The aircraft had flown thousands of short flights between the Hawaiian Islands, subjecting the fuselage to an unusually large number of compression–decompression cycles. (Retrieved from *http://www.warman.demon.co.uk/anna/incident.htm*; used with permission.)

low-cycle fatigue ($N < 10^4$ or 10^5). Typical S/N diagrams for nonferrous and ferrous metals are shown in Fig. 5.8. The S/N diagram shows that the fatigue strength decreases with increasing number of cycles. These curves allow engineers to design systems whose components operate below the threshold where failure is likely. For materials such as steel and Ti, the S/N curve becomes horizontal at a certain limiting stress. Below this limiting stress (the *fatigue limit*), the material can probably endure an infinite number of cycles without failure. Many other nonferrous materials, such as Al and Cu alloys, have an S/N curve that slopes gradually downward with increasing number of cycles. These materials have no fatigue limit because the S/N curve never becomes horizontal. Under these conditions, it is common practice to characterize the fatigue properties of the material by giving the fatigue strength at an arbitrary number of cycles: for example, 10^7 cycles. The tendency toward fatigue is enhanced significantly by the presence of a stress concentration such as a sharp notch, which if suitably located can lead to the initiation of a fatigue crack at its root. The cause of the fatigue limit of some metals has not been settled, but it is probably a strain-aging effect (Sec. 6.2).

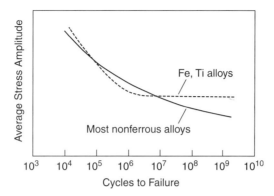

Fig. 5.8 Typical S/N curves for ferrous and nonferrous metals.

5.3.1 Stages of Fatigue

The fatigue process can be divided conveniently into three stages: crack initiation, crack propagation, and final fracture. Fatigue cracks often initiate at the surface or at voids or second-phase inclusions that serve as stress risers. Dislocations accumulate near such stress concentrations. As the dislocations harden the metal around the stress concentration during cyclic loading, minute cracks nucleate and start the fatigue process. It is difficult to determine the exact location of fatigue crack initiation on a failed specimen, even using powerful replication and inspection techniques. For materials with relatively homogeneous microstructures, this early stage of the fatigue process is normally shear stress dependent and acts on microscopic slip planes that are preferentially oriented to have high Schmid factors for the applied stress (hence, located at ∼45° to the direction of applied stress). Characteristically, this stage I crack growth starts at a point of discontinuity in a material, such as a machining mark, defect, or a change in cross section. The crack grows until it encounters a grain boundary, at which point it stops briefly until sufficient energy has been applied to the adjacent grain so that the process can continue. After traversing two or three grain boundaries, the direction of the crack propagation now changes into a stage II mode in which cracks typically lie perpendicular to the applied tensile stress. Dislocation cross-slip is the most important mechanism here. Stage II crack growth accounts for most of the crack propagation length through the bulk material. In this stage the extrusions formed on the surface are emitted, thus developing into fissures called *intrusions*, as shown in Fig. 5.9. In stage III the crack has grown to such a large size that the component is no longer physically able to carry the load for which it was designed. The remaining cross section of the component (i.e., the area still unaffected by the fatigue crack) subsequently fractures, due to overload in a conventional manner through the nucleation, growth, and coalescence of microvoids. Figure 5.10 shows the three stages of fatigue crack propagation. The relative proportion of the metal's service life involved with the three fatigue stages is dependent primarily on the test conditions and material composition. However, it is well known that crack initiation normally begins after the first 1% of the lifetime. The majority of crack life is occupied in stage II, the propagation phase. The final fracture phase, stage III, is relatively short. Fatigue failure requires the simultaneous action of tensile stress, cyclic stress, and plastic strain. If any of these factors is absent, the crack does not propagate.

5.3.2 Fatigue Crack Propagation

It is natural to assume that a material's fatigue crack growth rate should depend on the stress range applied. This dependence was quantified by Paris and Erdogan (1963), who proposed the equation

$$\frac{da}{dN} = C \, \Delta K^m \qquad (5.9)$$

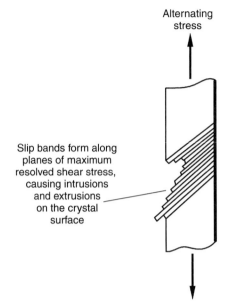

Fig. 5.9 Fatigue forms surface intrusions and extrusions.

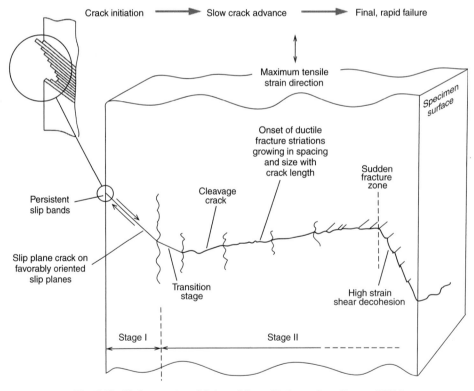

Fig. 5.10 Various modes of fatigue failure. (Redrawn from Provan, 2001.)

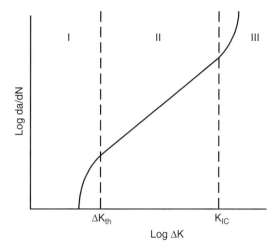

Fig. 5.11 The three stages of fatigue crack propagation.

If the fatigue crack growth rate da/dN (da is the change in crack length and dN is the change in number of cycles) is plotted versus the range of stress intensity factor ΔK on a log-log plot (Fig. 5.11), the slope of the curve is the constant m in equation (5.9) and C corresponds to the point where an extension of the curve will intersect with $\Delta K = 1$ MPa$\sqrt{\text{m}}$.

When plotted in a log-log plot, experimental fatigue data may not always follow a straight line, particularly at low and high fatigue crack growth rates. Thus, it is common to divide crack propagation into three regions (Fig. 5.11): (I) threshold region, (II) stable crack growth region, and (III) unstable crack growth region. The Paris law [equation (5.9)] is valid only in region II. At very low crack growth rates (region I) the fatigue behavior of a material is controlled by the threshold value ΔK_{th}, which is a material property. Below this threshold, cracks either remain dormant or grow at undetectable rates. The fatigue threshold ΔK_{th} is an important parameter for damage tolerance design because it allows estimation of the critical crack size in a structure, or oppositely, by knowing a crack size in a structure, the critical load levels can be predicted. Region II is also sometimes called the *Paris regime*, in recognition of the researcher who first correlated stress intensity factor range with fatigue crack growth rate. In the final region, the maximum stress intensity factor K_{max} approaches the fracture toughness K_C, which causes rapid crack growth.

Increasing the mean stress in the fatigue cycle ($R = K_{min}/K_{max} = \sigma_{min}/\sigma_{max}$) increases the crack growth rates in all regions of the curve. The effect of increasing R is generally less in region II than in regions I and III. The influence of R on the Paris law was given by Forman et al. (1967):

$$\frac{da}{dN} = \frac{C\,\Delta K^m}{(1-R)K_{crit} - \Delta K} \qquad (5.10)$$

where C and m are material specific constants different from the Paris coefficients and $K_{crit} \approx K_{IC}$. Similar to the Forman equation is the correlation developed by Walker (1970):

$$\frac{da}{dN} = \frac{C\,\Delta K^m}{(1-R)^{m(1-\gamma)}} \qquad (5.11)$$

Once again the constant coefficients C and m are not related to the previous expressions, and the γ term adjusts for R effects. Of the two previous expressions [(5.10 and 5.11)], the addition of the stress ratio R to the equations is highly significant. The da/dN versus K data for a single

74 FRACTURE AND FATIGUE

material vary with stress ratio R. Due to this dependence, a plot of da/dN versus K cannot truly be considered a material property.

5.3.3 Characteristics of Fatigue Fracture

Crack initiation/nucleation leads to microcrack development. Under repeated applied stress, this microcrack will grow in size, and as the crack expands, the load-carrying cross section of the sample is reduced, with the result that the stress on this section rises. Eventually, the point is reached where the remaining cross section is no longer able to carry the load, and the part fractures. This behavior causes two distinct regions to form on the fracture surface (Fig. 5.12). The first is a relatively smooth or burnished area appearing in the region where the crack grew slowly. This area usually has concentric marks about the point of origin of the crack, which correspond to the positions at which the crack was stationary for some period of time. The rest of the fracture surface exhibits a typically rough or granular appearance where the failure has been swift and catastrophic. The entire process of fatigue failure is a combination of fatigue and either brittle or ductile fracture. During the crack propagation period, the fatigue mechanism dominates, but the final fatigue fracture may be either ductile or brittle, depending on the material, temperature, and loading rate.

5.3.4 Variables Affecting Fatigue Life

Even when design engineers carefully follow good design practices to minimize fatigue, fatigue problems are sometimes introduced into the structure after the design step. Fatigue failures can result from geometrical or strain discontinuities, poor workmanship, improper manufacturing techniques, material defects, or the introduction of residual stresses that may add to existing service stresses. Factors affecting fatigue include:

1. *Surface finish*. Since fatigue cracks most often initiate from a preexisting defect at the component's surface, the surface quality greatly influences the probability of a crack initiating. Most engineering fatigue test samples have a mirror finish and therefore achieve the longest fatigue lives, but real components rarely have such smooth (and expensive) surface finish.

Fig. 5.12 Typical fatigue fracture surface. (From Smallman, 1970, p. 511; with permission.)

Consequently, it is often possible to improve fatigue life by improving surface finish quality on parts that will be loaded cyclically. Surface finish has a more important effect on the fatigue of components subjected to low-amplitude stress cycles.

2. *Grain size.* Generally, fatigue strength increases as grain size is reduced. Grain boundaries are good obstacles to the propagation of a fatigue crack, just as they are for the propagation of a brittle crack.

3. *Residual stress.* Residual stress (mean stress) also influences the rate of fatigue damage. Residual stresses are those that remain in a part even when it is not subjected to an external force. When a residual tensile stress is present in a part experiencing stage II crack propagation, the crack is being forced open, and any stress cycles applied will therefore lower the fatigue limit. On the other hand, if a mean compressive stress is present, the crack will be forced shut and any stress cycle would first have to overcome this residual compressive stress before any growth can proceed. Critical components subject to fatigue (e.g., aircraft landing gear) are often shot-peened to introduce residual compressive stresses at the surface to improve their fatigue resistance.

4. *Environment.* Corrosive attack by a liquid can produce etch pits on the metal's surface that act as notches and stress raisers, but when the corrosive attack is simultaneous with fatigue stressing, the damaging effect is much greater than just a notch effect. Corrosion reduces the amount of metal present to carry the imposed load, thereby increasing the actual stress. Corrosive environments often cause faster crack growth and/or crack growth at a lower tension level (Sec. 21.5.6). Even a mildly corrosive environment can reduce the fatigue strength of Al structures by 75 to 25% of their fatigue strength in dry air. Corrosion fatigue can be reduced either by lowering the cyclic stresses or by limiting contact between the metal and the corrosive environment.

5. *Frequency of stress cycle.* In most metals, the frequency of stress cycle has little effect on the fatigue life (at ordinary temperature), although lowering the frequency usually results in a slightly reduced fatigue life. The effect becomes greater at higher temperatures; in that case, the fatigue life tends to depend on the total time of testing rather than on the number of cycles.

6. *Temperature.* Generally, materials lose ductility as temperature decreases and yield and fatigue strength increase. Fatigue strength increases faster than tensile yield strength, however. Conversely, an increase in temperature generally reduces fatigue strength.

ADDITIONAL INFORMATION

References, Appendixes, Problem Sets, and Metal Production Figures are available at
ftp://ftp.wiley.com/public/sci_tech_med/nonferrous

6 Strain Rate Effects and Creep

> They must often change who would be constant. ...
> —Confucius, *The Analects*, 500 B.C.

6.1 INTRODUCTION

The simpler explanations of metallic deformation treat the metal as a static entity whose only change with time is the motion of dislocations and/or cracks. This assumption holds well for metals deformed at low temperature (e.g., 77 K), but for deformation at or above room temperature, diffusion actively alters the lattice (and sometimes the dislocations) while deformation progresses. This complicates deformation and causes strain aging and creep phenomena that pose challenges for the design engineer.

6.2 YIELD POINT PHENOMENON AND STRAIN AGING

A *yield point* (Fig. 13.7) can occur in metals containing C, O, or N interstitial impurity atoms. Interstitial atoms induce compressive strain in the lattice (Fig. 13.5), and dislocations may induce tensile strain in portions of the nearby lattice. When interstitial impurities diffuse to dislocations, these strain fields partially cancel one another, lowering the total energy of the crystal. Atmospheres of interstitial atoms collect around dislocations and "anchor" them because additional energy is required to increase lattice strain again when they separate. This increases the stress required to set dislocations in motion (i.e., the upper yield point stress). Once the dislocations leave their atmospheres, it is relatively easier for them to move, and the lower yield point is the stress to move dislocations freed from their atmospheres. Metal can be overstrained to remove the yield point (e.g., to eliminate stretcher strains on finished products), but if the metal is unloaded and then permitted to age before re-straining, the yield point returns (Fig. 13.7). This is known as *strain aging*, in which interstitial impurities diffuse to dislocations during aging, forming new atmospheres of interstitials that anchor the dislocations again.

Another type of strain aging can occur *during* deformation. This *dynamic strain aging* (DSA) is common in BCC metals and arises from the same attractive interaction between interstitial solute atoms and dislocation strain fields. DSA occurs at higher temperatures, where solute atom diffusion rates are sufficiently rapid to diffuse interstitials to dislocations while plastic deformation occurs, causing serrated yielding (Fig. 13.7), believed to result from dislocations alternately breaking away from and reacquiring solute atom atmospheres.

6.3 ULTRARAPID STRAIN PHENOMENA

Metals are subjected to high strain rates during such processes as high-speed machining, rapid crack propagation, forging, impacts, and explosions. The split Hopkinson bar technique is widely

---

Structure–Property Relations in Nonferrous Metals, by Alan M. Russell and Kok Loong Lee
Copyright © 2005 John Wiley & Sons, Inc.

Fig. 6.1 Anisotropy effects in exploded canisters of (*a*) single-crystal and (*b*) polycrystalline Al. (From Glass et al., 1961, p. 122; with permission of AIME.)

used to test metals at strain rates that range from about 10^2 to 10^4 s^{-1}. In this technique a short metal specimen placed between two bars is loaded beyond the elastic limit by impact with the incident bar (Fig. 13.10). Upon loading, part of the wave is reflected back to the incident bar and part is transmitted to the output bar. The stress, strain rate, and strain in the plastically deforming specimen are determined by strain gauges measuring both bars' strain, which remains elastic throughout the test.

In general, dislocation velocities increase with deformation rate. Conventional elasticity theory predicts that subsonic dislocation speed cannot exceed the speed of sound in the metal because the energy required to do so would be infinitely large. Finite-element models using a thin strip subjected to a homogeneous simple shear show that dislocations can move faster than the speed of sound if they are created as transonic or supersonic dislocations at a strong stress concentration (e.g., a crack tip, dislocation pile-up, or sharp indenter) and are subjected to high shear stress. This behavior is vital for understanding deformation processes such as mechanical twinning and martensitic transformation, but its practical importance is limited because high stresses are required for sustained motion at such high velocities. However, if the loading rate is sufficiently rapid, dislocations cannot respond fast enough to deform the metal plastically, and the metal fails by brittle fracture, much as ceramics do at low strain rates. In fact, ceramics have higher fracture energy per kilogram of material, and modern armor is often made of strongly bonded ceramic plates and tiles (e.g., Al_2O_3 or B_4C) rather than metal. Ultrahigh strain rate loading drastically alters mechanical behavior. The fracture strengths of metals under explosive loading are relatively insensitive to common metallurgical factors, such as purity, grain size, strain, and age hardening. Figure 6.1 shows single-crystal and polycrystalline specimens deformed by explosion. The effect of anisotropy on final sample shape is apparent.

6.4 CREEP

Under some service conditions, metals must sustain steady loads for long time periods at elevated temperature (e.g., $T > 0.4T_{m.pt.}$). Such conditions are common for furnace parts, turbine blades, and filaments in vacuum tubes. Under these conditions, metals gradually deform at loads well below the yield strength. This time-dependent strain under stress is called *creep*. The gradual loosening of bolts, sagging of long-span cables, and slow deformation of engine parts are examples of creep. In many cases such deformation eventually stops because the deformation eliminates the force causing creep. Creep extended over a long time ultimately ruptures the metal, but the more common problem is that dimensions gradually change until parts become unusable. Understanding and managing creep have become more important as engineers continually raise operating temperatures to improve performance and efficiency.

6.4.1 Stages of Creep

Creep tests are normally carried out by loading a specimen, either in tension or compression, inside an inert-gas-filled furnace where temperature can be controlled accurately over a long period. Usually, load and temperature are both held constant, and the extension of the sample is measured over time. Such tests produce creep curves of the general shape seen in Fig. 6.2. The curve is usually split into three regimes: the primary, steady-state, and tertiary stages. *Primary creep* occurs early in the test and is characterized by a fast, decreasing creep rate occurring over a relatively short time. The steady-state (or *secondary*) *stage* begins when the creep rate has stabilized, and throughout this stage the creep rate remains fairly constant. The somewhat misleading term *steady state* refers only to the creep rate; the metal's microstructure degenerates throughout the entire creep process and is never "steady" at any instance. Knowledge of the creep rate allows the design engineer to estimate a component's service life.

When a component's dimensions creep beyond allowable tolerances, it must be replaced. The jet engine (Fig. 16.14) provides a good example of this. The gap between the outer tip of the rotating blades and the engine casing must be small to maximize engine efficiency. If the gap is large, substantial amounts of gas flow through the engine without acting on the blades. However, if a blade elongates too much by creep, it will scrape against the inner wall of the casing. Engines are designed to tolerate a certain depth of such scraping in abradable seal designs, but serious damage occurs if the blade cuts too deeply. However, if the creep rate of the metal is known, blades can be replaced at predictable intervals before irreparable damage occurs. In the final *tertiary stage*, a continually increasing creep rate usually spells the beginning of the end of the specimen's test life. The stress is no longer constant because the specimen necks and decreases its cross-sectional area. At the onset of tertiary creep, macroscopic evidence of creep damage (e.g., voids and cracks opening in the grain boundaries) begins to appear.

6.4.2 Variations in Creep Behavior

Creep behavior varies with metal composition and temperature. The ratios between the durations of the three stages of creep may differ substantially from those shown in Fig. 6.2. In some creep experiments, individual stages may be lacking entirely. Composites reinforced with continuous fibers normally constitute an extreme, upper bound of strength (Fig. 6.3*a*), and their behavior is viscoelastic, with the matrix forming the viscous part, in parallel coupling with the elastic fibers. Composites with short aligned fibers (Fig. 6.3*b*) have initial primary creep (viscoelastic)

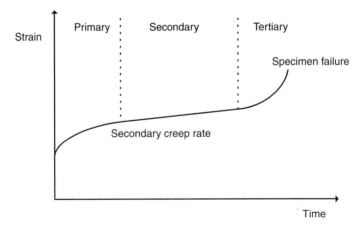

Fig. 6.2 Schematic representation of a typical creep deformation curve during testing under a constant applied load.

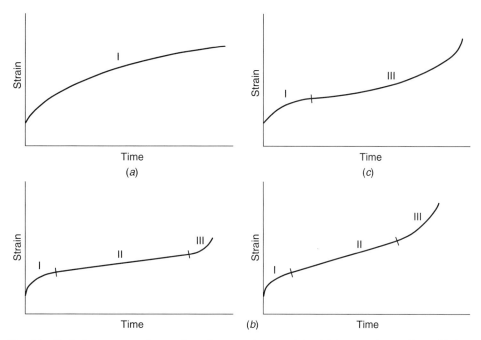

Fig. 6.3 Typical creep curves for metal-matrix composites containing (*a*) long continuous fibers, (*b*) short fibers, or (*c*) particles. I, Primary creep; II, secondary creep; III, tertiary creep.

and steady-state creep where sections of the matrix that are not supported by fibers (e.g., between ends of fibers) creep with a constant creep rate. Composites reinforced with particles rather than fibers have virtually no steady-state creep (Fig. 6.3*c*), and these particles cause early initiation of creep damage (e.g., voids) in the matrix, leading to a long tertiary creep regime. The creep curves are determined by the fiber arrangement (length/diameter ratio, orientation, volume fraction), stress, and temperature.

6.4.3 Mechanisms of Creep

At low temperatures ($T < 0.4T_{m.pt.}$), plastic deformation is caused primarily by dislocation glide (slip) and to a lesser extent by twinning. At high temperatures ($T > 0.4T_{m.pt.}$), there are two main deformation mechanisms, dislocation creep and diffusion creep, although both are related to diffusion mechanisms.

6.4.3.1 Dislocation Creep Mechanisms. When a creep specimen is loaded initially, an elastic strain occurs that eventually gives way to some plastic component of strain. This plastic component represents the region of dislocation multiplication, where dislocation sources generate dislocations that glide on slip planes with favorable Schmid factors. The dislocations eventually become blocked, and a new Frank–Read source is needed for slip to proceed. Eventually, the flow stress will become greater than the applied stress, and any time-dependent deformation will rely on localized thermal fluctuations assisting with the movement of jogs (Sidebar 4.1 and Fig. 4.18).

For $T > 0.4T_{m.pt.}$, recovery processes occur as deformation progresses. Secondary creep is a balance between work hardening and thermal softening. It can be defined by two terms: (1) a coefficient of work hardening ($h = \delta\sigma/\delta\varepsilon$), where σ is the flow stress at a strain, ε, obtained from the slope of the stress–strain curve in the absence of recovery, and (2) the rate of recovery

80 STRAIN RATE EFFECTS AND CREEP

($r = -\delta\sigma/\delta t$), which is obtained from the decrease in flow stress with time when annealing the work-hardened specimen. The internal yield stress, σ, of deformation elements can change by an amount $(\delta\sigma/\delta t)\,dt$ with time and by an amount $(\delta\sigma/\delta\varepsilon)\,d\varepsilon$ with strain, so that

$$d\sigma = \frac{\delta\sigma}{\delta t}\,dt + \frac{\delta\sigma}{\delta\varepsilon}\,d\varepsilon \qquad (6.1)$$

During steady-state creep, the flow stress remains constant so that in equation (6.1), $d\sigma = 0$. Therefore,

$$\frac{d\varepsilon}{dt} = \frac{-\delta\sigma/\delta t}{\delta\sigma/\delta\varepsilon} = \frac{r}{h} \qquad (6.2)$$

Experiments have shown that this expression is more or less correct, and many mechanistic theories of recovery creep use this model to account for steady-state creep.

The earliest models of dislocation creep assumed that the dislocation climb rate controlled the recovery process and that dislocations generated from active sources were confined to their glide planes and moved until they became blocked by obstacles. Eventually, an elastic back stress grew strong enough to prevent the dislocation source from emitting any new dislocations. The leading dislocations could then climb out of their slip planes by vacancy emission. These dislocations were free to glide in a new slip plane or to be annihilated by a dislocation of opposite sign. The climb represents the recovery stage since a new dislocation can be released from the source as creep continues. From this model, one can determine the values for n [as in equation (6.3)] and Q_c (the Arrhenius-type activation energy for creep) which agree with the values observed for creep. However, formations of piled-up dislocations are not often seen in practice, so the model is largely unsupported by microstructural examination, although piled-up dislocations may disperse during thinning of TEM samples, before they can be observed.

For intermediate to high stress levels at temperatures above $0.5T_{m.pt.}$, the strain rate is given by

$$\dot{\varepsilon} = A\frac{D_{\text{eff}}Gb}{kT}\left(\frac{\sigma}{G}\right)^n \qquad (6.3)$$

where $\dot{\varepsilon}$ is the axial strain rate, A a dimensionless constant, D_{eff} the effective diffusion coefficient, G the shear modulus, **b** the Burgers vector, k the Boltzmann constant, T the absolute temperature, and σ the applied stress. In this case the effective diffusion coefficient, D_{eff}, is a function of both the lattice diffusion and diffusion through dislocation cores. Values for the stress exponent, n, between 3 and 8 indicate that dislocation creep is the dominating mechanism. Typically, $n \approx 3$ for glide-controlled creep in metals below $0.5T_{m.pt.}$, $n \approx 5$ for creep controlled by high-temperature dislocation glide plus climb or lattice diffusion, $n \approx 7$ for creep controlled by core diffusion at lower temperatures, and $n \approx 8$ for creep controlled by lattice diffusion through a constant substructure.

6.4.3.2 Diffusional Creep. The diffusion that drives creep requires vacancies to operate. An atom can move to a site of an adjacent vacancy, provided that it has adequate thermal energy to jump from its original site (Fig. 6.4). At any temperature, the average thermal energy of an atom is $3\,kT$, where k is Boltzmann's constant (1.38×10^{-23} J/atom·K). As atoms vibrate about their mean positions, they collide with their neighbors, transferring energy continually from one to another. As a result, the thermal energy is not homogeneously distributed among the atoms in the crystal, and at any instant, any single atom has more or less energy than the average value. Even if the atom has sufficient energy to jump into a neighboring lattice site, it can move only if a vacancy exists on that lattice site to allow the move to occur. The likelihood of the two independent processes above occurring simultaneously is the product of

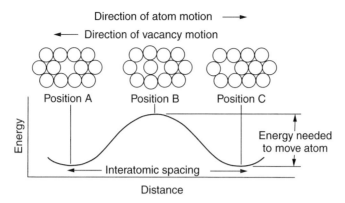

Fig. 6.4 Diffusing atom moving to a site of adjacent vacancy. (From Evans and Wilshire, 1993, p. 28; with permission.)

the individual probabilities. The rate of diffusion in a solid also depends on the frequency with which individual atoms move.

Surfaces, grain boundaries, and dislocation cores provide "easy paths" for diffusion, since atomic arrangements are less regular there than in a perfect lattice. Activation energies are lower for these easy path processes than for diffusion through the bulk of the crystal. At high temperatures when lattice diffusion is rapid, diffusion along the easy path makes a negligible contribution to the overall diffusion rate because the volume of the relatively disordered regions of the boundaries and dislocations is very small compared to the total volume of the polycrystalline sample. However, as the temperature is decreased, the bulk diffusion rate decreases exponentially [equation (2.1)], and diffusion along easy paths comprises a greater fraction of total diffusion because the rate of the process with the lower activation energy changes less as temperature decreases.

At low stress the crystal resistance almost inhibits dislocation motion. However, time-dependent plastic deformation at low strain rates takes place by a diffusional flow of atoms. At moderately elevated temperatures, this diffusional flow occurs as grain-boundary diffusion (Coble creep). At high temperatures diffusion through the lattice becomes dominant (Nabarro–Herring creep). The strain rate due to diffusion creep is given by

$$\dot{\varepsilon} = \frac{7\sigma D_1 b^3}{kTd^2} \tag{6.4}$$

where D_1 is the lattice diffusivity, b the atomic spacing, k is Boltzmann's constant, T the absolute temperature, and d the grain diameter. The creep rate depends strongly on grain size, because grain boundaries are the main source of vacancies. As grain size decreases, the area of the grain boundaries increases, so the creep rate increases.

Figure 6.5 depicts a square grain subjected to tensile stress along the vertical axis and compressive stress along the horizontal axis. Such stresses provide a driving force for the diffusion of vacancies from the horizontal to the vertical grain boundaries, because such diffusion lowers strain energy. The lattice and the grain boundaries provide two independent paths for vacancy diffusion; these processes are termed Nabarro–Herring and Coble diffusion creep, respectively. *Nabarro–Herring diffusion creep* may be represented by

$$\dot{\varepsilon}_{\text{NH}} = \frac{A_{\text{NH}} D_L G b}{kT} \left(\frac{b}{d}\right)^2 \frac{\sigma}{G} \tag{6.5}$$

82 STRAIN RATE EFFECTS AND CREEP

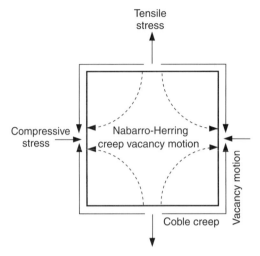

Fig. 6.5 Schematic illustration indicating vacancy transport through the lattice (dashed lines; Nabarro–Herring diffusion creep) and along grain boundaries (solid lines; Coble diffusion creep).

where the subscript NH represents Nabarro–Herring creep, A is a constant, and D_L is the lattice diffusion coefficient. In a similar manner, *Coble creep* may be expressed as

$$\dot{\varepsilon}_{\text{Co}} = \frac{A_{\text{Co}} D_{\text{gb}} G b}{kT} \left(\frac{b}{d}\right)^3 \frac{\sigma}{G} \tag{6.6}$$

where the subscript Co represents Coble creep and D_{gb} is the grain boundary diffusion coefficient.

Nabarro–Herring and Coble diffusion creep processes operate independently, so the faster one is rate controlling. Analysis of the processes indicates that Coble creep is dominant in fine-grained metals deformed at intermediate temperatures, whereas Nabarro–Herring creep is rate controlling in coarser-grained metals deformed at higher temperatures. There is a continuous transition between dislocation creep and diffusional creep and between dislocation glide and dislocation glide plus climb; the relative rates of these processes depend strongly on metal parameters. In practice, engineers assure safe performance by operating a metal well below its creep limit, which is the stress to which a metal can be subjected without having creep exceed a specified amount of strain for a given time and temperature (e.g., a creep limit of 0.01 strain during 100,000 hours at 600°C).

6.5 DEFORMATION MECHANISM MAPS

Several more or less independent mechanisms hide behind the term *creep*. Each of these mechanisms can be the dominating deformation mechanism for a given metal and set of creep parameters. Two-dimensional maps have been developed to provide a convenient guide to which type of deformation behavior is dominant in a particular regime. There are three basic types of deformation mechanism maps: stress–temperature maps, grain size–stress maps, and grain size–temperature maps.

In a *stress–temperature map*, the modulus-normalized stress is plotted as a function of homologous temperature at a constant grain size (Fig. 6.6), calculated using fundamental atomistic parameters. The upper limit of this type of diagram is set by the fact that conventional plastic flow behavior is observed at very high stresses. Within the creep region, at any specified temperature and stress, the contribution of each independent mechanism could be determined from

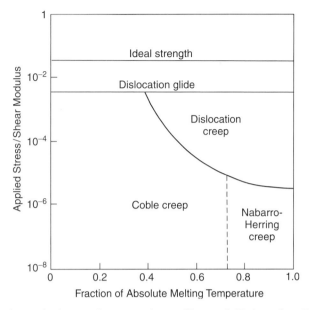

Fig. 6.6 Deformation mechanism map for a pure polycrystalline metal. (Redrawn from Evans and Wilshire, 1993, p. 48.)

creep data or from calculations of the secondary creep rate expected from the appropriate theoretical equations for a given process. The map indicates the stress–temperature regions within which a given mechanism is likely to be dominant. It is also possible to superimpose contours of constant creep rate on the map, since a metal shows a fixed creep rate at any particular temperature and stress. Thus, much useful information is presented in a compact form.

There are, however, definite limits to the usefulness of such diagrams. First, the diagram cannot be more accurate than the strain equations from which it was constructed; equations that describe only the steady-state creep phase may contain poorly determined constants. Second, the mechanisms shown have different dependencies on microstructural features (texture, precipitates, grain size) that are not expressed directly in the diagram and can change the size of the respective areas significantly. Third, although creep experiments are performed in an area where one specific mechanism is rate controlling, significant contributions to the overall strain can still occur from other mechanisms. Despite these shortcomings, deformation mechanism maps are quite useful for roughly defining the different regimes.

6.6 SUPERPLASTICITY

Superplastic metals exhibit extremely high ductility when plastically deformed. Metals normally fail at elongations of less than 100%, but superplastic metals routinely reach elongations of greater than 1000%, and elongations as high as 8000% (Fig. 6.7) have been observed. To achieve superplasticity, the strain rate during deformation must not exceed the rate of recovery. Superplastic metals are not a class of metals, but rather, a type of deformation behavior possible in many metals if they are deformed within a narrow range of temperature and loading rate. Not all metals exhibit superplasticity, but this behavior occurs at some specific loading conditions in a wide range of materials, including intermetallic compounds, ceramics, composites, and nanocrystalline metals. Although superplasticity is generally considered a recent discovery,

Fig. 6.7 Superplasticity in Cu–Al alloy (8000% elongation). Tensile test specimen shapes before (upper) and after (lower) deformation. (Redrawn from Hori et al., 1991.)

it has been suggested that the famous Damascus swords, made with steels from India, were manufactured several centuries ago under conditions favoring superplastic deformation.

For superplastic behavior a metal must possess a fine, equiaxed grain structure that remains stable during deformation. The grain size of superplastic metals should be as small as possible, normally in the range 2 to 10 μm. Given a suitable microstructure, superplasticity occurs over a narrow range of temperatures, generally above $0.5T_{\text{m.pt.}}$. The most important mechanical characeristic of a superplastic metal is its high strain rate sensitivity of flow stress, referred to as m and defined by

$$\sigma = K\dot{\varepsilon}^m \tag{6.7}$$

Superplastic behavior usually requires that $m \geq 0.5$, and for the majority of superplastic metals m lies in the range 0.4 to 0.8. The presence of a neck in a metal subject to tensile straining leads to a locally high strain rate, and for a high value of m, to a sharp increase in the flow stress within the necked region. Hence the neck undergoes strain hardening that inhibits further necking. When $m = 1$ the flow stress is directly proportional to the strain rate, and the metal behaves as a Newtonian viscous fluid, such as hot glass drawn from the melt into fibers without the fibers necking down. Confusion sometimes arises between superplasticity and creep since both are strain rate effects. Superplasticity is mainly a high-strain-rate phenomenon (strain rate between 10^{-1} and 10^{-4} s^{-1}) and occurs in a limited number of alloys (Sec. 19.2.6), whereas creep is a slow phenomenon (strain rate between 10^{-5} and 10^{-13} s^{-1}) and occurs in most metals that are mainly strain hardening in a tensile test.

Superplastic metals exhibit large ductility at relatively low flow stresses, so they can be formed into complex shapes. Superplastic forming technology is used extensively (Sec. 19.2.6.1), particularly in the aircraft and automobile industries. A conventionally formed aircraft component may actually be made up of several simple parts that can be fastened together mechanically. The joining steps require time, and the fasteners increase the overall weight of the component. Using superplastic metal-forming methods, the component can be made as a single piece, thereby eliminating the labor and weight associated with the fasteners.

ADDITIONAL INFORMATION

References, Appendixes, Problem Sets, and Metal Production Figures are available at
ftp://ftp.wiley.com/public/sci_tech_med/nonferrous

7 Deviations from Classic Crystallinity

There are no quasicrystals, only quasi-scientists.
—Linus Pauling, from a speech at the 1983 ACS Meeting

Ride bene chi ride l'ultimo. (He who laughs last, laughs best.)
—Italian Proverb

7.1 INTRODUCTION

Metals have a strong tendency to crystallize (Sec. 1.2) as they solidify, but for certain combinations of composition and processing history, their crystallinity and grain structures can be altered or completely suppressed. Metals with altered crystallinity possess unusual properties that can be exploited in engineering systems. In this chapter we describe four deviations from "normal" crystallinity: nanocrystalline, amorphous, quasicrystalline, and radiation-damaged metals.

7.2 NANOCRYSTALLINE METALS

Ordinary cast or recrystallized metals typically have grain sizes between 5 μm and 5 mm. However, special processing methods can produce grains as large as hundreds of millimeters (Sec. 21.2.6) or as small as a few nanometers. In many ways, exceptionally small grain sizes are the more interesting of these two extremes because strength, ductility, wear resistance, formability, magnetic behavior, and fatigue behavior all change substantially when average grain size is reduced to tens of nanometers. In *nanocrystalline metals*, up to 50% of the atom population can be within a few atom spacings of a grain boundary. Early investigators postulated that the property changes in nanocrystalline metals resulted from less uniform packing of atoms within ~1 nm of nanocrystalline metal grain boundaries. However, high-resolution TEM micrographs (Fig. 7.1) and large-scale molecular dynamics computations indicate that atoms immediately adjacent to nanocrystalline metal grain boundaries have nearly perfect atomic order. Even without a disordered region near the grain boundaries, there seems little doubt that the small volume of the grains is the key factor in altering nanocrystalline metals' properties.

7.2.1 Methods of Producing Nanocrystalline Metals

Grain boundaries are crystal defects that add free energy (hundreds of mJ/m^2 of grain boundary) to the metal. To produce nanocrystalline metals, energy must be added to the metal to form the exceptionally large grain boundary area. Three methods are used to produce nanocrystalline metals: (1) crystallizing amorphous metals, (2) producing isolated nanoparticles that are subsequently sintered into a bulk specimen, and (3) increasing the free energy of the system by introducing high concentrations of lattice defects that recrystallize the metal to an exceptionally small grain size.

Structure–Property Relations in Nonferrous Metals, by Alan M. Russell and Kok Loong Lee
Copyright © 2005 John Wiley & Sons, Inc.

86 DEVIATIONS FROM CLASSIC CRYSTALLINITY

Fig. 7.1 High-resolution TEM image from nanocrystalline Cu produced by gas-phase condensation and compaction. Note that the atomic order in the two grains is preserved right up to the boundary; no disordered region appears at the boundary. (From Kumar et al., 2003.)

7.2.1.1 Crystallizing Amorphous Metals. Certain metal compositions (Sec. 7.3.1) can be cooled rapidly enough to preserve the amorphous structure of the liquid phase in a metastable solid at room temperature. Amorphous metal can also be produced as electroplated alloys from aqueous solutions. Annealing these amorphous metals nucleates large numbers of exceedingly small crystallites in the amorphous phase that convert the amorphous metal into nanocrystalline metal.

An example of this process is a Ni–25 at% W alloy electroplated as an amorphous layer on a Cu cathode. The electrodeposited metal can be crystallized by annealing at 500°C to produce an FCC Ni–W solid solution with 5- to 8-nm crystallites. Of course, nanocrystalline metals can be produced with this method only in alloys that can be made amorphous. However, increasing numbers of alloys can be processed into bulk (i.e., millimeters thick) amorphous metals at moderate cooling rates (Sec. 7.3), making this method a feasible means to produce bulk nanocrystalline alloys of several different compositions.

7.2.1.2 Formation and Sintering of Nanoparticles. Nanoscale metal particles can be produced by condensing metal vapors (Sec. 18.2.6.2), spark-eroding bulk metal, or by precipitating metals from solutions. These fine particles can then be assembled and sintered into bulk material. Since the particles are so small, sintering temperatures are lower than those usually required in powder metallurgical processing. The materials so formed are often surprisingly resistant to grain growth (Fig. 7.2), which may be partially attributable to the presence of oxides and porosity in the sintered metal. There are indications, however, that nanocrystalline materials are inherently resistant to grain coarsening (Sec. 7.2.2.3).

7.2.1.3 Converting Coarse-Grained Metals to Nanocrystalline Metals. If sufficient energy is added to polycrystalline metal of normal grain size, it can recrystallize to a nanocrystalline microstructure. Several processing methods can introduce large numbers of crystal defects into the lattice, including *ball milling* (sometimes called *mechanical alloying*), high-strain-rate deformation, explosive detonation, and irradiation with high-energy neutrons. These methods all disrupt the normal lattice so severely that it recrystallizes from many closely spaced nucleation sites, making the newly recrystallized grains exceptionally small. These processing methods

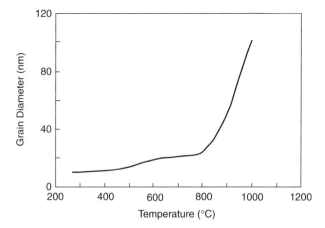

Fig. 7.2 Grain growth in TiAl particles formed by vapor condensation and sintered into bulk material. Note that breakaway grain growth did not occur until 60% of the homologous melting temperature was reached; TiAl melts at 1500°C. (From Cahn and Haasen, 1996, Vol. 3, p. 2481.)

have the advantage that they can be used on relatively massive pieces of metal; however, they often produce lower-purity metals with 100 to 1000 nm average grain size and a wide dispersion of grain sizes. These factors tend to give such metals properties intermediate between normal metal and nanocrystalline metal.

7.2.2 Properties of Nanocrystalline Metals

7.2.2.1 Mechanical Properties. Metals with an average grain size under 100 nm generally have much higher yield strength, ultimate tensile strength, and hardness than normal metals, but their ductilities are low (Fig. 7.3). Several nanocrystalline metallic elements and alloys have

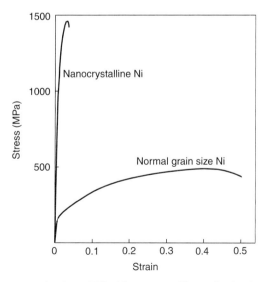

Fig. 7.3 Tensile test stress–strain plots of Ni with nanocrystalline grain size (average grain size 40 nm) and normal grain size (average grain size 60 μm). Strain rate was 3×10^{-4} s^{-1} for both tests. (Adapted from Kumar et al., 2003, and Betteridge, 1984.)

been tested by tension, compression, and indentation testing. FCC metals (particularly Cu, Ni, and Pd) have received the most study, although some data are available on BCC and HCP metals. Maximum strength typically occurs for grain sizes around 15 to 25 nm and decreases for still smaller grain sizes (Fig. 18.20).

Deformed nanocrystalline metals show no evidence of Frank–Read dislocation sources or dislocation pile-ups near grain boundaries. Computer simulations predict that dislocation sources are inoperable within such small grains, and they further predict that deformation will occur by partial dislocation emission and reabsorption at grain boundaries, by grain boundary sliding, and/or by Coble creep (Sec. 6.4.3.2). TEM observation of deformation in such small grains is difficult, so confirmation of computer simulations predictions is not entirely conclusive. TEM does show that shear bands appear in the grains after only a small amount of plastic deformation. Shear band motion leads to rapid void formation and crack growth at grain boundaries. Failure produces dimpled fracture surfaces with average dimple sizes greater than the metals' grain size (Fig. 7.4). Some evidence also exists for twin formation during plastic strain of nanocrystalline metals.

Nanocrystalline metals' low ductilities raise concern about their suitability for structural use. These concerns have been allayed somewhat by fracture toughness measurements, indicating relatively high plane stress fracture toughness (e.g., 25 to 120 MPa$\sqrt{m}$ in nanocrystalline Ni). For comparison, the equivalent fracture toughness in normal-grain-size Ni is about 60 MPa$\sqrt{m}$. Fatigue behavior of nanocrystalline metals has had limited study, but it appears that fatigue cracks may advance about 10 times faster in nanocrystalline metals than in normal metals during intermediate-cycle fatigue. However, nanocrystalline metals display generally better high-cycle fatigue strength than that of normal metals. At higher temperatures, Coble creep becomes active and works in tandem with grain boundary sliding to permit large creep strains. The presence of so many grain boundaries allows superplastic forming (Sec. 19.2.6) of nanocrystalline metals, sometimes at room temperature. Strains as large as 100 to 1000% have been achieved at rather high strain rates (about 1 s^{-1}). Nano-sized grains allow superplastic forming of otherwise brittle intermetallic compounds at elevated temperature, and even Al_2O_3 can be deformed to hundreds of percent elongation at 1800°C if the grain size is below 100 nm.

7.2.2.2 Magnetic Properties. In nanocrystalline metals with a grain size of 10 to 50 nm, the magnetic exchange length is longer than the grain size. This makes the domain wall width larger than the grain size, so each grain is a single domain. The most widely used nanocrystalline soft magnetic material, $Fe_{73.5}Si_{13.5}B_9Nb_3Cu_1$, can be solidified rapidly to produce an amorphous structure containing small clusters of Cu atoms in the glassy metal. When the metal is annealed, the Cu clusters act as nucleation sites, producing a nanocrystalline material with some amorphous regions left between the crystallites. This microstructure allows an exchange interaction to occur between nanocrystals across the amorphous regions, which gives the metal exceptional magnetic properties, including high relative permeability (approximately 10^5), coercivities below 1 A/m (0.0125 Oe), high saturation magnetization (10^6 A/m, or 13 kG), high electrical resistivity (1.15×10^6 Ω·m), and low magnetostriction (approximately 10^{-6}).

7.2.2.3 Thermal Stability. In metals with normal grain size, the driving force ($\Delta\mu$) for grain growth is expressed by the Gibbs–Thompson equation,

$$\Delta\mu = \frac{2\gamma V_a}{d}$$

where γ is the interfacial energy, V_a the atomic volume, and d the grain size. This expression predicts rapid grain growth in nanocrystalline metals at room temperature, because their d values are so small. Surprisingly, however, this rapid grain growth does not occur. Nanocrystalline metals are generally stable against grain growth until they reach about $0.4T_{m.pt.}$ (Fig. 7.2), which is similar to the behavior of normal metals. Stranger still, nanocrystalline metals with grain sizes

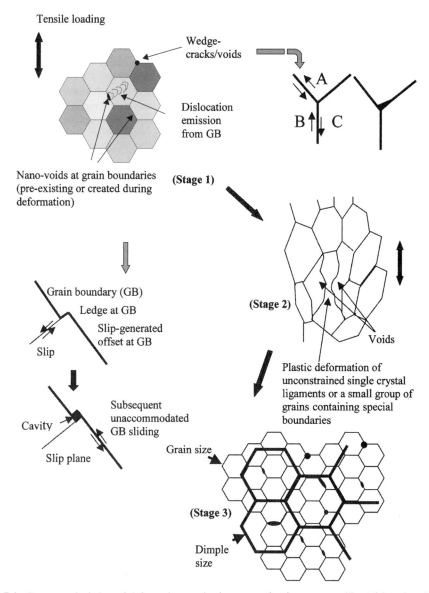

Fig. 7.4 Summary depiction of deformation mechanisms operating in nanocrystalline Ni based on TEM observation during straining and on SEM observations of fracture surfaces. (From Kumar et al., 2003; with permission.)

below 10 nm are sometimes observed to be *more* resistant to grain growth than nanocrystalline metals with somewhat larger grain sizes!

Grain growth in metals with normal grain size is generally governed by the kinetics of *grain boundary diffusion*, which has a lower activation energy than bulk diffusion. However, activation energies for grain growth in some nanocrystalline metals closely match the activation energy for *bulk diffusion*, which implies that some volume diffusion process within the grains may be the rate-controlling mechanism for grain growth in nanocrystalline metals. Although the issue is complicated by the effects of pores and impurities acting to stabilize the grain boundaries, it may be that lattice strain in nanograins becomes more severe as grain size decreases, and bulk diffusion within the grains is necessary to relieve these strains before grain growth can begin. In this model, interface diffusion is necessary but not sufficient, and since bulk diffusion has a higher activation energy, it limits grain growth kinetics in nanocrystalline metals. Some theorists have suggested that a critical grain size exists below which the rate-limiting factor in grain growth is migration and rearrangement of triple junctions and accommodation of the excess volume located in the annihilated boundary to other sites in the metal or to the surface.

7.2.3 Engineering Applications for Nanocrystalline Metals

Nanocrystalline alloys are the dominant materials for hard disk magnetic data storage in computers (Sidebar 16.1 and Fig. 16.9). Another engineering use is the Finemet $Fe_{73.5}Si_{13.5}B_9Nb_3Cu_1$ nanocrystalline alloy employed in a variety of soft magnetic applications. Finemet is more stable at elevated temperature than totally amorphous soft magnetic materials and is used as sheets built into the walls of low-field, magnetically shielded rooms, as well as in transformers, switch systems, and induction devices.

Applications exploiting the mechanical properties of nanocrystalline metals have been few, due primarily to high production costs and difficulty producing thick sections. The performance of WC–Co cermet cutting tools (Sec. 16.2.3.3) has been improved by reducing the WC and Co phase sizes to tens of nanometers. Hardness increases nearly 50% vis-à-vis coarse-grained WC–Co cermets, and wear resistance and toughness are improved similarly. Advances in bulk amorphous metals (Sec. 7.3) makes formation of nanocrystalline metals and mixed amorphous–nanocrystalline microstructures more feasible. Wear-resistant nanocrystalline surfaces could be applied to tougher, more ductile substrate metals. Nanocrystalline metal at surfaces and high stress locations could also improve high-cycle fatigue performance of normal-grain-size substrate components.

7.3 AMORPHOUS METALS

Glassy ceramics have been used for millenia. But the first amorphous metal, a rapidly solidified Au–Si alloy, was not produced until 1960. The first metallic glasses required high cooling rates (about 10^5 K/s or faster) to preserve the disordered atomic structure of the liquid metal as a metastable solid. Such high cooling rates are possible only in small volumes of metal, which limited amorphous metal production to powders, ribbons, and thin films. However, amorphous metals' mechanical and magnetic properties and their corrosion resistance are so exceptional that they were used in commercial coatings, brazing foils, and transformer core laminates as early as the 1970s. Newer alloys with stronger glass-forming abilities have allowed production of glassy metal parts as thick as 0.1 m. This greatly expanded their potential engineering use, and amorphous metals are now employed in a diverse range of products that exploit their high strength, hardness, corrosion resistance, modulus of resilience, and easy formability.

7.3.1 Methods of Producing Amorphous Metals

Pure metalloid elements such as Si can be made amorphous because their bonding is partially covalent, but computations indicate that normal pure metals such as Ag, Cu, Ni, and Pb crystallize so rapidly that they need cooling rates of 10^{12} to 10^{13} K/s to retain the amorphous structure of the liquid as a glassy solid. Such cooling rates are achievable only in tiny volumes of metal by methods such as condensation of sputtered metal atoms onto chilled surfaces. Cooling rates up to about 10^{14} K/s are achieved by striking small surface regions of massive metals with nanosecond or picosecond laser pulses. This melts the metal momentarily, and it is immediately quenched by the cold metal surrounding the heated spot. But producing substantial amounts (grams to kilograms) of amorphous metals requires alloys rather than pure metallic elements.

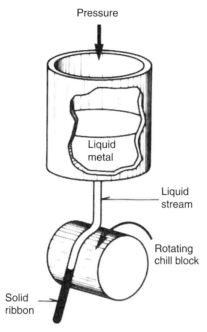

Fig. 7.5 Melt-spinning technique for producing amorphous metal alloys. The molten metal is released under pressure from a crucible suspended over a water-cooled spinning cylindrical chill block, producing a thin, rapidly cooled ribbon of metal. (From Gessinger, 1984, p. 52; with permission from Elsevier.)

Amorphous metal ribbons of some simple binary alloys are made by quenching the liquid onto rapidly rotating, chilled metal wheels (*melt spinning*) (Fig. 7.5) at cooling rates of 10^4 to 10^6 K/s. Higher cooling rates (10^6 to 10^8 K/s) are possible by gas atomization of liquid metal onto chilled metal substrates (*splat cooling*) to produce thin films. Several other methods have been developed to make amorphous alloys, including irradiation with high-energy particles, ball milling, spark erosion, thermal spraying, electroplating, and even solid-state diffusion reactions between polycrystalline metal foils with widely differing atomic radii.

The rapid heat transfer needed to achieve cooling rates of 10^4 K/s or more requires at least one dimension of the product to be no thicker than 10 to 100 μm. This severely limits dimensions for engineering parts. More massive amorphous metal products require alloys that can remain amorphous at lower cooling rates. Alloys are now available (Fig. 7.6) that remain amorphous when cast in sections nearly 0.1 m thick because their critical cooling rates are only 10^3 to 10^{-1} K/s. This reduction in the critical cooling rate increases the time available to cool the metal from milliseconds to minutes. The keys to improving the glass-forming ability of alloys are (1) formulating multicomponent alloys (i.e., three or more elements), (2) using elements with atomic radii that differ by more than 12%, (3) using compositions close to a deep eutectic, and (4) selecting elements with a negative heat of mixing. These four factors allow the metal to form a viscous, supercooled liquid that persists well below the equilibrium freezing temperature, thereby delaying crystallization. Multicomponent alloys often have complex equilibrium structures that require considerable atom movement for crystallization. The thermodynamic relationship $\Delta G = \Delta H_f - T \Delta S_f$ (where H_f and S_f are the enthalpy and entropy of fusion) indicates that a lower free energy results when ΔS_f is large. Since multicomponent systems have greater numbers of microscopic states, their entropy of fusion is larger. Elements with differing atomic radii achieve higher packing densities in the liquid state, which impedes atomic rearrangement during solidification and cooling. Negative heats of mixing raise the energy

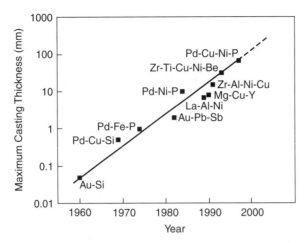

Fig. 7.6 Increase in maximum possible casting thickness achieved in amorphous metal alloys during the late twentieth century. Discovery of alloy compositions with progressively stronger glass-forming activities increased maximum casting thickness by three orders of magnitude during this period. (Redrawn from Telford, 2004, p. 38.)

barrier at the solid–liquid interface during freezing, which makes atomic rearrangement to form the equilibrium crystal structure more difficult. This extends the liquid temperature range over which the liquid metal can remain as a supercooled liquid. Finally, using compositions near a deep eutectic forms a stable liquid at lower temperatures, reducing the amount of heat that must be extracted to reach the glass transition temperature.

Alloys specifically formulated to maximize glass-forming ability can be cast in normal molds to form an amorphous structure. This permits thick sections to be produced easily without resorting to melt spinning or other costly processing methods. The resistance to crystallization of amorphous metals with high glass-forming ability confers a great advantage in forming the metal to the dimensions of a final product; the amorphous metal can be heated to lower its viscosity (just like a ceramic glass), molded with low forces while soft, and then cooled to room temperature, all without crystallizing the metal.

7.3.2 Properties of Amorphous Metals

7.3.2.1 Mechanical Properties. The mechanical properties of amorphous metals vary markedly from those of crystalline metals. Amorphous metals tend to have higher tensile yield strengths and higher elastic strain limits than crystalline metal alloys; however, their ductilities and fatigue strengths are considerably lower (Table 7.1). The absence of long-range crystalline order frustrates dislocation formation and motion. Thus, amorphous metals have higher yield strengths and a larger elastic limit than those for crystalline metals with similar compositions. As strain increases to about 2% elongation, however, amorphous metals begin to form shear bands at angles with the greatest resolved shear stress. All subsequent deformation is confined to these shear bands (inhomogeneous flow), which are only 10 to 20 nm thick and form extremely rapidly. The metal in the shear bands becomes less dense and weaker than the surrounding metal (work softening) and behaves much like a viscous liquid. Continued strain leads to rapid fracture along shear bands. Thus, amorphous metals have low tensile ductility. At low strain rates and/or moderately elevated temperatures, amorphous metals can be deformed extensively by creep, which occurs by pure viscous flow. Since viscous flow increases the density of the material, viscosity rises as creep progresses; thus, amorphous metals display a progressively decreasing creep rate under constant load and temperature. Amorphous metals' fatigue strength

TABLE 7.1 Comparison of Mechanical Properties of Amorphous Vitreloy 1 ($Zr_{41.2}Ti_{13.8}Cu_{12.5}Ni_{10}Be_{22.5}$) with Widely Used Crystalline Metal Alloys

Property	Vitreloy 1	Al Alloys	Ti Alloys	Steels
Tensile yield strength (MPa)	1900	100–630	180–1320	350–1600
Elastic strain limit[a] (%)	~2	~0.5	~0.5	~0.5
Fracture toughness, K_{IC} (MPa$\sqrt{m}$)	20–140	23–45	55–115	50–154
Density (g/cm^3)	6.1	2.6–2.9	4.3–5.1	7.8
Specific strength (MPa/g·cm^3)	320	<240	<310	<210

Source: Telford (2004); with permission.
[a]Elastic strain limit is the strain at which substantial plastic deformation begins. Since the transition from elastic to plastic strain is somewhat gradual, these values are approximate. Small permanent deformations occur at strains below these values. Note that this value is not equivalent to the 0.2% strain construction used to define yield strength in many metals.

is generally poor; cyclic tensile loading creates shear bands that lead to more rapid fatigue failure than crystalline metals would experience under the same loading.

The low ductility and fatigue strength of amorphous metals can be improved by partially devitrifying the material to form a composite structure that is a nanoscale mixture of amorphous and crystalline metal phases. The crystalline phase acts as the equivalent of a "crack stopper" to inhibit shear band propagation, thereby enhancing ductility and toughness. The second phase can also be introduced extrinsically (e.g., C or W fibers in an amorphous matrix). These composites are expected to maintain the high yield strength of the amorphous phase while making the material more reliable and fracture resistant.

7.3.2.2 Corrosion Resistance. Two characteristics of amorphous metals make them more corrosion resistant than crystalline metals. On the metal's surface, grain boundaries frequently serve as active sites for corrosion to begin, and microscale composition variations in crystalline metals act as galvanic cells. The complete absence of grain boundaries and the compositional homogeneity of metallic glasses improve corrosion performance, sometimes quite dramatically. Crystalline metals that are rapidly attacked by mineral acids are often unaffected by the same acids when amorphous.

7.3.2.3 Magnetic Properties. Several Fe- and Co-based amorphous alloys have excellent soft magnetic performance, mechanical strength, and corrosion resistance. For example, the bulk amorphous alloy $Fe_{60}Co_8Zr_{10}Mo_5W_2B_{15}$ remains amorphous below 800 K and has an extremely narrow BH hysteresis loop, Vickers hardness of 1360, compressive fracture strength of 3800 MPa, and can withstand an hour's immersion in aqua regia acid with no weight loss. Amorphous metals also serve as useful precursors for production of nanocrystalline soft magnetic materials (Sec. 7.2.3) and hard permanent magnets. The bulk magnetic material with the highest work product (Sec. 16.2.3.4 and Fig. 16.4) is $Nd_2Fe_{14}B$. This intermetallic compound is widely used in commercial applications (e.g., electric motors and speakers), and it is often produced by melt spinning to generate an amorphous ribbon which can then be devitrified into a nanocrystalline permanent magnet. The fine grain size (14 to 50 nm) achieved by forming the intermediate amorphous phase improves coercivity.

7.3.2.4 Thermal Stability. Metallic glasses are metastable. There is always a combination of crystalline phases with a lower free energy than the amorphous structure. Given enough time at a sufficiently high temperature, any metallic glass will crystallize. Figure 7.7 shows a typical differential scanning calorimeter trace produced by heating amorphous $Pd_{77.5}Cu_6Si_{16.5}$. The small drop in the curve at 370°C marks the endothermic glass transition (T_g); the large exothermic peak at 423°C is the first major crystallization event (T_{x1}); and the smaller peak at 436°C (T_{x2}) marks the transformation of the metastable crystalline phase formed at T_{x1} to the equilibrium

94 DEVIATIONS FROM CLASSIC CRYSTALLINITY

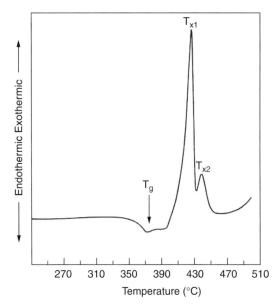

Fig. 7.7 Differential scanning calorimetry trace for heating a 6-mg sample of amorphous $Pd_{77.5}Cu_6Si_{16.5}$ at 20 K/min. The endothermic event at T_g is the glass transition temperature. T_{x1} marks an exothermic transformation to a metastable crystalline phase. At T_{x2} that metastable phase transformed to the equilibrium crystalline phase. (Redrawn from Cahn and Greer, 1996, p. 1786.)

crystal phase. The potential for crystallization places an upper limit on metallic glass's use temperatures, but crystallization often occurs several hundred degrees above ambient.

7.3.2.5 Formability. Ceramic glasses are easily formed into intricate shapes by heating them until their viscosity drops low enough to allow plastic flow. This same approach can also be used for bulk metallic glasses, giving them outstanding formability. When a metallic glass with a high glass-forming ability is heated, the following transitions occur: amorphous metal → glass transition → *supercooled liquid* → crystallization → melting. In the supercooled liquid temperature range, metallic glass deforms by simple viscous fluid flow rather than by shear bands. Thus, large strains are easily accomplished without cracking the metal. Figure 7.8 shows the relation between flow stress and strain rate in amorphous $La_{55}Al_{25}Ni_{20}$ alloy in its supercooled liquid temperature range. Each line in Fig. 7.8 has a slope of 1, so these metals behave as classic Newtonian fluids with high viscosity. Viscosity is an exponential function of temperature, so the metal's strength can be "tuned" by adjusting its temperature. For example, a Zr-based metallic glass alloy, Vitreloy-1 (Table 7.1), becomes quite malleable at about 400°C, allowing the supercooled liquid to be press forged or extruded (Secs. 8.4.3–4). For amorphous alloys with high glass-forming ability, these processes are easier to perform than conventional casting or forming operations are for normal metals because the working temperature and forces required are both lower, the shrinkage of the amorphous metal is much less, no porosity forms, and the surface finish is typically mirror smooth, eliminating the need for subsequent machining or polishing. The metal becomes almost as easy to form as a thermoplastic polymer, and the formed part remains totally amorphous during forming and subsequent cooling to 20°C.

7.3.3 Engineering Applications for Amorphous Metals

The high yield stress and large elastic strain limit of amorphous metals give them a modulus of resilience (Sec. 12.2.2.3 and Fig. 12.7) as much as 25 times greater than that of crystalline

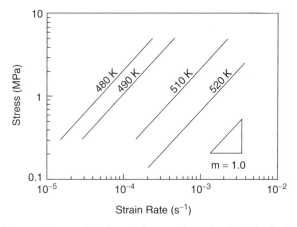

Fig. 7.8 Relation between stress and strain rate in amorphous $La_{55}Al_{25}Ni_{20}$ in its supercooled liquid temperature range. The slope of each line is approximately 1 ($m = 1.0$), indicating ideal Newtonian fluid flow (i.e., ideal superplasticity). (Redrawn from Inoue and Hashimoto, 2001, p. 40.)

alloys made of the same elements. This greatly enhanced ability to store and release elastic strain energy makes amorphous metals ideal for springs and sports implements (Sidebar 7.1). The high hardness and easy formability of amorphous metals makes them well suited to use in surgical tools, knives, computer, and instrument housings. Their hardness and excellent corrosion resistance make them useful for jewelry and for coatings on tanks, pumps, valves, filters, and electrodes used in corrosive fluids.

When molten, alloys with high glass-forming ability are two to three orders of magnitude more viscous than conventional alloys, making it easier to produce foamed metals by bubbling gases through the liquid and injecting the resulting froth into a mold. Amorphous metals with 99% porosity have been manufactured with this technique. Diffusion at moderately elevated temperature is much slower through amorphous metals than through crystalline metals because the dominant diffusion mechanism at such temperatures is grain boundary diffusion, and amorphous metals have no grain boundaries. This allows amorphous metals to be used as diffusion barriers in microelectronic devices. The limited deformability of amorphous metals makes them

Fig. 7.9 Golf club striking a ball. An amorphous metal club head has a higher modulus of resilience, so it can store and release more elastic strain energy than can a crystalline club head.

96 DEVIATIONS FROM CLASSIC CRYSTALLINITY

> **SIDEBAR 7.1: USE OF AMORPHOUS METALS IN SPORTS IMPLEMENTS**
>
> An amorphous metal ball dropped onto a rigid surface will bounce for more than 2 minutes before finally coming to rest. By contrast, a crystalline metal ball typically stops bouncing within 6 or 8 seconds, because its various crystal defects dampen the elastic rebound, rapidly converting the elastic energy to heat. The homogeneity of the amorphous structure allows more than 99% of the collision energy to be returned to the surface to accelerate the ball upward at the start of each bounce. The rebound of amorphous metal is not only highly efficient, it also has a much greater capacity to store elastic energy, due to its high yield strength and unusually large elastic strain limit (about 2% vis-à-vis about 0.5% in crystalline metals).
>
> Amorphous metals' high moduli of elastic resilience (Sec. 12.2.2.3 and Fig. 12.7) can be exploited in making more effective sports implements. When an amorphous metal golf club strikes the ball (Fig. 7.9), it can store more elastic energy from the collision than crystalline metal. It then imparts a greater force to the ball during the later (rebound) stage of the collision as it returns its stored elastic energy, accelerating the ball to higher velocity. The result is a 25-m longer flight for a golf ball hit by an amorphous metal driver. Similar benefits accrue to athletes using amorphous tennis rackets, baseball bats, and hunting bows.

effective armor-penetrating projectiles. An amorphous bullet shears into self-sharpening slivers on impact rather than flattening as Pb alloys do. When reinforced with W, amorphous metal ammunition provides performance approaching that of environmentally controversial depleted-U ammunition (Sidebar 1.3 and Sec. 24.4.4.3).

There are two major barriers to wider use of amorphous alloys. The first is their limited ductility, which poses problems in many structural applications. Continuing research on improving ductility by forming amorphous-crystalline composites (Sec. 7.3.2.1) may allow greater structural use. The second challenge is reducing the current high costs of these alloys, which are typically priced around $35 per kilogram. High-purity metals are required to produce alloys with high glass-forming ability. Development of alloys that are less sensitive to impurity content would substantially lower costs.

7.4 QUASICRYSTALLINE METALS

The crystallographer's world was a well-defined orderly place in 1981. Everyone "knew" that crystals were comprised of unit cells that could be stacked side by side in any desired number, filling up all available space with exactly repeating copies of the atom positions of the unit cell. To fill space completely with these repeating unit cells, the cells must have two-, three-, four-, or sixfold rotational symmetry. Crystals with five- or n-fold symmetry (where n is an integer greater than 6) cannot fill space completely and are therefore forbidden structures in perfect crystals.

This tidy picture suffered a severe jolt in 1982 when Shechtman generated a diffraction pattern from an Al–Mn alloy similar to the one in Fig. 7.10. The alloy had the symmetry of an icosahedron, possessing six axes of fivefold symmetry, 10 axes of threefold symmetry, and 15 axes of twofold symmetry. The finding was so controversial that Shechtman was forced to wait two and a half years to receive peer review approval to publish his findings in a scientific journal. Even after the findings were published, many scientists were reluctant to concede the existence of such crystals, and statements such as the quotations at the beginning of this chapter exemplify the level of skepticism then extant.

In the ensuing decades, many more such crystals with forbidden symmetries have been discovered, and the term *quasicrystal* is used to identify the entire family of such materials. Although quasicrystals cannot be defined by a stacking of unit cells that fill space completely,

Fig. 7.10 (a) Electron diffraction pattern taken from as-cast, quasicrystalline $Al_{65}Ni_{30}Ru_5$ alloy along its 10-fold axis. Note the five- and 10-fold rotational symmetry. Such symmetry is forbidden by conventional crystallographic models of repeating unit cells stacked side by side to fill all space within the crystal. (b) Fivefold symmetry is also prominent in this optical photograph of an icosahedral $Ho_9Mg_{34}Zn_{57}$ crystal. [(a) From Sun and Hiraga, 2002; (b) from Canfield and Fisher, 2001; with permission of Elsevier.]

the debate about their existence ended long ago. Their crystal structure is most easily described by unit cells that interpenetrate one another (Fig. 7.11). The typical quasicrystal is a binary or higher-order alloy, often containing 60 to 70 at% Al. Quasicrystals are typically hard, brittle materials with unusually low conductivities and coefficients of friction (Table 7.2). They are used sparingly in current engineering design, finding applications in cookware, surgical tools, and electric razors. Their unusual combination of high electrical resistivity (400 to 600 $\mu\Omega \cdot$ cm), high hardness, and low coefficient of friction makes them potentially useful as wear-resistant coatings for parts subjected to sliding contact, thermal barrier coatings, and reinforcing phases in metal matrix composites. Some quasicrystals can dissolve large amounts of H, making them potentially useful in H fuel storage systems. They can be applied as coatings by thermal spray, and bulk parts can be fabricated by powder metallurgy.

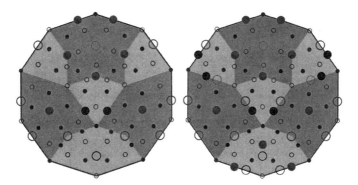

Fig. 7.11 Model of the decagonal quasi-unit cell of $Al_{72}Ni_{20}Co_8$. Large circles represent Ni (open) and Co (black) atoms; small circles represent Al atoms. Circle sizes do not correlate with actual atom sizes. The structure has two distinct layers along the periodic c axis, which is perpendicular to the page. Solid circles represent the atoms on the $c = 0$ level, and open circles represent the atoms on the $c = \frac{1}{2}$ level. The image on the right is the same decagonal quasi-unit cell showing additional atoms present from the "overlap" of neighboring quasi-unit cells. (From Steinhardt, 1996, pp. 613–614; with permission.)

TABLE 7.2 Comparison of Physical and Mechanical Properties of Quasicrystalline Icosahedral $Al_{71}Pd_{20}Mn_9$ and $Al_{64}Cu_{23}Fe_{13}$ with Pure Cu Metal

Property	i-$Al_{71}Pd_{20}Mn_9$	i-$Al_{64}Cu_{23}Fe_{13}$	Cu
Hardness (Vickers hardness no.)	700–900	800–1000	40
Young's modulus (GPa)	200	62	110
Poisson's ratio	0.38	—	0.34
Surface energy (mJ/m^2)	25[a]	—	—
Surface coefficient of friction	—	0.05–0.20	0.42
Fracture toughness, K_{IC} (MPa·m$^{0.5}$)	0.3	1	30

Source: Dubois (1997).
[a] For comparison, Teflon-type chlorofluorocarbons have surface energies of ~17 mJ/m^2.

7.5 RADIATION DAMAGE IN METALS

High-speed particles (ions, neutrons, electrons, protons) and high-energy photons (γ rays) can knock atoms out of their equilibrium positions in a crystal. In high-radiation environments, a crystalline metal's normal atomic lattice can be disrupted severely by such collisions. When a fast-moving particle knocks a metal atom out of position, a cascade of subsequent collisions occurs, leaving the lattice riddled with defects fanning out from the initial collision site (Fig. 7.12). If the irradiation is intense and prolonged, every atom in the metal can be displaced many times. Most displaced atoms come to rest on normal lattice sites, but a substantial minority become vacancies or interstitials, producing defect concentrations several orders of magnitude higher than equilibrium levels. The accumulation of defects in radiation-damaged metal impedes dislocation motion, making the metal stronger and less ductile (Sec. 12.3.5.2 and Fig. 12.23). Diffusion is proportional to the number of vacancies in the lattice, so radiation damage increases diffusion rates. It also causes the metal to swell and distort, posing clearance and fit problems with mating parts in assemblies.

If pure metal is irradiated at low temperature (e.g., $T < 20$ K), the lattice disruption is so drastic that the metal can become largely amorphous. Alloys can become amorphous during

Fig. 7.12 Computer simulation of the defects in a section of a Cu grain struck by a high-energy neutron. The arrow indicates the direction of the incident neutron. The black spheres identify vacancies caused by the collision cascade; the light gray spheres identify positions in the crystal lattice where displaced atoms came to rest as interstitial defects. Atoms in correct lattice sites are not shown. (Courtesy of H. L. Heinisch, Pacific Northwest Laboratory, U.S. Department of Energy; with permission.)

irradiation at much higher temperatures than pure metal, even above 20°C in some cases. Temperature "spikes" occur in the crystal volume around each new collision cascade, but their heating effect is brief and localized. The surrounding cold metal cools the collision cascade atoms rapidly, and their disruptive effects are greater than their annealing effect. Vacancies require roughly 1 eV of energy for both their creation and for a "jump" into a neighboring lattice site (Fig. 6.4). Interstitials have much higher creation energy (about 5 eV) because they strain the surrounding lattice more severely; however, the jump energy for interstitials is quite low (about 0.1 eV). Thus, interstitial defects are more difficult to form than vacancies, but once formed, interstitials move through the lattice more easily. For this reason, most metals have a temperature range where interstitial motion is relatively rapid but vacancies are nearly immobile.

At higher irradiation temperatures, both vacancies and interstitials diffuse through the lattice. This allows some vacancies and interstitials to recombine, annihilating both defects and releasing their defect energies as heat in the lattice. Defects can also lower the strain energy in the crystal by moving to other defect sites, such as impurity atoms, dislocations, grain boundaries, or the free surface. Interstitials are attracted to the dilated lattice just below the edge of the half-plane in edge dislocations, and vacancies are attracted to the strain fields around both edge and screw dislocations. Screw dislocations containing edge-type jogs (Sidebar 4.1) are normally difficult to move, but the large numbers of mobile interstitials and vacancies in radiation-damaged metals allow jogged screw dislocations to move, accelerating creep (Sec. 6.4). If the irradiated metal is at or moderately above room temperature, vacancies and interstitials congregate by diffusion into dislocation loop-stacking fault defects (Fig. 4.13) that are much less mobile than individual defects. Planar defects of coalesced vacancies can grow into three-dimensional voids that persist to $\sim 0.5 T_{m.pt.}$, where full annealing occurs. Metal irradiated between about 20 K and $0.5 T_{m.pt.}$

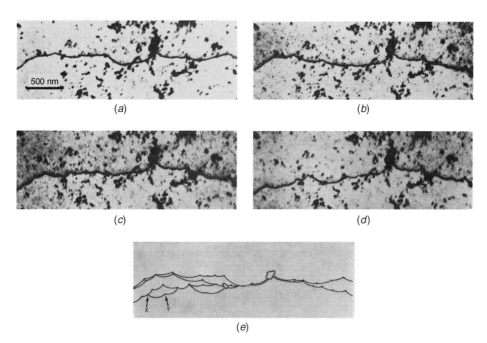

Fig. 7.13 (a–d) Time-lapse sequence of four TEM images of dislocation–defect interaction in Cu irradiated by $3 \times 10^{17} \alpha$ particles/cm^2 ($E = 38$ MeV). The dark black line is a dislocation moving downward under an applied shear stress. The irregular black spots are radiation-induced defect clusters. The successive positions of the dislocation line are shown superimposed in part (e) with the defect clusters removed for clarity. Bowed dislocation segment XY is labeled for reference in Problem 4.1. (From Barnes, 1964.)

SIDEBAR 7.2: RADIATION DAMAGE IN THE *GALILEO* SPACECRAFT

Irradiating structural metals does not significantly alter their mechanical properties unless the radiation is intense and prolonged, as in the core of an operating nuclear reactor. However, semiconducting and magnetic devices used in modern electronic systems are much more susceptible to radiation damage because vacancies, interstitials, dislocation loops, and voids in semiconductor devices degrade their performance and can even cause complete failure. For this reason, engineers designing computers for use in Earth orbit deliberately avoid using modern microprocessors; their small feature size makes them more likely to fail from solar flare radiation and cosmic rays. The *Space Shuttle* computers fly with 1980s-era processors, because their larger feature size (fewer transistors/mm^2) makes them more robust against radiation damage.

Jupiter poses an especially difficult radiation exposure challenge to spacecraft because its radiation levels are more intense than Earth's. The *Galileo* spacecraft orbited Jupiter continuously from 1995 to 2003, studying the planet and its satellites (Fig. 7.14). Fly-bys of the moon Io were especially hazardous for *Galileo*'s electronics because Io ejects a huge torus of charged particles from its volcanoes. Io's orbital motion around Jupiter through this magnetized plasma torus generates 2 trillion watts of fast-moving electrons and ions between Io and Jupiter. Some of *Galileo*'s electronic systems absorbed more than 1 Mrad of radiation even though these devices were heavily shielded to minimize radiation damage. (By comparison, the lethal radiation dose for a human being is 0.000 5 Mrad.) Each passage through this torrent of high-energy particles further damaged *Galileo*'s electronic systems. The spacecraft continued to function throughout its eight-year mission, but several systems' capabilities (e.g., computer memory, tape recorder, LEDs, and various science instruments) were substantially impaired by mission's end.

Fig. 7.14 Artist's depiction of the *Galileo* spacecraft during a close fly-by of the Jovian satellite Io. The spacecraft's electronics were subjected to intense radiation while in orbit around Jupiter (1995–2003). (Courtesy of NASA.)

typically shows large numbers of planar defects and voids that interact with dislocations to harden and embrittle the metal (Fig. 7.13).

Neutron irradiation not only displaces atoms but can also change their nuclear structure. Neutrons have no electrical charge, so they can be absorbed by the nuclei of some of the metal atoms, transmuting them into other isotopes and/or other elements. Thus, a pure metal gradually becomes an alloy if subjected to sustained, intense neutron irradiation. V offers an example of such a transmutation, gradually becoming a V–Cr solid solution alloy by the reaction $^{51}_{23}\text{V} + ^{1}_{0}\text{n} \rightarrow ^{52}_{23}\text{V}$ (unstable) $\rightarrow ^{52}_{24}\text{Cr} + \beta^{-1}$. Since many of the isotopes formed by neutron irradiation are radioactive, the metal becomes dangerous to handle, requiring heavy shielding to protect workers from radiation injury.

Radiation fluences are especially high in nuclear fission reactors, making their structural metals particularly vulnerable to radiation damage. More than 450 commercial power reactors produce a total of 360,000 MW, 16% of the world's total electric power generation. An additional 280 reactors are used for research and isotope production, and 150 ships are powered by nuclear reactors. Steels and Zr alloys (Secs. 12.3.2 and 12.3.3) are the most widely used metals in reactor cores. Radiation effects in these metals have been studied extensively to allow reactor designers to predict the property and dimensional changes of the metals as they are irradiated. Reactor operating temperatures are typically about 300°C, so part of the damage anneals away as it occurs in Fe and Zr. Future fusion reactors (Sidebar 10.2) will pose greater radiation damage challenges, since they will generate higher-energy neutrons and higher total dosages to their structural materials. In addition, these reactors will generate large amounts of $^{1}_{1}\text{H}$, $^{2}_{1}\text{H}$, $^{3}_{1}\text{H}$, and He, which can form gas-filled voids in irradiated materials that exacerbate swelling. Radiation damage problems also occur in space vehicle electronics (Sidebar 7.2) and particle accelerators.

ADDITIONAL INFORMATION

References, Appendixes, Problem Sets, and Metal Production Figures are available at
ftp://ftp.wiley.com/public/sci_tech_med/nonferrous

8 Processing Methods

> If you don't like something, change it.
>
> —Mary Engelbreit and Patrick Regan, *Mary Engelbreit:*
> *The Art and the Artist Hardback*

8.1 INTRODUCTION

It is a rare event, indeed, when an object found in nature is directly useful in its native form. Almost every inorganic object used in our daily lives is processed in some way to make it more usable. For example, the Pb in a car battery began its processing as PbS mixed with other minerals. It was crushed, separated, and roasted in air to form PbO, reduced with C to produce Pb metal, refined, alloyed with Ca, cast into plates, rolled to sheet form, expanded by piercing, coated with PbO, and immersed in an acid bath to produce the battery (Sec. 21.5.3.1). Every other part of that automobile has its own processing history. Processing alters the structure of the material, and the engineer must monitor and anticipate these structural changes to optimize the material's performance for its intended task. Chapters 10 to 25 contain brief descriptions of processes that convert minerals into metallic elements. In this chapter we describe the procedures that reshape and modify those metals for engineering use.

8.2 CASTING

One particularly straightforward way to reshape metal is to melt it and cast (pour) the liquid metal into a mold whose interior has the desired shape. The solidified metal is shaped like the inner contours of the mold. Casting is one of the oldest metal-forming processes. Its apparent simplicity belies a surprisingly complex interplay between casting methods and the final properties of cast metal.

8.2.1 Furnaces and Crucibles

The most common furnace design to melt metal for casting is a refractory-lined steel tank heated by electric resistance heating elements or by hydrocarbon combustion. The crucible containing the solid casting metal is placed in the hot central zone of the furnace until the metal melts. Some furnaces heat the casting metal with rapidly alternating electromagnetic fields; this induces eddy currents in the metal that melt it by ohmic resistance heating. Still other casting operations melt the metal by striking it with an electric arc. The liquid metal surface may be covered with a floating liquid slag of molten salts to avoid oxidation from contact with air. For particularly reactive metals (e.g., Ti), slags are inadequately protective, and melting is done in inert gas or vacuum. Recycled scrap metal and alloying additions are often added to the molten metal to achieve the desired composition in the casting.

Structure–Property Relations in Nonferrous Metals, by Alan M. Russell and Kok Loong Lee
Copyright © 2005 John Wiley & Sons, Inc.

Crucibles are typically steel or cast iron with a refractory ceramic interior lining that does not dissolve or react with the molten metal (e.g., SiC bonded with C for Al). Some lower-melting alloys (e.g., Mg, Pb) have little reaction with ferrous metals, so bare steel or cast iron can be used with no refractory lining. Most gases are much more soluble in liquid metal than in solid metal (Fig. 20.7); thus, it is often necessary to remove dissolved gases from the liquid metal to avoid gas bubble porosity in the final casting. This can be done by bubbling other gases through the molten metal with a submerged tube to entrain the dissolved gases (e.g., Cl_2 or N_2 removes H dissolved in molten Al). Some metals pose special challenges to crucible materials, either because they are so reactive that they attack almost all crucible materials (e.g., Ti) or because they are so high melting (e.g., Re, W, Ta) that no crucible materials have sufficient strength and inertness to hold the molten metal.

8.2.2 Moving Molten Metal into a Mold

Molten metal for casting is moved into the mold either by tilting the crucible to pour the metal out the top or by tapping it out the crucible's side or bottom. The pouring method must avoid entraining slag into the casting metal, and it should minimize contact with air and other materials that could add impurities or dissolved gases (Sec. 20.2.8.3). Channels are arranged to direct the molten metal into the mold cavity (Fig. 8.1). The pouring basin funnels molten metal down the sprue, which in turn directs the molten metal into one or more runners to fill both the mold and auxiliary molten metal reservoirs called *risers*. The molten metal in the risers flows into the mold to compensate for volume shrinkage as the metal in the mold cools and solidifies. Metal from the risers minimizes formation of shrinkage cavities (empty regions) in the center of the casting. After the final casting cools, the solidified risers and runners are cut away. The assembly is designed to allow molten metal to flow from the risers into the mold before the runners solidify. Risers are often surrounded by insulating material so they will freeze last. The acceptable geometry for the assembly varies with the fluidity of the molten metal and with the thermal conductivities of the mold material and casting metal.

8.2.3 Casting Defects and Segregation

The ideal casting has no porosity, slag inclusions, or entrained mold material. Real castings sometimes fall short of this ideal and contain one or more physical defects. *Hot tears* are

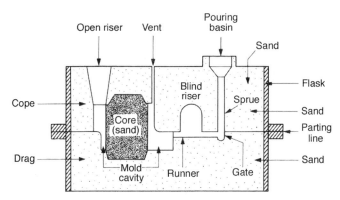

Fig. 8.1 Molten metal is poured into a sand mold through a sprue. The sprue channels the molten metal into one or more runners to fill both the mold and auxiliary reservoirs for molten metal called *risers*. Molten metal in the risers flows into the mold to compensate for the volume shrinkage during cooling and solidification. The metal from the risers minimizes formation of empty regions (shrinkage cavities) in the center of the mold. Solidified risers and runners are cut from the casting after it is removed from the mold. (From Kalpakjian, 1989, p. 302; with permission.)

separations that occur in the semisolid metal caused by molds that do not allow the metal to contract as it shrinks during cooling. *Porosity* is caused by dissolved gases evolving in the solidifying metal or by shrinkage that was not compensated by additional flow from risers. *Misruns* are "missing portions" of the intended casting, caused when the mold fails to fill completely with metal. *Cold shuts* occur when liquid metal streams entering from more than one gate fail to fuse when they meet inside the mold (caused by excessive cooling or oxide skins). *Blows, blisters, scabs*, and *scars* are surface defects caused by dissolved gas in the liquid metal, entrained flux, or air trapped in the mold as it filled with liquid metal. *Wash* is an unwanted protrusion from the casting surface caused by mold wall erosion. Defects can sometimes be ignored if they are merely cosmetic, other defects can be repaired by welding or grinding, but in some cases defective castings must be rejected.

Castings without macroscopic physical defects will still usually contain both microstructural and macrostructural segregation. High-purity metals and eutectic compositions freeze at one discrete temperature, but other alloys freeze over a temperature range. This generally leads to nonuniform compositions over a range of micrometers to hundreds of micrometers, resulting from constitutional supercooling and dendritic freezing (Sec. 24.5.6.1 and Figs. 8.2 and 24.9). When an alloying element is rejected from the solid at the advancing solid–liquid interface, the casting develops macroscopic segregation because the last metal to freeze contains a higher concentration of that element. This causes a castings composition near the mold walls to differ from the composition near the center. Still another segregating effect is the tendency of light phases to rise and heavier phases to sink before the casting freezes. Microscopic segregation can be reduced by diffusion in homogenizing anneals, but macroscopic segregation cannot be remedied.

Castings usually have nonuniform grain structure. As Fig. 8.3 shows, the metal near the mold walls freezes to form small, randomly oriented solid nuclei, giving this region equiaxed grains. As freezing progresses, the grain structure tends to become columnar, with long, narrow grains oriented parallel to the direction of heat flow. Finally, the last metal to freeze in the casting's center has an equiaxed grain structure again, although the grain size is often larger than at the mold wall. The large grain size and strong preferred orientation in the columnar region makes this metal anisotropic, weaker, and less ductile.

8.2.4 Continuous (Strand) Casting

In traditional foundry practice, plate and sheet were cast in large, vertical molds, then hot rolled after solidifying to reduce ingot thickness. The cast ingot usually had macrosegregation, and its

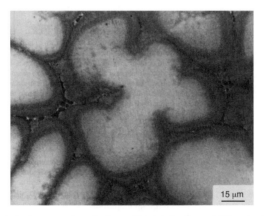

Fig. 8.2 Back-scattered electron scanning electron micrograph of a cast Ni superalloy showing W coring. The W-rich regions in the centers of the dendrites appear bright because heavy W nuclei back-scatter electrons more efficiently than do lighter elements. The dark border surrounding each dendrite is γ-phase depleted in W. (From Willemin and Durand-Charre, 1988; with permission.)

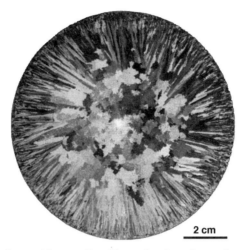

Fig. 8.3 Grain structure in cast Ni superalloy. The grains formed near the outer mold wall are relatively small and equiaxed. The last grains to freeze in the center of the casting are coarser equiaxed grains. The intermediate region shows radial, columnar grains lying parallel to the freezing direction. (From Durand-Charre, 1997, p. 54; with permission.)

upper end had to be cut away and remelted to remove a shrinkage pipe or porosity. Cost savings are possible if this process is replaced by one that freezes the metal continuously to form long plates (Fig. 8.4) with more uniform composition, less waste, and less rolling required to achieve the desired thickness. The continuously cast bar is withdrawn at the same rate as liquid metal is poured into a water-cooled mold; the metal exits the mold with a solid skin around a liquid core. Water is sprayed on the metal as it leaves the mold, to hasten solidification. To prevent the solidifying metal from sticking to the mold, molds are usually covered with specially formulated coatings to minimize friction, and the mold may be vibrated to reduce adherence to the casting.

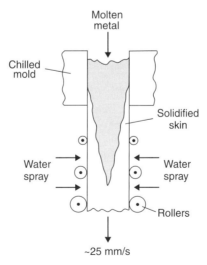

Fig. 8.4 Continuous casting of molten metal. Water-cooled molds solidify the outer skin of the casting, and the water sprays accelerate the cooling and solidification of the center of the casting. (From Kalpakjian, 1989, p. 167; with permission.)

8.2.5 Sand Casting

Casting molds can be made from mixtures of sand, clay, binders, and water packed around a pattern in a steel housing (flask). The pattern has the shape of the desired casting but is dimensioned slightly larger to allow for metal shrinkage during cooling. The mold is usually split horizontally into lower (*drag*) and upper (*cope*) sections (Fig. 8.1), although more than one parting line can be used. The sand binds to itself firmly enough to hold the desired shape as the mold is split along the parting line to remove the pattern. Sand cores may be inserted into the mold to block liquid metal from entering certain regions of the mold, thus forming hollows or holes in the cast metal. Vents allow gases to escape as liquid metal flows into the mold. After the metal freezes and cools, the sand is shaken loose from the metal on vibrating racks. The metal frozen within risers, vents, and runners is cut away from the casting and reused.

Although dimensional accuracy and surface finish quality are generally lower in sand castings than those available from other mold materials, sand casting can produce remarkably complex shapes (e.g., internal combustion engine blocks) at low equipment cost. Sand casting is more labor intensive than casting with some other mold types because a new mold must be prepared for every casting. Consequently, it is more cost-effective when used for smaller production runs.

8.2.6 Shell Mold Casting

Shell molds are thin-walled sand molds held together by a thermosetting binder. The two mold halves are formed around metal patterns heated to 175 to 375°C by spraying a sand-thermosetting binder mix onto the pattern. The hot pattern makes the binder-covered sand grains stick to the pattern and to one another. This sand-coated pattern is cured in an oven to harden the polymer, forming a rigid sand–polymer shell 5 to 10 mm thick around the pattern. The two halves of the mold are assembled (with cores, if needed) and placed in a steel box. The space between the mold and the box is filled with steel shot to support the thin, relatively weak shell mold against the weight of the casting metal. The mold walls are sufficiently porous to allow gases to escape during casting.

Shell molds have smoother surfaces than conventional sand molds, allowing the liquid metal to flow more easily through the mold with less turbulence and lower resistance. This allows shell mold castings to have thinner sections and sharper corners, and it produces a smoother finish on the cast metal. The limited strength of shell molds restricts this technique to castings under about 100 kg, and most shell mold castings are only a few kilograms. The dimensional accuracy is as good as that of other high-precision casting methods (Sec. 8.2.7), and costs are about the same as for conventional sand mold casting.

8.2.7 Precision Casting (Plaster Mold and Ceramic Mold Casting)

When fine detail and dimensional accuracy are needed for a cast product, sand molds can be replaced by plaster of paris molds. The mold is formed by pouring a water-based slurry of fine gypsum, talc, and silica powders over the pattern, waiting about 15 minutes for it to set, and then removing the pattern. The two mold halves are dried at 120 to 260°C and assembled to form the complete mold. The mold is heated to 120°C before the molten metal is poured. Since the mold has low permeability to gas flow, the casting is done either in vacuum or under pressure, so the gases released by the cooling metal will not cause defects. Plaster mold casting is generally used for smaller castings (less than about 10 kg). It can produce sections as thin as 1 mm, with a smoother surface finish than a sand casting.

Plaster of paris molds fail above about 1100°C. For higher-melting metals, precision cast molds use more refractory ceramic powders [e.g., finely ground zircon sand ($ZrSiO_4$), silica, and Al_2O_3 mixed with gelling agents] that work with ferrous metals, Ni alloys, and Ti alloys. The mold material slurry is poured over the pattern and allowed to gel (Fig. 8.5). It is then lifted off the pattern, briefly air dried, and ignited with a torch to burn off volatile bonding

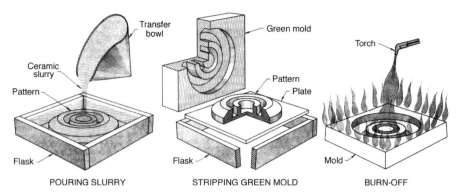

Fig. 8.5 Preparation steps for a one-piece ceramic mold precision casting. (From ASM, 1985, p. 23.27; with permission.)

agents. Finally, the mold is baked at 500 to 1000°C before casting. Parts as large as 700 kg are cast by ceramic-mold methods with good dimensional accuracy and an excellent surface finish. Precision casting is more expensive than sand mold casting, and it produces slower cooling rates for the freezing metal.

8.2.8 Investment Casting (Lost Wax Casting)

Most mold-making processes are complicated by the need to remove the pattern from the mold without damaging the mold. *Investment casting* avoids this problem by using expendable pattern materials that melt and flow out of the mold. For example, patterns can be made by injecting wax or thermoplastic polymer into a metal die. This pattern is then coated (often repeatedly) with a slurry of mold powders and binders to produce a one-piece mold with no parting lines. The mold is then dried in air and heated to 90 to 175°C, so the pattern will melt and flow out the opening. After the pattern material has drained from the mold, the mold is fired at 650 to 1050°C to harden it prior to casting. Since the mold materials are quite refractory, high-melting metals can be investment cast.

Preparing an investment casting mold is an expensive, multistep process best suited to making smaller parts (35 kg or less), such as gears, cams, rotors, and intricate or irregular shapes (e.g., dental implants). Investment cast parts have good surface finish, close tolerance control, and usually require little or no trimming or grinding. Evaporative pattern casting (lost foam casting) is a variation on this method involving low-density foam polymer patterns that are vaporized by the hot incoming molten metal during casting.

8.2.9 Permanent Mold and Pressure Die Casting

In the casting processes described in the preceding sections, the mold is broken apart to remove the casting. Although the mold material is generally reused, the labor and time spent shaping the mold are lost with each casting. Permanent mold casting uses metal or graphite molds designed to produce many castings (Fig. 8.6). Although it is expensive to make a permanent mold, one mold produces thousands of castings, making this the most economical casting method for large production runs. Since the mold is exposed repeatedly to the liquid metal, the mold and casting materials must not react or dissolve in one another. Steel or cast iron molds are the most common, which limits the casting metals to lower-melting alloys (e.g., Sn, Pb, Zn, Mg, Al, and Cu)(Sec. 19.2.3.3). A release agent is usually applied to the inner mold surface to assure easy mold-casting separation and to minimize reaction between the mold and the casting metal. If sand cores are used in combination with a permanent mold, the process is called *semipermanent casting*.

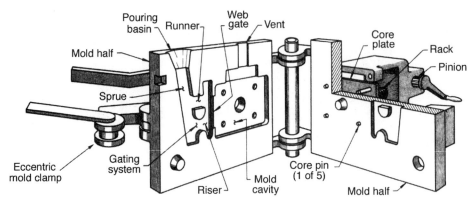

Fig. 8.6 Book-type permanent mold made of steel. (From ASM, 1985, p. 23.28; with permission.)

Permanent molds can be filled by pouring molten metal into the die or by pulling it into the die with a vacuum, but the most common approach is to pump the liquid metal to force it into the mold under high pressure. This *die casting* or *pressure die casting* process moves metal rapidly into the mold, allowing high production rates. It also permits narrower die passages and more intricate contours because the metal fills the mold so rapidly that cold shuts or misruns rarely occur. Many pressure die casting systems are highly automated and pump metal into the die at pressures as high as 700 MPa, allowing as many as 1000 injections/h. The chamber may be immersed in the molten metal reservoir (hot chamber die casting) or separate from the molten metal (cold chamber die casting). Water or oil is usually pumped through cooling channels inside the dies to accelerate freezing and allow more castings to be produced per hour. For low-melting casting alloys (e.g., Sn and Zn alloys), a die may produce 500,000 parts before significant wear occurs; with higher-melting metals (e.g., Cu alloys), die lifetimes are much shorter. Semisolid die casting is a widely used variation on conventional die casting in which the metal injected into the mold is partially liquid and partially solid (Sidebar 8.1).

8.2.10 Centrifugal Casting

Some casting systems use centrifugal force to shape the metal. A hollow cylindrical die can be partially filled with molten metal and rotated rapidly around its centerline to cast the metal into the shape of a pipe. Closely related processes called *centrifuge casting* and *semicentrifugal casting* rotate the mold on a rotating table and rely on centrifugal force to move the liquid metal into the die portions farthest from the center of rotation.

8.2.11 Single-Crystal Casting

Certain applications require single-crystal products. Ni superalloy combustion zone turbine blades deliver superior creep strength if all grain boundaries are eliminated, thereby preventing grain boundary sliding (Sec. 16.3.3.6). Strictly speaking, these turbine blades are not true single crystals since other phases are present in the matrix, but the γ-phase matrix is cast to produce a monocrystal (Fig. 16.26).

The *Czochralski crystal pulling method* produces single-crystal Si for microprocessors. The ultrapure Si must be free of grain boundaries and dislocations to operate properly (Sec. 21.2.6). This can be achieved by immersing a seed crystal into the surface of a pool of liquid Si under high vacuum and rotating it slowly (about 1 revolution/s) while lifting it vertically (about 10 μm/s). About 80% of Si single crystals are grown by the Czochralski method to produce boules up to 380 mm in diameter. An alternative process, the float zone method, simultaneously

> **SIDEBAR 8.1: SEMISOLID DIE CASTING**
>
> There are definite advantages to injecting a solid–liquid mixture into a casting mold rather than injecting an all-liquid charge. Metal shrinkage and dissolved gas evolution accompany solidification, but in a semisolid charge, these problems are lessened because part of the metal has already solidified before injection. In addition, the temperature of the solid–liquid mixture will be lower, which reduces die wear. In an ordinary gravity-fed casting, use of semisolid metal is difficult because its high viscosity tends to clog runners and result in misruns or coldshuts. However, in die casting, the high pressure of the injection can overcome these flow problems, making it possible, for example, to inject metal in the condition shown on Fig. 20.11 for 356 Al–Si alloy at about 590°C.
>
> Two methods have evolved to inject semisolid charges into the die. In *thixocasting*, a solid billet with a fine-grained equiaxed microstructure is partially melted to put it into a semisolid state just before injection. The other process, *rheocasting*, begins with completely molten metal that is cooled below the liquidus temperature while being stirred to produce a semisolid slurry. This slurry is then injected into the die. Both processes inject metal that is 50 to 95% solid. Rheocasting offers certain advantages over thixocasting because it does not require a specially prepared billet, and rheocasting can accommodate remelted scrap. Both methods increase die life and produce a product with lower porosity than the traditional, all-liquid injection method. Semisolid casting is particularly useful for Al alloys due to Al's large liquid → solid shrinkage (Sec. 20.2.2.5) and tendency to evolve H gas upon freezing (Sec. 20.2.2.6).

purifies the polycrystalline starting rod and converts it into a single crystal by slowly (20 μm/s) moving an induction coil along the length of the rod to melt a small zone immediately under the coil. The coil is started at the intersection of the polycrystalline Si rod and a monocrystal Si seed crystal; as the coil moves up and away, the Si left behind refreezes by "patterning itself" on the seed single crystal. Eventually, the entire rod is converted to a single crystal with the same crystallographic alignment as the seed crystal. Impurities in the polycrystalline rod are preferentially rejected into the liquid phase at the solid–liquid interface, so the last few centimeters at the end of the new single crystal contain most of the original rod's impurities.

8.3 POWDER METALLURGY

Metal parts can be fabricated by compacting metal powders into the desired shape and then sintering them into a solid object. For refractory metals this powder metallurgical (P/M) approach is the only workable fabrication method because their high melting points make casting difficult (Sec. 14.4.4). P/M also provides cost or performance advantages for complex part shapes that make it preferable to casting, forging, or machining methods.

8.3.1 Making Metal Powders

Several methods can produce metal powders. A liquid metal stream can be atomized by gas or water jets that break it into small droplets (Fig. 8.7 and Sidebar 8.2), using the same precept as that of an ordinary can of spray paint. This method produces round particles with minimal oxide scale that flow well and are easily pressed and sintered. In the rotating electrode process, a metal rod is rotated at high speed in an inert-gas chamber around its centerline while a W electrode strikes an arc with the rod end (Fig. 8.9). As the arc melts the metal, droplets are flung off to the sides by centrifugal force and freeze. Metal powders can also be produced by reducing fine metal oxide particles with H_2 or CO gas. Certain metals (e.g., Ni, Sec. 16.3.5) will react

SIDEBAR 8.2: SKULL MELTING AND GAS ATOMIZATION OF Ti ALLOYS

The high reactivity and high melting points of some early transition metals make them difficult to contain in crucibles. Liquid Ti, for example, aggressively attacks common crucible refractory materials, which not only damages crucibles but also contaminates the Ti with unwanted impurities. Induction skull melting allows the molten alloy to be held by a thin solid shell (the skull) of the Ti alloy being melted. Inside the vacuum induction furnace (Fig. 8.8), the skull is held on a water-cooled Cu crucible that keeps the skull from melting. The stirring action of the induction field homogenizes the melt.

To start the atomization, a second induction coil melts a small region at the bottom of the skull. The liquid metal pouring through the hole in the skull is blasted by high-pressure inert-gas jets. These jets break the molten stream into tiny droplets that freeze into powder particles as they fall through the tower. The resulting powder is spherical and cools so rapidly that little segregation occurs.

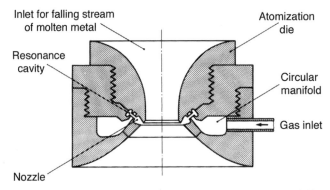

Fig. 8.7 Atomization die for producing metal powder. A molten metal stream falling into the inlet is broken up into small droplets by inert gas fed at high pressure through the nozzles on the side of the die. These droplets fall from the bottom of the die and freeze rapidly due to their small size. Powder is collected below, as shown in Fig. 8.8. (From Gessinger, 1984, p. 53; with permission of Elsevier.)

with CO to produce carbonyl compounds that decompose at higher temperatures to produce fine, high-purity metal powder. In yet another method, metals can be deposited electrolytically to form powder particles rather than massive metal platings on the electrode. Vaporized metal can be collected on a cold surface to form particles as small as a few nanometers in diameter. Brittle metals can be powdered by milling them in a vial with agitated hard steel or WC balls, a process that can also be used to alloy ductile metals on a submicrometer grain/phase scale. This process also works with ductile metals if they are first embrittled with H, then milled to fine powder, and finally, heated to release the H again (Sidebar 5.2). Certain metals (e.g., Au) can be precipitated from aqueous solution to produce fine metal powders.

8.3.2 Mixing Powders and Compaction

Metal powders may be coated with thin layers of wax or lubricants to make them pour more easily and hold together after they are pressed into a green body. The loose powders can be compacted at high pressure in a die (cold pressing) or by loading them into a soft polymer mold, sealing the mold, and then subjecting it 300 to 1000 MPa of pressure (cold isostatic pressing). Different metal powders can be blended to produce alloys or metal–metal composites (Sec. 18.3.3.1) that are then compacted into green bodies.

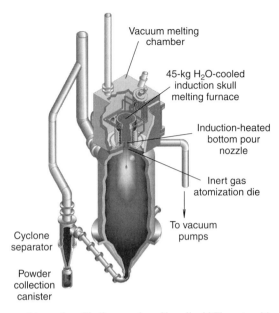

Fig. 8.8 Gas atomizer used to produce Ti alloy powders. Since liquid Ti reacts with conventional refractory crucible materials, it is melted in a "skull" of Ti that is kept solid by a water-cooled Cu containment. A stream of inert gas is sprayed at high pressure through the falling liquid Ti (see Fig. 8.7) to break it into small droplets that are collected by the separator at the lower left. (Courtesy of E. S. Bono, Crucible Research LLC, *http://www.crucibleresearch.com*.)

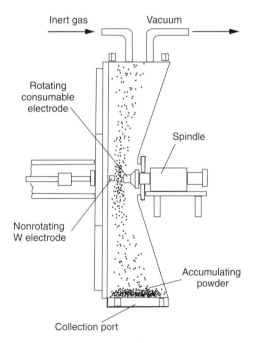

Fig. 8.9 Powder production using a rotating consumable electrode. An arc is struck between the W electrode and the rotating consumable electrode. As the consumable electrode melts, the molten metal is thrown off by centrifugal force and collected at the bottom of the vacuum chamber. (From Kalpakjian, 1989, p. 502; with permission.)

8.3.3 Sintering

Cold-pressed P/M green bodies are strong enough to be handled and can even be lightly machined, but they must be sintered prior to engineering use. Sintering at 70 to 90% of the homologous melting temperature diffusion-bonds the powder particles to one another. Sintering is usually performed in a controlled atmosphere (e.g., H_2, N_2, vacuum) to minimize oxidation for 0.5 to 8 hours. Any wax or lubricants present in the green body will vaporize during sintering. The sintered part's strength and ductility are typically slightly lower than they would be in a wrought part due to the presence of various contaminants on the powder particles' surfaces and a few percent porosity in the sintered metal. If one of the powder constituents melts at the sintering temperature, the sintering is said to be liquid-phase assisted, and it is often possible to achieve near 100% density in such parts. To achieve the highest strength and fracture toughness in a sintered part, the green body can be hot isostatically pressed in a sealed refractory metal can (e.g., Mo, Ta). The metal can then be sintered under high pressure to collapse porosity, producing about 100% density; however, the process is expensive.

The need for pressing the green body limits P/M to smaller part sizes; and P/M parts are generally more costly than competing methods. P/M parts have the advantage that they can be produced to near netshape without subsequent machining or grinding, which saves production costs for complex shapes such as gears, particularly when the metal involved is hard or high melting. P/M allows unusual combinations of metals that would not be possible by melt processing (e.g., W–K for light bulb filaments, Sec. 14.4.5), and it can also produce metals that have intentional porosity to hold lubricants or act as filters (e.g., U isotopic enrichment with porous Ni membranes, Sec. 24.4.2.1).

8.4 FORMING AND SHAPING

Metals can deform without breaking. We are so familiar with metal's malleability that it seems unremarkable, but it is essential for most metal-forming processes. For example, a 15-μm-thick Al metal foil is so easy and inexpensive to produce that most kitchens contain a large roll of it. In contrast, producing a similar-sized panel of 15-μm-thick Al_2O_3 is exceedingly difficult. In many cases, metals are chosen for a given engineering application primarily because they can, quickly and inexpensively, be bent, hammered, rolled, or stretched into the needed shape. Most metal products are plastically deformed at some point in their preparation for service.

8.4.1 Flat-Rolling Plate and Sheet

Metal plate, sheet, and foil are used to fabricate such diverse products as ships' hulls, aircraft skin, and x-ray windows. Flat metal plates or sheets can be fabricated rapidly by rolling a metal ingot between two parallel steel cylinders (Fig. 8.10). Rolling makes the ingot thinner and longer. Single upper and lower rolls can have undesirably large elastic deflection while delivering the large forces needed to roll metal, but the four-high design shown in Fig. 8.10 greatly reduces such deflection by providing each roll with a "backup" roll to support it along its length.

Rolling can be performed on either hot or cold metal. Hot rolling reduces the force required to deform the metal, and it can break down and recrystallize the large, nonuniform grains often present in castings, producing the fine, equiaxed grain structure shown in Fig. 8.10. However, hot rolling can produce an undesirably thick oxide layer, and it is more difficult to achieve accurate dimensions in hot-rolled metal. Cold rolling requires more energy, but it can produce excellent surface finish and dimensional accuracy. Cold rolling deforms the metal below its recovery and recrystallization temperature, so it work hardens and often develops residual stresses. Small roll diameters or small reductions in thickness per pass tend to deform the metal near the surface more heavily than the metal at the sheet's center, producing residual compression in the surface and residual tension in the core (Fig. 8.11a); large roll diameters or large reductions in thickness

FORMING AND SHAPING   **113**

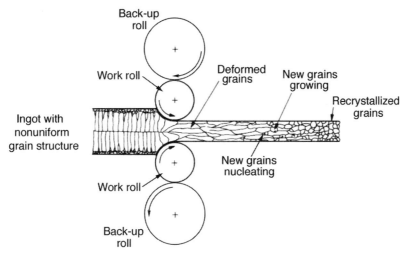

Fig. 8.10 Four-high rolling mill. Grain structure changes are shown for hot-rolling as-cast metal to achieve hot breakdown and recrystallization. (Adapted from Kalpakjian, 1989, p. 367; with permission.)

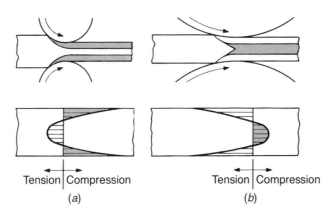

Fig. 8.11 Residual stresses in cold rolling (*a*) with small rolls or small reductions in thickness per pass and (*b*) with large rolls or large reductions in thickness per pass. (From Kalpakjian, 1989, p. 369; with permission.)

per pass have the reverse effect (Fig. 8.11*b*). Although some friction is necessary to pull the workpiece through the rolls, excess friction increases the force needed for rolling and wastes power. Most hot work is done dry or with graphite or MoS_2 lubricant; cold work allows use of organic oils and emulsions as lubricant.

8.4.2 Shape-Rolling Operations

Flat rolling is the most common rolling operation, but rollers with variable diameters along their lengths can form rods, I-beams, corrugated sheet, and threaded surfaces. Threads can be produced more rapidly by rolling than by cutting, and rolled threads have residual compressive stresses on their surfaces that increase fatigue strength. Rolling has even been adapted to produce thick-walled seamless tubing by piercing a solid rod with a mandrel while rollers deform the rod's exterior.

8.4.3 Forging

Forging presses or hammers a metal billet between two surfaces (Fig. 8.12). A hot metal workpiece is more ductile and requires less forging force, but the billet's surface may oxidize. Cold forging can be controlled more accurately, but it requires larger forces, work-hardens the metal, and may crack the billet. Friction forces between the die and billet tend to give the forged workpiece a barrel shape (Fig. 8.12), but lubricants can minimize this effect. A family of equations exists to predict the forging force necessary to achieve a given deformation level in the workpiece. For example, the force (F) needed to reduce the height (h) of a cylindrical billet of diameter d can be determined from

$$F = \frac{\pi \sigma_{\text{flow}} d^2}{4} \left(1 + \frac{\mu d}{3h}\right)$$

where σ_{flow} is the metal's maximum flow stress over the range of true strains reached during the forging and μ is the coefficient of friction between the die and the workpiece. Simple shape changes (e.g., converting a flat billet to a roughly cylindrical shape) can be fabricated in a flat die forge by repositioning the billet between deformation steps. By using dies with contoured surfaces, many complex shapes can be forged, producing objects such as coins, pistons, and crankshafts. In a swaging mill, tapered, cylindrical dies rotate around a slowly advancing cylindrical workpiece to reduce its diameter.

Forging works best on highly ductile metals. Al and Mg can be hot forged at 250 to 550°C, and at these temperatures wear rates on steel dies are quite low. Harder, higher-melting metals pose a greater challenge for both the forging press and the dies. Metals such as Mo and Ni superalloys require hot forging temperatures well above 1000°C, and the brief, rapid strokes of a hammer forge are generally preferred to the more slowly applied force of a forging press to minimize heat transfer to the dies and chilling the workpiece. Lubricants applied to the dies (hot forging) or workpieces (cold forging) improve the workpiece's surface finish, reduce the chances of cracking the workpiece, and prolong die life.

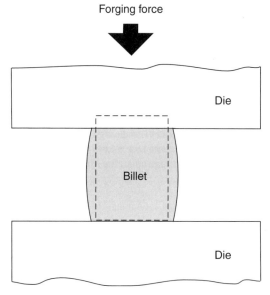

Fig. 8.12 Simple flat-die forging of a metal billet. Frictional forces at the billet–die interfaces retard deformation near the billet top and bottom ends, causing barreling of the billet.

8.4.4 Extrusion and Drawing

Extrusion pushes a ductile metal through a die to reshape it (Fig. 8.13). Extrusion imposes severe demands on workpiece ductility, so it is often performed at elevated temperature. Metal can be extruded as simple cylindrical rods or in more complex tubular or irregular shapes. Extrusion dies are usually made of hardened steel, and the interface between the extrusion die and the workpiece is lubricated to minimize die wear and ram force. Molten glass, MoS_2, hexagonal BN, or metals near their melting points make effective high-temperature lubricants. Organic lubricants may be used at lower temperatures. The extrusion ratio is defined as $A_0 : A_f$, where A_0 is the workpiece's initial cross-sectional area and A_f is cross-sectional area after extrusion. Extrusion ratios are typically $4:1$ to $10:1$ for less ductile metals and may be as high as $100:1$ for a highly ductile workpiece. For many Al alloys (e.g., 6061), extrusion and heat treatment are combined in one process. The extrusion preheating solutionizes the alloy, the metal is quenched immediately after extrusion to preserve a metastable solid solution, and post-extrusion aging precipitation hardens the metal.

Small variations in frictional forces between the workpiece and the die cause most extruded metal to become slightly twisted or bowed; this can be corrected by stretching the final extruded product forcefully to straighten it. Extrusion can be used to clad one metal onto another (e.g., Ti over Cu) by extruding metal sealed inside a can comprised of the desired cladding material. Seamless tubing can be produced by extrusion (Sidebar 20.5 and Figs. 20.20 and 20.21). Hydrostatic extrusion (Sidebar 14.2) reduces friction losses and permits higher extrusion ratios or extrusion of less ductile metals with less risk of workpiece cracking.

Extensive plastic flow during extrusion heats the metal, so metal preheated near its solidus temperature can melt partially as it passes through the die. This causes intergranular tearing and cracking similar to hot shortness. Surface flaws can result from intermittent sticking of the workpiece to the die, producing *bamboo defect*, a series of circumferential cracks. A more insidious extrusion defect is *centerburst* or *center cracking*, in which hydrostatic tensile stress in the workpiece's center produces internal voids that weaken the metal but show no visible sign on the part's exterior. The last portion of the extruded product to pass through the die often contains a funnel-shaped void along the centerline called *tailpipe defect*; this must be cut away, reducing product yield.

Like extrusion, *drawing* reduces the workpiece's diameter and modifies its contours as it passes through a die. But drawing *pulls* the metal through the die (Fig. 8.14), whereas extrusion *pushes* it through. Drawing requires that the metal work harden as it passes through the die so that the smaller-diameter portion of the workpiece can transfer the pulling load to the larger-diameter portion without necking or breaking. Consequently, drawing cannot be performed above the recovery/recrystallization temperature, and the drawing reduction in cross-sectional area is

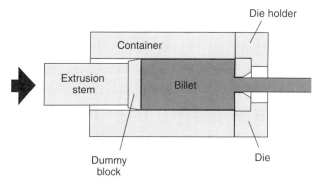

Fig. 8.13 Basic elements of direct extrusion. The extrusion stem and dummy block push the billet to the right, forcing it to flow through the die to emerge as a smaller-diameter cylinder.

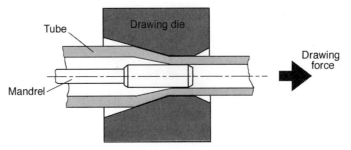

Fig. 8.14 Drawing a tubular workpiece over a mandrel (section view).

usually limited to a maximum of 45% to avoid deforming or breaking the metal under the pulling stress. Drawing dies vary from several centimeters in diameter to as fine as 2 μm and are made of tool steels, carbides, or diamond. Organic lubricants are generally used for drawing. Drawing is a cold-working process, so it can impart residual stresses in the drawn metal. Light reductions tend to produce residual compressive stress near the surface, while heavier drawing ratios produce a residual tensile stress near the surface similar to the effect seen in cold rolling (Fig. 8.10). Wire is by far the most common drawing product, and automated multistage drawing machines form tens of millions of tons of wire per annum for electrical and structural use.

8.4.5 Sheet Metal Forming

Sheet metal is produced by rolling and given subsequent processed by bending, punching, deep drawing, spinning, or stretching the metal. More exotic sheet metal–forming techniques include superplastic, explosive, peen, and magnetic-pulse forming. The amount of deformation that sheet metal will tolerate without cracking or fracturing varies from one metal to the next. The minimum bend radius (R) provides a simple metric to predict sheet metal's ability to deform in bending (Fig. 8.15). R can be estimated from the metal's tensile test ductility,

$$R \approx T \left(\frac{50}{r} - 1 \right)$$

where T is sheet metal thickness and r is the metal's percent reduction in area in a standard tensile test. A highly ductile metal (e.g., pure Al) has $r > 50\%$ and can be folded over upon itself completely like a sheet of paper; thus, its minimum bend radius is zero. For less ductile metals, $R > 0$. Annealed commercial purity Ti has $R = 0.7T$. In annealed Mg, $R = 5T$; for fully work-hardened Mg, $R = 13T$, which indicates that the metal can be bent only slightly before it cracks. When sheet metal is bent, it deforms by both elastic and plastic bending. When the bending load is removed, the elastic deformation relaxes, causing a certain amount

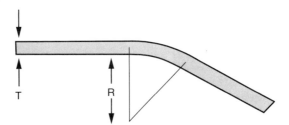

Fig. 8.15 Bend radius (R) of sheet metal with thickness (T).

of springback (Sec. 12.2.2.3). It is possible to compensate for springback by overbending the metal or by bending at high temperature to reduce σ_y and the amount of springback.

An amazing array of techniques and machines exists to bend sheet metal and metal tubing; most of these are variations on the basic theme of a die to hold the metal and a punch to deform it (Fig. 8.16). Spinning (Fig. 8.17) bends the sheet around a mandrel by tracking a tool in a helical path to force the sheet metal gradually to conform to the mandrel's shape. The sharp edge of sheared metal is usually hemmed, roll folded, or beaded. These steps protect the consumer from being cut by a sharp edge, and they also increase the part's bending moment and improve its appearance.

More exotic sheet metal–forming techniques exist for special processing needs. *Superplastic forming* (Sec. 19.2.6) exploits grain boundary sliding in fine-grained metal and produces severe

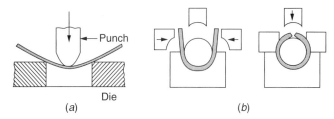

Fig. 8.16 Two sheet metal bending methods: (*a*) air bending; (*b*) bending in a four-slide machine. (From Kalpakjian, 1989, p. 466; with permission.)

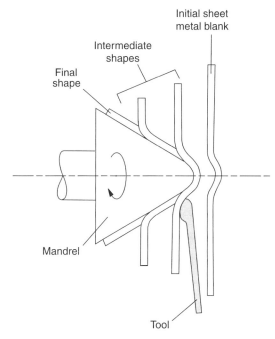

Fig. 8.17 Sheet metal forming by spinning. Four stages of forming are shown here. The rightmost image of the workpiece shows the sheet metal in its initial shape. Moving from right to left, the shape of the workpiece is forced to conform to the shape of the mandrel as the tool pushes the metal down and left across the surface of the mandrel. The final shape of the workpiece is achieved when it is in full contact with the mandrel.

shape changes without cracking. The standard metal punch can be replaced by other methods of applying force to sheet metal. *Hydroforming* uses a rubber membrane to transfer high fluid pressure to the workpiece; *peen forming* replaces the punch with high-velocity steel balls; *explosive forming* uses a detonation shock front in a water tank to push the sheet metal into the die; and *magnetic-pulse forming* employs a magnetic repelling force generated by induction to push the sheet metal over the die.

8.5 MATERIAL REMOVAL

The forming methods described in Secs. 8.2 to 8.4 shape metal to the approximate dimensions needed, but they are often augmented by techniques to remove metal to produce special features such as holes; sharp corners; smooth, flat surfaces; and precisely controlled contours. Material removal processes allow mating parts to be fabricated with accurate control of their clearance (or interference) with mating parts in assemblies. Metal can be removed from the workpiece by cutting it with sharp tools, grinding it away with abrasives, dissolving it with reactive chemicals, or vaporizing it with lasers or electrical discharges.

8.5.1 Cutting Machines and Cutting Tools

Saws, lathes, mills, and drills shear metal from the workpiece surface. These machines apply large forces to a small region of the metal with a sharp-edged tool (Fig. 12.5) to cut it away from the workpiece. These metal-cutting systems produce closely controlled dimensions in the workpiece, but they require a ductile workpiece and consume large amounts of energy. Cutting may also work harden or heat the metal sufficiently to alter its properties.

A major metal-cutting challenge is selecting and fabricating cutting tool materials with sufficient hardness, toughness, and chemical stability to slice through the workpiece without dulling or breaking the cutting edge too rapidly. If the workpiece metal is hard (e.g., steel, Ti), the cutting tool must be harder still to cut it effectively. Many workpiece metals can be cut with tempered martensite tool steels containing substantial amounts of V, Cr, Nb, Mo, and/or W carbides (Sec. 13.3.2.1). These high-speed steels are inexpensive, tough, and easily fabricated into tool shapes. However, for especially hard workpiece metals (Mo, W) or for high cutting speeds (which can heat tools by several hundred degrees), high-speed steel tools wear rapidly, forcing frequent resharpening or replacement. In these difficult cutting applications, WC–Co cermet cutting tools (Sec. 16.2.3.3) are harder, although they have lower fracture toughness than that of high-speed steel. All-ceramic tools (e.g., Al_2O_3, TiC, TiB_2, cubic-BN, and diamond) have excellent hardness, but their fracture toughness is low, making them vulnerable to chipping or fracture, particularly in interrupted cuts. Cutting tools are often coated with multiple layers of hard materials (e.g., TiN, diamond) to minimize wear and reaction with the workpiece metal. Lubricants can be sprayed onto the workpiece and cutting tool to lower temperatures and minimize friction. Although lubricants cannot penetrate the actual cutting zone where the tool edge shears through the metal, capillary action can move fluid onto the tool face, cooling and lubricating the contact with the chip to reduce wear. Flooding the workpiece with lubricants can be counterproductive with ceramic tools making interrupted cuts (e.g., milling) since the thermal shock caused by alternating hot cutting and cold lubricant can crack brittle tool materials.

8.5.2 Tool–Workpiece Interactions

Chemical reactivity between the tool and the workpiece is a major factor in selecting cutting tool materials. Diamond tools, for example, provide extreme hardness but wear rapidly when cutting Ti or Fe at high speed by dissolving into the workpiece metal to form carbides. Since the tool edge slides underneath the oxide and adsorbed gas layers on the metal's surface, it is exposed to reactive, atomically clean metal at elevated temperatures. Both the tool flank and

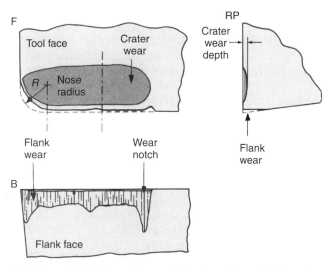

Fig. 8.18 Front (F), right profile (RP), and bottom (B) orthographic projections of a cutting tool for a lathe, showing flank and crater wear. Dashed lines indicate original tool dimensions before wear. Feed direction is leftward as shown in the front and bottom views. Figure 12.5 may be used with the RP view to help visualize the workpiece position. (From Kalpakjian, 1989, p. 599; with permission.)

the tool face are in direct contact with this metal, and they can dissolve into the metal as cutting proceeds, producing flank wear and crater wear (Fig. 8.18). As cutting speed increases, temperature rises and wear becomes much more rapid. At high cutting speeds, the heating rate of the chips removed from the surface can be as high as 10^5 K/s, and chip-forming strain rates can reach 10^5 s^{-1}. Metals with high work-hardening rates (e.g., Ni, Ta) are difficult to machine because the surface becomes much harder as cutting proceeds. Pure Re work-hardens so drastically that it is essentially unmachinable (Sec. 15.5.2.2). Tools cutting through soft, ductile metals (e.g., annealed Cu) sometimes accumulate a mass of workpiece material that welds temporarily to the tool face. This built-up edge breaks away and re-forms many times during a machining operation, changing the tool geometry continuously and degrading workpiece surface finish quality.

8.5.3 Grinding

Normal metal-cutting methods work poorly or fail completely on hard, brittle metals (e.g., Ru, Os). They are also less effective at polishing metal or producing exceptionally small reductions in material thickness. For these tasks, grinding can be used to remove material with abrasives. *Grinding* drags hundreds of small, hard crystals across the workpiece surface, and each crystal acts like a miniature cutting tool, shaving tiny chips off the surface (Fig. 8.19). Abrasive disks, wheels, and belts used for grinding are composites containing many particles of a hard material (e.g., SiC, WC, B_4C, or cubic BN) bound together in a resin, glass, or metal matrix. As grinding proceeds, individual hard particles cut the workpiece until they become dull or pull out of the matrix. As the grinding wheel wears, new particles are exposed to continue the cutting action on the workpiece surface. If the abrasive particles are friable, they may fracture, exposing a sharp, new edge to cut with renewed effectiveness. Large abrasive particles remove more metal from the workpiece and change workpiece dimensions rapidly; finer abrasives remove small amounts of metal and polish the surface. Grinding wheels can become glazed when the particles are worn flat, or they can become loaded with chip material lodged in the gaps between abrasive particles. Glazed or loaded wheels grind poorly and heat

120   PROCESSING METHODS

the workpiece excessively, so the wheel must be dressed periodically by having it cut briefly through a superhard material that breaks away flattened sections of abrasive grains and unclogs loaded material.

Grinding can alter the workpiece metal in several ways. The frictional heat raises the metal's temperature, which may oxidize the surface or inadvertently alter heat treatments (e.g., solutionizing precipitate particles). Since temperature rise is high at the surface and less in the interior, residual stresses can develop in the workpiece, particularly in metals that are poor thermal conductors (e.g., Ti). If residual stresses are tensile at the surface, the metal's fatigue strength can be degraded substantially. To reduce heating problems, grinding fluids are often used to lubricate and cool the workpiece. Fluids can also play a more active role in machining. Abrasive jet machining entrains fine abrasive particles in a high-speed stream of liquid or gas. The abrasives' impacts cut the workpiece, and the fluid flow cools the metal and clears chips from the work zone. For some applications high-speed water alone cuts the workpiece without abrasive particles. Fluids are also used in ultrasonic machining, in which a rapidly vibrating tool works an abrasive particle slurry against the workpiece surface. The fluid cools the workpiece and clears away chips and worn abrasive particles.

8.5.4 Chemical and Electrochemical Machining and Grinding

Machining operations are not limited to metal-cutting processes. Chemical and electrochemical methods can dissolve metal to controlled depths in patterns across the workpiece surface. Complex patterns can be chemically milled by using simple lithographic masks to selectively block the chemical action on some surface regions (Sidebar 8.3). Chemical milling relies entirely on the reaction between the metal and a suitable reagent (e.g., acid). Electrochemical milling imposes a galvanic current on the workpiece to accelerate material removal, with the cathode acting as a cutting tool that approaches but does not touch the anodic workpiece; this method is sometimes described as "electroplating in reverse." These processes are more expensive than conventional cutting, so they are used primarily for hard, brittle metals; thin parts that cannot withstand the tool pressure of conventional cutting; workpieces where no temperature rise is tolerable; and machining tasks where complicated patterns need to be machined with excellent surface finish.

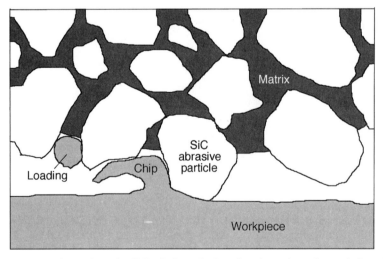

Fig. 8.19   Close-up of a portion of a SiC grinding wheel cutting the surface of a workpiece. Each SiC particle shears a tiny chip away from the workpiece surface as the wheel rotates clockwise. Chips often lodge in spaces between SiC particles, loading the wheel and reducing cutting efficiency.

SIDEBAR 8.3: MEMS AND LIGA DEVICES

Lithographic processing methods are used to mask, etch, and dope extremely small areas of ultrapure Si single crystals to produce microprocessors for electronic devices. Lithographic techniques are also used to machine Si to produce microelectromechanical systems (MEMS). Such systems include sub-millimeter-sized gyroscopes, valves, pumps, filters, ink jet printheads, accelerometers for automobile airbag deployment, electric motors, and optical switches (Fig. 8.20). The advantages of the lithographic etching technique include small device features and the ability to produce thousands or even millions of parts rapidly and inexpensively. The closely related lithography electroforming (*Galvanoformung*) molding (*Abformung*) (LIGA) technique allows metals other than Si to be fabricated at exceptionally small dimensions. LIGA uses photolithographic chemical machining to produce cavities in polymer blanks; these cavities are then filled with metals such as Ni or Cu by electrodeposition to produce sub-millimeter-sized parts.

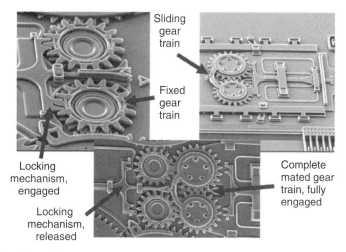

Fig. 8.20 Chemical machining was used to fabricate the weapons system safety lock MEMS device shown in these micrographs. When the correct code is entered into the device, the MEMS device releases a lock that aligns small mirrors, completing an optical path to arm the weapon. The gear teeth in this image are 7 μm wide. (From Romig et al., 2003, p. 5849; with permission.)

8.5.5 Electrical Discharge Machining

Electrodischarge machining (EDM) erodes the workpiece by striking a spark between a tool and a workpiece that are both submerged in a dielectric fluid (Fig. 8.21). This technique can machine any electrically conductive material and works well to fabricate hard, brittle, fragile parts to complex or difficult-to-cut shapes. The cutting electrode is usually graphite, Cu, brass, or Cu–W composite and may have the shape of the desired product or may be a wire. Wire EDM is often used to cut hard or brittle metals and carbides. EDM produces an exceptionally smooth surface finish if the material removal rate is kept low.

8.5.6 Laser- and Electron-Beam Machining

The energy of laser beams or focused electron beams can be used to saw and machine metal by melting or vaporizing it. Laser beams can operate in air. Electron beams require a vacuum

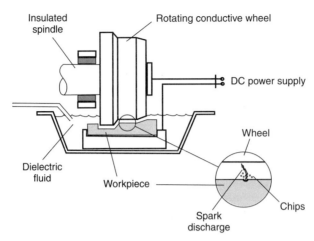

Fig. 8.21 Electrical discharge machining erodes the workpiece material by creating a spark between the tool and the workpiece.

chamber since air rapidly attenuates electron beams. These processes are particularly useful for cutting narrow slots and drilling small holes (down to about 10 μm in diameter). Material removal rates are relatively low in both processes, and the workpiece near the cut area experiences large temperature spikes, so alteration of prior heat treatments and formation of residual thermal stresses are common to both processes. Oxide films may thicken near laser machined regions. In laser-beam machining, workpiece reflectivity is important in determining how effectively the beam will heat the metal and may pose a safety hazard to workers.

8.6 JOINING

Many assemblies require methods to join separate pieces permanently or temporarily. A simple tubular bicycle frame, for example, is rather difficult to produce as one piece but can be fabricated by connecting individual lengths of tubing. These connections can be made by welding, brazing, gluing, or fastening the individual parts, and these various methods are collectively called *joining methods*.

8.6.1 Traditional Welding

Two parts can be joined by melting juxtaposed regions and allowing the liquid metals to mix and fuse into a single part as the liquid solidifies. The metal may be heated with a torch fueled with a hydrocarbon–oxygen flame or an electric arc. Flame temperatures as high as 3300°C are produced by burning acetylene gas (C_2H_2) with pure O_2. By altering the torch's C_2H_2 : O_2 ratio, the flame can be made reducing, neutral, or oxidizing. A reducing flame minimizes metal oxidation in the weld zone. An oxidizing flame can cut metal by a combination of melting, oxidizing, and blowing the debris out of the kerf. The weld zone is typically supplied with a filler metal that melts into the joint to provide additional metal to aid the fusion process. Filler rods often contain metal salts (fluxes, Sec. 19.2.3.5) to dissolve oxide films and to shield the molten metal from oxidation. In arc welding, a high-amperage electric arc heats the metal (Fig. 8.22). In some types of arc welding, the electrode (which often contains a flux; Sec. 15.3.2.5) is gradually consumed to provide filler metal to the weld zone; in other methods the electrode is W and is not consumed. An inert shielding gas (Ar, He, CO_2) is often passed over the work zone to minimize oxidation of the hot metal.

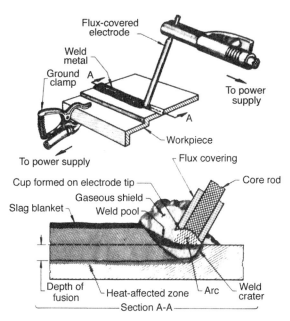

Fig. 8.22 Gas metal-arc welding. The arc struck between the workpiece and the electrode wire melts the metal. A conductive wire clamp is placed on the workpiece to complete the electrical circuit for current flow. A continuous flow of inert gas covers the molten metal to exclude air from the weld zone, thereby reducing oxidation of the molten metal. (From ASM, 1985, p. 30.12; with permission.)

Electric power can also be used to weld without creating an arc. Two metal parts clamped forcefully together can be fused by running 3000 to 40,000 A through the clamped joint at low voltage to melt the metals together by ohmic heating (Fig. 8.23). This *resistance welding* or *spot welding technique* is used primarily to join sheet metal sections. A typical automobile contains several thousand spot welds made by robots. If the electrodes are configured as wheels rolling along longer workpiece sections, the process is called *seam welding*.

The metal in the weld fusion zone melts, and the metal adjacent to the fusion zone is heated well above ambient temperature. This melting and heating profoundly alter the metal's

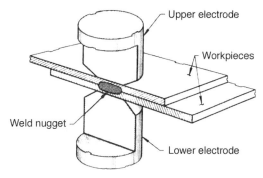

Fig. 8.23 Resistance welding. The two Cu alloy electrodes (shown here in partial section view for clarity) are pushed forcefully together, compressing the two workpieces while electric current flows from one electrode to the other. The current heats the metal sufficiently to form a nugget of fused metal that joins the two workpieces. (From ASM, 1985, p. 30.45; with permission.)

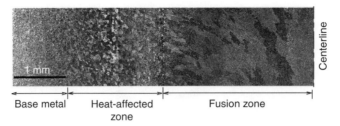

Fig. 8.24 Photomicrograph montage of the effect of welding on grain structure of as-cast Al–4 wt% Cu alloy (2024) welded by shielded gas tungsten arc welding. The weld centerline is the right edge of the image. The fusion zone is the metal that melted during welding; note the large columnar grains in this region. The metal in the heat-affected zone experienced grain growth but did not melt. The base metal retains the original fine grain structure that all the metal had prior to welding. (From Kang et al., 2003; with permission.)

microstructure (Fig. 8.24). Prior treatments such as precipitation hardening or work hardening are lost in the fusion and heat-affected zones. The enlarged grain size in these zones lowers strength and ductility. Since welds usually cool rapidly, there are often residual tensile stresses in the weld and compressive stresses near the weld that lower weld strength and/or warp the assembly. Gas porosity and/or flux inclusions in the weld metal may also compromise strength, particularly fatigue strength. In some cases the welded assembly can be heat treated after welding to improve properties, but this is often not feasible. Materials properties changes are a key consideration in design of welded structures.

8.6.2 Other Joining Processes

In addition to the widely used torch, arc, and resistance welding methods, other joining techniques are available to meet more specialized needs. Some of these methods join metals without melting them. Simple cold deformation can join two ductile metals by rolling or forging one piece stacked atop the other. The deformation expands the interface area, exposing atomically clean metal that cold welds to the other workpiece metal across the interface. This *roll bonding* method is used to clad a pure Cu midlayer between two 75% Cu–25% Ni alloy faces for the U.S. 25-cent coin. The method can be used with most ductile metals, although intermetallic compounds can sometimes form brittle interlayers at the joint that compromise strength. *Friction welding* rotates one workpiece against the face of the other, heating the interface and fusing the surfaces together. The oxides and absorbed gases usually present on metal surfaces (Sec. 8.7.1) are scraped loose, allowing freshly exposed metals to weld at the interface. Friction welding is often used to connect wire or tube ends or to fasten bolt heads to bolt shafts. A variation of friction welding is *ultrasonic welding*, in which an oscillating tool (10 to 75 kHz) causes a shearing action between the workpieces that heats them and breaks loose the oxide scale, bonding the metals without melting them. Another variation on friction welding is *friction-stir welding*, in which a rapidly rotating tool is pushed through the metals at the joint between the juxtaposed workpieces. Plastic flow of metal around the rotating tool joins the metals.

Explosives can join dissimilar metals that cannot be fusion welded due to formation of brittle intermetallic compounds or large differences in melting points (Sidebar 8.4). Diffusion bonding can join metals without melting them, and this method is particularly useful in joining Ti sections (Sec. 12.2.2.2). Difficult welding problems can sometimes be solved by using laser beams or electron beams to heat the metal. As with laser and electron beam metal machining (Sec. 8.5.6), these methods can weld refractory metals and weld in narrow passages that conventional equipment cannot reach. Electron-beam welding can join plates up to 150 mm thick, forming a fusion zone that is about 10 times narrower than conventional fusion welding. Joining thin metal parts by arc or torch welding is quite difficult, but electron-beam systems permit

JOINING   125

SIDEBAR 8.4: EXPLOSION WELDING

During World War I, soldiers observed that shrapnel from exploding artillery shells would sometimes weld itself to steel objects. The two pieces of metal would bond so strongly that they would not separate even when hammered, pried, or chiseled. This concept was exploited under more controlled circumstances after World War II for joining dissimilar metals over areas too large to permit fusion welding or in systems that form brittle intermetallic compounds if co-melted. When two metal plates are stacked with a sheet of explosive (Fig. 8.25), the blast front from the detonating explosive travels 3000 m/s across the top of the flyer plate, slamming it down onto the base plate and welding the two plates together. The process does not actually melt significant volumes of metal, so little or no equilibrium intermetallic forms. Plates as large as 6 m × 2 m have been clad in this manner, allowing large steel structures to be joined to Al, Ti, and Zr plates for corrosion protection or to connect Al superstructures to steel ships' hulls.

---

precise power control that achieves reliable welding of thin sheets and foils. Laser welding also produces an unusually high ratio of fusion zone depth/width (up to 30:1), but it is limited to thinner workpieces (about 25 mm maximum thickness). Both methods produce such rapid heating and cooling that centerline cracking from thermal shock is a concern; appropriate feed rate and power-level adjustments can usually prevent cracking.

8.6.3 Brazing and Soldering

Often, metal needs to be joined in situations where melting the metals is either unacceptable or unnecessary. In such situations, brazing and soldering are often used to join metals by melting

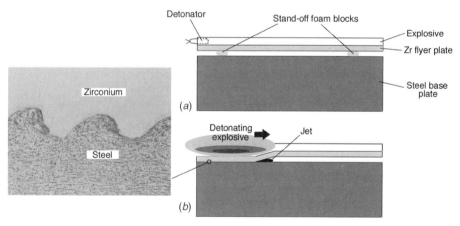

Fig. 8.25 Explosive welding is used to bond corrosion-resistant metals (e.g., Ti, Zr) to steel surfaces to improve corrosion performance. (*a*) Prior to detonation, a Zr plate is set atop light polymer foam blocks on the steel surface to form a gap between the two metal plates. A sheet of explosive is pressed onto the Zr plate and rigged with a detonator. (*b*) The detonator ignites the explosive, which burns from left to right, collapsing the Zr plate forcefully onto the steel plate and ejecting a jet of shattered oxide layers, vaporized polymer blocks, and compressed air to the right. This cleans the surface immediately before the Zr contacts the steel. The high energy of the impact between the plates produces a rippled interface (shown in the micrograph at the left), bonding the two metals. Although Fe and Zr form brittle intermetallic compounds at equilibrium, the explosive bonding is too rapid to allow intermetallic formation. (Courtesy of A. Nobili, T. Masri, and M. C. Lafont, Nobelclad Europe SA; with permission.)

TABLE 8.1 Commonly Used Brazing Alloys

Metal to Be Joined	Filler Metal	Working Temperature (°C)
Al alloys	Al containing 6–13% Si, 0.2–5% Cu, 0.8% Fe	570–620
Cu alloys	Cu containing 5–7% P or 40% Zn or 6–92% Ag	700–925
Ni alloys	Ni containing 6–20% Cr, 1.5–3.5% B, 4–10% Si, 0.5–5% Fe	890–1130

a lower-melting alloy between the two workpieces. The distinction between these two methods is somewhat arbitrary, but the term *brazing* usually implies that higher temperatures were used and significant diffusion occurs near the joint; by contrast, soldering is a lower-temperature procedure. Brazing can produce high-strength joints but requires accurate temperature control, because the workpiece metals are often heated quite near their melting points. Soldered joints are usually weaker but simpler to produce. Some common brazing alloys are listed in Table 8.1; common soldering compositions are described in Secs. 21.4.3.4 and 21.5.3.5 and Sidebar 21.4. For both processes, the workpieces and filler metals are heated by torch, furnace, induction, or electric resistance. Fluxes are usually needed to dissolve oxide films from the workpieces, assuring a stronger bond. Soldering's low temperatures usually cause little change in the workpieces' microstructures and properties, but brazing causes grain growth, loss of work hardening, loss of precipitation hardening, and possible oxidation.

8.6.4 Adhesive Bonding

Organic adhesives can join metal parts. A wide range of thermosetting polymers (e.g., phenolics and epoxies) are available with shear strengths as high as 40 MPa. Adhesive joining methods involve either low heat or no heat to form the bond, so the microstructure of metal near the joint is unaffected. To maximize adhesive bond strength, the joint should be designed to increase contact area and minimize possible lifting and peeling forces on the joint. Adhesive materials weaken rapidly at elevated temperature and char and decompose above about 250 to 300°C, so their use is limited to components operating at or near room temperature.

8.6.5 Mechanical Fasteners

The various joining methods described in the preceding sections all produce joints that are permanent or quite difficult to separate. Many applications require that parts can be disassembled for shipping, repair, or routine maintenance. In those cases, welded, brazed, or glued joints are inappropriate, and removable fasteners are specified. Bolts, nuts, washers, splines, keyways, clips, cotter pins, and related devices used as temporary fasteners are familiar to everyone, but their use involves certain potential mechanical and corrosion problems that are often less well known. The holes required for bolted joints often load the surrounding metal biaxially or triaxially, lowering fatigue life and/or contributing to stress corrosion cracking. Water may penetrate the gaps between bolted parts, but narrow gaps often have poor oxygen exchange with the water on the metal surfaces outside the gap. This can lead to oxygen-concentration galvanic cells that cause crevice corrosion (Sidebar 12.3 and Fig. 12.9). Poorly designed fastener–joint assemblies may place dissimilar metals in electrical contact, causing rapid corrosion when wet.

8.7 SURFACE MODIFICATION

8.7.1 Nature of Metal Surfaces

To the casual glance, metal surfaces often appear clean and smooth. However, closer examination usually reveals a less tidy surface (Figs. 8.26 and 20.2), contaminated with microscopically thin layers of adsorbed gas, oxide films, residual skin oil and lubricating oil, or adherent dust and other debris. In addition, many metal-working processes (e.g., grinding, polishing) produce a work-hardened metal surface layer tens or hundreds of micrometers deep. The apparent smoothness of the surface is also deceptive. Even metal polished to a mirror finish contains asperities (high points), scratches, and waviness lying hundreds of nanometers above or below the mean surface height.

The complex layered structure of most metal surfaces changes their performance in several ways. For example, oxide and adsorbed gas layers inhibit corrosion of the underlying metal. These surface layers also prevent the spontaneous cold welding that would occur if two atomically clean metal surfaces touch. Without these contaminant layers, metals parts would weld lightly to one another on contact, and pieces pushed forcefully together would bond so well that they would be difficult to separate. We are so familiar with handling contaminated metals that spontaneous cold welding sounds bizarre; however, metal parts can weld to one another when sliding friction scrapes away contaminants. Sometimes this is deliberate, as in friction or ultrasonic welding (Sec. 8.6.2). At other times it occurs inadvertently when lubrication fails and a bearing seizes. The friction and wear associated with unintended surface interactions can be alleviated by interposing a lubricating barrier (e.g., oil, graphite, chlorides, or sulfides), by making one of the contacting surfaces hard and the other soft, or by coating the surfaces with low-friction materials such as Cr, quasicrystals, or chlorofluorocarbons.

8.7.2 Mechanical Surface Treatment

Metals' surface properties can be improved by various mechanical treatments. *Shot-peening* bombards the part with high-velocity steel, glass, or ceramic shot to work-harden the surface and induce residual compressive stresses near the surface to retard fatigue. Aircraft landing gear struts are often shot-peened. *Roller burnishing* is used to increase fatigue resistance by polishing the surface with a hard, smooth burnishing roller to remove scratches and other flaws that could start fatigue cracks. Roll bonding (Sec. 8.6.2) is sometimes used to add a metal layer to the part's surface to improve wear or corrosion performance.

8.7.3 Coatings

Thermal treatments are also used to alter metal surfaces. *Hard facing* applies hard metal coatings to surfaces using welding techniques (also called *weld overlay*). *Thermal spray* coats the surface by melting the coating material and spraying it at high speed (about 100 m/s) onto the metal.

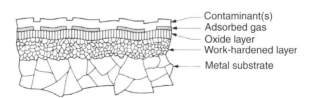

Fig. 8.26 Various layers commonly found on metal surfaces. Common contaminants include skin oil, cutting fluids used in machining, and lubricating oils. (From Kalpakjian, 1989, p. 916; with permission.)

The coating particles flatten from impact and freeze into an adherent (but somewhat porous) layer. In plasma-arc coating systems, the metal or ceramic coating material is heated to extreme temperatures (e.g., 15,000 K) and sprayed onto the metal surface as a plasma. Ceramic coatings provide oxidation resistance, wear resistance, and thermal insulation, but such coatings often suffer from weak bonding at the metal–ceramic interface, and differences in elastic moduli and coefficients of thermal expansion can crack or spall the coating.

Physical vapor deposition (PVD) coats a substrate with vaporized metal in a vacuum chamber. The hot vapor condenses as a solid coating when it contacts the cooler workpiece (Sidebar 20.1). The simplest way to vaporize the coating metal is by electrical resistance heating, but several alternative methods are also used, such as striking an arc to the coating metal to produce a plasma vapor. The coating material can be "sputtered" into vapor form by bombarding it with high-energy Ar^+ ions. In reactive sputtering, the inert gas is replaced by O, N, or C to deposit metal oxides, nitrides, or carbides. PVD is primarily a "line-of-sight" coating process that covers exposed areas but produces much thinner coatings or none at all in deeply recessed regions or the interior walls of tubing and holes.

Chemical vapor deposition (CVD) coats a hot substrate by allowing gaseous, metal-bearing compounds to contact the workpiece in a controlled atmosphere. The high temperature of the workpiece and/or reactions in the chamber atmosphere decompose the chemical, leaving metals or compounds on the workpiece surface. This technique is often used to produce layered semiconductor devices. For example, gaseous $Ga(CH_3)_3$, $Al(CH_3)_3$, and AsH_3 can be reacted to coat a substrate with GaAs, AlAs, and AlGaAs semiconducting layers. CVD is also a useful purification method for several metallic elements (Sidebar 8.5).

Ion implantation injects selected impurity atoms a few micrometers beneath the workpiece surface. It is somewhat analogous to carburizing steel parts, but in the ion implantation technique the metal's surface is bombarded by accelerated charged ions of the impurity atom. This technique is used to add O to the surface of Ti alloy surgical implants to raise the metal's hardness (Sec. 12.2.2.1 and Fig. 12.3) to improve wear resistance. Using high-O Ti for the entire part would provide unacceptably low ductility and fracture toughness, but increasing O content

---

SIDEBAR 8.5: METAL PURIFICATION BY CHEMICAL VAPOR DEPOSITION AND QUARTZ–HALOGEN LIGHT BULBS

Zr metal can be purified by reacting low-purity Zr metal with I_2 to 200°C in a vacuum chamber to form ZrI_4 vapor. Many of the impurities in the crude Zr do not react with I_2 and are thus excluded from the ZrI_4 vapor. An ohmically heated Zr wire in the chamber is maintained at about 1300°C, and when the ZrI_4 vapor contacts the hot wire, it decomposes into Zr metal and I_2 vapor. The Zr metal thus formed coats the hot filament, gradually accumulating a thick layer of high-purity Zr crystals. This process (the van Arkel–de Boer process) yields 99.98% pure Zr with low interstitial impurity content. Similar decomposition CVD processes are used to produce high-purity Ti, Hf, Cr, Th, V, Nb, Ta, and Mo.

A variation of this process allows more efficient conversion of power to visible light in quartz–halogen light bulbs. In these bulbs, a small amount of I_2 is sealed with the W filament inside a small quartz bulb. In an ordinary incandescent bulb, the hot W filament sublimates W atoms that condense on the inner bulb wall. The loss of W eventually leads to filament burn-out. Inside the quartz–halogen bulb, sublimated W atoms react with I_2 vapor inside the bulb to form W iodide. This iodide vapor eventually contacts the hot W filament, which decomposes the iodide, returning W atoms to the filament and reforming I_2 molecules to repeat the cycle. This allows the filament to operate considerably hotter than in normal bulbs, which produces more visible light and less infrared light per watt of electricity. A quartz bulb is needed rather than soda-lime glass to allow the lamp to operate hot enough to maintain W iodide in the vapor phase.

only at the surface gives the part good wear resistance while retaining high toughness and formability. The O atoms also induce residual compressive stress in the surface that improves fatigue strength.

PVD and CVD produce accurately controlled thicknesses of amorphous, nanocrystalline, or crystalline coatings, but they require vacuum chambers or controlled atmospheres, which raise costs and make it difficult to coat large areas or large numbers of parts. In such situations, *electroplating* is often the preferred method to coat large parts or large numbers of parts inexpensively. Electroplating deposits metal by making the workpiece the cathode in an electrolyzed solution. Electroplating base metals with Ni (Sec. 16.3.3.2), Cu, Cr (Sec. 14.2.3.3), Ag (Sec. 18.3.3.4), Cd (Sec. 19.3.3.2), Sn (Sec. 21.4.3.1), Zn (Sec. 19.2.3.1), or Au increases corrosion resistance, wear resistance, and contact electrical conductivity. It may also improve a product's appearance.

Unfortunately, electroplating tends to produce thicker metal plating at sharp corners and edges and thinner coatings on flat areas, recessed surfaces, and interior surfaces. Electroless plating is a variation on electroplating that produces metallic coatings by chemical reactions rather than electric current. Electroless plating reduces a metal salt such as $NiCl_2$ with NaH_2PO_2 to deposit Ni metal on all available surfaces, even those that are nonconducting and located on interior or recessed surfaces. Anodizing is a "reverse electroplating" operation performed to increase oxide layer thickness by making the metal part the cell's anode. It is used most widely to improve corrosion resistance in Al, but it can also color the surface by using electrolytes containing dyes so that dye molecules are entrained in the growing oxide layer.

Metal coatings can be applied to steel parts by dipping them in liquid Zn (Sec. 19.2.3.1), Pb, or Sn (Sec. 21.4.3.1). Vitreous metal oxide coatings (*porcelain enameling*) can be applied to hot metal (425 to 1000°C) to fuse glass onto the surface. Although this process is most commonly used to coat steel (Sec. 15.3.2.5), it is also used on Al and Ni superalloys.

8.7.4 Painting

Painting coats metal surfaces with blends of organic and inorganic materials suspended in a liquid carrier. Paint is applied by spraying, brushing, or dipping. During drying or curing, the liquid carrier evaporates or reacts with other components in the paint to solidify the coating. As with all coating operations, the metal surface must be cleaned of debris, oils, and loose oxide scale by scraping, blasting, or brushing or by dissolving them chemically. Heat is often used to accelerate paint carrier drying and polymerization reactions needed to form a strong, adherent coating. Pigments in the paint impart the desired color to the painted surface. Although spray painting is the fastest way to coat large metal parts, as much as 70% of paint can be wasted as "overspray" that misses the target and pollutes the air. Electrostatic painting operations impose opposite charges on the paint droplets and the metal to draw the droplets onto the surface by electrostatic attraction, greatly reducing overspray. Evaporation and disposal of organic liquid carriers present hazards to workers and potential environmental problems; consequently, many painting operations have converted to *dry coating* methods that spray a dry mixture of fine pigment and resin particles onto an electrostatically charged surface. These particles then react to form a smooth coating in a curing oven.

ADDITIONAL INFORMATION

References, Appendixes, Problem Sets, and Metal Production Figures are available at
ftp://ftp.wiley.com/public/sci_tech_med/nonferrous

9 Composites

> It might be said now that I have the best of both worlds: a Harvard education and a Yale degree.
>
> —Harvard alumnus John F. Kennedy, upon being awarded an honorary degree from Yale University, June 1962

9.1 INTRODUCTION

All materials have shortcomings. Most conventional metal alloys are ductile and reasonably strong, but for some applications they are too heavy, insufficiently rigid, or weaken unacceptably at elevated temperature. Ceramics are light, strong, stiff, and corrosion resistant, and they retain those characteristics well at high temperatures, but ceramics are difficult to form, and their brittleness and low fracture toughness make them unreliable for many applications. Combining two or more materials with different properties, such as metals and ceramics, into a new material combines their strengths and reduces their shortcomings. Such dissimilar materials combinations are called *composites*. Their superior properties have led some observers to call them "materials of the future"; however, the composite concept is not new. Thousands of years ago, builders learned the advantages of blending straw into the mud used for brick making. The twelfth-century Mongols made composite archery bows. Tendons formed the bow's tension side, bamboo acted as the core, and horn lay on the bow's compression side in a laminate held together by silk soaked in pine resin. Mongol bows were 80% as strong as modern composite bows.

9.2 COMPOSITE MATERIALS

The ideal composite material exploits the favorable properties of each component while mitigating each component's shortcomings. Composite materials can have strength, stiffness, and lightness superior to those of any single (monolithic) component. By varying a composite's composition, its properties can be tailored to meet the particular needs of an application. For example, within broad limits, it is possible to specify strength and stiffness in one direction, coefficient of expansion in another, and so on. This is seldom possible with monolithic materials.

9.3 METAL MATRIX COMPOSITES

The variety of composites reflects the continuing quest for materials that can withstand ever greater loads in increasingly aggressive environments. These improvements are achieved by embedding a second-phase fiber or particle into a matrix that holds the entire composite together. Composites may be divided into three categories: ceramic matrix, polymer matrix, and metal matrix composites (MMCs). Compared to monolithic metals, MMCs have higher strength-to-density and stiffness-to-density ratios, lower coefficients of thermal expansion, better abrasion resistance, and superior creep and fatigue properties. MMCs have higher conductivity

---

Structure–Property Relations in Nonferrous Metals, by Alan M. Russell and Kok Loong Lee
Copyright © 2005 John Wiley & Sons, Inc.

and toughness than those of most ceramic matrix composites, and MMCs have better high-temperature performance than that of polymer matrix composites. Typical fiber and matrix materials for MMCs are shown in Table 9.1. MMCs are commonly classified by the shape of the reinforcements used (Fig. 9.1):

1. *Dispersoids* are roughly spherical particles (often, metal oxides) about 0.01 to 0.1 μm in diameter. Ideally, these are homogeneously distributed and have a relatively low volume fraction (1 to 20%).

2. *Particulates* are approximately equiaxed particles about 10 to 20 μm in diameter. These are often SiC or Al_2O_3, but TiB_2, B_4C, SiO_2, TiC, WC, BN, W, and others have been used as well. The particulate volume fraction is about 10 to 30% when used for structural applications, but can be as high as 80% when used for electronic packaging and switches. Dispersoid and particulate-reinforced composites are isotropic.

3. *Discontinuous fibers* are manufactured as continuous strands, then chopped into short lengths. Fiber lengths depend on a material's intended application. The fibers may be whiskers, extraordinarily strong fibers that are typically single crystals with a diameter of 0.25 to 1 μm and aspect ratios (length/diameter) up to several hundred. Whiskers were originally expensive, but cheaper production methods for SiC whiskers have now been developed. Randomly oriented fibers produce isotropic composites, but extrusion, drawing, or rolling orients fibers and makes

TABLE 9.1 Typical Fibers and Matrixes for Metal Matrix Composites and Characteristic Values of Relevant Properties

	Density (g/cm³)	Thermal Expansion (10^{-6} K^{-1})	Stiffness (GPa)	Strength (MPa)
Fiber				
Glass	2.5	5	70	2500
Aramid	1.4	−6/+75	140	2500
Carbon/graphite	1.8	−1/+15	200–700	2500
Boron	2.6	5	400	7000
Al_2O_3	4.0	7	390	2800
(Saffil) short	4	8	300	1500
(FP) long	4	8	370	2100
SiC	3.2	5	390	
(Whisker) short	2.6	4	300	1500
(Nicalon) long	2.6	4	200–600	3000–8000
W	19.3	4	400	4000
Matrix				
Mg	1.7	26	45	180
Al	2.7	23	70	35
Ti	4.5	9	120	170
Cu	8.9	17	130	100
Ni	8.9	13	220	120

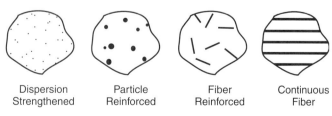

Fig. 9.1 Different types of reinforcements used in metal-matrix composites.

the composite anisotropic. Fiber composite manufacture requires rigorous safety precautions because submicrometer-diameter fiber fragments can easily become airborne, and their inhalation injures lung tissue.

4. *Continuous fibers* are normally about 2 to 30 μm in diameter. Continuously reinforced MMCs are designed to orient the fibers in the direction of the applied load. They are stronger and stiffer parallel to the fibers than perpendicular to them. These composites have typical fiber volume fractions of about 10 to 50% and are highly anisotropic.

As recently as the late 1980s, MMCs could claim only a few significant applications. However, dramatic advances in material and process development have led to much wider utilization today. MMC applications are quite specialized, and they are rarely selected solely for their mechanical strength. Their excellent specific strength and stiffness make them appealing materials for aerospace systems. Their good hot strength and oxidation resistance make them useful in gas turbine engines. MMCs' combination of high stiffness and high strength are well suited to use as automotive driveshafts and bicycle frames. Their stiffness and lightness are well suited for tennis racquets, baseball bats, and golf clubs. MMCs' high creep resistance makes them attractive materials for heat exchangers, pistons, and engine cylinder blocks.

MMCs now have a proven track record as successful "high-tech" materials for many applications. MMCs deliver significant performance benefits (improved productivity, component lifetime), environmental benefits (fewer airborne emissions, lower noise levels), and economic benefits (energy savings or lower maintenance cost). In the early years of the twenty-first century, MMCs were experiencing a market growth rate of 15 to 20%, with use in rail and automotive vehicles, electronic packaging, and thermal management systems. Growth in the rail and automotive industry is driven by increasing pressure for low cost, light weight, fuel economy, and high reliability. In the electronic packaging and thermal management sectors, increased MMC use occurred in response to high demand for wireless communications and network installations.

9.4 MANUFACTURING MMCs

For a given MMC application, the fabrication method is the key factor that determines both properties and cost. There are two principal challenges in MMC fabrication: (1) finding a cost-effective way to distribute the reinforcement phase in the desired configuration in the matrix, and (2) achieving a strong bond between the matrix and reinforcement to allow good load transfer between phases without failure. MMCs can be produced by a variety of fabrication techniques, including liquid-, solid-, and vapor-state processes. Selection of the processing route depends on the type and level of reinforcement loading and the degree of microstructural integrity required for a given application.

9.4.1 Liquid-State Processing by Stir Casting

Perhaps the most obvious method to distribute ceramic reinforcing particles into a metal matrix is to stir them into the molten matrix metal and solidify the mixture (Fig. 9.2). This can be done rapidly and in large quantities using conventional metallurgical processing equipment. Successful stir casting (sometimes called *vortex mixing*) requires high turbulence in the molten metal to ensure good wetting between the particulate reinforcement and the liquid matrix melt. This can be achieved by melting the matrix metal in a crucible, stirring vigorously to form a vortex at the surface of the melt, and introducing the ceramic particles at the side of the vortex. Stirring is maintained for a few minutes before the slurry is cast. Normally, it is possible to incorporate up to 40% ceramic particles in the size range 2 to 100 μm in a variety of molten matrix metals.

Stir casting is the simplest and least costly method to produce MMCs, but it has some drawbacks. Particle agglomeration and sedimentation can occur in the melt and also during

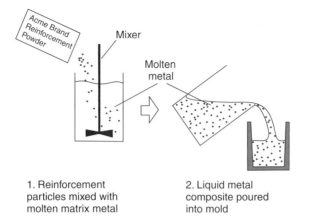

Fig. 9.2 Basic principle of stir casting. (Redrawn from Hashim et al., 2002.)

solidification. Nonuniform reinforcement distribution in cast MMCs can result from interaction between the ceramic particles and the advancing solid–liquid interface during freezing. Rapid solidification reduces the problem, because it refines the scale of the structure and because there is a critical solid–liquid interface velocity, above which solid particles are enveloped rather than pushed. Stir casting generally involves prolonged liquid matrix–ceramic particle contact, which can cause chemical reactions between the two phases. For example, in Al–SiC composites, Al_4SiC_4 or Al_4SiC_3 form at the Al–SiC interface. Formation of Al_4C_3 can degrade the final MMC properties by raising the viscosity of the slurry, making subsequent casting difficult. The rate of reaction can be reduced in several ways: (1) using a suitable coating on particles, (2) using Al alloys with a high Si content, or (3) using preoxidized silicon carbide particles.

9.4.2 Solid-State Processing by Powder Blending and Consolidation

Some stir casting problems can be avoided by blending metallic powder with ceramic short fiber/whisker particles and sintering the blended powders to produce the MMC. Blending can be carried out dry or in liquid suspension. This is normally followed by cold compaction, canning, degassing, and high-temperature consolidation by hot isostatic pressing (HIP) or extrusion (Fig. 9.3). Attaining a uniform mixture can be difficult, particularly with short fibers. Powder metallurgical (P/M) material usually contains platelike oxide particles 10 to 50 nm thick (0.05 to 0.5 vol%) that originally formed on the metal powder. These fine oxide particles act as unintended dispersion-strengthening agents and often strongly influence matrix properties (Sec. 9.8), particularly at high temperature. P/M production of MMCs has several advantages. P/M allows microstructural control of the two phases (difficult to achieve by liquid-phase processing). P/M processing uses lower temperatures and therefore, in principle, offers better control of interface kinetics. P/M also makes it possible to use matrix alloy compositions and microstructural refinements that are available with rapidly solidified powders. P/M processing allows use of existing metal-working equipment, which has contributed to fast growth in P/M processing of MMCs even though it is a somewhat costly production method.

9.4.3 Vapor-State Processing by Spray Deposition

Spray deposition can produce Al matrix composites by injecting ceramic particles/short fibers/whiskers into a liquid metal spray. The liquid metal droplet stream originates from either a molten bath (Osprey process) or by continuous feeding of cold metal into a zone of rapid heat injection (thermal spray process). Typical droplet velocities are 20 to 40 m/s. A thin layer of

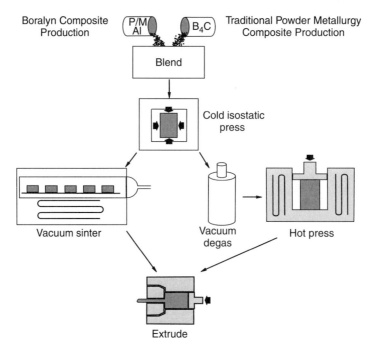

Fig. 9.3 Processing steps for manufacture of powder metallurgy composites. (Redrawn from Harrigan, 1998.)

liquid, or semisolid, is normally present on the sprayed surface of the ingot as it develops. This method often results in inhomogeneous particle distribution and 5 to 10% porosity in the final product. Spray deposition costs are intermediate between stir casting and P/M methods.

9.5 MECHANICAL PROPERTIES AND STRENGTHENING MECHANISMS IN MMCs

Adding hard, brittle reinforcement particles (e.g., SiC, Al_2O_3) to a ductile metal matrix can strengthen the metal substantially. When a tensile stress is applied parallel to the fiber direction in a continuous fiber MMC (stage 1 in Fig. 9.4), both fiber and matrix deform elastically. As higher stresses are applied (stage 2), the matrix deforms plastically while the fibers (which have a higher yield strength than the matrix) are still in the elastic regime. If the fibers possess some ductility, a stage 3 phase can occur where both matrix and fibers deform plastically, but generally, fibers break without undergoing plastic deformation. As the fibers fracture, the load is transferred back to the weaker matrix, and the composite fails swiftly. MMCs exhibit higher strength because they are able to transfer much of the load to the strong reinforcement particles, reducing the stress borne by the matrix. Strengthening also results from modifications to the matrix microstructure caused by the presence of reinforcement.

9.5.1 Load Transfer

A key factor in a composite's mechanical behavior is its ability to share the load between the matrix and the reinforcing phase. The stress varies greatly from point to point within the microstructure (especially with short fibers or particles as reinforcement), but the proportion of the applied load carried by each phase can be obtained from volume averaging the load within

MECHANICAL PROPERTIES AND STRENGTHENING MECHANISMS IN MMCs 135

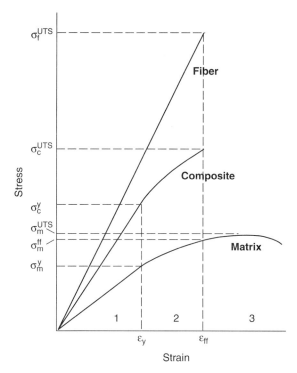

Fig. 9.4 Generic stress–strain curve of a continuous fiber MMC.

each of them. At equilibrium, the external load is equal to the volume-averaged loads borne by the two constituents

$$\sigma_f v_f + \sigma_m v_m = \sigma_A \tag{9.1}$$

where v_f is the reinforcement volume fraction, v_m the matrix volume fraction, σ_m and σ_f the mean stress in the matrix and reinforcement, respectively, and σ_A the stress on the composite. Thus, for a simple two-phase composite under a given applied load, a certain proportion of the load will be carried by the reinforcement and the remainder by the matrix.

The fraction of the load carried by the reinforcement will be independent of the load applied for composites operating in the elastic regime. Since nearly all engineering use of materials is intended to occur in the elastic regime, this condition covers nearly all design situations. If a large fraction of the total applied load is carried by the reinforcement, more stress must be applied to produce large-scale yielding in the matrix, and hence the composite is stronger than the unreinforced metal (Fig. 9.5). The main function of MMC reinforcement is to carry most of the applied load. The main functions of the matrix are to bind the reinforcement together and to transmit and distribute the external load to the individual reinforcement particles or fibers. The degree of load transfer from the matrix to the reinforcement depends on such factors as the relative stiffnesses of the phases, the volume fraction of the reinforcement, the aspect ratio of the reinforcement, the strength of the interfacial bond, and whether the deformation of the composite is purely elastic or contains some degree of plasticity.

9.5.2 Matrix Strengthening

It is easiest to analyze and predict composite properties by assuming that the matrix properties are the same as those of the unreinforced metal; however, this is not the case in actual composites.

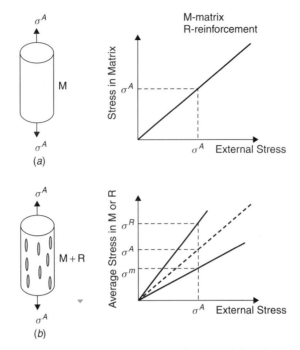

Fig. 9.5 Stress distribution of an applied external stress in (*a*) unreinforced metal and (*b*) an MMC.

The presence of the reinforcing phase significantly alters the matrix microstructure, dislocation motion in the matrix, and the nature of plastic flow. Consequently, the matrix of a MMC behaves differently from the unreinforced metal in several ways.

9.5.2.1 Dislocation Strengthening. Dislocation density is higher in a composite's matrix than in unreinforced metal with the same history. In a composite, dislocations can be generated either by straining in response to an applied load or through straining to relax residual thermal stresses caused by *coefficient of thermal expansion* (COTE) mismatches between matrix and reinforcement. In the latter case, Arsenault and Shi (1986) have calculated the change in dislocation density due to matrix-reinforcement COTE mismatch by estimating the effect of "prismatic punching" of dislocations around reinforcement particles. A similar approach was taken by Miller and Humphreys (1990), who made the assumption that the particles were cubic, to obtain

$$\Delta\rho = 12\frac{\Delta\alpha\,\Delta T f}{bd} \tag{9.2}$$

where $\Delta\rho$ is the increase in dislocation density, $\Delta\alpha$ the COTE mismatch, ΔT the temperature difference, b the Burgers vector length, f the reinforcement volume fraction, and d the particle size. From these expressions, the change in matrix yield strength can be estimated:

$$\Delta\sigma \approx Gb\sqrt{\Delta\rho} \tag{9.3}$$

$\Delta\sigma$ is the change in yield strength of the matrix and G is the shear modulus of the matrix. These expressions show that the dislocation density, and hence the matrix strengthening, increase with decreasing reinforcement particle size and increasing reinforcement volume fraction.

9.5.2.2 Orowan Strengthening. Dislocation motion is impeded by closely spaced hard particles of reinforcing phase on a matrix slip plane. The shear component of the applied stress

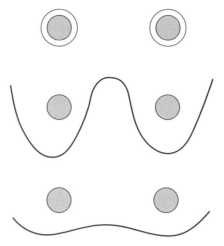

Fig. 9.6 Orowan mechanism.

bows the dislocations around the particles (Fig. 9.6). At the yield point a critical value of the bowing angle is exceeded, and the dislocation passes by the obstacle, leaving it surrounded by a dislocation loop. This strengthens the composite by making subsequent dislocation motion more difficult. Although important in dispersion-strengthened systems, Orowan bowing and looping of dislocation lines does not significantly strengthen other MMCs because their reinforcing particle size and interparticle spacing are both large. Moreover, MMC reinforcement particles are frequently located on grain boundaries, where Orowan-type strengthening will not occur. However, Orowan strengthening can be significant in P/M discontinuously reinforced Al composites. The oxide film on the Al particles' surfaces breaks up and aligns during extrusion, leaving oxide stringers in the matrix parallel to the extrusion direction. These oxide stringers hinder dislocation motion in the matrix, producing a dispersion-hardening condition for both the MMCs and for unreinforced metal, given the same P/M processing.

9.5.2.3 Grain Size Refinement. MMCs usually have much finer grain size than that of unreinforced matrix metal. Miller and Humphreys (1990) observed that the Hall–Petch relationship [Sec. 3.2 and equation (3.1)] implies that the beginning of plastic flow in the matrix depends on the local magnification of stress at grain boundaries resulting from dislocation pile-ups. A small grain size results in fewer dislocations in the pile-ups, and hence the applied stress must be higher to cause yielding.

9.6 INTERNAL STRESSES

Internal stresses arise in MMCs from mismatches of matrix and reinforcement COTE and Young's moduli and from plastic flow in the matrix. Thermal residual stresses arise during the manufacturing processes or from temperature changes during service. When the composite is cooled to room temperature, the matrix cannot contract fully to its stress-free condition since the reinforcement contraction is smaller (assuming that $COTE_{reinforcement} < COTE_{matrix}$); the matrix will therefore have a residual tensile stress while the reinforcement is in residual compression. The thermal residual stresses are independent of applied load and have no influence on the composite's elastic modulus. During elastic deformation of a composite, the stress field due to the mechanical loading is simply superimposed on the thermal stress distribution. Hence, the bulk composite strain is a function of the applied load only. However, these thermal residual stresses have undesirable effects, such as promoting cavitation and reducing the initial yield stress. This is illustrated in Fig. 9.7, where the stress–strain curve of the composite is moved downward along the elastic line by an amount equivalent to the thermal stress. The relative positions of the various parts of the curves for matrix and composite are no longer the same.

Plastic flow will occur locally when the stresses generated in the matrix satisfy a yield criterion during loading. This will change the original shape of the hole containing the reinforcement. Therefore, if stress relaxation processes (Fig. 9.8) do not occur, the reinforcement must respond

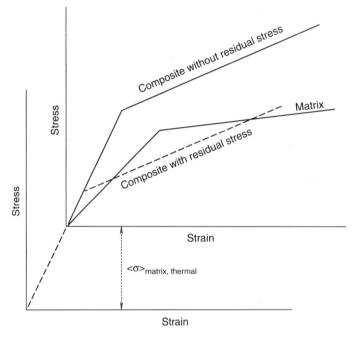

Fig. 9.7 Effect of internal thermal stress on the stress–strain curve of a composite. (Redrawn from Lilholt, 1991.)

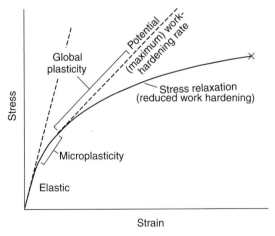

Fig. 9.8 Typical stress–strain curve for an MMC exhibiting a reduced work-hardening rate from the observed maximum potential rate due to the activation of relaxation mechanisms. (Redrawn from Clyne and Withers, 1993, p. 95.)

by elastic deformation, which creates an internal stress field. This would result in more load being transferred to the reinforcement as plastic flow in the matrix increases.

9.7 STRESS RELAXATION

9.7.1 Driving Force for Relaxation

There is a distinction between relaxation mechanisms and global plastic flow in the matrix. During global plastic deformation, load transfers to the reinforcement, which reduces the load carried by the matrix. Thus, the stress in the matrix is lower than it would be in the pure metal. This *load shifting* to the reinforcement increases the free energy of the system because the plastic misfit increases between the ductile matrix and the elastically strained reinforcement by generating Orowan loops around the reinforcement. Misfit between two phases could arise from a temperature change and also during mechanical loading because a stiff reinforcement deforms less than the surrounding matrix. This process will not continue indefinitely because relaxation mechanisms act to reduce the high stresses in the reinforcement and hence reduce the overall free energy of the composite. Reduction in free energy is the driving force for stress relaxation, not simply the reduction of local matrix stresses. Figure 9.8 shows how stress relaxation mechanisms cause a large reduction in the work hardening of the composite from the maximum possible work-hardening rate.

Stress relaxation occurs only if it will reduce the free energy of the system and only if a mechanism is available to act. The relaxation mechanisms that operate will depend on the reinforcement size and aspect ratio, the temperature, matrix ductility, and the relative strengths of interfacial bond strength. Mechanisms of stress relaxation can be classified into two types, those that reduce the reinforcement stress by microstructural modification, and those that introduce microdamage into the composite.

9.7.2 Microstructural Modification

9.7.2.1 Dislocations. Thermally generated misfit strains near the matrix–reinforcement interface can be relaxed by punching out and moving dislocations (Fig. 9.9). This reduces the strain in the region around the reinforcement (i.e., it reduces the misfit). Dislocation relaxation is particularly important in Al matrix composites because Al's high stacking fault energy limits the separation of partial dislocations, making cross-slip easy (Sec. 20.2.2.1).

9.7.2.2 Diffusion. The matrix stress distribution generated around the reinforcement is not uniform. Thus, stress gradients exist near the reinforcement that promote diffusion from regions of compressive stress to regions of tensile stress. The diffusion path will depend on the temperature. Volume diffusion will dominate at high temperatures, while lower-energy diffusion paths such as dislocation cores, grain boundaries, or matrix–reinforcement interface will be preferred at intermediate temperatures (Sec. 6.4.3.2 and Fig. 6.5). Of course, all diffusion processes are stymied at high strain rates and low temperatures because diffusion is too slow to act in the time available.

9.7.2.3 Recrystallization. The increased dislocation density near the reinforcement promotes recrystallization, and the inclusion misfit strain can be relaxed at temperatures high enough to permit recrystallization. However, oxide dispersion phases present in the matrix (Sec. 9.4.2) hinder recrystallization by retarding grain boundary migration. Prior thermomechanical history of the material can also affect how readily recrystallization can occur. Recrystallization will soften the matrix by reducing dislocation density. As with diffusional relaxation, recrystallization will be significant only at elevated temperatures.

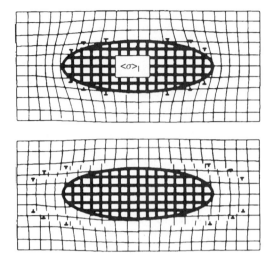

Fig. 9.9 Thermally generated misfit strains are relaxed by punching out the dislocations originally present at the matrix–reinforcement interface. This reduces the strain around the reinforcement. (From Clyne and Withers, 1993, p. 100; with permission.)

9.7.3 Microdamage Processes

Some stress relaxation mechanisms actually damage the composite microstructure by fracturing the reinforcement particles or separating the matrix from the reinforcement. These microdamage mechanisms weaken the composite and influence its failure mechanisms.

9.7.3.1 Reinforcement Fracture. For discontinuously reinforced composites, stress relaxation seldom fractures reinforcement particles. Reinforcement fracture is important only when the reinforcement has a high aspect ratio and a low volume fraction, factors that increase stress in the reinforcement phase. For fairly long fibers, a condition for reinforcement fracture can be calculated using the *shear lag model*. This model assumes that load transfer to the reinforcement during straining is due solely to shear stresses at the matrix–reinforcement interface. The critical reinforcement aspect ratio s_* is the ratio of reinforcement length/diameter above which reinforcement fracture may be expected:

$$s_* \approx \frac{\sigma_{I*}}{2\tau_y} \approx \frac{\sigma_{I*}}{\sigma_y} \tag{9.4}$$

where σ_{I*} is the failure stress of the reinforcement and τ_y and σ_y are the shear and tensile yield stresses of the matrix. Although this model is insufficient for describing composite behavior, it does show that increasing the aspect ratio of the reinforcement raises the stress in the reinforcement for the same applied stress.

Reinforcement cracking tends to dominate in composites with low-ductility matrixes and high interfacial bond strengths. Studies on Al–SiC composites with different SiC particulate sizes found that particle cracking became a significant damage mechanism only when the particle diameter was larger than 20 μm (Fig. 9.10). Larger particles are more prone to crack since they have a higher probability of containing a critical flaw. Reinforcement clustering also promotes reinforcement cracking. Sectioned samples that had reached instability show that most of the damage occurred in clusters of particles. Matrix fracture, together with any particle fracture within the cluster, allows a crack to grow across reinforcement clusters. Once this process occurs, there can be fast fracture through the matrix between nearby clusters, resulting in failure

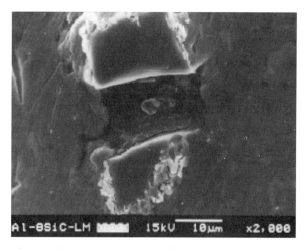

Fig. 9.10 Scanning electron micrograph of a fractured SiC particle in a failed tensile specimen of Al–SiC composite. (From Tham et al., 2001; with permission.)

of the part. In general, reinforcement cracking is promoted by high aspect ratio, high interfacial bond strength, and large particle size.

9.7.3.2 Cavitation and Debonding. Composite failures usually begin near the matrix–reinforcement interface. The two main interface failure mechanisms are cavitation and debonding. In particulate or short-fiber-reinforced MMCs with a strong interface bond, ductile failure occurs by nucleation, growth, and coalescence of cavitation voids near the reinforcements or within the matrix. For a weakly bonded interface, the reinforcements tend to debond from the matrix in a brittle manner when the local stress (or strain energy release rate) exceeds the interface bond strength.

Two void nucleation modes are frequently seen in discontinuously reinforced metal matrix composites: particle cracking and particle–matrix interface debonding. As the stress applied to the composite increases, the plastic constraint imposed by the reinforcing particles on the matrix increases proportionally. This causes void nucleation in the matrix followed by particle cracking or interface debonding. For short-fiber reinforcements, failure normally occurs by formation and coalescence of microvoids within the matrix. Nucleation and growth of these voids ultimately causes interfacial debonding. Voids generally nucleate near reinforcement particles due to the high triaxial stress state and the increased level of matrix work hardening near the interface. Figure 9.11 is a TEM micrograph of a region just below the fracture surface of an Al–SiC composite tensile test specimen. Voids (marked by arrows) 20 to 30 nm in diameter nucleated on the fiber end corner. In other fibers, these voids had grown toward the center as well as away from the fiber end. The growth and coalescence of multiple voids at fiber ends normally produces single, equiaxed voids of approximately the same diameter as the fiber. Void nucleation is a fairly stable process that needs considerable plastic strain to develop; it does not lead instantly to sudden failure of the composite.

The stress state needed for cavitation will depend on the exact nature of the void nucleation event. For a weakly bonded interface, hemispherical voids are most likely to form at the fiber ends along the loading direction. For a strongly bonded interface, voids tend to grow from the matrix. The major factors promoting cavitation are high matrix flow stress, large reinforcement size, large imposed plastic strain, particle clustering, high particle aspect ratio parallel to the stress direction, and small grain size. Knowledge of void nucleation, growth, and coalescence in composites is vital for predicting their tensile behavior and creep deformation since they are related to composite failure.

Fig. 9.11 Void initiation in tensile fracture specimen of Al–SiC composite at corners of the SiC fiber end (shown by arrows) with intense strain in the matrix. As strain increases, these voids typically grow toward the center of the fiber end and coalesce into one large void at the fiber end. (From Nutt and Needleman, 1987; with permission.)

9.8 HIGH-TEMPERATURE BEHAVIOR OF MMCs

Many compositions of metal matrix composites are available that deliver high strength and impact resistance and excellent wear and corrosion behavior. Their greatest value, however, probably lies in their superior high-temperature performance. Composites are not only stronger than conventional monolithic metals at room temperature, they retain their higher strength at elevated temperatures better than does unreinforced metal (Fig. 9.12). MMC mechanical behavior changes at high temperature, and creep becomes particularly important. A schematic representation of typical creep deformation for unreinforced metal and a MMC is shown in Fig. 9.13. The loads and temperatures seen in normal high-temperature MMC use are too low to cause appreciable creep of the reinforcement. Therefore, the MMC's creep behavior is dictated by the amount of matrix creep. A continuous-fiber MMC normally has a very low creep rate since the fibers carry most of the load. However, if the fibers break, the composite will then begin to behave like a discontinuous MMC, reinforced by long fibers.

An MMC generally exhibits short primary creep, long steady-state creep, and short tertiary creep (Sec. 6.4.1 and Fig. 6.3). In many MMCs, however, the secondary stage is rather brief and may be more appropriately termed a *minimum creep rate*. A well-defined steady-state creep rate can usually be obtained in compression but not in tensile creep. This behavior can be explained in terms of damage mechanisms. Damage can occur very early during tensile deformation and will build up throughout the deformation; consequently, tensile creep behavior never reaches a true steady state. The damage is not normally catastrophic, and it does not cause immediate failure. These damage mechanisms do not operate in compression, and therefore true steady-state behavior can be achieved in compression.

MMC's load transfer to the reinforcement phase lowers the load borne by the matrix, and since only the matrix is vulnerable to creep, the creep life and the creep strength of the MMC can be several orders of magnitude higher than they would be for the unreinforced matrix metal.

HIGH-TEMPERATURE BEHAVIOR OF MMCs   143

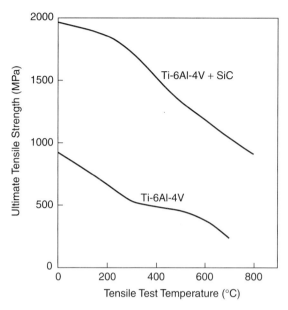

Fig. 9.12 Comparison of high-temperature strength of Ti–6% Al–4% V alloy with and without SiC reinforcing fibers.

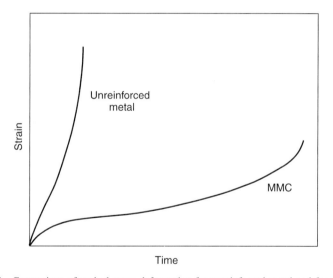

Fig. 9.13 Comparison of typical creep deformation for unreinforced metal and for a MMC.

Figure 9.14 shows that in Al–SiC composites with the same applied stress, the creep rate for the composites is considerably lower than for the unreinforced material. Figure 9.14 also shows that whisker reinforcement gives a lower creep rate than does particulate reinforcement. The creep resistance of P/M Al–30 vol% SiCp/6061 composite is one order of magnitude higher than that of unreinforced 6061 Al alloy tested under the same conditions. Higher creep resistance results from partial load transfer to the reinforcement, with a corresponding reduction in the effective stress acting on the matrix.

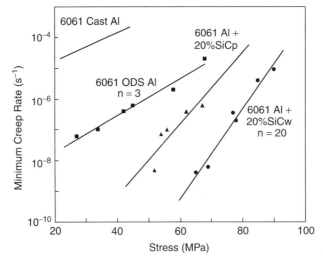

Fig. 9.14 Comparison of cast and powder route unreinforced material with particulate reinforced (p) and whisker reinforced (w) composites tested at 288°C. (ODS, oxide dispersion strengthened matrix.) (Redrawn from Nutt and Needleman, 1987.)

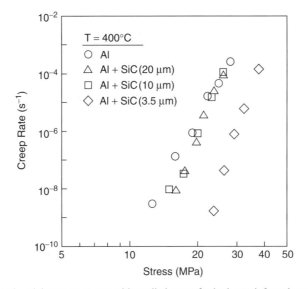

Fig. 9.15 Variation in minimum creep rate with applied stress for both unreinforced pure Al and 10 vol% SiCp–Al composites at 400°C. Note that the larger SiC particles deliver no better creep performance than that of pure metal. (From Tjong and Ma, 1999; with permission.)

However, composite creep strength is not always superior to that of ordinary metals. Experiments on the creep behavior of P/M 10 vol% SiCp/2124 Al (Sec. 20.2.4.4) composite and monolithic 2124 Al alloy under the same testing conditions showed that the creep strength of these two materials was essentially the same at high strain rates because the SiC particles debonded from the 2124 Al matrix. Similar behavior was also found in the creep properties of P/M Al–SiC composites and pure Al at 300 to 400°C. Figure 9.15 shows that incorporating small SiC particulates (3.5 μm) into Al decreases the creep rate of the Al matrix by about three orders of magnitude, whereas large SiC particulates (10 to 20 μm) provided no creep strengthening at all. The creep properties of both reinforced and unreinforced Al are controlled by fine oxide particulates introduced during P/M processing of the materials (Sec. 9.4.2). The large SiC particulates do not improve the creep properties of pure Al because the interparticle spacing for SiC particulates was about two orders of magnitude larger than that for fine oxide particles.

Matrix flow stress decreases with increasing temperature, so at high temperatures the local stresses are too small to fracture the particles; consequently, most of the strain has to be accommodated by the matrix. The usual failure mechanism in samples tested at high temperature is coalescence of voids nucleating at the reinforcement particles. Research work on Al–20 wt% SiC whiskers showed that the failure resulted from void nucleation at the ends of whiskers. When the reinforcement fraction was raised to 30 wt% SiC, the failure initiated from voids in the matrix. In another study of Al–SiC particle composites, the majority of creep deformation arose from the nucleation of cavities at low applied stresses, whereas at higher stresses the cavitation level was reduced considerably. This behavior was due to activation of dislocation glide mechanisms at the higher stress levels.

Experiments have also shown that increasing the test temperature enhances grain boundary sliding in Al–4 wt% $MgAl_2O_4$ composites, forming voids at the grain boundaries. These voids then grew and coalesced due to vacancy diffusion and grain boundary sliding. As the temperature increased, nucleation and void growth also increased, which eventually led to crack generation. The differences between the damage mechanisms at room and high temperature are due to the increased plasticity of the matrix and the operation of thermally activated processes such as dislocation climb (Sec. 6.4.3.1).

ADDITIONAL INFORMATION

References, Appendixes, Problem Sets, and Metal Production Figures are available at
ftp://ftp.wiley.com/public/sci_tech_med/nonferrous

10 Li, Na, K, Rb, Cs, and Fr

10.1 OVERVIEW

Lithium, sodium, potassium, rubidium, cesium, and francium (Li, Na, K, Rb, Cs, and Fr) are called *alkali metals* from *al-qili*, the Arabic word for "plant ashes," an early source of alkali salts. The alkali metals' outer electron configuration is (rare gas core) $+ s^1$. The filled p shell of the rare gas core shields the nucleus from the outer s electron so well that the outer s electron is only weakly bonded to the atom. This electron configuration gives the alkali metals a host of "extreme" characteristics. They are easily ionized and have low melting and boiling temperatures. Their unusually large atomic radii, low densities (Sec. 1.6), and high chemical reactivity all result from this electron structure.

The terrestrial abundances of the group 1 metals vary enormously. Na and K are among the most common elements in Earth's crust, comprising 2.3 and 1.8 wt%, respectively. On a volume basis (cost/cm^3), Na is the least costly of all metals. Both elements are essential to plant and animal life. Rb (78 wt ppm), Li (18 wt ppm) (Sec. 11.3.1), and Cs (2.6 wt ppm) are far less abundant. Fr has no stable isotopes, and the radioactive isotopes have short half-lives. However, Fr is present in the decay chain of some actinide elements (Sec. 24.6.1), so tiny amounts are present in U ores. It has been estimated that the entire Earth's crust contains only ~25 g of Fr. K and Rb also contain naturally occurring radioactive isotopes, but their half-lives are so long that their radioactivity levels are low.

The alkali elements' low melting temperatures (28 to 180°C) mean that any measurement of properties performed at room temperature (Table 10.1) is really a determination of high homologous temperature behavior (0.65 to $0.98 T_{m.pt.}$). At room temperature, alkali elements are so soft that they can be cut with a dull knife. Their reactivity and low strength limit their use in metallic form to certain specialty applications and to alloying additions to other metals. However, their compounds play a major role in many engineering processes.

10.2 HISTORY, PROPERTIES, AND APPLICATIONS

> Observation has been made of a sodium system operating at 1000°F where a vertical stream of sodium $\frac{1}{4}$-in. diam and roughly 18-in. high was being forced from the system. The first 3 to 4 in. of the leaking stream retained the bright metallic luster characteristic of sodium. From 3 to 4 in. to approximately 9 in., the surface was a dark-black color, and beyond 9 in., oxidation of the alkali metal was occurring with smoke and burning.
>
> — Mausteller et al. (1967)

10.2.1 History

Ancient civilizations were familiar with many alkali element compounds, most notably sea salt, which is mostly NaCl. But the alkali elements' low electronegativities made it difficult for early chemists to reduce these compounds to metallic form. The first production of metallic Na

---

Structure–Property Relations in Nonferrous Metals, by Alan M. Russell and Kok Loong Lee
Copyright © 2005 John Wiley & Sons, Inc.

TABLE 10.1 Selected Room-Temperature Properties of Alkali Metals with a Comparison to Cu

Property	Li	Na	K	Rb	Cs	Cu
Valence	+1	+1	+1	+1	+1	+1, +2
Crystal structure at 20°C	BCC[a]	BCC[b]	BCC	BCC	BCC	FCC
Density (g/cm^3)	0.54	0.97	0.86	1.53	1.90	8.92
Melting temperature (°C)	181	98	63	39	28	1085
Thermal conductivity (W/m·K)	85	140	100	58	36	399
Elastic (Young's) modulus (GPa)	4.9	10	3.4	2.4	1.7	130
Coefficient of thermal expansion (10^{-6} m/m·°C)	46	71	83	90	97	16.5
Electrical resistivity (μΩ·cm)	9.5	4.9	7.4	13.1	20.8	1.7
Cost ($/kg), large quantities	95	0.38	90	High[c]	High[c]	2

[a] Transforms to α-Sm below 70 K.
[b] Transforms to α-Sm below 40 K.
[c] No regular market trade occurs in Rb or Cs metal; large orders are infrequent and metal costs of about $800/kg are cited. Small quantities are typically priced at several dollars per/gram.

occurred in 1807 by electrolysis of fused NaOH, and fused salt electrolysis is still used to produce the lighter alkali metals. Rb and Cs are produced in small quantities either by reduction of their oxides or by electrolysis of their fused salts. Fr has no stable isotopes; it was first identified in 1939 from ^{227}Ac decay. Even today it is available only in extremely small quantities, since its longest-lived isotope, ^{223}Fr, has a half-life of just 21 minutes. Of all elements with atomic number less than 100, Fr is the most difficult to accumulate and study.

10.2.2 Physical Properties

The alkali metals have fairly high electrical and thermal conductivities, low melting temperatures, and very low strength. All have BCC (cI2) crystal structure at room temperature; Li and Na transform to the close-packed Sm crystal structure (hR3) at cryogenic temperatures (Appendix A, ftp://ftp.wiley.com/public/sci_tech_med/nonferrous; and Sec. 10.4). The metals are reactive and must be stored under mineral oil or inert gas to avoid rapid oxidation. Alkali metals even attack glass at moderately elevated temperatures, reducing the SiO_2 to Si.

10.2.2.1 Color, Photoelectric Effect, and Thermionic Effect. The alkali elements are well known for intense ionization emission colors (red for Li and Rb, yellow for Na, and blue for K and Cs). Objects that have been handled by bare hands will emit the characteristic yellow color of Na when held over a flame because traces of Na from perspiration coat the surface. Pure Cs metal has a pale golden tinge; it is one of only three pure metallic elements, the others being Cu and Au, with inherent color (Sec. 18.2.2.1).

Because of the especially weak attraction of the outer s electron in both Rb and Cs, both metals display photoelectric behavior. When exposed to light in a vacuum, the metal surface emits electrons that are dislodged from the outer s band of the metal by the energy of the photons. These electrons can be collected on a positively charged grid and directed to flow in a circuit between the grid and the alkali metal. Although this effect can be used to construct a photocell or a light meter, it is more expensive and cumbersome than semiconductor devices. Closely related to the photoelectric effect is the thermionic effect, in which thermal energy ejects electrons from the metal surface (see Sidebar 10.1).

10.2.2.2 Plasticity. At room temperature alkali metals slip in the ⟨111⟩ direction, which has densest linear packing in BCC crystals. The slip planes are {110}, {211}, or {321}. All alkali metals are highly ductile at room temperature since 20°C is $0.65T_{m.pt.}$ or greater for these elements. The variation of critical resolved shear stress (τ_{CRSS}) with temperature is large, which

SIDEBAR 10.1: Cs IN THERMIONIC GENERATORS

Heating a metal provides sufficient energy for some of its electrons to escape the surface (the thermionic effect). Cs has the lowest work function of any metal (1.89 eV for an electron to escape), and Cs is often coated onto refractory metals to enhance its thermionic emission. Without a Cs coating, most transition metals have a work function of 4 eV or higher, which requires high temperatures to achieve useful current flow. Thermionic devices have efficiencies of only 10 to 15%, but they have no moving parts and can reliably convert thermal energy to electricity for years. This makes them appealing for specialty applications such as spacecraft power supplies (Fig. 10.1).

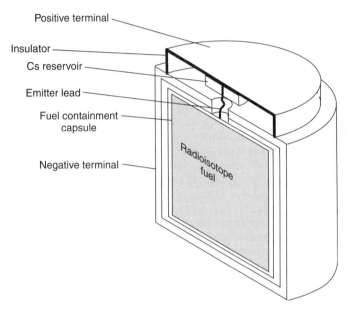

Fig. 10.1 In a thermionic power supply, the decay energy of a radioactive isotope heats a cesiated W emitter. Thermionic electrons leaving the emitter generate electric current flow to power spacecraft systems. The system has no moving parts and converts the radioactive decay thermal energy to electricity with 10 to 15% efficiency.

is typical of BCC metals (Fig. 10.2). In many BCC metals, τ_{CRSS} becomes so high at cryogenic temperatures that cleavage fracture becomes a lower-energy failure mode than plastic flow (e.g., the familiar ductile–brittle transition seen in steel). This strong temperature dependence is caused by the low mobility below about $0.15T_{m.pt.}$ of screw dislocations moving in the $\langle 111 \rangle$ direction (Sec. 14.4.2.2). Near 0 K, alkali metals remain ductile. Although the curves in Fig. 10.2a and b have similar shapes, the τ_{CRSS} values for BCC Ta are more than two orders of magnitude larger than those of Na and K because alkali metals' bond strengths are much weaker than in refractory Ta.

10.2.2.3 Corrosion. Despite their reputation for chemical reactivity, high-purity Li, Na, and K actually form thin, transparent, protective oxide layers in ultradry, high-purity O_2 (partial pressure of $H_2O \leq 0.00002$ mmHg). Even prolonged exposure in dry O_2 at elevated temperature does not attack these metals appreciably. However, in normal room air, H_2O vapor, O_2, CO_2, and N_2 all combine to form complex mixtures of hydroxides, oxides, carbonates, and bicarbonates

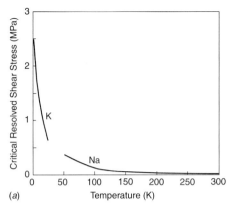 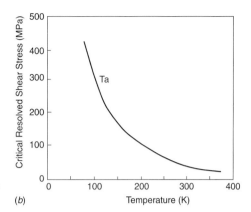

Fig. 10.2 (*a*) Variation of τ_{CRSS} with temperature for BCC Na and K. Na transforms at 40 K to the α-Sm structure; all solid K is BCC. (*b*) Equivalent curve for the refractory BCC metal Ta ($T_{m.pt.} = 2996°C$) is shown for comparison. The τ_{CRSS} of all these BCC metals vary greatly with temperature, although the values for Ta are much higher due to its greater bond strength.

(and for Li, a nitride) that are nonprotective. Na and K corrode rapidly in normal air, and even thick sections will be completely converted to nonmetallic compounds within days. Li does not corrode as rapidly as the heavier alkali metals, but it blackens and deteriorates with linear kinetics and cannot tolerate long-term atmospheric exposure. The heavier alkalis are still more reactive and oxidize even in ultradry, high-purity O_2. Cs forms a series of suboxides, one of which, Cs_7O, lies near a eutectic point in the Cs–O phase diagram at $-2°C$. A partially liquid oxide layer does little to inhibit further oxygen transport. In addition, the heat evolved by the initial oxidation reaction can warm the underlying metal to its $28°C$ melting point and further accelerate oxidation. When exposed to normal air, Rb and Cs can erupt spontaneously in vividly colored flames with the characteristic spectral emission colors of Rb (red) and Cs (blue).

Alkali metals react violently with water, evolving H_2 gas and considerable heat. The reaction of Na is typical:

$$2Na + 2H_2O \rightarrow 2NaOH + H_2 \qquad (\Delta H_{298} = -45.7 \text{ kcal/mol})$$

The reaction heat can ignite the H_2 gas, melt the Na, and lead to an explosion that scatters a dangerous mix of liquid Na metal and hot, caustic lye solution. One of the authors (A.R.) still regrets his boyhood decision to hurl a $\frac{1}{2}$-kg brick of Na metal into the water hazard of a local golf course. Although the resulting flames and explosions festively accented the evening twilight, the loud concussions attracted an undesired level of attention from the neighbors. Moreover, the pond's small population of muskrats and ducks were probably never again completely at ease.

10.2.2.4 Conductivity. Alkali metals have fairly high electrical and thermal conductivities (Table 10.1 and Fig. 10.3). Their reactivity and low strengths preclude conventional use as electrical conductors, but there is considerable interest in using liquid alkali metals as coolants in nuclear reactors. Their good conductivity allows the liquid metal to be pumped magnetically. At high temperature, other coolants (e.g., water) must be kept under pressure to remain liquid, but pressure is unnecessary for alkali metal coolants, and their high thermal conductivity provides efficient heat transfer.

10.2.3 Applications

Because of their reactivity and low strength, alkali metals have no structural uses; but they are employed as reactants in numerous chemical processes, as coolants, as alloying additions to

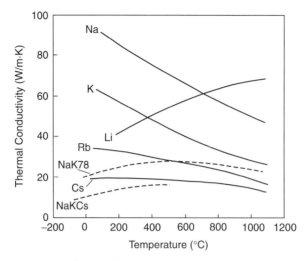

Fig. 10.3 Thermal conductivity of liquid alkali metals. NaK78 is the eutectic Na–78 wt% K alloy; NaKCs is the ternary eutectic (Sidebar 10.3). For comparison, the thermal conductivity of solid Cu at 20°C is 399 W/m·K.

other metals, and even as propellants for advanced rocket engines. Na metal is by far the most heavily used alkali metal, with an annual world production of 340,000 tons; Li ranks a distant second at 15,000 tons of metal per annum. K metal production is only a few hundreds of tons per year. World production of Rb and Cs is so small that no regular market exists for these metals. All five metals (excluding Fr) have engineering applications.

10.2.3.1 Lithium. Several decades ago, Li was used primarily in compounds such as lithium stearate, an additive for lubricating oil; lithium hydroxide for CO_2 absorption in rebreather scuba gear, submarines, and spacecraft; and lithium oxide in specialty glasses and ceramics. Li has more recently been used in high-performance batteries and as an alloying addition to other metals. Metallic Li is produced by electrolysis of fused LiCl and KCl. It has excellent heat transfer and nuclear properties. In addition, its low electronegativity makes it useful as a degassing additive to Ni and Cu alloys. A large and rapidly growing demand for Li batteries is becoming a major use for Li metal.

Al–Li precipitation-hardening alloys (Sec. 20.2.4.7) have lower density and higher elastic moduli than those of other Al alloys, making them useful aerospace alloys. Al–Li alloys typically contain only about 2 wt% Li, but they are produced in sufficient quantities to make this a significant factor in Li demand. Li additions to Mg reduce density and convert Mg from an HCP metal with limited ductility to a highly ductile BCC metal. Additions of 7 to 11 wt% Li produce duplex alloys (part BCC, part HCP), and Li contents above 11 wt% produce single-phase BCC alloys. At high Li contents, Mg–Li alloys densities are less than 1 g/cm^3, so they float on water. Mg–Li alloys saw limited use years ago as lightweight armor, but they are not competitive with modern polymer armor. In general, their cost and corrosion performance are inferior to those of other Mg alloys.

Li reacts with both the O_2 and the N_2 in air, forming Li_2O and Li_3N on exposed Li surfaces. Li–Ca alloys are useful degassing additives for Ni and Cu castings because they not only deoxidize the liquid metal but also remove dissolved N and S. Li metal is also a minor alloying addition to some solder and bearing alloys. The low atomic weight of Li (6.94 g/mol) gives it the highest heat capacity (in terms of J/kg·K) of any metal. Its high heat capacity and relatively good thermal conductivity combine to make Li an appealing high-temperature heat transfer material. Since ^6_3Li can produce tritium (^3_1H) by neutron capture, designers of fusion power reactors have

SIDEBAR 10.2: Li USE IN FUSION POWER REACTORS

Several experimental fusion reactors have been built in the past 50 years to develop use of nuclear fusion power generation. The reactor fuels are deuterium (^2_1H) and tritium (^3_1H). Deuterium is a naturally occurring isotope that can be extracted from ordinary water by electrolysis, but tritium is radioactive ($t_{1/2} = 12.7$ years) and must be produced artificially. The ^6_3Li isotope can produce tritium fuel for thermonuclear fusion power reactors by capturing some of the neutrons emitted by the fusion reaction: $^1_0n + ^6_3Li \rightarrow ^4_2He + ^3_1H$. The fuel is heated to extreme temperatures and confined magnetically as a plasma inside a large toroidal vacuum chamber where the fusion reaction occurs (Fig. 10.4). The newest fusion test reactor is the ITER (International Thermonuclear Experimental Reactor). ITER will be the largest fusion reactor ever built (500 MW of thermal power with a 15 million ampere plasma current). A research-scale Li breeding operation will be performed with ITER, but commercial fusion reactors will need to position Li "blankets" alongside the reaction chamber to breed tritium. Information acquired from ITER operation will guide design and operating procedures for future fusion electric power generation.

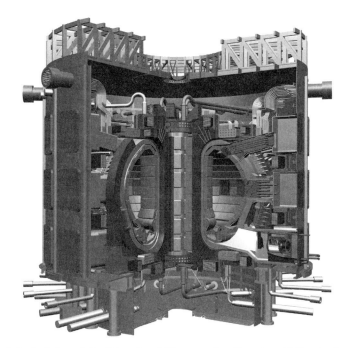

Fig. 10.4 Artist's depiction of the International Thermonuclear Experimental Reactor for fusion power research. This reactor will breed tritium fuel from Li on a small scale for research; however, a commercial fusion reactor will need to transmute sufficient Li into tritium (^3_1H) to fuel the fusion reaction. (Courtesy of ITER, *http://www.iter.org/*; with permission.)

made Li an integral part of every plan for future fusion power generation (Sidebar 10.2). Li could be used as a combination coolant and tritium breeding feedstock, or solely as a tritium breeding material.

10.2.3.2 Sodium. Sodium compounds are essential for plant and animal life and are widely used in industry, medicine, and agriculture. Most glass contains about 15% Na_2O. Caustic soda

(NaOH) is used in petroleum refining and the manufacture of soaps, paper, and textiles. Sodium borate is used in manufacturing ceramics, soaps, and water softeners. Sodium sulfate is used in detergents. Metallic Na represents only a small percentage of total Na use; it is produced by the Downs process, electrolysis of a fused NaCl–CaCl$_2$ eutectic mixture. The CaCl$_2$ lowers NaCls 800°C melting temperature to 600°C at the eutectic composition.

Numerous chemicals are manufactured using Na metal, including sodium hydride, sodium peroxide, and various dyes, herbicides, and pharmaceuticals. Metallic Na is also used to produce tetraethyl Pb and tetramethyl Pb antiknocking agents for gasoline; although these chemicals have been largely phased out in Europe and North America (Sec. 21.5.4), they are still widely used in some nations. Na metal is also used to reduce metal halides and other compounds to produce metallic Ti, Zr, Hf, Ta, Si, Mg, and Ca. Liquid Na is an effective mill scale removal agent for stainless steel and Ti. Na is also used in both low- and high-pressure Na lamps for outdoor illumination. These lamps emit a characteristic yellow light and are highly efficient (180 lu/W).

Na finds several uses as a liquid coolant and heat transfer medium. Na-filled exhaust valves dissipate heat in high-performance internal combustion engines, and liquid Na is used as a heat transfer agent in die-casting equipment. Perhaps the most exotic use of Na coolant is in liquid metal fast breeder fission reactors (LMFBRs). These reactors simultaneously produce power by fission of mixed U and Pu fuel and convert an abundant but less useful U isotope, $^{238}_{92}$U, to fissionable $^{239}_{94}$Pu by capture of excess neutrons in the reactor core: $^{238}_{92}$U + $^{1}_{0}$n → $^{239}_{92}$U. The $^{239}_{92}$U formed by this reaction quickly decays to $^{239}_{94}$Pu, which can fuel the reactor. For optimal efficiency, LMFBRs need a coolant that will neither capture too many neutrons nor reduce neutron velocities too greatly in collisions with coolant atom nuclei. Na meets both of these requirements. At 1 atm pressure, Na has a liquid range of 98 to 892°C, which covers the operating temperature range of a LMFBR core.

The reactive nature of liquid Na coolant requires that it be kept in a sealed system, isolated from contact with air or water to avoid fire and explosion hazards. Despite early concerns about the safety of Na coolant use in LMFBRs, the operating records of several small experimental LMFBRs and one large prototype LMFBR (France's Super-Phenix reactor) indicate that the safety record of Na coolant is about the same as that of high-pressure water and steam in conventional light-water-cooled fission reactors. During operation, nuclear fission heats the Na to about 500°C, and the primary coolant loop Na (which becomes intensely radioactive) transfers the heat to a second, nonradioactive Na loop, which in turn heats H$_2$O in a third coolant loop. Finally, the H$_2$O is flashed to steam to drive turbines for electricity generation. Experience has shown that pure Na will not attack the reactor core and pipe materials used in LMFBRs, but the presence of even small amounts of impurities, especially O, greatly accelerates corrosive attack. For this reason, impurity content of the Na is monitored continuously to hold O levels below 10 to 30 ppm. LMFBRs can produce about 20% more fuel than they consume, utilizing nearly 75% of the energy of the natural U rather than the 1% used in ordinary light-water fission reactors. LMFBRs have proven their technical feasibility, but their viability in the marketplace will finally be decided by their economic competitiveness with conventional reactors and possible concerns about unauthorized Pu diversion for production of nuclear weapons.

10.2.3.3 Potassium. K compounds are essential nutrients for plant and animal life, and they are also used heavily for industrial processes; however, little use is made of K metal. Potassium chloride, sulfate, nitrate, and carbonate are used in fertilizers and fireworks, and potassium permanganate, KMnO$_4$, is a widely used industrial oxidizing agent. Potassium phosphate is commonly used in liquid detergents. Medical patients on low-sodium diets often use KCl to season food. K metal has few engineering uses because Na metal has similar properties and is much less costly. There is no simple, efficient process (e.g., the Downs process for Na) to electrolyze K salts to produce the metal. K is produced by electrolysis of KOH (a process with several difficulties) or by reduction of KCl with Na, followed by extensive distillation to remove Na from the product. The complex production methods and the low demand for K metal result in a surprisingly high market price for such an abundant element.

K metal is used in small quantities as a grain boundary pinning agent in W filament wire for incandescent light bulbs (Sec. 14.4.5). The large K atom (atomic radius = 0.231 nm) is insoluble in W (atomic radius = 0.137 nm) and segregates to grain boundaries. At the 3000-K operating temperature of the W filament, the K vaporizes, forming small gas bubbles that pin the grain boundaries and prevent grain growth that would weaken the filament. K metal comprises 78 wt% of eutectic NaK alloy ($T_{m.pt.} = -12.5°C$), which is occasionally used as a heat transfer liquid. An even-lower-melting Na–K–Cs alloy exists ($T_{m.pt.} = -79°C$) (Sidebar 10.3), but this alloy is expensive and highly reactive.

10.2.3.4 Rubidium and Cesium. Rb and Cs are relatively rare, and the difficulties and expense involved in their purification limits their use to a few specialty applications, most of which exploit their unusually low ionization energies. Rb and Cs compounds are produced primarily as by-products of Li mining, so their availability fluctuates with Li production. Cs is used as a catalyst in the hydrogenation of a few organic compounds, and it can also be used in photoelectric cells

---

SIDEBAR 10.3: THE COLDEST LIQUID METAL

The lowest melting point of any metal alloy occurs in a ternary eutectic of 47.4 at% K–40.8 at% Cs–11.8 at% Na (Fig. 10.5). This alloy freezes at $-79.2°C$, substantially lower than the freezing point of Hg ($-38.8°C$). Additions of Hg and Rb have been made to this K–Cs–Na eutectic in attempts to lower the freezing temperature further, but these efforts were unsuccessful, suggesting that this ternary eutectic may be the "ultimate" low-freezing alloy. This alloy could serve as a heat transfer fluid operating over an exceptionally low temperature range, but it is expensive, reactive, and has a lower thermal conductivity than that of the pure alkali metals (Fig. 10.3), so it remains a laboratory curiosity.

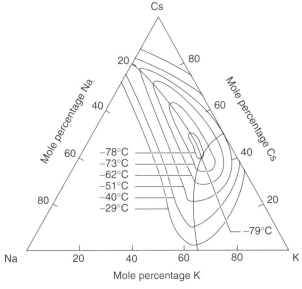

Fig. 10.5 Liquidus surface of the Cs–K–Na ternary phase diagram (Sec. 10.2.3.3). The 47.4% K–40.8% Cs–11.8% Na ternary eutectic freezes at $-79.2°C$, the lowest known freezing temperature of any metallic alloy.

and to capture O inside vacuum tubes. Cesium formate is used as a specialty drilling fluid in very deep, high-pressure, high-temperature oil wells.

The outer s electron of Cs will fluoresce when excited by microwave radiation, and the natural resonance frequency of the Cs atom is used to define the second. "Cs fountain" atomic clocks keep time with an uncertainty of less than 2×10^{-15}, which is equivalent to an error of 0.1 ns/day. Cs has also been used as a propellant for experimental electrically powered ion rocket engines that accelerate ions to speeds of 100,000 km/h, achieving a small thrust that can be sustained for long periods. Ion rocket efficiencies are 10 times greater than those of conventional combustion rockets. Cs, Xe, and Hg have all been used as the propellant in ion rockets; Xe has the advantage that it is already gaseous at ambient temperature, but its ionization energy is high. Cs requires energy to vaporize it, but its ionization energy is low. Rb can be used for many of the same applications as Cs (e.g., photocells, O gettering, and ion rocket propellant), but it performs them somewhat less effectively. Rb has also been used to produce certain specialty electronic porcelains with high insulating capacity. $RbAg_4I_5$ has the highest room-temperature electrical conductivity of any known ionic crystal, which may allow its use in thin-film batteries.

10.3 SOURCES

Na metal is produced from NaCl, which is present in large quantities in salt domes, salt flats, saltwater lakes, and the oceans. The primary factors in siting Na production facilities are access to low-cost electric power, proximity to salt deposits rich in NaCl, and proximity to customers. Na metal from the Downs process is usually pumped as a liquid into railroad tank cars and remelted at its destination so it can be pumped or drained from the tank. Major Li deposits are found in Australia, Canada, Chile, and Zimbabwe; both hard rock mining and Li-bearing brines are used. Germany and the United States dominate the manufacture of Li metal and chemicals. The best Rb and Cs sources are found in Canada.

10.4 STRUCTURE–PROPERTY RELATIONS

10.4.1 Low-Temperature Phase Changes and Deformation in Li

When deformed at or above 180 K, Li is at 40% or more of its homologous melting temperature, and deformation is essentially hot work [i.e., dominated by thermally activated processes (Sec. 4.8) and rapid recovery and recrystallization]. Above 180 K, Li's strain rate sensitivity is high, its work-hardening rate is low, and work softening actually occurs for strain rates less than 10^{-5} s^{-1}. When deformed below 180 K, a stress-induced martensitic transformation occurs in small regions of the metal, forming the low-temperature α-Sm (hR3) crystal structure (Sec. 23.3.1). The martensitic transformation occurs by distortion of the BCC unit cell, transforming the (011)$_{BCC}$ plane into a close-packed plane. Stress at dislocation pile-ups causes embryonic nuclei of α-Sm to form, but the nuclei are stable only in the presence of the stress field close to a dislocation, so they can grow no larger than a few atom diameters. Although the equilibrium temperature for Li's BCC → α-Sm transformation is 70 K, stress fields around dislocation pile-ups trigger partial transformation at higher temperatures. This second phase strengthens the metal. At 100 K, Li is at $0.22T_{m.pt.}$, and its critical resolved shear stress (τ_{CRSS}) is about 1 MPa. By comparison, Cu is at $0.22T_{m.pt.}$ at room temperature, and its τ_{CRSS} is also about 1 MPa.

Even when no external stress is applied, Li's BCC ↔ α-Sm transformations are complex and hysteretic. When Li is cooled to 4.2 K and reheated, a complex sequence of transformations occurs (Fig. 10.6). During cooling, the volume percent of the low-temperature α-Sm phase increases steadily with decreasing temperature; the M_s temperature is 70 K, but even at 4.2 K,

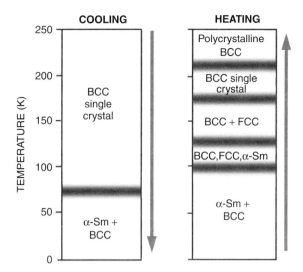

Fig. 10.6 Phase changes in an unstressed Li single crystal cooled from room temperature to near 0 K, then warmed back to room temperature.

the transformation is incomplete, so there appears to be no M_f temperature. Upon heating, the reverse transformation begins at about 90 K and finishes at 180 K. During heating the α-Sm phase gradually transforms first into the closely related FCC structure, which then transforms to BCC as heating continues. As the (now BCC) single-crystal material is heated back toward room temperature, it recrystallizes into a polycrystal. The recrystallization is induced by defects formed during the transformations, but recrystallization cannot occur until the temperature is high enough (about 200 K) for substantial diffusion.

This martensitic transformation causes peculiar mechanical properties in Li below 180 K. Li's τ_{CRSS} value normally does not vary with crystal orientation, but in Li below 180 K it changes substantially as the angle between the (011) plane normal and the tensile axis varies; in compression, τ_{CRSS} also varies with the crystal orientation, but in a different manner than the tensile variation. In addition, the strong temperature dependence of τ_{CRSS} found in most BCC metals is absent in Li.

ADDITIONAL INFORMATION

References, Appendixes, Problem Sets, and Metal Production Figures are available at
ftp://ftp.wiley.com/public/sci_tech_med/nonferrous

11 Be, Mg, Ca, Sr, Ba, and Ra

11.1 OVERVIEW

Beryllium, magnesium, calcium, strontium, barium, and radium (Be, Mg, Ca, Sr, Ba, Ra) are the alkaline earth metals. The outer electron configuration of an isolated alkaline atom is (rare gas core) $+ s^2$, but in metallic form the outer electrons hybridize to an $s^1 p^1$ configuration that provides two bonding electrons. They have large atomic radii and low densities (Fig. 1.13), which makes their properties somewhat similar to those of the alkali metals, although less extreme. Mg and Be are the lightest metallic elements with useful structural strength. The other alkaline metals are weaker and too reactive for use in normal-humidity air, but they are used as alloying additions to other metals, and their compounds have a wide range of applications.

Terrestrial abundances of the group 2 metals vary enormously. Ca and Mg are among the most common elements, comprising 4.7 and 2.8 wt% of Earth's crust, respectively. Both elements are essential to plant and animal life. Sr (384 wt ppm) and Ba (390 wt ppm) are less abundant but still relatively easily obtained, whereas Be (2.0 wt ppm) is rare and expensive. Ra has no stable isotopes, but some of the radioactive isotopes have moderately long half-lives. U decay replenishes the Ra lost to decay, giving Earth's crust a steady-state abundance of about 10^{-7} wt ppm Ra.

11.2 HISTORY AND PROPERTIES

Humankind has used alkaline element compounds for thousands of years. Ca compounds are major constituents of two of the most important structural materials in human existence: bone and concrete. Alkaline metals, however, are reactive and they were first reduced from their compounds only two centuries ago. Most of the alkaline elements are simple metals with purely metallic bonding (Be being the exception, Sec. 11.3.2.2). The lighter alkaline metals possess fairly high electrical and thermal conductivities, low densities, and sufficient strength for structural use (Table 11.1). The heavier elements are poor conductors and are too weak and reactive for structural use. All the alkaline metals have simple crystal structures (HCP, FCC, or BCC).

11.3 BERYLLIUM

> Beryllium is used as a reflector in modern, light-weight fission warheads and thermonuclear triggers. It has special value for triggers since it is essentially transparent to thermal radiation emitted by the core. It is a very efficient reflector for its mass, the best available.
>
> —Carey Sublette, *http://nuclearweaponarchive.org/Nwfaq/Nfaq4.html*, 1999

11.3.1 History of Be

Be is rare, both on Earth and throughout the observable universe. The stellar fusion reaction of two $^4_2$He nuclei ($^4_2$He $+ ^4_2$He $\to ^8_4$Be) might be expected to make Be quite common. However, $^8_4$Be

---

Structure–Property Relations in Nonferrous Metals, by Alan M. Russell and Kok Loong Lee
Copyright © 2005 John Wiley & Sons, Inc.

TABLE 11.1 Selected Room-Temperature Properties of Alkaline Metals with a Comparison to Cu

Property	Be	Mg	Ca	Sr	Ba	Ra	Cu
Valence	+2	+2	+2	+2	+2	+2	+1, +2
Crystal structure at 20°C	HCP[a]	HCP	FCC[b]	FCC[c]	BCC	BCC	FCC
Density (g/cm^3)	1.85	1.74	1.55	2.63	3.51	5.00	8.92
Melting temperature (°C)	1287	650	842	777	727	700	1085
Thermal conductivity (W/m·K)	190	160	200	35	18	19	399
Elastic (Young's) modulus (GPa)	296	45	20.7	17.2	12.4	—	130
Coefficient of thermal expansion (10^{-6} m/m·°C)	11.3	25.4	22.8	22.5	20.6	—	16.5
Electrical resistivity (μΩ·cm)	3.7	4.5	3.4	13	35	100	1.7
Cost ($/kg), large quantities	800	2.00	4.80	4.40	100[d]	High[e]	2

[a] Transforms to BCC at 1263°C.
[b] Transforms to BCC at 455°C.
[c] Transforms to HCP at 231°C and transforms to BCC at 623°C.
[d] Also available as Fe–60% Si–22% Ba alloy, in which the Ba basis cost is $3.60 per kilogram.
[e] Ra metal is strongly radioactive and rarely produced; Ra salts are available in gram quantities or smaller.

is unstable, decaying with a half-life of only 2×10^{-16} s into two $^4_2$He nuclei. Relativistic mass shifts stabilize $^8_4$Be in the cores of massive stars when temperature exceeds about 10^8 K, but $^8_4$Be decays when its velocity is no longer an appreciable fraction of the speed of light. Thus, Be's only stable isotope, $^9_4$Be, is thought to form by collision of heavier nuclei in interstellar space. For example, a high-speed collision between $^4_2$He and $^{56}_{26}$Fe sometimes produces a $^9_4$Be nucleus in the collision fragments. Such *spallation reactions* may be the only means of synthesizing the nuclides between He and C (i.e., $^6_3$Li, $^7_3$Li, $^9_4$Be, $^{10}_5$B, and $^{11}_5$B), making these three elements somewhat rare.

Be metal was isolated in 1828 by reducing BeCl$_2$ with K metal. Be metal was largely ignored until the early twentieth century, when the fledgling aviation industry realized that its low density and high melting temperature could be useful in airframe structures. This produced a flurry of research activity during and shortly after World War I. However, when it became apparent that Be was difficult and costly to produce and had marginal ductility, aviation engineers quickly lost interest. The well-known Cu–Be precipitation-hardening alloys (Sec. 18.2.3.8) came into wide use in the 1920s, when a simple reduction reaction of mixed CuO and BeO was developed to produce master alloy directly. Cu–Be alloys remain a major use for Be. Fluorescent Be compounds saw their first widespread use in fluorescent light bulbs in the 1930s, but this use was discontinued when it was discovered that about 5% of persons exposed to Be powders develop a debilitating and even fatal allergic reaction (Sec. 11.3.4). Be's toxicity necessitates careful handling of powdered Be. Early nuclear weapons used Be–Po neutron sources to initiate the fission of U and Pu, and the low neutron cross section of Be attracted interest for use as fuel cladding, moderators, and reflectors in fission power reactors. Other materials largely replaced Be in nuclear reactor applications in the later twentieth century. Be metal first became available for general commercial use in the late 1950s. Due to its low atomic number ($Z = 4$), it possesses a low x-ray absorption cross section, making it useful in x-ray tubes. Ultrathin Be foil is used in x-ray lithography for microprocessor fabrication. Be is used as a structural material for high-speed aircraft, missiles, spacecraft, and communication satellites. Its lightness, stiffness, and dimensional stability make it useful in gyroscopes, computer parts, and instruments. BeO has a high melting point and an unusual combination of high thermal conductivity and low electrical conductivity.

11.3.2 Physical Properties of Be

The physical properties of Be are summarized in Fig. 11.1.

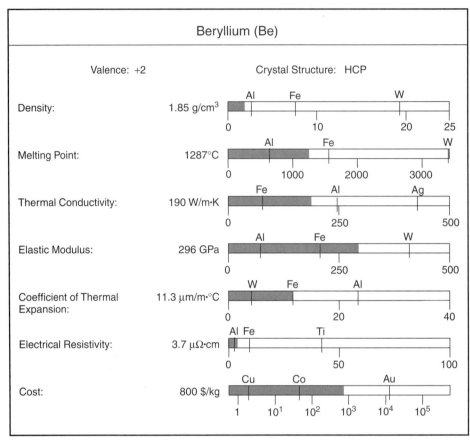

Fig. 11.1 Physical properties of Be.

11.3.2.1 Corrosion. Be reacts strongly with O_2, forming a surface layer of BeO about 10 nm thick immediately upon exposure to air that protects the metal from further oxidation up to about 600°C. Above that temperature Be diffusion through the oxide layer thickens the oxide layer with parabolic kinetics. Prolonged exposure to high-temperature air (>800°C) will eventually lead to failure. C impurity in Be can degrade oxidation resistance by forming Be_2C precipitates that slowly hydrolyze to BeO and CH_4 in humid air. This not only accelerates oxidation around the carbide particle, but produces loose BeO powder that poses a toxicity hazard (Sec. 11.3.4).

11.3.2.2 Plasticity. Be's low density, high modulus of elasticity, oxidation resistance, high conductivity, high heat capacity, and high melting temperature make it an excellent material for aerospace vehicles and other transportation applications. Unfortunately, Be's high cost, toxicity, and limited ductility restrict its use to specialty applications. The cause of Be's low ductility is its proximity to the nonmetallic elements in the upper right corner of the periodic table, which results in mixed metallic and covalent bonding. In HCP Be (Fig. 11.2), bonding *within* the (0001) basal plane is metallic, but bonding *between* (0001) planes is partly metallic and partly covalent. This causes large differences in the critical resolved shear stresses (τ_{CRSS}) among Be's three active slip systems (Fig. 11.3) and gives Be the lowest *c/a* ratio (1.568) of any HCP element. The Poisson ratio of Be is only 0.02, much lower than the 0.25 to 0.40 Poisson ratios of most metals.

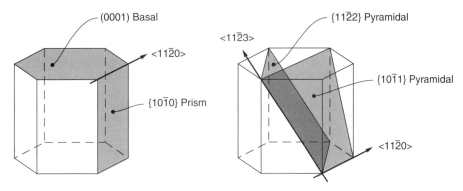

Fig. 11.2 Slip planes and directions in the HCP unit cell.

The von Mises criterion for high ductility in a polycrystalline metal requires five active, independent slip systems to accommodate the complex shape changes imposed by neighboring grains upon one another (Sec. 4.6). If four or fewer independent slip systems are active, the metal will elongate only a few percent in tension before fracturing from strain incompatibilities at grain boundaries. At 20°C, Be slips readily on the basal plane (0001) in the $\langle 11\bar{2}0 \rangle$ direction, which provides two independent slip systems. First-order prism slip $\{10\bar{1}0\}$ occurs at a much higher τ_{CRSS} value, which provides two more independent slip systems, but the Burgers vector for prism plane slip is also the $\langle 11\bar{2}0 \rangle$ direction. Only when second-order pyramid planes $\{11\bar{2}2\}$ slip will the necessary fifth independent slip system be provided with a Burgers vector $(1/3\langle 11\bar{2}3 \rangle)$ that does not lie in the basal plane. The τ_{CRSS} for Be pyramidal slip is about 1000 times greater than for basal slip (Fig. 11.3), which impairs ductility in polycrystalline Be below about 200°C. At room temperature Be slips preferentially on the basal plane and fails by brittle fracture on the (0001) plane after only a few percent elongation. This poses serious design limitations on Be structural use (Sidebar 11.1). In the 1950s and 1960s numerous investigators attempted to improve Be's ductility by improving purity, but these efforts served only to confirm that the ductility limitations of Be are inherent in its bonding. Reducing impurity atom concentrations yields only modest ductility improvement.

In some metals, ductility is enhanced by twinning reactions that reorient the crystals for more favorable dislocation slip to occur. Unfortunately, this is not the case in Be. Be twins on $\{10\bar{1}2\}$

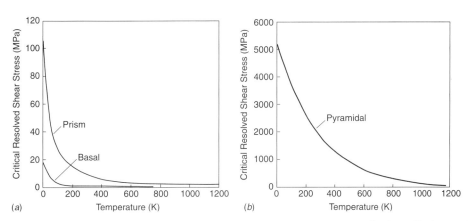

Fig. 11.3 Critical resolved shear stress (τ_{CRSS}) for (a) basal and prism slip in HCP Be, and (b) pyramidal slip in HCP Be. The high τ_{CRSS} for pyramidal slip is a major contributor to Be's limited room-temperature ductility.

> **SIDEBAR 11.1: IMPORTANCE OF DUCTILITY IN ENGINEERING DESIGN**
>
> Nearly all engineering parts are designed for use only in the elastic regime. In fact, after design safety factors are applied, most components never exceed 30 to 50% of their yield strength. Thus, it may seem odd that ductility is considered important in selecting materials for parts where no plastic flow is intended. Ductility is important because small regions of the part are likely to exceed the yield stress even when the *overall* stress is well under the yield strength. For example, in a threaded assembly, normal dimensional variations within dimension tolerances and in response to temperature changes can cause as much as 3 to 4% plastic flow in small volumes near thread roots. If the material cannot deform plastically 3 to 4%, the threaded assembly may crack, even though the macroscopic load is well below the material's yield stress. Even in unthreaded parts with simple shapes, these same dimensional factors require material with at least 2% tensile elongation to accommodate loading imbalances and size variations without cracking. Design engineers generally consider 15% tensile elongation the criterion for "true ductility" because parts with stress-intensifying geometries (e.g., notches, corners, or flaws) can require as much as 15% tensile elongation to assure that the material will bear the design load reliably.
>
> Ductility is also vital for safety-critical components that must "fail gracefully" if they are accidentally overloaded. It is costly to retire an aircraft landing gear assembly bent by a bad landing, but that is a better outcome than a fracture that causes the aircraft wing to strike the runway at 200 km/h. Ductility is also needed in less dramatic circumstances that occur routinely in service. Something as mundane as rapid thermal expansion from a quick temperature change or a wrench dropped accidentally onto a part will simply bend or dent a ductile component, but the same incident may crack a brittle material, causing subsequent fracture in service.

planes in the $\langle 10\bar{1}1 \rangle$ direction with a τ_{CRSS} value of 40 MPa; the shear from twin formation is large (0.189). Formation of twins at room temperature leads to nucleation of cleavage cracks along the (0001) and twin boundaries, so the nonbasal shear motion from twinning does not enhance ductility in Be.

Be's low-ductility problem is exacerbated by its tendency to form large columnar grains when cast. Grains as large as 10 mm are common in as-cast Be ingots. Such large grains accumulate dislocation pile-up stresses that lead to even lower ductility, and wrought Be must be given multistage hot-working treatments to achieve a finer, more equiaxed grain structure via recrystallization. A frequently used alternative approach is to form Be components by powder metallurgy. This assures small, equiaxed grains, although it introduces the new problem of BeO particles at the grain boundaries that can serve as crack initiation sites. Recent trends in Be fabrication favor very fine grain sizes (10 μm or smaller) achieved through P/M processing of electrolytically refined Be powder with low O content (to minimize BeO formation). This provides Hall–Petch strengthening (Sec. 3.2) and improves ductility. Tensile elongations of 5% with yield strengths of 500 MPa can be achieved by this approach.

It should be noted that large improvements in tensile elongation in one or two dimensions are possible when the metal is highly textured. Extruded Be rod, for example, often has more than 10% elongation when tensile tested parallel to the extrusion axis, but ductility in the transverse direction is quite low in such material. Similarly, textured Be sheet can achieve higher ductility in the plane of the sheet, but ductility normal to the sheet is near zero. Consequently, this approach does not improve the overall performance of a part subject to triaxial loading, which occurs in most engineering assemblies.

Be fracture is complex. Calculated surface energy is lowest on the (0001) basal planes at 2300 mJ/m^2, but cleavage sometimes occurs on the $\{11\bar{2}0\}$ second-order prism planes (surface energy = 6300 mJ/m^2) and on the $\{11\bar{2}2\}$ twinning plane (surface energy = 11,900 mJ/m^2).

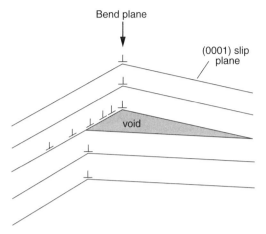

Fig. 11.4 Depiction of bend plane splitting in Be. The (0001) basal planes are seen here in edge view. Pile-ups of dislocations on the basal planes cause voids to form that lead to fracture.

Fracture typically occurs by the *bend plane splitting* mechanism illustrated in Fig. 11.4. Dislocation arrays form low-angle grain boundaries lying parallel to the c axis within a grain. These occur near obstacles such as twins or grain boundaries. When additional dislocations approach such an array, they generate forces normal to the basal plane that cannot be relieved by slip at an angle to the basal plane due to the very high τ_{CRSS} value for pyramidal slip. The result is cleavage along the basal plane, as shown in the optical micrograph in Fig. 11.5. Be does, however, possess some of the fracture toughness of a classic metal. Specimens that are sectioned after plastic deformation but before failure usually show many microcracks that have halted at grain boundaries, a clear indication that neighboring grains have served as crack stoppers.

The Stroh relation models Be's bend plane splitting mechanism and predicts that refining grain size should improve both yield strength and ultimate tensile strength:

$$\sigma_N = \frac{4\gamma G \cos\theta}{\pi D \tau_s}$$

where τ_s is the effective resolved shear stress, σ_N the applied normal stress, γ the surface energy of new surface created by cleavage, G the shear modulus, θ the angle between the basal plane

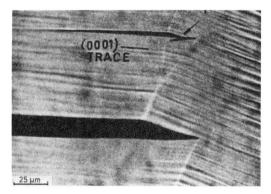

Fig. 11.5 Optical micrograph of basal plane cracks formed by bend plane splitting in Be. The (0001) basal planes are seen here in edge view. Compare to Fig. 11.4. (From Damiano et al., 1969.)

and the tensile axis, and D the dislocation slip distance (grain size in polycrystalline specimens or specimen diameter in single crystals). Studies with high-purity Be show that yield strength rises from 175 MPa for a grain size of 20 μm to 650 MPa for a grain size of 2 μm. Ductility improves substantially with decreasing grain size as well, but the behavior below about 10 μm grain size is complicated by other factors, such as BeO inclusion content.

11.3.2.3 Conductivity and Heat Capacity. Despite a low electron density at the Fermi energy level (Fig. 1.10*b*), Be is an excellent thermal and electrical conductor. Below 200 K, Be actually conducts electricity better than Cu. Be's low atomic weight gives it the highest heat capacity per unit mass (1825 J/kg·K) of all structural metals; this value is much higher than that of other highly conductive metals (e.g., Cu, 386 J/kg·K or Al, 900 J/kg·K). This combination of high conductivity and high heat capacity makes Be an outstanding heat sink in applications where low mass is important, such as nuclear weapons and spacecraft (Sidebar 11.2).

11.3.2.4 Impurities. The unusual bonding within the HCP Be unit cell and Be's small atomic radius make it difficult to dissolve other elements into Be. Only a few metallic elements have more than 1 at% solubility in Be at equilibrium. At high temperature, Be can dissolve 10 at% Cu (Fig. 18.12), 5 at% Ni, and 2 to 3 at% Co, Mn, Pd, and Ag. Elements at the right end of the transition elements (e.g., Cu and Ag) have the best solid solubility because the Be 2*s* electron can be accommodated in the transition metal *d* band, and this action becomes easier when the *d* band is nearly full. Elements with completely filled *d* bands (Zn, Cd, and Hg) have essentially zero solubility in Be. Interstitial solubility is also low in Be, since Be's octahedral and tetrahedral interstitial sites are smaller than O, C, and N atoms; when these elements are present in Be, they form BeO, Be_2C, and Be_3N_2. H fits in the Be octahedral site, but its solubility is low. For this reason, pure Be is the most commonly used form of the metal, although some use has been made of Al–Be metal–metal composites and Be–BeO composites.

Theorists estimate that 8 at% or more monovalent solid solution impurity atoms in Be could sufficiently alter the bonding in Be to achieve high ductility at room temperature. Since even Cu has solid solubility of less than 3 at% at room temperature (Fig. 18.12), there is no prospect of achieving such an alloy at equilibrium at room temperature. High-fluence ion implantation studies have been performed to produce nonequilibrium solid solutions with as much as tens of percent Ag, Ti, Zn, and Cu in Be, but these efforts produced only thin films that are difficult to test for ductility, and the prospects for achieving ductile Be in such metastable alloys remain unexplored.

Commercial Be's principal impurities are O (as BeO), Fe, Al, and Si. Al is insoluble in solid Be and segregates to grain boundaries, where it lowers ductility, particularly above 400°C. At these temperatures Be normally has high ductility, so Al impurity produces behavior akin to hot shortness. Fe has limited solid solubility in Be, and Fe strengthens Be by solid solution hardening and formation of fine precipitates of $FeBe_{11}$, which can induce a yield point drop in Be (Sec. 6.2). Be producers usually adjust Al and Fe impurity contents to maintain 2(Fe wt%) = (Al wt%). This composition provides enough Fe to "purge" Al metal from the grain boundaries

---

SIDEBAR 11.2: PROJECT MERCURY HEAT SHIELDS

The first U.S. manned spacecraft used on a 1961 Project Mercury suborbital mission had a Be heat shield to protect astronaut Alan Shepard from atmospheric reentry heat. The capsule's designers had experience using Be's high heat capacity to build missile warhead reentry structures, and the heat shield for the 1961 *Freedom 7* flight was the largest Be forging ever made at the time. Shortly after the suborbital Mercury flights ended, it was found that glass-fiber and Al honeycomb structures were more effective at dissipating reentry heat than Be, because they ablated during reentry, allowing particles of incandescently hot material to fly off the surface as reentry proceeded, carrying heat away with them.

by preferentially forming an AlFeBe$_4$ precipitate at grain boundaries. AlFeBe$_4$ is high melting and does not degrade mechanical properties. Other transition metal impurities (e.g., Ni, Mn, Co) can partially substitute for Fe in this precipitate. Si is a valuable sintering aid in P/M Be products, lowering hot pressing times and temperatures. However, P/M parts produced with a Si sintering aid have lower creep strength, due to formation of small pockets of low-melting Al–Mg–Si ternary mixtures at grain boundary triple points. Small amounts of BeO (0.5 to 1%) can usually be tolerated, and the BeO particles can be beneficial by retarding undesired recrystallization. However, as BeO content rises above 1%, the oxide particles act as cleavage initiation points, degrading ductility and fracture toughness. BeO is sometimes raised to high levels deliberately to enhance strength and stiffness for applications where ductility is not essential. Vacuum hot pressing of electrolytically refined Be powder is used for the majority of engineering applications, although wrought Be is employed as well.

11.3.3 Applications of Be

11.3.3.1 X-ray Equipment Use. Absorption of x-rays is lowest for elements with low atomic numbers:

$$I = I_0 e^{-C\rho d}$$

where I is the x-ray intensity transmitted through material of thickness d (cm), I_0 the incident x-ray intensity, C the absorption cross section (cm^2/g), and ρ the material density (g/cm^3). Values of C are lowest for elements with low atomic numbers (Z), so low-Z elements are used to make windows transparent to x-rays in x-ray generation tubes, x-ray lithography systems, and electron microscopy analytical devices. Of all the elements with $Z < 12$, only Be ($Z = 4$) has good structural strength and reasonably good fracture toughness. Consequently, Be foils are the dominant material for this application.

11.3.3.2 Nuclear Use. In the 1940s, Be shifted from a laboratory curiosity to a strategic metal because of its value in nuclear weapons and fission reactors. Early atom bombs used a $^{210}_{84}$Po radioactive source to bombard $^{9}_{4}$Be with alpha particles ($^{4}_{2}$He); this generates neutrons to start the fission chain reaction in Pu or U metal ($^{9}_{4}$Be + $^{4}_{2}$He → $^{8}_{4}$Be + $^{4}_{2}$He + $^{1}_{0}$n). This neutron emission reaction takes place in only one of every 12,000 collisions, so a strong alpha source such as $^{210}_{84}$Po is required to achieve the necessary neutron flux to trigger the explosion. The short half-life of $^{210}_{84}$Po (138 days) necessitated frequent replacement, and other neutron sources based on $^{2}_{1}$H–$^{3}_{1}$H fusion have replaced the early Be–Po triggers.

Be also serves as a neutron reflector to surround the core of both fission and fusion weapons and reflect neutrons back into the reaction during the crucial first microsecond of the fission explosion (Sec. 24.4.2.2). Although several materials can serve as reflectors (and the closely related function of inertial tamper), Be's low mass is an advantage for reflectors in weapons delivered by aircraft or missiles. During the 1950s and 1960s, Be was also used as a nuclear fuel cladding in fission reactors. Its low neutron absorption cross section and self-protective oxide made it well suited to this use. However, its limited ductility and its tendency to generate damagingly large He concentrations within the metal from neutron reactions put it at a competitive disadvantage to Zr alloys in reactors with long burn periods (i.e., essentially all commercial power reactors).

11.3.3.3 Aerospace Structural Applications. Aerospace engineers demand exceptional materials performance in both airframe and engine applications. Specific strength (σ_y/density) and specific stiffness (E/density) are critically important in aircraft performance and efficiency, and Be excels by both of these measures. For particularly demanding performance requirements, Be is often selected, despite its high cost, limited ductility, and the special handling requirements required by its toxicity. It is used in applications such as military aircraft disk brake rotors, gyroscopic inertial guidance systems, space-based telescope mirrors and frames, and structural elements in communications satellites, the Space Shuttle, and the International Space Station. For most of these applications, pure Be is used. Some use has been made of Al–Be metal–metal composites for applications that do not involve high temperatures; these composites have cost, density, and mechanical properties intermediate between those of Al and Be metal.

11.3.3.4 Electronic Applications. Be–BeO composites known by the trade name E Materials can solve thermal management problems in electronic packaging. These composites contain as much as 70 vol% BeO; their low density, high thermal conductivity, and tailorable coefficient of thermal expansion make them useful substrates for circuits with high heat output. In these applications ductility is of little importance.

11.3.4 Toxicity of Be

Inhalation of Be-laden dust or fumes can cause lung inflammation and related medical problems known as *berylliosis*. Only 2 to 6% of the population is susceptible to berylliosis, but the risk demands that all Be powder handling be done in accordance with stringent safety procedures. Bulk Be metal poses little risk, although it becomes hazardous if it is corroded or abraded to generate Be-bearing powders. Acute berylliosis results from a high dose of inhaled Be dust and is marked by a sudden onset of coughing, shortness of breath, and lung inflammation and rigidity, sometimes accompanied by eye and skin injuries. Acute berylliosis can be treated by using ventilators to assist breathing and administering corticosteroids to minimize lung inflammation. Although there have been fatalities, acute berylliosis patients are usually severely ill for 7 to 10 days but recover with treatment with no lasting ill effects.

Chronic berylliosis is an allergic reaction to long-term Be dust exposure. The victim's lungs form abnormal tissue and lymph nodes enlarge. In some cases, other parts of the body are damaged. Chronic berylliosis symptoms are essentially the same as those of acute berylliosis, but they develop over months or years. Diagnosis can be difficult since the symptoms are similar to those of other pulmonary diseases and may not appear until 20 years after exposure. The disease is incurable, causes progressive deterioration of lung function, and may lead to premature death by heart failure or cancer. Corticosteroid therapy is sometimes used, but its ability to slow the progression of the disease is uncertain, and it does not alleviate scarring of lung tissue. Drugs that bind with Be and allow it to be excreted (chelation therapy) are currently under study. One of the authors (A.R.) had a mentor and colleague who battled berylliosis for much of his adult life and finally died from the disease. It was a painful sight to watch this once vibrant man struggle for breath after the simple act of climbing a single flight of stairs.

11.3.5 Sources of Be and Refining Methods

Total world Be use typically varies between 200 and 300 tons/yr. Although the metal's low density makes tonnage figures somewhat deceiving (a ton of Be is more than half a cubic meter of metal), Be usage by any measure is tiny compared to that of major commercial metals. Be is present in over 30 minerals; the most commercially important of these are bertrandite, $4BeO \cdot 2SiO_2 \cdot H_2O$ and beryl, $Be_3Al_2(SiO_3)_6$. All Be-bearing minerals are relatively scarce. Aquamarine and emerald are familiar gemstones comprised of beryl. Refining processes are complex and sometimes even involve hand-sorting of ore particles. The final reduction step is reduction of BeF_2 by Mg. World Be mine production is less than world consumption because Be is assiduously recycled, and substantial government stockpiles exist.

11.3.6 Structure–Property Relations in Be

11.3.6.1 Cross-Slip and Dislocation Locking in Be. As in most close-packed metals, Be dislocations on both (0001) basal and $\{10\bar{1}0\}$ prism planes frequently separate into two partial dislocations separated by a stacking fault (Sec. 4.5):

$$\tfrac{1}{3}[11\bar{2}0] \rightarrow \tfrac{1}{3}[01\bar{1}0] + \tfrac{1}{3}[10\bar{1}0]$$

Be's stacking fault energy is high; partial dislocations are spaced only about three Burgers vectors apart on the basal plane, which is too close for resolution in the transmission electron microscope. Stacking fault energy is higher still on prism planes (Appendix B1, *ftp://ftp.wiley.com/public/ sci_tech_med/nonferrous*), and this difference causes a cross-slip dislocation locking phenomenon in Be. Near room temperature (170 to 320 K), thermal fluctuations allow partial dislocations on prism planes to collapse the stacking fault and constrict into one perfect dislocation. When this occurs, the perfect dislocation can then cross-slip onto the basal plane. However, as it begins to move on the basal plane, it again dissociates into two partial dislocations, and since stacking fault energy is lower on the basal plane, the separation of the partials is greater on the basal plane than it was on the prism plane. In fact, the spacing on the basal plane is too great to allow the dissociated dislocation on the basal plane to constrict to a perfect dislocation by thermal fluctuations. Thus, the cross-slipped dislocation is now effectively "trapped" on the basal plane, unable to cross-slip back to a prism plane (Fig. 11.6). This causes the screw portions of prism plane dislocation loops to become pinned after cross-slipping onto the basal plane, while the edge component of the loop (edge dislocations cannot cross-slip) continues to move freely on the prism plane. At temperatures above 320 K (about 45°C), the partial dislocations trapped on the basal plane have sufficient thermal energy to constrict to a perfect dislocation. This allows them to cross-slip back to prism planes again, effectively "unlocking" them.

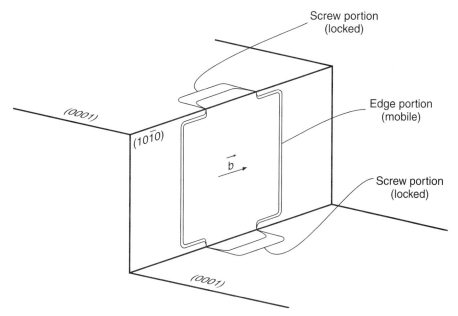

Fig. 11.6 Screw portions of a dissociated dislocation loop on the plane in Be become locked by cross slipping onto the (0001) plane, which has a lower stacking-fault energy. After the dislocation cross-slips, it dissociates into partial dislocations again, and the spacing between those two partials is too great to allow thermal fluctuation of atom positions to constrict the partials back into a perfect dislocation. Although the screw portions of the original loop become locked in their new slip plane, the edge portions remain mobile on their original slip plane because edge dislocations cannot cross slip.

11.4 MAGNESIUM

> And whoever aspires to the heights and wants to fly must cast off much that is heavy and make himself light—I call it a divine capacity for lightness.
>
> —Friedrich Nietzsche, *Sämtliche Werke: Kritische Studienausgabe*, Vol. 10

11.4.1 History of Mg

The first Mg metal was a tiny, low-purity bead produced by Sir Humphry Davy by electrolyzing mixed MgO and HgO. It was considerably later in the nineteenth century when Bunsen's electrolytic reduction of $MgCl_2$ led to the first commercial Mg production in Germany. Larger-scale production began in the United States in 1916, when Dow Chemical Co. began electrolyzing $MgCl_2$ from well brine in Michigan, and thermal processes followed in the mid-twentieth century. Today several Mg production processes are available to reduce Earth's enormous quantities of Mg-bearing salts and minerals, but none match the efficiencies of the Hall process (Sec. 20.2.7) used to produce Al. Consequently, Mg is more costly and less heavily utilized than its light metal rival, Al. Although Mg has a substantial density advantage, Al's lower cost, better corrosion resistance, and greater strength and ductility prevent Mg from seriously challenging Al's dominance of the light metals marketplace. Global Al production is about 40 times greater than Mg production.

11.4.2 Physical Properties of Mg

Mg is the lowest-density metallic element with sufficient strength and corrosion resistance for general structural use (Fig. 11.7). It is relatively inexpensive; easy to cast, machine, and weld; and available in essentially inexhaustible quantities in Earth's crust and oceans. It has an HCP crystal structure with rather limited room-temperature ductility and good hot ductility. Its conductivity and heat capacity are relatively high. It also provides high damping capacity to suppress noise and vibration. Its high chemical reactivity is often exploited for use in reducing the compounds of other metals and in sacrificial anode corrosion protection systems.

11.4.2.1 Corrosion. Mg is positioned near the bottom of the standard electromotive force emf series (Table 11.2), making it anodic to all commonly used commercial metals. MgO forms rapidly on exposed Mg metal, and the oxide is only partially protective of the underlying metal. Mg was once thought to have inherently poor corrosion resistance, but it is now clear that its corrosion rate is unusually sensitive to certain impurity elements, particularly Fe, Ni, and Cu. In the 1980s, higher-purity Mg alloys became commercially available, and these have

TABLE 11.2 Mg Is the Most Anodic Element Compared to Other Common Metals in the Standard Electromotive Force Series

Electrode Reaction	Standard Electrode Potential (V)
$Ag^+ + e^- \rightarrow Ag$	+0.800
$Cu^{2+} + 2e^- \rightarrow Cu$	+0.340
$Pb^{2+} + 2e^- \rightarrow Pb$	−0.126
$Ni^{2+} + 2e^- \rightarrow Ni$	−0.250
$Fe^{2+} + 2e^- \rightarrow Fe$	−0.440
$Zn^{2+} + 2e^- \rightarrow Zn$	−0.763
$Al^{3+} + 3e^- \rightarrow Al$	−1.662
$Mg^{2+} + 2e^- \rightarrow Mg$	−2.363

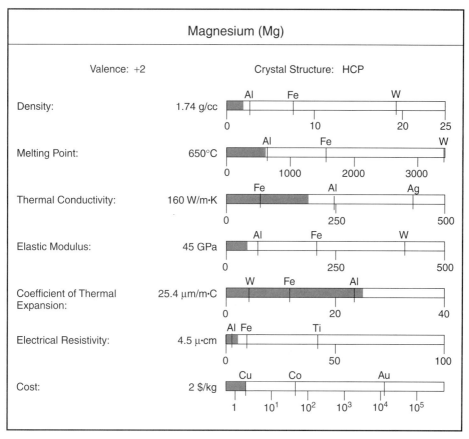

Fig. 11.7 Physical properties of Mg.

sharply reduced corrosion problems. Most Mg products are cast, usually from steel crucibles into steel molds, which makes some Fe contamination inevitable, even when the Mg starts with excellent purity. However, $MnCl_2$ additions to the liquid metal remove the Fe from solution by precipitating a benign AlMnFe intermetallic compound. Adding sufficient Mn to achieve a ratio of Fe/Mn ≤ 0.032 assures nearly complete precipitation of Fe from the alloy. This dramatically improves corrosion resistance (Fig. 11.8) for the most widely used Mg alloy, AZ91 (Sidebar 11.3). High-purity alloy AZ91D has overall corrosion performance somewhat better than that of ordinary steel. The corrosion performance of these new-generation, high-purity Mg alloys is similar to that of Al alloys in salt spray tests. Several anodizing treatments, chromate coatings, and polymeric coatings have been developed to further protect Mg from corrosion, even from abrasive environments and in hot air and hot lubricating oils.

Zr can also be added to Mg to reduce the harmful effects of Fe, Si, and to some degree Ni on corrosion performance. Y, Nd, and Dy promote a more strongly protective oxide coating and also deliver better elevated temperature creep strength (Sec. 11.4.3.1). Despite these advances, however, Mg's low position on the electromotive force scale must still be borne in mind in designing Mg assemblies; even the best Mg alloys are vulnerable to galvanic corrosion if they are in electrical contact with a more noble metal such as Fe. Rapid corrosion results when galvanic couples (e.g., a steel bolt used to join Mg plates) are exposed to liquid water. Even a thin film of water condensed from humid air is sufficient to cause serious corrosion in such galvanic couples.

SIDEBAR 11.3: ALLOY DESIGNATIONS FOR Mg

Mg alloys are designated by a mixed alphanumeric code. The first two symbols are letter abbreviations for the two alloying additions present in the highest concentrations:

$$
\begin{array}{ll}
A = Al & N = Ni \\
B = Bi & P = Pb \\
C = Cu & Q = Ag \\
D = Cd & R = Cr \\
E = \text{rare earth} & S = Si \\
F = Fe & T = Sn \\
H = Th & W = Y \\
K = Zr & Y = Sb \\
L = Li & Z = Zn \\
M = Mn &
\end{array}
$$

The next two symbols are numbers that indicate the weight percentages of these two alloying additions. The fifth symbol is a letter describing a standard alloy within the broad composition range indicated by the first four symbols. Finally, the temper designation is indicated by the same coding system as that used to designate temper in Al (Sidebar 20.2). An example of a complete Mg alloy designation is QH21A-T6. This alloy contains approximately 2 wt% Ag and 1 wt% Th. The "A" designation specifies a particular composition of one standard version of this alloy; the exact composition would have to be obtained from a handbook or Web site. The "T6" code indicates that the alloy has been solutionized, quenched, and aged artificially.

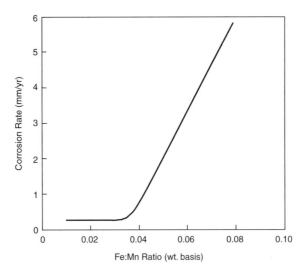

Fig. 11.8 Salt spray test corrosion rate of Mg alloy AZ91 (Mg–9% Al–1% Zn) with low Ni and Cu content and varying Fe/Mn ratios. When sufficient $MnCl_2$ is added to molten AZ91 (Fe/Mn $\leq$ 0.032), Mn precipitates the Fe as intermetallic compounds in the melt, removing Fe from solid solution, where it accelerates corrosion.

Mg liquid or finely divided solid can ignite and burn with an intense, white-hot flame. Both the O_2 and N_2 in air react when Mg burns, forming MgO and Mg_3N_2. In the early years of Mg use, metal fires gave Mg a reputation as a dangerous metal. Safe handling procedures for Mg are now used, making Mg fires rare, although some engineers still bear an unreasonable prejudice against Mg use for this reason. Use of SF_6 cover gas in Mg casting operations reduces liquid metal fire risk and also improves the corrosion resistance of the cast alloys. SF_6 allows Mg to be cast without a liquid flux; this eliminates flux inclusions in the cast metal that can accelerate Mg corrosion. (Unfortunately, SF_6 is a greenhouse gas, so Mg foundries strive to reduce SF_6 escape from the casting process or to replace it with SO_2, to minimize global warming.)

11.4.2.2 Mechanical Properties. Mg has limited room-temperature ductility. Deformation is essentially limited to (0001) slip and twinning on the $\{10\bar{1}2\}$ plane. At room temperature Mg's τ_{CRSS} value for basal slip is 0.6 MPa, whereas τ_{CRSS} for pyramidal slip is 44 MPa and for $\{10\bar{1}0\}$ slip is 60 to 100 MPa. Mg tolerates 5 to 10% reduction by cold rolling, but heavy cold work fractures the metal. Above 250°C, Mg is highly ductile because the $\{11\bar{2}0\}$, $\{10\bar{1}0\}$, and $\{10\bar{1}1\}$ slip planes become active (Fig. 11.2); consequently, most Mg deformation is performed as hot work. Adding solutes to Mg (e.g., Zn, Al) lowers prism plane τ_{CRSS} and raises basal plane τ_{CRSS}, so solutes have the somewhat unusual effect of improving ductility in Mg. Mg's elastic modulus is low (45 GPa), but its density is so low that it provides much greater stiffness in a structural member than does an equivalent mass of steel, Ti, or Al. Thus, if the design of an assembly has space for the bulkier Mg component, Mg can deliver much greater bending resistance than that of the same mass of a heavier metal.

Twinning plays a major role in Mg plastic deformation. Because τ_{CRSS} for pyramidal slip (44 MPa) is so much higher than for basal slip (0.60 MPa), slip is difficult in grains unfavorably oriented for basal slip, much as it is in Be (Sec. 11.3.2.2). However, in Mg, twinning reorients the lattice to facilitate basal slip on the newly oriented grains, and twinning occurs so frequently that a significant percentage of Mg's volume is reoriented by twinning during plastic deformation. The result is a substantial ductility advantage of Mg over Be, even though both metals are largely constrained to basal slip at room temperature.

11.4.2.3 Castability. Mg's ductility limitations have not greatly restricted its engineering use, because most Mg components are produced by casting. Mg casting can be divided into three types: die casting, sand casting, and thixomolding. Fabricating a part such as the safety housing for a lawn mower blade would require severe deformation of plate for most metals, but Mg has such excellent casting properties that it can be cast into that shape. Mg and its alloys have exceptionally low viscosity, allowing the metal to fill long, narrow mold cavities that are difficult to use with most other metals. Its relatively low melting temperature (650°C for pure Mg, somewhat lower for alloys) allows use of hot-chamber die casting, and Mg's minimal reactivity with steel below 700°C permits use of inexpensive steel crucibles and molds. Although die casting is the most frequently used Mg casting method, sand mold casting is used for low-volume production of larger parts, and thixocasting (Sidebar 8.1) has gained wider use in recent years.

Mg sand-mold casting poses a special challenge. Mg metal reduces SiO_2, and the alloys will attack the sand in the mold, producing Si metal and MgO. To prevent this, the sand is usually coated with an inhibitor such as potassium fluoroborate or sodium silica fluoride to prevent the reaction. Mg alloys are also well suited for thixomolding, a casting process in which partially solid, partially liquid metal is injected into molds to achieve thin-walled, high-density components with complex shapes. Thixomolded components are used in the automotive, electronics, power tool, and computer industries; they have less shrinkage and a lower incidence of porosity than those of other castings. Examples of thixomolded parts include television cases, laptop computer housings, eyeglass frames, hand drills, and steering wheels.

A persistent problem in Mg alloy casting has been the variability of the strength and ductility of castings produced under seemingly identical conditions. Since Mg alloy ductility at room temperature is always somewhat low, Mg is especially susceptible to the deleterious effects of shrinkage porosity, gas (air) porosity, and entrained oxide films and flux (Sec. 8.2.3). Failure tends to occur by fracture along paths connecting regions of high casting defect density, and the variability of defect concentrations and location causes the variation in mechanical properties often seen in Mg castings. For this reason, tight control of casting parameters is particularly important in Mg castings to assure predictable product performance.

11.4.2.4 Conductivity and Heat Capacity. The thermal and electrical conductivity of Mg and its alloys are fairly high, and it has a heat capacity of 1020 J/kg·K, which exceeds that of most other good conductors (e.g., Cu, 386 J/kg·K, or Al, 900 J/kg·K). These factors combine to make Mg an efficient, lightweight heat sink with good vibration and noise reduction capability.

11.4.3 Applications of Mg

11.4.3.1 Cast Mg Alloys. Mg's excellent casting behavior makes this the preferred forming process for many Mg parts; one-third of all Mg production is used to make Mg casting alloys. Relatively few elements have extensive solid solubility in Mg due to Mg's low electronegativity and large atomic radius (0.160 nm). However, several elements have enough solubility to be useful alloying additions, including Ag, Al, Cd, Ga, Li, Pb, Pu, rare earth metals, Sc, Th, Tl, Y, Zn, and Zr. Unfortunately, with the exception of Al and Zn, all these metals are expensive and/or toxic. Mn is also present in many Mg alloys, although its solid solubility is less than 1%. Mn has little direct effect on mechanical properties but is used to remove Fe from solution (Sec. 11.4.2.1). Si solubility is nil, but Si forms Mg_2Si precipitates that increase creep strength by reducing grain boundary sliding.

AZ (Al–Zn) Alloys. The most frequently used Mg alloy, AZ91, contains Al (9.0%) and Zn (0.7%) with sufficient Mn added to remove Fe impurity (Sec. 11.4.2.1). Al enhances Mg's castability and improves strength and corrosion resistance. The Mg–Al system (Fig. 11.9) presents seemingly

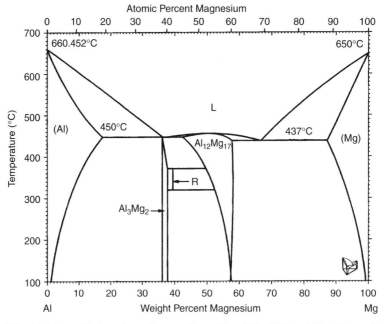

Fig. 11.9 Al–Mg equilibrium phase diagram. (From Massalski, 1986, p. 160; with permission.)

Fig. 11.10 Transmission electron micrograph of peak-aged $Mg_{17}Al_{12}$ intermetallic precipitates in AZ91 Mg alloy. The alloy was solutionized at 425°C, water quenched, and aged 70 hours at 175°C. Continuous, incoherent precipitates lying parallel to the (0001) primary slip plane are visible on the left side of the micrograph, and cellular, discontinuous precipitates at a grain boundary appear at the right. $Mg_{17}Al_{12}$ precipitates raise hardness only slightly, and when they form on grain boundaries, they reduce creep strength. (From Zheng et al., 2003; with permission.)

ideal phase fields for solutionizing, quenching, and aging to produced a precipitation-hardened alloy (Sec. 3.5). Unfortunately, the precipitate that forms from such heat treatment is a relatively coarse, incoherent $Mg_{17}Al_{12}$ intermetallic compound that forms parallel to the (0001) slip plane (Fig. 11.10). Zn refines the precipitate, making it a somewhat more effective strengthening particle, but even so, the precipitation-hardened alloy (designated AZ91E-T6) has a yield strength of only 145 MPa and 275 MPa ultimate strength. Die-cast AZ alloys usually have relatively fine grain size, since die castings cool rapidly. In sand-mold AZ alloy castings, the slower cooling rate can lead to unacceptably large grain sizes, so C-based grain refining additives such as CaC_2 or hexachloroethane are often added for grain refinement. The Zr grain refining additives used in some other Mg alloys (see below) cannot be used in AZ alloys because Zr and Al form undesirable intermetallics.

Other Mg–Al alloys, such as AM60 (Mg–6% Al–0.15%Mn), provide better ductility than AZ91, but the strength is lower. AM60 has a yield strength of 125 MPa, ultimate strength of 210 MPa, and 8 to 13% elongation. The enhanced ductility of AM60 makes it more fracture resistant; this alloy is often used for cast "mag" wheels on high-performance automobiles.

ZK (Zn–Zr) Alloys. Zn can precipitation harden Mg by formation of platelike GP zones (Sec. 20.2.4.4) on Mg $\{11\bar{2}0\}$ and (0001) planes, followed by formation of two metastable transition precipitates, β'_1 ($MgZn_2$ needles) and β'_2 ($MgZn_2$ plates). The final equilibrium β phase (MgZn) appears in overaged ZK alloys. The strengthening effect apparently results from dislocations shearing through underaged and peak-aged precipitates or from Orowan looping around overaged precipitates (Fig. 20.16). ZK61A-T6 (Mg–6% Zn–0.8% Zr) has a yield strength of 195 MPa, ultimate tensile strength of 310 MPa, and 10% elongation. However, the ZK alloys are somewhat more difficult to cast than AZ alloys; cast ZK alloys are vulnerable to microporosity, and the alloy cannot be welded because the welding heat vaporizes the Zn. Addition of 1.5 to 3 wt% Cu to these alloys [e.g., ZC63 (Mg–5.9% Zn–3.2%Cu)] produces a lamellar eutectic at the grain boundaries that reduces the problems of microporosity but increases density slightly (1.87 g/cm^3) and lowers strength.

Fluorzirconate is often added to the melt of Mg–Zn alloys to produce about 0.6 wt% Zr in the final casting. Mg reduces fluorzirconate, releasing Zr metal into the alloy, and small Zr crystals begin to nucleate and grow in the liquid metal. Mg and Zr have nearly identical HCP lattice parameters, and the solid Zr particles in the molten alloy provide nuclei for Mg crystallization, increasing the number of Mg grains and reducing their average size. Although the presence of Zr actually lowers the strengthening effect of the precipitates, the benefits of grain refinement on strength and ductility more than compensate for that loss.

Creep-Resistant Alloys (AS41, EZ33, ZE41, HK32, WE43, QE22). The low density of Mg improves fuel efficiency and performance in motor vehicles and aircraft. However, for many transportation applications, the low creep strength of AZ and ZK Mg alloys causes problems in components such as gearboxes, where temperatures often exceed 130°C, especially in crankcases and engine blocks, where 170 to 200°C is typical. A transmission housing or valve cover, for example, requires a specific torque to tighten the bolts used in assembly. If the Mg creeps from long-term exposure to engine heat, the torque on the bolts gradually diminishes, causing fluid leaks and increased noise and vibration. Several Mg alloys have been developed to improve strength and creep resistance at or above 200°C. These alloys use rare earth, Th, or Ag solutes to form precipitates stable at these temperatures; they are significantly more expensive than most other Mg alloys.

Although Al has several salutary effects in Mg alloys, high Al content promotes formation of low-melting $Mg_{17}Al_{12}$ eutectic at grain boundaries, making them more vulnerable to grain boundary sliding. Thus, $Mg_{17}Al_{12}$ at grain boundaries sharply degrades creep strength. Addition of about 1% Si to Mg–Al alloys produces the more refractory Mg_2Si precipitate at grain boundaries that partially offsets the deleterious creep effects of $Mg_{17}Al_{12}$. Si additions in alloys such as AS21 and AS41 improve creep resistance at low cost.

Better creep resistance is possible by addition of rare earth metals (e.g., RE = Nd or mixed rare earth metals) to Mg–Zn alloys. This allows precipitation of Mg–Zn–RE intermetallic compounds at the grain boundaries, which suppresses microporosity. The room-temperature strength of alloys such as EZ33 (Mg–2.6% Zn–3.2% RE–0.7% Zr) and ZE63 (Mg–5.7% Zn–2.5% RE–0.7% Zr) is reduced by the RE additions, because the RE takes some of the Zn out of solution to form the grain boundary intermetallics. These alloys are precipitation hardenable, forming sequences of GP zone structures and nonequilibrium phases that precipitate on the $\{10\bar{1}0\}$ planes. These precipitates resist coarsening at elevated temperatures and give the alloys substantially better creep strength than AZ91. Th acts similarly to RE additions, precipitating platelike intermetallics on the $\{10\bar{1}0\}$ planes. HK31A-T6 (Mg–3.3% Th–0.7% Zr) and HK32A-T5 (Mg–3.3% Th–2.1% Zn–0.7% Zr) have been used at temperatures as high as 350°C.

Y also confers good creep resistance to Mg; alloy WE43-T6 (Mg–4.0% Y–3.4% RE–0.4% Zr) has rather ordinary room temperature mechanical properties: a yield strength of 180 MPa and an ultimate strength of 250 MPa with 2% elongation. However, it maintains these strength values when retested at room temperature after thousands of hours at 200°C, and the creep strength is excellent. It has been used in helicopter and other aerospace components with operating temperatures as high as 250°C.

Ag has high solid solubility in Mg (15.3 wt% at 472°C, diminishing to about 2% at 20°C). Ag produces an excellent age-hardening response in Mg; quenched QE22 (Mg–2.5% Ag–2.0% RE–0.6% Zr) forms ternary ellipsoidal and rod-shaped zones parallel to the (0001) Mg planes which evolve into rod-shaped and equiaxed hexagonal precipitates as aging progresses. The precipitates are uniformly distributed throughout the grains, giving good dislocation blocking. This alloy also forms massive $MgRE_9$ precipitates at grain boundaries.

The cost and increased casting difficulties associated with Mg alloys containing Ag, Th, and RE metals makes these alloys substantially more expensive to use than AZ and ZK alloys. In the aerospace industry the superior performance justifies the added expense, but in the cost-sensitive automotive industry, these more expensive alloys have seen relatively limited use. To address this problem, Mg casting alloys are now under development and in early stages of commercial application that achieve high-temperature strength and creep resistance at lower cost (Sec. 11.4.5.1).

11.4.3.2 Wrought Mg. Wrought Mg is much less widely used than castings, and a relatively limited number of the AZ, ZK, and ZM alloys are used for most wrought products. AZ31B is widely used in plate and sheet products. Since room-temperature ductility is limited, heavy deformation must be performed above 250 to 300°C. The degree of subsequent cold work and annealing time and temperature can be controlled to achieve the strength and ductility best suited for the final product. Mg is somewhat more expensive than Al, and Al is much more easily deformed; these two factors give Al a decisive cost advantage over Mg for wrought products; consequently, wrought Mg products see relatively limited use.

11.4.3.3 Mg as an Al Alloying Addition. Nearly half of all Mg metal production is used as alloying additions to Al alloys. Many Al alloys contain 0.5 to 5% Mg (Sec. 20.2.4). Mg has solid solubility in Al of nearly 15% at 450°C, but this approaches zero at room temperature (Fig. 11.9). Al alloys often contain substantial amounts of Mg in metastable solid solution. Mg provides an impressive list of benefits to Al alloys. It lowers density, provides solid solution hardening, increases the work-hardening rate, accelerates the precipitation-hardening reaction of Cu and Zn, and forms Mg_2Si precipitates in 6000 series Al alloys.

11.4.3.4 Mg as a Deoxidation/Desulfurization Agent. Mg has a strong affinity for O and S, and Mg is frequently added to the melt of other alloys to "getter" these impurities. In gray cast iron (Sidebar 11.4), Mg forms insoluble MgO and MgS that float to the surface and join the slag layer. Only ~0.1% Mg is needed to deoxidize and desulfurize cast iron. This removes O and S from solution, allowing graphite nodules to grow in both the a- and c-axis directions of the hexagonal graphite unit cell. (O and S inhibit growth in the c-axis direction in cast iron.) This results in roughly spherical graphite nodules of ductile cast iron separated from one another by tough, ductile ferrite or pearlite. This microstructure gives nodular iron high ductility (15 to 25% elongation). By contrast, in conventional gray cast iron the graphite forms thin, interconnected plates that allow easy crack propagation, making the metal brittle (1 to 2% elongation). Mg competes with Ce and other REs in this application; although the cost of Mg is lower, the REs have much lower vapor pressures at the temperature of liquid iron and are thus easier to use than Mg. Mg is also used as a deoxidation/desulfurization agent in other metals, such as Cu and Ni.

11.4.3.5 Mg as a Chemically Active Agent. The reactivity and low cost of Mg allows its use as a reducing agent to produce reactive metals such as Be, Ti, V, Zr, Hf, Th, and U from their halides and B from B_2O_3. For more difficult reduction reactions such as those of the rare earth halides, Ca metal is often preferred to Mg. These metallothermic reduction reactions are often referred to as "bomb" reductions (Sidebar 11.5). The halides and Mg are placed in a sealed container and ignited to initiate the strongly exothermic reduction reaction. The Mg–halide slag floats to the top; the product metal sinks to the bottom. After the reactants cool, they break apart easily at the slag–metal interface.

SIDEBAR 11.4: Mg INOCULATION IN CAST IRON

Dissolving Mg into molten cast iron is difficult because molten iron's 1400°C temperature is higher than Mg's 1100°C boiling point. In addition, Mg is much lighter than iron. Simply dropping a chunk of Mg into a crucible of liquid Fe causes the Mg to float atop the Fe, boiling and burning violently with little Mg entering the iron. Mg alloys, such as 30% Ca–55% Si–15% Mg or 18% Mg–65% Si–2% Ca–0.6% Ce can be plunged below the iron's surface in a refractory "bell." Alloys are preferred to pure Mg to control the rate of dissolution and to slow the loss to boiling. Pneumatic injection can also be used to entrain salt-coated Mg powder into Ar or N_2 gas that is injected below the surface of the molten Fe with a lance. Still another tactic is placing Mg alloy slugs in casting sprues so that the liquid Fe dissolves the Mg while flowing into the mold.

SIDEBAR 11.5: HAZARDS OF BOMB REDUCTION REACTIONS

High heat and pressure are generated inside bomb reduction containers. A ruptured container can send hot metal flying with great destructive force. These hazards prompted the name *bomb* for the reaction vessels (Fig. 11.11). During World War II, the U.S. Manhattan Project used metallothermic reduction of UF_4 to produce U metal for nuclear weapons production. This work was performed in secrecy in small wood-framed buildings on the campus of Iowa State University. During one production run, two technicians were lifting a bomb from its recessed well in the building floor when the container failed. The top of the steel container and a jet of molten slag and U metal shot between the workers' heads, narrowly missing both. When the hot material hit the building's roof, both the wood rafters and the liquid metal were instantly set ablaze. The city fire department responded to reports of a fire, but armed security guards were under strict orders to deny access to anyone without a security clearance. The fire crews lacked the necessary clearance, so they were detained outside the perimeter fence, helpless to do more than watch as the building burned to the ground. Only several years later did the city and university communities learn the true origin of the fire.

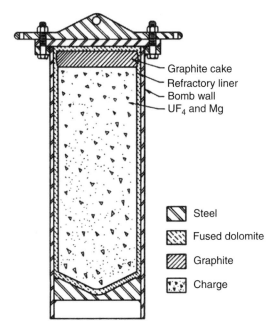

Fig. 11.11 Section view of a bomb reaction assembly for reducing UF_4 to U metal. (From Katz et al., 1986, p. 226; with permission.)

Mg is widely used to prevent corrosion in buried pipelines, storage tanks, and other structures by acting as a sacrificial anode. The Mg metal is electrically connected to the structure, forming a galvanic couple (Fig. 11.12). Mg lies lower on the electromotive force scale than steel and all other structural metals (Table 11.2), so Mg acts as the anode, leaving the cathode passivated against corrosion. Protection lasts as long as some of the Mg remains in metallic form, but the Mg is eventually consumed and must be replaced periodically to maintain corrosion protection. The chemical reactivity of Mg is also useful in flares and pyrotechnic devices; the intense white heat of Mg combustion produces a bright light that is visible for safety and military

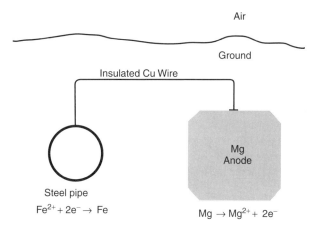

Fig. 11.12 Mg serving as a sacrificial anode for a buried steel pipeline. Since Mg is lower than Fe on the electromotive force scale, the Mg corrodes preferentially in this galvanic couple, protecting the steel from corrosion. Since eventually, the Mg is completely oxidized by this action, it must be replaced periodically to maintain corrosion protection for the pipe.

countermeasure flares even in full sunlight. When combined with other alkali or alkaline metals, Mg produces colored flares and fireworks.

11.4.4 Sources of Mg and Refining Methods

Total world Mg production is typically about 500,000 tons/yr. The metal's low density makes comparison of tonnage figures with those of heavier metals somewhat deceptive. Mg production is roughly 15 times smaller (weight basis) than production of metals such as Cu or Zn, although on a volume basis the ratio is nearer to four or five times. World reserves of Mg could be described as "essentially unlimited" because the metal can be extracted economically from large saltwater lakes, brine wells, and seawater. The most common process for large-scale Mg production electrolyzes a mixture of alkali metal chlorides (e.g., KCl) and $MgCl_2$ at about 700°C. This process produces Mg with purities as high as 99.8%. An unavoidable 300 to 400 ppm of Fe is typically present, due to dissolution of steel components in the electrolysis cell. For higher-purity Mg production that is cost-competitive on a smaller production scale, the silicothermic processes (e.g., Pidgeon process, Magnetherm process, and Bolzano process) use ferrosilicon to reduce MgO in a molten slag at temperatures at or above 1200°C to evolve Mg vapor. This vapor is condensed outside the main furnace, then remelted, refined, and cast. Silicothermic Mg production can produce purities as high as 99.95%. About one-third of annual Mg demand is met by recycled old and new Mg scrap.

11.4.5 Structure–Property Relations in Mg

11.4.5.1 Lower-Cost Creep-Resistant Mg Alloys. Creep resistance is important for many Mg alloy uses, but conventional creep-resistant alloys use Ag, Th, Y, and RE additions, all of which are costly. These alloys are also more difficult to cast, which drives costs higher still. Newer-generation Mg alloys are entering commercial service that use alkaline metals such as Ca and Sr to partially replace the more expensive elements while achieving the needed high-temperature strength and creep resistance. These alkaline elements sharply decrease creep deformation, as shown in Table 11.3 and Fig. 11.13.

These new alloys maintain Al content to assure good fluidity for castability. However, Al forms eutectic $Mg_{17}Al_{12}$ intermetallic at grain boundaries that lowers creep strength, so Ca and/or Sr are added to preferentially react with the Al, suppressing $Mg_{17}Al_{12}$ formation. Ca

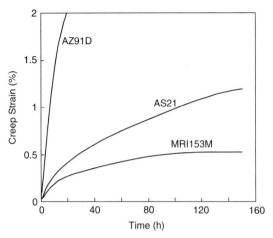

Fig. 11.13 Relative creep resistance under 85 MPa stress at 135°C of AZ91D and MRI153, a Mg alloy containing Ca and Sr alloying additions. (Drawn from data in Hryn, 2001, p. 123.)

TABLE 11.3 Creep Behavior of MRI 151 and MRI 153 Alloys Compared to AS21

Alloy	Stress Exponent, n^a (135°C, 85–110 MPa)	Activation Energy, Q^a (kJ/mol) (90 MPa, 130–150°C)
MRI 151	7.0	175
MRI 153	7.6	181
AS21	19.5	166

Source: Hryn (2001), p. 129; with permission.

[a] For $E' = A\sigma^n \exp(-Q/RT)$, where E' is the steady-state creep rate, A is a constant, σ the applied stress, R the gas constant, and T the absolute (K) temperature.

and Sr have stronger affinity for Al than for Mg, so they form Al_2M intermetallics (M = Ca or Sr) rather than $Mg_{17}Al_{12}$ or Mg_xM_y. In addition, the Al_2M intermetallics have melting points above 1000°C, so they do not contribute to grain boundary sliding. The Al_2Ca and Al_2Sr intermetallic compounds replace $Mg_{17}Al_{12}$ at the grain boundaries (Fig. 11.14) while preserving the benefits of Al (i.e., castability, solid solution strengthening, and corrosion resistance). The MRI (Magnesium Research Institute) series alloys use small amounts (<1%) of Ca and/or Sr along with reduced RE content. Costs are kept low by using mischmetal rather than pure REs (Sec. 23.4.1). These alloys show some tendency to produce hot tears and other casting defects, but their defect rates are equal to or better than those of other creep-resistant alloys. The MRI alloys perform well in thixomolding operations, and their cost is substantially lower than that of older creep-resistant Mg alloys.

11.4.5.2 Extended Solid Solubility of Transition Metals Through Nonequilibrium Methods. As mentioned in Sec. 11.4.3.1, few elements have suitably large atomic radii and low electronegativities to satisfy the Darken and Gurry criteria for extensive solid solubility in Mg. Ultrarapid quenching (Sec. 7.3.1) and physical vapor deposition (PVD) (Sec. 8.7.3) have been used to produce metastable Mg solid solutions of several transition metals (e.g., Ti, Zr, Al, V, and others) at much higher solute concentrations than their equilibrium solubilities. Single-phase solid solutions with compositions such as Mg–46 wt% Ti, Mg–17 wt% V, and Mg–10.5 wt% Zr can be produced by PVD. Ultrarapid quenching has produced Mg–23 wt%

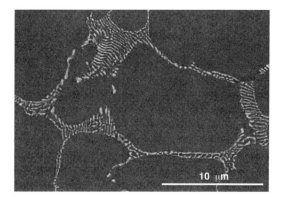

Fig. 11.14 Back-scattered electron image of Mg–5% Al–1.2% Sr alloy. Note the presence of Al–Sr intermetallic (white phase) in a lamellar microstructure at the Mg (dark phase) grain boundaries. No $Mg_{17}Al_{12}$ was present in this alloy because Sr reacts preferentially with Al, preventing formation of low-melting $Mg_{17}Al_{12}$ and thereby improving creep strength. (From Hryn, 2001, p. 123; with permission.)

Al, Mg–23 wt% Y, and Mg–6 wt% Mn. All these compositions are far beyond the equilibrium solubility limits. These materials are metastable, but tests show that they persist to about 280°C. When the solid solution breaks down, precipitates of pure transition metal usually form.

Entirely new metastable FCC phases have also been made by rapid cooling of Mg–18% Sn and Mg–23% Pb. Metallic glasses (Sec. 7.3) form in rapidly cooled Mg–Ni, Mg–Zn, and Mg–Cu alloys. When rapid solidification is applied to conventional Mg alloys such as ZE62 and ZK60, the grain size refinement boosts strength and ductility. Hardness frequently doubles in such materials after rapid solidification processing.

These metastable Mg alloys have better corrosion resistance than equilibrium Mg alloys. Mg–10.5% Zr and Mg–44% Ti, for example, display much lower corrosion rates in NaCl solution spray tests than pure Mg or conventional Mg alloys. It appears that the Zr or Ti form more protective oxide layers. Large solid solution concentrations of transition metals can also change the c/a ratio of Mg, lowering it in the case of Mg–Zr alloy. Since active slip systems tend to change with c/a ratios in HCP metals, this raises the possibility that one or more additional slip systems may be activated in this material. This has already been observed in equilibrium HCP Mg–Li alloys (Sec. 10.2.3.1). Rapid cooling of other, more corrosion-resistant solutes present the intriguing (but as yet undemonstrated) possibility of producing corrosion-resistant, highly ductile wrought Mg.

11.5 HEAVIER ALKALINE METALS

11.5.1 Calcium

Ca metal is produced by calcining limestone ($CaCO_3$) to lime (CaO) and then reducing the lime by mixing it with Al powder and heating the mixture to produce Al_2O_3 and Ca vapor. The vapor is then condensed outside the furnace and remelted. World production of Ca metal is about 2500 tons/yr. Ca–Si alloys are used to deoxidize and desulfurize iron and steel and to control graphite morphology in cast iron (Sec. 11.4.3.4); Ca is a more effective desulfurizer than Mg. Ca is also used to remove Bi from Pb and as an alloying addition to battery grid Pb in lead–acid batteries (Sec. 21.5.3.1). Ca is quite ductile but relatively weak. Its partially protective oxide and nitride coatings allow it to be machined and handled in air. The low density and FCC crystal structure of Ca might make it an appealing alternative to Mg for structural use were it not for its low elastic modulus and its slow reaction with water, corroding the metal and emitting H_2 gas. Ca metal is used as a reducing agent in reactive metal production (Sec. 11.4.3.5), and it is an effective alloying agent for Al, Be, Cu, Pb, and Mg (Sec. 11.4.5.1) alloys.

11.5.2 Strontium

Sr metal is produced by electrolysis of fused $SrCl_2$ mixed with KCl; global production of Sr metal is small (about 400 tons/yr). The main producers of Sr minerals are Spain, Mexico, and China. The principal use for Sr metal is in Al and Mg alloys (Sec. 11.4.5.1). Bulk Sr metal oxidizes rapidly in air and reacts with water; Sr metal powder ignites spontaneously on contact with air. Sr has an unusual electronic band structure; the electron density of states is very low at the Fermi energy, making it nearly a semiconductor at room temperature. Sr carbonate is used in the glass for television and computer monitor CRTs, accounting for 70% of all global Sr consumption. Sr salts emit an intense red color at high temperature, which is exploited in fireworks, flares, and pyrotechnics. Sr is present in some ferrite magnet compounds.

11.5.3 Barium

Ba metal reacts strongly with both air and water. There is little demand for pure Ba metal, due to its softness and reactivity; annual world production is ∼25 tons/yr. Ba is used as an alloying additive to Ni to produce high-emissivity spark plug wire. Ba alloyed with Si and Fe is a more effective (and more expensive) deoxidizer and desulfurization agent in Fe and steel than the competing Mg–Ca–Si alloys, and it can be used to produce ductile iron. Baryte (barite), a natural mineral containing $BaSO_4$, is used as a weighting agent in well drilling fluids. Annual world demand for baryte in petroleum and natural gas drilling operations is about 6 million tons, and this is by far the largest use for Ba. $Ba(CO_3)_2$ and other Ba compounds are used in polymer and rubber production. All water-soluble Ba compounds are toxic, and $Ba(CO_3)_2$ is used in rat poison. $Ba(OH)_2$ is an additive in glass manufacture. Insoluble $BaSO_4$ serves as a medical imaging agent for x-ray diagnostic work on the gastrointestinal tract. Ba's high atomic weight absorbs x-rays strongly, allowing radiologists to see the outlines of the stomach and intestines. The sulfate and sulfide are used as white pigment in paint. $Ba(NO_3)_2$ adds a green color to signal flares and pyrotechnics. Ba ferrite is used as a permanent ferromagnetic material, and $BaTiO_3$ is a ferroelectric material used in transducers.

11.5.4 Radium

Marie Curie discovered Ra in U ore in 1898. Metallic Ra was first produced by electrolysis of $RaCl_2$ solution, using a Hg cathode to form an amalgam with the Ra. The Ra was then isolated by distilling off the Hg. Ra salts can be purchased in gram quantities, but they are strongly radioactive, emitting α radiation (plus some β and γ rays) and producing radioactive Rn gas. Consequently, Ra salts and Ra metal pose severe safety hazards. Ra salts sometimes contaminate drinking water taken from U-bearing soils and wells. Ra is a "bone seeker"; the body metabolizes Ra similarly to Ca and incorporates Ra into bone tissue, which elevates cancer risk. Isotope $^{223}_{88}Ra$ has an odd and rare (one decay event out of 10^9) radioactive decay mode in which a $^{14}_{6}C$ nucleus is emitted: $^{223}_{88}Ra \rightarrow {}^{209}_{82}Pb + {}^{14}_{6}C$.

Ra metal rapidly forms a black nitride upon exposure to air, and Ra reacts strongly with water. There is no commerce in Ra metal. Ra has been used as a source of Rn gas for treatment of cancer and other diseases. In proximity to Be, it can serve as a neutron source (Sec. 11.3.3.2) for research and cancer radiation therapy. In recent years, Ra has been largely replaced by other, safer materials for medical applications. Ra salts were once widely used to produce self-luminous paints for watch dials, but radiation safety concerns have nearly ended this use as well. The longest-lived and most common isotope, $^{226}_{88}Ra$, has a half-life of 1600 years and acts as a bone seeker when ingested, elevating bone cancer risk.

ADDITIONAL INFORMATION

References, Appendixes, Problem Sets, and Metal Production Figures are available at
ftp://ftp.wiley.com/public/sci_tech_med/nonferrous

12 Ti, Zr, and Hf

12.1 OVERVIEW

The elements of group 4, titanium, zirconium, and hafnium (Ti, Zr, and Hf) are reactive, electropositive metals. But unlike the neighboring rare earth and alkaline elements, Ti, Zr, and Hf form adherent, diffusion-resistant oxide layers that protect them from both aqueous corrosion and high-temperature oxidation. The s^2d^2 electronic structure of these metals provides four bonding electrons per atom, giving them relatively high melting temperatures. These attributes, combined with good ductility and abundance, make Ti, and to a lesser extent Zr, important industrial metals.

Ti, Zr, and Hf oxides and other compounds are difficult to reduce to metal. Consequently, the metals saw little engineering use until after World War II. Ti and Zr are often selected for use in seawater, hot acids, sour gas petroleum wells, and high-temperature exposure to air. Ti alloys have high strength/weight ratios, particularly at elevated temperatures, and they are used widely in aircraft structures and jet engines. Zr and Hf have useful nuclear properties. All three metals are HCP at room temperature, transforming to a BCC phase at high temperature. Interactions between the inner (core) electrons and moving conduction electrons make them poor electrical and thermal conductors.

Ti ranks among the 10 most abundant terrestrial elements, comprising 0.63% of Earth's crust. Zr is among the 20 most abundant terrestrial elements (162 ppm). Hf is somewhat scarce (2.8 ppm). Production of Ti, Zr, and Hf metals of reasonably high purity poses several challenges for extractive processors that have been solved only by costly production methods not required for less electropositive metals. For this reason all three metals are more expensive than their abundances would suggest.

12.2 TITANIUM

> The AL-37FU engines are configured for thrust vector control with titanium rather than steel units to reduce the weight of the nozzle.... When the Su-37 was shown at Farnbrough in 1996, it stole the show, performing an astounding aerobatic display. The Su-37's most impressive maneuver is the kulbit (somersault). With an entry speed of 350 km/h the aircraft flipped onto its back (a full 180°) facing the opposite direction, inverted, and practically stationary. After "pausing," thrust vectoring completes the kulbit (a 360° somersault) with a nose-down angle of 30° and an exit speed of 60 km/h.
>
> —John Pike, "Su-37 Super Flanker," GlobalSecurity.org,
> *http://www.globalsecurity.org/military/world/russia/su.htm*, October 2003

12.2.1 History of Ti

Ti is a relative newcomer to metallurgy. It was first produced in low-purity form in 1825, but reasonably pure Ti metal was not available until 1910. Ti remained a rare metal until after World War II. Even today, Ti remains difficult to refine (Sec. 12.2.2.1). However, Ti alloys'

Structure–Property Relations in Nonferrous Metals, by Alan M. Russell and Kok Loong Lee
Copyright © 2005 John Wiley & Sons, Inc.

high strength/weight ratio and excellent corrosion resistance are so compellingly attractive to engineers that large-scale production began in the 1950s for use in jet engines and military aircraft. Other applications soon followed, including marine components such as propellers and submarine hulls, surgical implants, and petroleum drill pipe. Current world demand for Ti metal is 75,000 tons/yr, and its high extraction and fabrication costs are the principal impediments to much greater use. TiO_2's refractive index is 2.63, even greater than that of diamond (2.42). TiO_2 is used as a white pigment in paints and polymers, and demand for TiO_2 (4.5 million tons/yr) greatly exceeds usage of Ti metal.

12.2.2 Physical Properties of Ti

Ti has sometimes been called the "prima donna" of metals. It can perform exceptional engineering feats but is difficult to refine and use. Engineers accustomed to working with a "forgiving" metal such as low-carbon steel find Ti an exasperatingly demanding material to refine and fabricate due to its unusual combination of physical properties (Fig. 12.1).

12.2.2.1 Reactivity. Ti is so reactive that its production and fabrication pose several daunting challenges. In the decades-old Kroll process, Ti ore (TiO_2 or $TiFeO_3$) is converted to $TiCl_4$ and reduced with Mg metal. A newer molten $CaCl_2$ electrolytic process (the Cambridge process) for Ti reduction is under development and shows promise of simplifying Ti reduction and lowering Ti production costs (Sidebar 12.1). The Kroll and Cambridge processes form a mixed Ti metal

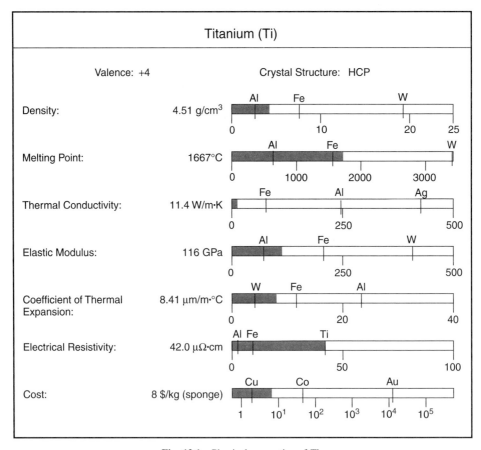

Fig. 12.1 Physical properties of Ti.

SIDEBAR 12.1: LOWER-COST Ti

Ti's corrosion resistance, high strength/weight ratio, and good hot strength could improve the performance of many engineering systems. However, Ti's uses are limited today by its high cost. When William Kroll developed the $TiCl_4$ reduction process in the mid-twentieth century, he predicted that it would be replaced within 15 years by a simpler, less costly electrolytic reduction process, analogous to the Hall process for Al. However, despite considerable effort, more than half a century passed before a successful electrolytic process was demonstrated on a laboratory scale. If scale-up work is successful, this new method may be able to lower the cost of Ti metal by as much as two-thirds, making Ti cost competitive with stainless steel.

Most previous attempts to produce Ti metal by electrolysis tried to dissolve TiO_2 into a molten salt, as is done in the Hall process (Sec. 20.2.7) for Al and the Downs process (Sec. 10.2.3.2) for Na. A University of Cambridge team took a different approach by immersing solid TiO_2 pellets into liquid $CaCl_2$ (Fig. 12.2). An electric current separates the O from the pellets, leaving Ti metal. TiO_2 pellets are held on a wire suspended in the molten salt. TiO_2 is normally an insulator, but at 900°C under an applied voltage (3.2 V) it converts to TiO_x ($x < 2$), which conducts well. The O ionizes, dissolves in the molten salt, migrates to the anode, and is discharged into the cell atmosphere. Pure Ti accumulates at the cathode. The process also allows Ti alloys to be produced directly by using mixed metal oxides instead of pure TiO_2.

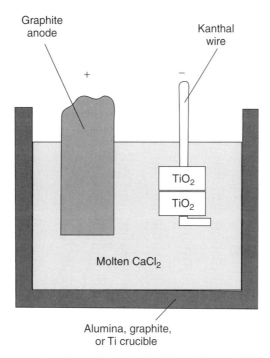

Fig. 12.2 Cambridge process electrolytic cell for reducing TiO_2 to Ti metal. TiO_2 pellets held on a Kanthal (Ni–Cr) wire suspended in the molten salt convert to TiO_x ($x < 2$), which conducts well. The O in the TiO_2 ionizes and migrates to the anode, where it is discharged into the cell atmosphere. Pure Ti accumulates at the cathode. (Redrawn from Chen et al., 2000.)

plus alkaline or alkali salt product called *sponge Ti*. Crushing sponge Ti allows the salt to be removed by leaching or vacuum distilling; the sponge Ti is then pressed into large briquettes (along with recycled Ti scrap and alloying additions) and melted into billets. Liquid Ti is so reactive that it dissolves all refractories; consequently, melting must be done by either arc melting into chilled molds or by a "skull" process in which liquid Ti metal is held in a cooled, solid-Ti metal shell. Ti cannot be melted in air because of the risk of fire and the certainty of O and N absorption that hardens and embrittles the metal (Fig. 12.3). Consequently, all Ti melting is done in vacuum or protective atmospheres. To achieve high ductility and minimize inclusion content, Ti usually receives two vacuum melting operations. The first consolidates the sponge and other feedstock metals and removes remaining salt and some of the interstitial O and N; the second reduces O and N content further and improves billet homogeneity. For highest ductility, three vacuum melting purification steps are performed. In the late 1990s, plasma arc melting and electron beam cold hearth melting began to compete with traditional arc melting methods, offering superior control of inclusion content.

After Ti ingots are produced, the metal's reactivity causes additional problems in hot working, grinding, joining, machining, and casting. Ti can be hot-forged and hot-rolled in air, but above about 500°C, a significant thickness of oxide mill scale accumulates that hardens and embrittles the surface, leading to cracking. Mill scale formation can be reduced by applying silicate-based coatings to form a glassy barrier to O, N, and H pickup. Coatings are reasonably effective up to about 800°C, but their protection diminishes at higher temperatures. Coatings and mill scale can be removed after use by any of three methods. Mixed HF–HNO$_3$ acid pickling solutions work best with the thin (10-μm) scales formed below 600°C. Abrasive grinding or grit blasting are used for the thicker oxides (as much as 200 μm in hot-rolled plate) formed above 600°C. Ti's low thermal conductivity necessitates low grinding speeds to avoid overheating and "smearing" the surface, which degrades fatigue performance and alters prior heat treatments near the metal surface. Thicker oxide scale can also be removed by immersion in oxidizing molten salt baths held at 200 to 480°C. At the higher end of this temperature range, the baths remove mill scale rapidly but may alter microstructure. Most descaling operations employ more than one of these three methods for optimal surface treatment, and some salt-bath descaling methods are augmented by electrolytic action to accelerate the process.

Ti's high reactivity requires that all joining operations (fusion welding, solid-state diffusion welding, and brazing) be performed in inert atmospheres or vacuum. Inert-gas shielded welding has been used on Ti, but weld quality is often marginal; glove boxes or collapsible glove bags yield more consistent joining quality (Sidebar 12.2). In addition, surface contaminants such as oxide scale and oil must be removed prior to joining to avoid adding interstitial impurities that

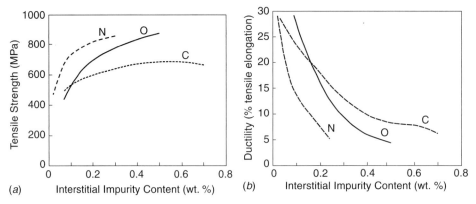

Fig. 12.3 (*a*) Hardening and (*b*) embrittling effect of O, N, and C on Ti. (Adapted from Donachie, 2000, p. 27.)

SIDEBAR 12.2: WELDING LARGE Ti SUBMARINE HULL SECTIONS IN AN INERT ATMOSPHERE

The high strength/weight ratio of Ti and its resistance to seawater corrosion makes it an excellent material for shipbuilding. However, ship hulls need to be welded in large sections, and welding Ti in air produces weld quality problems. In addition, the welded metal has better properties if it is heat treated after welding. Even when inert-gas shielded welding is used, the occasional contamination of the fusion zone with O, N, and H_2O makes for inconsistent weld quality. The Soviet Union addressed this problem by constructing large enclosures for submarine assembly that were sealed from the atmosphere and filled with a dry, oxygen-free, inert-gas atmosphere. Workers could weld large Ti alloy hull sections inside this space without risk of weld contamination and embrittlement. This required that workers wear the equivalent of a full-body space suit (Fig. 12.4) to isolate breathing air from room air and pass through air locks to avoid contaminating the inert atmosphere. The welded assembly was then heat-treated in a gigantic furnace. This extraordinary assembly method allowed many Soviet submarines built in the later twentieth century to be made of Ti alloys, a technological coup unmatched by other navies.

Fig. 12.4 Artist's depiction of welding operations on Soviet submarine hulls performed in a large assembly building with an inert-gas atmosphere. The workers must wear the equivalent of space suits to isolate their breathing air from the inert-gas atmosphere. After welding, the hulls were heat treated in enormous furnaces.

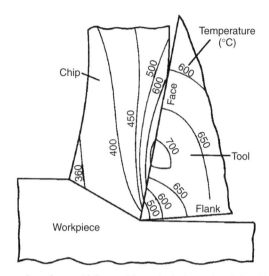

Fig. 12.5 The cutting action of a machining tool heats both the tool and the chip, sliding hot Ti with no protective oxide layer across the tool face. The highly reactive Ti dissolves some tool materials rapidly and leads to galling, smearing, or chipping of the workpiece surface. (From Kalpakjian, 1989, p. 506; with permission.)

embrittle the joined metal. Yet another challenge posed by Ti's reactivity is its tendency to attack cutting tools during machining. A cutting tool moving through the surface metal of a Ti workpiece heats both workpiece and chip and pushes Ti with no oxide layer against the hot cutting tool (Fig. 12.5). Ti's low thermal conductivity retards heat dissipation, and the reactivity of the hot, bare Ti dissolves the tool material and can gall, smear, or chip the workpiece surface. For these reasons, Ti machining requires carbide or high-speed steel cutting tools used at lower cutting speeds and higher feed rates than those typical for machining steel. Finally, the risk of Ti metal fire and tool wear considerations require use of cutting fluids (preferably Cl-free) to lubricate and cool the metal.

The difficulties in hot working, joining, and machining Ti alloys would seem to suggest that casting would be widely used to achieve near net-shape products requiring little subsequent fabrication work. Unfortunately, Ti's reactivity makes casting difficult. Refractory ceramics and metals are attacked aggressively by liquid Ti (Sidebar 8.2). Mold materials (usually, graphite or ceramic shells for investment castings) are attacked by the Ti, so patterns must be made oversize to account for subsequent removal of contaminated Ti near the mold wall. Inert atmospheres are essential in casting operations to avoid contaminating the Ti with O and N and to minimize fire risk. For safety-critical Ti castings (e.g., turbine components and surgical implants), a postcasting hot isostatic pressing operation is often used to close any voids and porosity in the casting. Despite these difficulties, castings as large as 1.5 m in diameter and 135 kg have been made; however, forging, stamping, and machining operations dominate Ti fabrication; less than 10% of the total Ti product weight is formed by casting.

12.2.2.2 Corrosion. When exposed to air, Ti immediately forms an oxide layer a few nanometers thick that protects the underlying metal from further oxidation. If this oxide layer is damaged, it re-forms in the presence of even trace amounts (a few ppm) of O or H_2O. The oxide is strongly adherent and stable over a wide pH range of corrosive solutions as long as moisture and O are present to maintain the protective oxide layer. The principal threats to oxide integrity are strongly reducing conditions, strongly oxidizing environments (e.g., red, fuming HNO_3), and F^{-1} ions. Strongly oxidizing or reducing environments with a complete absence of H_2O can cause violent exothermic attack of Ti.

At elevated temperature, Ti and O (and to a lesser extent N) atoms diffuse through the oxide layer, gradually thickening it. At oxide thicknesses similar to the wavelength of visible light, the metal appears blue or yellow; thicker oxides are gray. Exposure to air for 30 minutes thickens the scale to several micrometers at 700°C and to a few hundred micrometers above 1050°C. Thick oxide scales severely embrittle the metal's surface, so heavy scale cannot be tolerated in most applications. The surface of oxidized Ti is a complex multilayered structure of fully stoichiometric oxide on the outermost surface underlain by gradations of substoichiometric oxides (TiO_x, where $x < 2$). In addition, O has more than 30 at% solid solubility in Ti metal, and the metal near the oxide layers typically has much higher interstitial O content than the bulk metal.

The high solubility of O in Ti allows the metal to absorb its own oxide, making Ti one of the easiest metals to join by diffusion bonding. A diffusion joining operation begins by clamping two cleaned Ti workpieces together, heating them in a protective atmosphere, and holding them at about 900°C under stress for 1 to 6 hours. The thin oxide layers initially present diffuse into the metal, allowing the faying surfaces to bond. Since fine-grained Ti alloys can be formed superplastically (Secs. 6.6 and 19.2.6), combined superplastic forming and diffusion bonding techniques are used to fabricate large, complex structures from Ti plate and sheet (Fig. 12.6) that would be difficult to produce by other methods. Ti components can also be fabricated by powder metallurgy (P/M) (another process aided by Ti's ability to absorb its own oxide). However, low starting interstitial impurity concentrations and careful handling and processing of the powder are required to avoid final P/M parts with excessively high O and N content. Fatigue strength of P/M parts drops sharply for components with appreciable (several percent) porosity, so hot isostatic pressing is useful for safety critical components fabricated by P/M. Although P/M Ti components can perform nearly as well as those fabricated from wrought metal, only a few percent of Ti products are fabricated by P/M.

12.2.2.3 Elasticity and Plasticity. Although Ti alloys have strengths rivaling all but the strongest steels, Ti's Young's modulus is only 116 GPa, substantially lower than that of steel

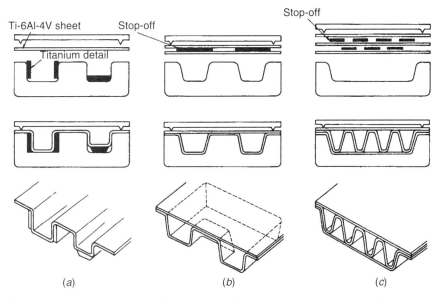

Fig. 12.6 Superplastically deformed, diffusion-bonded Ti alloy aerospace structures. (*a*) Reinforced sheet; (*b*) integrally stiffened structure with two diffusion-bonded sheets; (*c*) multiple sheets bonded to form sandwich structure. "Stop-off" is a diffusion barrier used to prevent diffusion in some locations. (From Donachie, 2000, p. 97; with permission.)

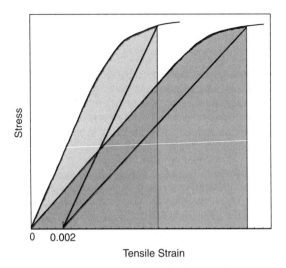

Fig. 12.7 Schematic comparison of two metals with equal 0.2% offset yield strengths but differing Young's moduli. Modulus of resilience is the elastic strain energy that the material can store (Energy $= \int_0^{\varepsilon_y} \sigma d\varepsilon$) shown here as the shaded area under the stress–strain curves. Note that the stored energy is considerably greater for the metal with the lower Young's modulus (darker shaded area).

(207 GPa). This lower modulus forces engineers to design for greater "flex" in Ti structures, but for some applications, this lower elastic modulus is actually beneficial, because it contributes to an exceptionally high modulus of resilience (Fig. 12.7) in high-strength Ti alloys. Ti golf club heads and tennis rackets can store large amounts of elastic strain energy as the implement strikes the ball (Sec. 7.3); this stored energy is released toward the end of the collision to deliver a greater impulse to the ball. Ti's high modulus of resilience causes a substantial "springback effect" in cold-forming operations, because the metal recovers the elastic portion of the deformation when the forming load is removed. Dies for Ti are dimensioned to "overform" the metal deliberately to compensate for the large springback. Ti is often hot formed to reduce the springback effect and stress-relieve the metal.

Pure Ti is one of the most ductile HCP metals. The principal slip system is the $\{10\bar{1}0\}\langle11\bar{2}0\rangle$ with the $(0001)\langle11\bar{2}0\rangle$ and $\{10\bar{1}1\}\langle11\bar{2}0\rangle$ systems acting as secondary slip systems (Fig. 11.2). The critical resolved shear stresses for these three slip systems change with interstitial impurity content. In Ti with (O + N content) < 1000 ppm, most slip occurs on the prism planes (τ_{CRSS} = 49 MPa) with some secondary slip occurring on the basal plane (τ_{CRSS} = 110 MPa). However, when interstitial impurity content is higher, slip becomes easiest on the pyramidal planes. Ti's ductility is enhanced by deformation twinning. Several twinning systems have been observed: $\{11\bar{2}3\}\langle\bar{1}\bar{1}22\rangle$, $\{11\bar{2}4\}\langle\bar{2}\bar{2}43\rangle$, $\{30\bar{3}4\}\langle20\bar{2}3\rangle$, and $\{10\bar{1}3\}\langle30\bar{3}2\rangle$.

High-purity polycrystalline Ti (O content < 500 wt ppm) has a 0.2% offset yield strength of 140 MPa and a tensile strength of 235 MPa with about 50% elongation. Strength rises and ductility decreases as interstitial impurity content increases (Fig. 12.3). Ti containing 0.1 to 0.4 wt% O is used widely. The reduced ductility and fracture toughness of metal with higher O content is acceptable in some applications, and less pure metal costs less. As with most metals, Ti's plasticity improves at higher temperatures, and hot rolling and hot forging are common fabrication methods. Much of the expense and complication of processing and fabricating Ti result from measures required to avoid adding interstitial impurities to the metal. O and H pickup are particularly troublesome. Ti will dissolve only about 0.1 at% H below 125°C, but solubility rises to 8 at% at 300°C. This poses the risk of H pickup at moderately elevated temperatures (e.g., during welding, heat treatment, or fused salt descaling operations) that will then form embrittling hydride precipitates as the metal cools. Ti is also subject to the Bauschinger effect (Sec. 12.2.6.2), a strain-hardening phenomenon in which 1 to 5% *tensile* elongation causes as much as a 50% drop in *compressive* yield strength. This behavior is most pronounced at room temperature and diminishes at elevated temperature, which provides another incentive to hot-form Ti.

12.2.2.4 Phase Transformations. HCP Ti (α-Ti) transforms to a BCC phase (β-Ti) at 882°C. This transformation is often exploited by heat treatments that strengthen Ti alloys by forming α + β microstructures (Secs. 12.2.3.3 and 12.2.3.4). With sufficient alloying additions (e.g., V, Mo) the BCC phase can be retained indefinitely as a metastable phase at room temperature. At high pressure, a third phase forms, hexagonal (hP3) ω-Ti (Fig. 12.8). The brittle ω phase can cause problems in Ti heat treatments by forming at low pressure in metastable β alloys; composition control and stabilizing anneals will suppress ω formation, and it does not pose a significant problem in Ti that is properly heat treated.

12.2.3 Alloys of Ti

Applications for Ti are based on two key characteristics: excellent corrosion resistance and a high strength/weight ratio, particularly at elevated temperature. Compared to competing alloys based on Fe, Ni, and Al, Ti has a higher initial cost and is also more expensive to fabricate into a finished product. Ti is not selected unless it can deliver higher performance, longer life, or reduced maintenance/operating costs. For many applications Ti delivers a lower total life cycle cost than competing materials, particularly in aerospace and marine structures, chemical processing equipment, surgical implants, and sports implements.

12.2.3.1 Pure Ti. Commercial-purity (CP) Ti comprises a significant portion of Ti consumption. The aqueous corrosion resistance of pure Ti is better than that of Ti alloys, and Ti components such as heat exchangers, tanks, and reactor vessels generally use unalloyed Ti.

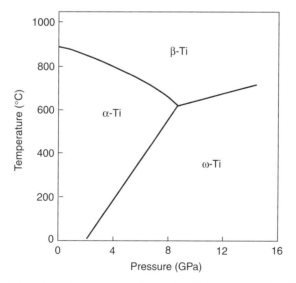

Fig. 12.8 Pressure–temperature phase diagram of pure Ti. At low pressure, HCP α-Ti transforms to BCC β-Ti at 882°C. The hexagonal ω-Ti phase (hP3) is stable at higher pressures and often appears as a metastable constituent in rapidly cooled Ti alloys. Transformations near the α–ω line are sluggish and hysteretic, particularly at lower temperatures.

Several grades of CP Ti exist, distinguished by their O (about 0.18 to 0.4 wt%) and Fe (about 0.2 to 0.5 wt%) content. CP Ti grades with larger O and Fe contents have higher strength and lower cost, but they also have somewhat lower ductility, fracture toughness, and corrosion resistance. CP Ti alloys containing 0.05 to 0.2% Pt group metals are available at substantially higher cost for maximum crevice corrosion resistance (Sidebar 12.3).

12.2.3.2 α-Ti Alloys. Additions of α-*stabilizing* elements such as Al, Sn, or O raise Ti's α → β transus temperature and increase strength by solid-solution hardening. Al additions are desirable for several reasons. Al is inexpensive, lowers alloy density, maintains alloy ductility better than do interstitial α-stabilizers, and improves high-temperature oxidation resistance. Al contents higher than 6 to 8% are usually avoided, however, since they can cause a sharp drop in ductility by forming α_2 phase, the brittle Ti_3Al intermetallic. Al additions to Ti tend to limit dislocation motion to $\{10\bar{1}0\}$ planes (Fig. 12.10), which strengthens the metal but lowers ductility. Alpha-stabilizing elements inhibit formation of β phase during heating for operations such as hot work. The most widely used α alloy is Ti–5% Al–2.5% Sn, which remains all α up to 950°C. Solid-solution strengthening makes Ti–5% Al–2.5% Sn much stronger than pure Ti, and the alloy also provides excellent weldability and good fracture toughness at cryogenic temperatures. Annealed Ti–5% Al–2.5% Sn has a yield strength of 880 MPa, ultimate tensile strength of 980 MPa, and 17% tensile elongation. When cold-drawn 15%, the alloy's yield strength rises to 1040 MPa with an ultimate tensile strength of 1200 MPa and 10% tensile elongation. Alpha-Ti alloys have better creep resistance than alloys that are partially or entirely β due to slower diffusion rates in the HCP lattice and the absence of phase boundaries.

Some α alloys contain small additions of β stabilizers (e.g., 1 to 2% V, Nb, or Ta) which allow small amounts of β formation by simple cooling or by solutionizing and artificial aging. These "near-α" or "super-α" alloys are still predominantly α phase, so they retain the weldability and toughness of the α alloys but with modestly higher strength. The most common annealing treatment for the frequently used near-α alloy Ti–8% Al–1% Mo–1% Nb is annealing at 790°C

SIDEBAR 12.3: IMPROVED CORROSION RESISTANCE IN Ti BY Pd AND Ru ADDITIONS

Ti's high corrosion resistance can be improved further by small additions of Pt group metals. Remarkably small amounts of these Pt group metals (0.05 to 0.20 wt% Pd or 0.10 wt% Ru) effect substantial improvement in corrosion resistance. The presence of Pd has been shown to decrease grain size in the surface oxide, which improves oxide layer adherence and inhibits stratification and spallation of the surface oxide. A second and probably greater contribution to corrosion resistance is the formation of Ti_2Pd precipitates or Ru-rich β-Ti dispersoids that act as tiny cathodic depolarization sites on the metal's surface. These cathodic sites shift the alloy corrosion potential in the positive (more noble) direction, improving H ion-reduction kinetics to suppress H evolution. This is particularly valuable in crevice corrosion (Fig. 12.9), where stagnant brine solutions in narrow gaps, under dirt or gaskets, or in surface cracks consume dissolved O in the brine faster than fresh O can diffuse into the crevice. The resulting O depletion produces a galvanic concentration cell between Ti in the crevice (anodic) and Ti exposed to fully oxygenated brine (cathodic) on the surface. This galvanic action releases positively charged Ti ions inside the crevice, which attract negatively charged Cl ions into the crevice to form unstable Ti chlorides that hydrolyze into dilute HCl acid. The acidity inside the crevice gradually increases over time and may reach pH values as low as 1 in an O-depleted environment, where Ti has difficulty maintaining its oxide layer. This can cause rapid corrosion that is hidden from view until the damage is extensive. The action of Pt group metals to suppress crevice corrosion can justify their high cost in environments that would corrode ordinary CP Ti.

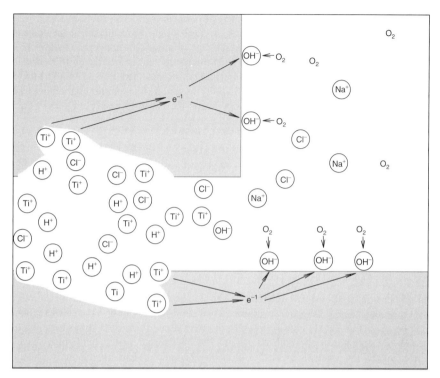

Fig. 12.9 Crevice corrosion mechanism. Stagnant brine in the crevice becomes depleted in dissolved O, forming a concentration cell that produces unstable Ti chlorides in the crevice. These chlorides hydrolyze to generate HCl acid in the O-depleted solution, attacking the Ti. (Redrawn from Fontana and Greene, 1967, p. 43.)

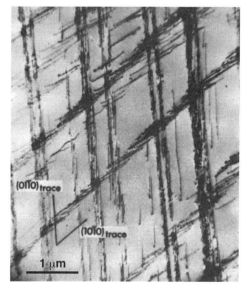

Fig. 12.10 Transmission electron micrograph of dislocation structures in Ti–6% Al alloy after 4% deformation. Dislocations are restricted to $\{10\bar{1}0\}$ planes. In pure Ti, both $\{10\bar{1}0\}$ and (0001) slip planes are active, and the typical dislocation cell structure of a ductile metal (Fig. 18.18a) would be present after 4% deformation. (From Blackburn and Williams, 1969, p. 308; with permission.)

for 8 hours followed by furnace cooling (termed *mill annealing*). This produces a microstructure that is predominantly α, with some β and $α_2$. In *duplex annealing* the mill-annealed metal is again heated to 790°C for 15 minutes, followed by air cooling. This converts much of the brittle $α_2$ present after mill annealing to α, and the final microstructure is a fairly uniform distribution of 2- to 5-μm-diameter β particles in an α matrix. Duplex annealed Ti–8% Al–1% Mo–1% Nb has a tensile yield strength of 950 MPa with 15% elongation. Precipitation hardening can push near-α alloys to still higher strengths, but this makes the metal susceptible to stress corrosion cracking in saltwater environments and is seldom used.

12.2.3.3 Metastable β-Ti Alloys. Additions of β-*stabilizing* elements (e.g., V, Mo, Cr, and Cu) lower the β → α transformation temperature. With sufficient β-stabilizer additions, a metastable, all-β crystal structure can be retained to room temperature by air-cooling thin sections or water-quenching thick sections. Metastable β alloys' BCC crystal structure has excellent formability at room temperature, but their strengths are low in the all-β condition. Extensive cold work at room temperature or heating to temperatures moderately above room temperature will form some α phase in the β matrix. This is often exploited to harden the alloy after forming; heating to 450 to 650°C will precipitate finely dispersed α particles in the β matrix, which produces yield strengths of about 1200 MPa with good fracture toughness (K_{IC} = 44 MPa$\sqrt{m}$).

Ti–13% V–11% Cr–3% Al and Ti–8% Mo–8% V–2% Fe–3% Al are representative of the several commercial β alloys. The relatively large percentages of heavier transition elements in β alloys raise densities to 5 g/cm^3 or higher. Their cold-forming properties are excellent, and they can be aged (Fig. 12.11) after forming to increase strength. Aged β alloys have the highest ultimate tensile strengths of all Ti alloys; $σ_{UTS}$ of 1500 MPa is attainable, but β alloys have lower creep strength and lower tensile ductilities in the aged condition than other Ti alloys. Specialty β

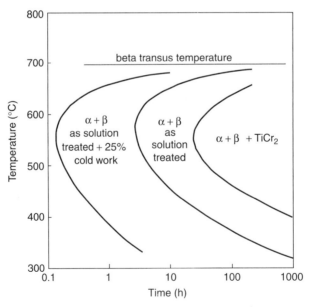

Fig. 12.11 Time–temperature transformation curves for aging metastable β–Ti–13% V–11% Cr–3% Al. The alloy was initially solutionized above the beta transus temperature, then cooled rapidly to room temperature to retain an all-β structure. The left curve indicates the start of the α → β transformation for metal that was cold-worked 25% in the all-β condition; the middle curve indicates the start of the α → β transformation for metal with no cold work; the right curve indicates the start of formation of TiCr$_2$ precipitates, which contribute further to age hardening. (Redrawn from Donachie, 2000, p. 256.)

alloys like Ti–8% V–5% Fe–1% Al deliver extraordinary strengths with aging ($\sigma_y = 1550$ MPa, $\sigma_{UTS} = 1700$ MPa), but their use is limited to a few "niche" applications (e.g., fasteners) because they require special melting procedures, are not weldable, and have low tensile ductility.

12.2.3.4 Alpha–Beta Alloys. The most widely used Ti alloys contain both α- and β-stabilizing elements to allow formation of mixed α + β microstructures with excellent strength, toughness, and corrosion resistance. One α–β alloy, Ti–6% Al–4% V (often called "Ti–6–4"), has dominated Ti metallurgy for decades, accounting for 45% of all Ti mill products. As Fig. 12.12 shows, α–β alloys can be solutionized at high temperature in the all-β phase field, then cooled to transform some of the β phase to α, resulting in an equilibrium α + β microstructure. A Widmanstätten structure results from slow cooling this alloy when α nucleates at the β grain boundaries and grows with the HCP (0001) α plane parallel to the BCC {110}β plane.

When α–β alloys are cooled from high temperature, most or all of the high temperature β phase transforms to α. This transformation can occur directly by diffusion-driven processes or

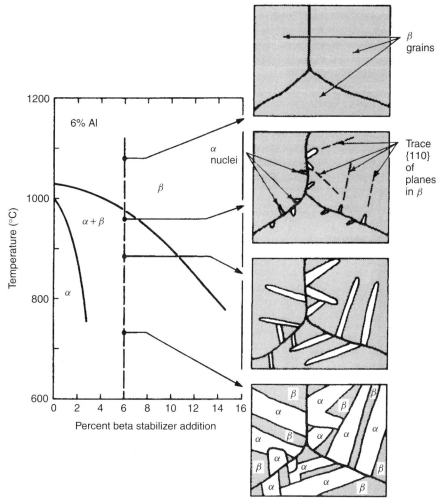

Fig. 12.12 Generic pseudo-binary α–β phase diagram. Alloys solutionized in the all-β phase field can be slow-cooled to form a Widmanstätten microstructure with α phase growing along {110} planes in the β grains. (From Donachie, 2000, p. 223; with permission.)

indirectly by martensitic transformations to metastable phases that eventually transform to α. Various heat treatments can be performed on α–β alloys to achieve the desired combination of properties for a particular application. Air-cooled Ti–6% Al–4% V transforms the β to acicular α (Fig. 12.13*b*). Water quenching transforms most of the β martensitically to α′ (Fig. 12.13*c*), a hexagonal, metastable phase; small amounts of the original β remain in the final structure. Although the microstructures in Fig. 12.13*a* and *b* look similar, their crystal structures differ. The martensite present in Fig. 12.13*c* is not as hard as the martensite that forms in steels. The hard, brittle body-centered tetragonal martensite in steel forms because interstitial C is much less soluble in the low-temperature BCC Fe than in the high-temperature FCC Fe. In Ti the interstitial impurities are actually *more* soluble in the low-temperature HCP phase than in the high-temperature BCC phase, so lattice strain from trapped interstitials does not harden Ti martensite. Ti martensite does, however, strengthen the metal somewhat since it is fine grained and has a high dislocation density from the martensitic shear transformation. Ti martensite is supersaturated with β-stabilizing elements (e.g., V, Nb, Mo), and it can be made somewhat harder by artificial aging to precipitate submicrometer-sized β phase within the martensite needles.

Cooling Ti–6% Al–4% V from within the α + β phase field (955°C on Fig. 12.13*a*) alters the final microstructure because primary α is present before cooling begins. Air cooling from

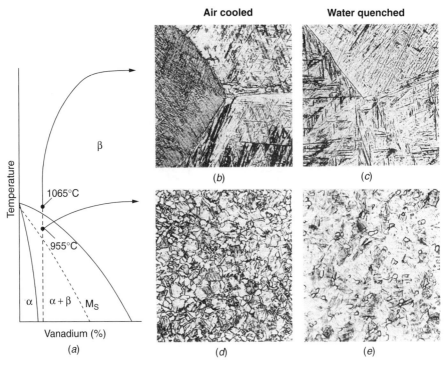

Fig. 12.13 (*a*) Ti–6% Al–4% V pseudo-binary phase diagram. Alpha–beta alloys can be solutionized in either the all-β phase field (1065°C) or at a lower temperature in the α + β phase field (955°C). (*b*) Air cooling from the all-β phase field produces acicular α (sometimes called *transformed* β) in the former β grains. Retained β appears as the dark phase between the needles of α. Prior β grain boundaries are still visible. (*c*) Water quenching from 1065°C produces a metastable α′; this martensitically formed hexagonal phase contains little or no β. Prior β grain boundaries are visible. Although the microstructures in (*b*) and (*c*) appear similar, they are crystallographically quite different. (*d*) Air cooling from the α + β phase field produces equiaxed primary α grains (light), retained β, plus some acicular α having shorter needles than the acicular α in part (*b*). (*e*) Water quenching from 955°C produces equiaxed primary α plus α′ martensite with little retained β. (Adapted from Donachie, 2000, p. 15; with permission.)

955°C converts the β that was initially present to transformed β, much of which is acicular α; the primary α remains unchanged (Fig. 12.13d). The acicular α has shorter needles than in material air cooled from 1055°C because the β grains are smaller at 955°C. If Ti–6% Al–4% V is water quenched from within the α + β phase field, the primary α remains unchanged, but the β phase transforms to α' (Fig. 12.13e).

Still another option is presented by water quenching Ti–6% Al–4% V from below the M_s temperature (e.g., 850°C). In that case, the lever law predicts that less β phase will be present as cooling begins, but the β that is present will have higher V content, high enough to stabilize the β phase throughout the quenching. This produces a microstructure of primary α plus untransformed β. The metastable, untransformed β may persist indefinitely at room temperature, but it can transform martensitically if strained.

Since a wide variety of microstructures is possible in α–β alloys, one might suppose that a wide range of mechanical properties is also available. That is not the case; the ranges of strengths and ductilities of α–β alloy microstructural variations are actually fairly narrow. Ti–6% Al–4% V tensile yield strengths range from 830 to 930 MPa in the annealed condition to a maximum of 1100 MPa for material quenched from 955°C and then aged at 550°C after quenching. Tensile ductilities vary from about 15% in the annealed condition to about 10% in the quenched and aged condition. The complexity of Ti heat treatments is compounded by the effects of processing operations (e.g., hot forging, casting) on microstructure.

12.2.4 Applications of Ti

The aerospace industry has always been the largest user of Ti alloys, and Ti use is especially high in military aircraft, where the greater costs of Ti can be justified by the more rigorous performance requirements of military aviation. Both military and commercial aircraft use Ti alloys extensively in gas turbine engine compressor blades. High strength/weight ratio, good toughness, and good oxidation resistance are essential in engine components, and Ti alloys have contributed greatly to the reliable, cost-efficient performance of jet engines. Use of Ti in airframe structures has been less pervasive, however, due to its high cost. The Boeing 757 has one of the highest Ti utilization rates in commercial aviation, but the empty weight of a 757 is only 5% Ti. If cost were not an issue, Ti would be used more extensively. Military aircraft make greater use of Ti outside the engines. The B1-B supersonic bomber is 22% Ti by weight, and the F-15 fighter plane is 34% Ti.

Ti's corrosion resistance makes it a valuable metal in the chemical processing and petroleum industries for pipe, reaction vessels, heat exchangers, filters, and valves. Biomedical devices also use Ti; the metal is fabricated into artificial joint prostheses, dental implants, eyeglass frames, endovascular stents (Sec. 12.2.6.1), heart valves, and wheelchairs. Marine applications include hulls for surface ships and submarines, propellers, and service water systems. Sporting goods manufacturers have also adopted Ti for golf clubs, bicycle frames, skis, scuba gas cylinders, lacrosse sticks, and even pool cues. Highly cost-sensitive industries such as manufacturers of automobile and architectural components have made limited use of Ti for exhaust valves, high-performance engine parts, and architectural trim. Ti is also used as an alloying additive to other metals, particularly steels. Ti additions minimize intergranular corrosion (sensitization) in stainless steel by reacting with C, thereby preventing undesirable Cr carbide formation that removes Cr from solid solution. Ti also improves the mechanical properties of high-strength low-alloy steel.

Ti has the potential to expand its utilization greatly. Many current users of stainless steel would prefer the lower density and superior corrosion resistance of Ti if the costs were equal. World stainless steel consumption is 20 million tons, hundreds of times greater than Ti consumption, but the price of stainless steel is only about $2/kg, much lower than the current cost of Ti mill products. Although Ti ore is readily available at low cost, greatly expanded use of Ti metal will occur only if the cost of refining and fabrication can be sharply reduced (Sidebar 12.1).

12.2.5 Sources of Ti

Ti is one of the most abundant elements in Earth's crust and is present in at least small concentrations in 98% of all rocks. Geological formations with high concentrations of the most useful Ti ores, rutile (TiO_2) and ilmenite ($FeTiO_3$), are found in Australia, Brazil, Canada, Finland, India, Malaysia, Norway, Portugal, Russia, Sierra Leone, South Africa, Sweden, and the United States. Since most Ti demand is for TiO_2 rather than the metal, rutile is preferred to ilmenite, but rutile is unfortunately somewhat less common than ilmenite. Ti ore reserves are large and will be adequate to meet anticipated demand far into the future.

12.2.6 Structure–Property Relations in Ti

12.2.6.1 Nitinol Shape Memory Alloy.

The NiTi intermetallic compound (usually called *nitinol*) has numerous applications that exploit its shape memory effect. NiTi undergoes a phase transformation near room temperature; the transformation temperature can be "tuned" to be either higher or lower than room temperature by making small changes to the Ni/Ti ratio. By analogy to the Fe–C system, the high-temperature phase is referred to as *austenite* and the low-temperature phase is called *martensite*. If the metal is cycled repeatedly through several temperature and deformation cycles, it establishes a shape pattern that will consistently return during future transformations. This allows the metal to be strained several percent below the transformation temperature, then return to its previous shape when it is heated sufficiently to transform back to austenite. The metal also displays a superelastic effect if it is deformed while its temperature is slightly above the transformation temperature, reshaping itself immediately to its predeformation shape. This occurs because the deformation stress momentarily transforms the metal to the martensite phase, but when the stress is removed, the metal reverts to the austenite phase and its prior shape.

The shape memory effect is used in nitinol eyeglass frames, which have a transformation temperature below room temperature. The frames can be "trained" to their desired dimensions, and if they are bent at room temperature, they revert to the correct shape as soon as the bending stress is removed. Accidents that would ruin ordinary metal eyeglass frames cause no permanent deformation in nitinol frames. In medicine, stents are used to hold open partially occluded arteries. A nitinol stent (Fig. 12.14) can be inserted in a collapsed form into an artery in the

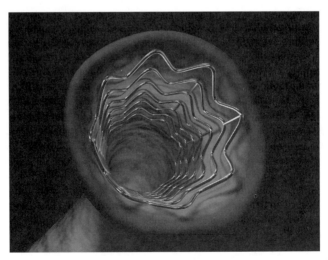

Fig. 12.14 Nitinol stent in its expanded shape configuration. The device can be collapsed for insertion through an arterial catheter, then expanded by the shape memory effect when in the proper location in the artery to relieve constrictions in blood flow.

thigh, then moved with a catheter to position it in a cardiac artery. Once in the correct location, the constraint on the catheter tip is released, and the stent "springs into shape," expanding to its "memory" shape to force the artery walls apart and relieve constriction in blood flow. Implanting a nitinol stent with a catheter is far less traumatic for the patient than heart surgery. Nitinol is also used to make orthodontic devices to straighten teeth. Body temperature triggers nitinol's reshaping behavior, so the wire exerts a steady force to move teeth gradually in the patient's jaw.

Nitinol devices have been in use for a number of years, but the crystal structure changes associated with the austenite–martensite transformation have been fully determined only recently. The high-temperature phase is cI2 (CsCl-type) (Figs. 25.3 and 12.15a), and this transforms martensitically upon cooling to the lower symmetry structure shown in Fig. 12.15b. First-principles calculations of bond energies indicate that the crystal structure in Fig. 12.15b is not actually the lowest-energy structure for NiTi at low temperatures. The structure depicted in Fig. 12.15c is the true equilibrium structure, but this does not form in NiTi martensite because stress in the lattice from twins and grain boundaries stabilizes the structure shown in Fig. 12.15b.

If the equilibrium structure shown in Fig. 12.15c were the martensite phase, the shape memory effect would be lost. The internal stresses originate at twin and grain boundaries and are present in both the austenite and martensite phases. The stress-based mechanism works for both the austenite $\rightarrow$ martensite and martensite $\rightarrow$ austenite transformations, because the twins reappear in the same place on cooling. "Information" about the martensite microstructure is stored in dislocation structures near the austenite grain boundaries. This explains why "training" nitinol by repeating temperature and deformation cycles gives it the ability to "remember" its low-temperature structure. The desired shape becomes the geometry for the part that minimizes total system energy (i.e., phase energy plus defect energy).

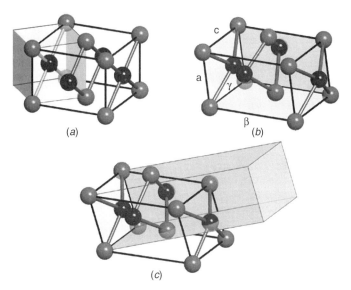

Fig. 12.15 Crystal structures of nitinol shape memory alloy. (a) The cubic (cI2) high-temperature austenite phase is shown by the shaded cube on the left side. The idealized, undistorted tetragonal low-temperature martensite crystal structure unit cell is outlined in bold black lines. Dark gray atoms are Ni; light gray atoms are Ti. (b) A distorted tetragonal unit cell is actually observed in the nitinol martensite phase. (c) The base-centered orthorhombic structure that is the true equilibrium structure at low temperature is shown by the tilted, shaded box. Although this is the lowest-energy crystal structure at low temperature, it is incapable of performing the shape memory effect, and the higher-energy crystal structure depicted in part (b) occurs instead; its presence is stabilized by stress in the metal induced by grain boundaries and twins in the nitinol martensite. (From Huang et al., 2003; with permission.)

12.2.6.2 Bauschinger Effect

Annealed metals have approximately equal tensile and compressive yield strengths. A metal plastically deformed in tension to a few percent strain displays the familiar work-hardening effect. However, the same tensile deformation that increases the tensile yield strength actually reduces compressive yield strength. This effect "works both ways"; that is, an initial compressive strain also causes a drop in the metal's tensile yield strength. This directionality of strain hardening, called the *Bauschinger effect*, occurs in all polycrystalline metals; however, it is especially pronounced in Ti and Ti alloys. Ti–6% Al–4% V, for example, develops more than a 50% difference between its tensile and compressive yield strengths after only a 2% tensile elongation. The Bauschinger effect diminishes as temperature increases, and it can be eliminated by either hot working the metal or giving it a subsequent stress-relief anneal after cold work.

In Fig. 12.16 an initial tensile deformation takes the test specimen past its yield stress (point 1) to a small plastic strain (point 2). When the load is released, the specimen contracts elastically along path 2–3. If the metal is then loaded in compression, its compressive yield strength would be point 4 on the figure. Note that the Bauschinger effect makes the absolute value of the compressive yield strength (4) considerably lower than the metal's tensile yield strength (2). The Bauschinger effect is caused by the tendency of dislocations to pile up in dislocation tangles (Fig. 18.18a). At point 2 in Fig. 12.16, the metal would have produced many dislocation pile-ups that make further dislocation motion in the same direction more difficult. However, if the specimen in then subjected to a compressive loading rather than continued tensile loading, the direction of the shear stress is reversed, and the dislocations will be driven to move in the opposite direction. This "backs them away" from the dislocation tangles that had formed during tensile elongation and allows them to move in the opposite direction, where the obstacles to dislocation motion are less numerous. Compressive loading from point 3 also has the additional effect of causing Frank–Read dislocation sources to begin generating new dislocations whose Burgers vectors have opposite sign to those produced in tensile loading. As these new loops

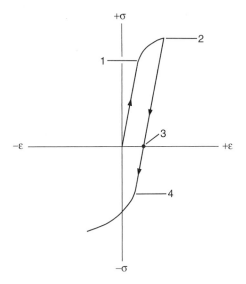

Fig. 12.16 Generic depiction of stress–strain behavior in a metal displaying the Bauschinger effect. In metals that have not been deformed plastically, the tensile (1) and compressive (4) yield stresses are approximately equal. However, a tensile deformation of a few percent elongation reduces the compressive yield strength. All metals display this behavior, but it is especially large in Ti. At room temperature Ti's Bauschinger effect can make the compressive yield strength less than half the tensile yield strength. This effect, sometimes referred to as *work softening*, is an important factor in cold-working Ti.

radiate outward, they will meet some of the dislocation loops formed during tensile loading, and the two loops will annihilate each other, further softening the metal in compression loading.

12.3 ZIRCONIUM

> Zirconium is just titanium on steroids. Zirconium is more corrosion resistant than Ti, higher melting, heavier, and more expensive.
>
> —T. W. Ellis, American Competitiveness Institute, Gold 2003 Conference: New Industrial Applications for Gold, Sponsored by the World Gold Council, Vancouver, BC, October 2003

12.3.1 Overview of Zr

Ellis's whimsically anthropomorphic statement succinctly captures Zr's properties (Fig. 12.17). Zr has an even more strongly adherent, protective oxide layer than that of Ti; the two metals have the same HCP and BCC crystal structures; and like Ti, Zr is highly reactive, making it difficult and costly to reduce, refine, hot work, join, descale, cast, and machine. Like Ti, Zr oxide is much more widely used than Zr metal. However, Zr lacks the high strength/weight ratio of Ti, so Zr's structural use is limited to corrosive environments, particularly in nuclear fission reactor cores. A typical 1000-MW nuclear power plant contains 150 km of Zr alloy tubing weighing

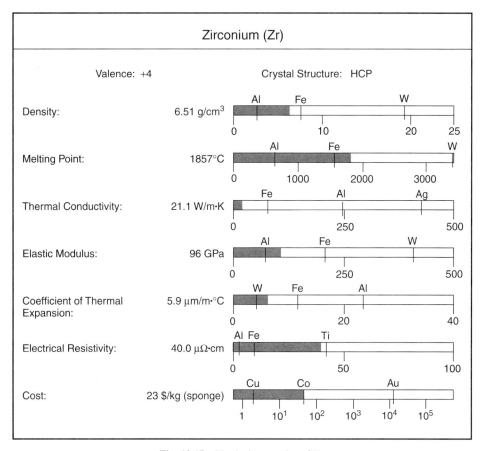

Fig. 12.17 Physical properties of Zr.

nearly 100 tons. Zr performs well in reactor cores due to its low thermal neutron capture cross section, its resistance to corrosion, and its good formability.

12.3.2 Physical, Mechanical, and Neutronic Properties of Zr

Zr is a relatively abundant element that is often present in many of the same mineral formations as Ti. The Kroll process for reduction of Ti (Sec. 12.2.2.1) has been adapted to produce sponge Zr metal. The reactivity of Zr makes reduction and purification challenging. Finely divided Zr is a fire hazard, so Zr machining operations need to minimize accumulation of chips and powder. The $\alpha_{HCP} \rightarrow \beta_{BCC}$ transformation in Zr occurs at 863°C with the BCC {110} planes forming parallel to the HCP (0001) planes and the HCP $\langle 11\bar{2}0 \rangle$ direction parallel to the BCC $\langle 111 \rangle$. Like Ti, Zr has an hexagonal ω phase (c/a ratio $= 0.617$) that is stable at high pressure. Formation of ω phase in Zr alloys produces yield strengths as high as 1200 MPa, but the ω phase embrittles the metal. Metastable ω can be formed by rapidly quenching pure Zr or by aging metastable Zr alloys. At room temperature and low pressure, the ω phase eventually transforms to α. Rapid quenching of Zr alloys from the β-phase field will produce an HCP α' martensite and in some cases, an orthorhombic α" phase. Beta-Zr can be retained to room temperature by rapid cooling of Zr alloys (e.g., Zr–Th).

Unalloyed Zr has a tensile yield strength of about 350 MPa and an ultimate tensile strength of 430 MPa with 30% elongation. α-Zr is ductile; fine-grained (10-μm grain size) polycrystalline Zr displays about a 50% reduction in area in tensile tests and 30% tensile elongation. The dominant slip system in Zr is the $\{10\bar{1}0\}\langle 11\bar{2}0 \rangle$; no basal slip is observed except in metal containing fine Zr hydride needles, which strongly suppress prismatic slip. With just one active slip system, the ductility of Zr relies heavily on twinning, particularly at low temperatures (Sec. 12.3.5.1). Interstitial O strengthens Zr and its alloys (Fig. 12.18) and lowers ductility.

Zr is a better thermal conductor than Ti, but its conductivity is still low. Zr has low electrical conductivity and an unusually low coefficient of thermal expansion (COTE). Since most ceramics also have low COTEs, this allows Zr metal to maintain ceramic coatings (such as its own oxide) over wide temperature ranges with minimal thermal stress accumulation. A protective ZrO_2 scale gives Zr excellent corrosion resistance in both aqueous solutions and at high temperature in air; corrosion resistance is a major factor for essentially all engineering uses of Zr. Certain interstitial

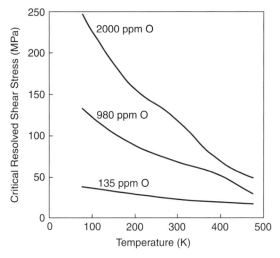

Fig. 12.18 Critical resolved shear stress for prism plane slip in Zr as a function of temperature and interstitial O impurity content. (Adapted from Soo and Higgins, 1968.)

TABLE 12.1 Comparison of Thermal Neutron Absorption Cross Sections of Several Elements[a]

Element	Thermal Neutron Absorption Cross Section (barns[b])	Element	Thermal Neutron Absorption Cross Section (barns[b])
Be	0.009	Cu	3.8
Mg	0.059	Ni	4.5
Zr	0.18	Ti	5.6
Zircaloy-4	0.22	Hf	113
Al	0.22	B	760
Sn	0.65	Cd	2,400
Nb	1.1	Gd	44,000
Fe	2.4		

[a] Thermal neutrons are neutrons released by fission whose kinetic energies have been moderated by repeated collisions with atoms near room temperature. Their average velocity is 2200 m/s at room temperature, or 3400 m/s at 400°C.
[b] A barn is 10^{-24} cm^2. The barn expresses the effective area the nucleus presents to the neutron for the absorption reaction. The larger the effective area, the greater the probability of reaction. The term originated from the phrase "as big as the side of a barn," used to describe a nucleus with a large cross section.

impurities, particularly N and H, degrade aqueous corrosion resistance, and measurement and mitigation of these effects has been a topic of intensive study in Zr and its alloys.

Zr's absorption cross section for thermal neutrons is low (Table 12.1). This makes the metal useful in nuclear reactor cores. Optimum reactor efficiency occurs when neutron absorption by ^{235}U nuclei in the core is high and nonproductive neutron capture is low. Zr alloys are used in nearly all nuclear reactors to isolate the U oxide reactor fuel and fission products from direct contact with the hot H$_2$O coolant while allowing the heat from fission in the fuel to be transferred to the H$_2$O. Zr's low thermal neutron capture cross section allows it to perform these mechanical and heat transfer tasks while absorbing few of the neutrons needed to sustain the fission chain reaction.

12.3.3 Commercial-Purity Zr and Zr Alloys

12.3.3.1 Pure Zr. Zr's corrosion resistance is exploited in various chemical processing operations. Pure Zr offers better corrosion resistance than Zr alloys against attack by hot acids and several other highly corrosive aqueous fluids. Zr occupies a market niche between Ta (which is more expensive than Zr but provides still greater corrosion resistance) and stainless steel and Ti (which are less costly but have somewhat lower corrosion resistance). Pure Zr performs well in hot mineral acids, organic acids, and alkali solutions. It is not resistant to concentrated sulfuric acid, Fe and Cu chloride solutions, and F^{-1} ions. Pure Zr's relatively high strength allows tubing in heat exchangers to be made with thin walls to improve heat transfer and lower cost. Pure Zr is also used in pumps, valves, and pipe for handling hot, corrosive fluids. Zr's market share in chemical processing applications has diminished in recent years as performance of polymeric materials in corrosive environments continues to improve.

12.3.3.2 Zr Alloys for Nuclear Reactor Core Components. A few nonnuclear uses exist for Zr, such as alloying additions in steel, Mg, Ni superalloys, and Ti alloys and as a "getter" in high-pressure Na light bulbs. But most Zr is used in nuclear power reactors. Certain alloys of Zr are much more corrosion resistant in superheated H$_2$O (up to 340°C) than pure Zr. Zr alloys are also stronger than pure Zr, so alloys are used in reactor cores for pipes, calandria tubes, and fuel cladding. Zr alloy options for nuclear applications are limited by Zr's large atomic radius, which means that relatively few elements have atoms large enough to dissolve extensively in Zr. Alloying options are further constrained by the desire to use alloying elements with low-thermal-neutron-absorption cross sections. About 50 ppm N is present in Kroll-processed Zr, and Sn reacts with the N, mitigating its deleterious effect on Zr's corrosion resistance. Sn also provides solid-solution hardening. Fe, Cr, and Ni improve the adherence of the oxide scale, although only small amounts are added because they reduce ductility and have rather high-neutron-absorption cross sections. Two commercial alloys have seen wide use in reactors designed in the United States and Europe: Zircaloy-2 (Zr–1.5% Sn–0.15% Fe–0.10% Cr–0.05% Ni) and Zircaloy-4 (Zr–1.5% Sn–0.20% Fe–0.10% Cr). Reactors designed in Canada and the Soviet Union/Russia often use Zr–2.5% Nb alloy, which is stronger than the Zircaloys but requires more complicated heat treatment to achieve adequate corrosion performance. In an operating reactor, these alloys rapidly form an adherent black oxide scale. The oxide is generally protective but is vulnerable to some corrosion situations involving complex interactions of radiation effects (Secs. 7.5 and 12.3.5.2), H absorption (Sidebar 12.4), and stress corrosion cracking.

12.3.4 Sources of Zr and Refining Methods

World production of zircon sand ($ZrSiO_4$) totaled 1.07 million tons in 2001; Australia and South Africa are the largest zircon producers. Zircon can be used directly or processed to produce ZrO_2. These materials are used in ceramicware, refractory brick, foundry sand, and pottery glazes and opacifiers. Less than 1% of the zircon mined each year is reduced to metal; approximately 5000

SIDEBAR 12.4: HYDRIDE FORMATION IN ZIRCALOY TUBING

The Zircaloy alloy fuel rod cladding of an operating nuclear reactor is unavoidably exposed to H from small amounts of H_3O^+ in the cooling water and from the corrosion reaction $Zr + 2H_2O \rightarrow ZrO_2 + 4H$. Zircaloy's oxide layer retards, but does not prevent, H diffusion into the metal. Fortunately, Zr is fairly tolerant of H at reactor operating temperatures (about 300°C); H concentrations of a few hundred ppm accumulate as a solid solution in the alloy without seriously lowering ductility. Although much of the H remains in solid solution in the metal, ZrH_x ($1.59 < x < 2$) can precipitate. In metal bearing only low stresses, the hydride lamellae form on grain boundaries, at dislocation tangles or cell walls, and at intermetallic precipitates in the alloy, and these hydrides cause no major changes in the alloy's mechanical properties. (At very high H concentrations, the hydrides sharply lower ductility by forming a continuous, interconnected network along the grain boundaries; however, such high H concentrations do not occur during normal reactor operation.)

Hydrides' biggest threat to cladding integrity is their habit of forming oriented platelets in metal with high elastic strains. In strained metal, hydride lamellae orient themselves perpendicular to the direction of a tensile stress (Fig. 12.19) or parallel to the compressive stress direction. If Zircaloy tubing is under high stress during hydride formation, the hydrides can align themselves parallel to the tube's radial direction. This arrangement poses a threat of cracking the tubing either from residual fabrication stresses or from pressure on the tube in service.

This problem has been addressed in three ways. First, Ni absorbs H strongly, and the newer Zircaloy-4 alloy has a maximum Ni content of only 70 ppm, compared to 300 to 800 ppm Ni in Zircaloy-2. For this reason, H absorption in Zircaloy-4 alloy is reduced by two-thirds vis-à-vis Zircaloy-2. Zr–Nb alloys also absorb less H than Zircaloy-2 because they lack the H absorbing $Zr(Fe,Cr)_2$ and $Zr_2(Fe,Ni)$ intermetallics present in Zircaloys. Second, the threshold stresses for hydride reorientation have been determined to be 180 to 220 MPa for Zr–2.5% Nb alloy and 80 to 110 MPa for Zircaloy. By restricting the operating pressure of the reactor coolant to stay below these values, the formation of radially oriented hydrides is avoided. Third, research has shown that control of texture formation and residual stress levels during fabrication and annealing of Zr reactor components (particularly in tube-forming operations) can reduce the tendency to form stress-oriented hydrides. These changes have reduced Zr alloy failures in commercial power reactors to near zero.

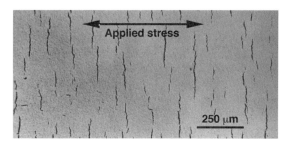

Fig. 12.19 Effect of applied tensile stress on the orientation of hydride platelets in Zircaloy-2. The darker gray vertical bands are Zr hydride precipitates. A stress of 165 MPa was maintained in the direction indicated while the hydride precipitates were forming. The precipitates lie perpendicular to the stress direction. (From Marshall, 1967; with permission.)

tons of Zr metal was produced in 2002, with nuclear applications accounting for 90% of metal sales. Zr metal production data are not publicly disclosed, but substantial capacity for Zr metal production exists in the United States and France, with smaller capacity in India. The primary Zr ore is zircon sand ($ZrSiO_4$), which often occurs mixed with Ti-bearing sands. The two major producers are Australia and South Africa; the Ukraine and the United States produce smaller amounts. Zircon sand typically contains about 2% Hf, which is a problematic impurity for nuclear use of Zr. Hf has a high-neutron-absorption cross section (Table 12.1) and is difficult to separate from Zr. The two elements behave quite similarly in most chemical reactions because they have the same valence and almost exactly equal atomic radii (Sec. 1.6). Despite their similarities, a solvent extraction method has been developed to separate the elements from their mixed tetrachlorides. Many sequential extractions are needed to achieve the low Hf content (Hf < 100 ppm) required for nuclear-grade Zr. The process also yields Hf containing about 4% Zr as a by-product. A newer, pyrometallurgical method of extractive distillation of molten Zr (Hf) chloride shows promise to lower separation costs.

12.3.5 Structure–Property Relations in Zr

12.3.5.1 Role of Twinning in Zr Plasticity. In most metals twinning plays a secondary role in plastic deformation, serving to reorient portions of a grain to facilitate further dislocation motion. However, in some metals, twinning plays a more important role, and Zr provides a striking example of a ductile metal that deforms almost solely by twinning under some conditions. The $\{10\bar{1}0\}\langle 11\bar{2}0\rangle$ is the only active slip system (Sec. 12.3.2) in Zr under most circumstances. Basal slip is seen only in metal containing Zr hydride particles (Sidebar 12.4).

By itself, the $\{10\bar{1}0\}\langle 11\bar{2}0\rangle$ slip system is insufficient to satisfy the von Mises criterion (Sec. 4.6) for extensive ductility in polycrystalline metal, yet pure HCP Zr displays good ductility at all temperatures from 77 K to the BCC transformation at 1136 K. The source of this surprising ductility is Zr's ability to twin on four different twinning planes: $(11\bar{2}1)\langle \bar{1}\bar{1}26\rangle$, $(10\bar{1}2)\langle \bar{1}011\rangle$, $(11\bar{2}2)\langle 11\bar{2}\bar{3}\rangle$, and $(11\bar{2}3)\langle \bar{1}\bar{1}22\rangle$. As Fig. 12.20 shows, $(11\bar{2}1)\langle \bar{1}\bar{1}26\rangle$ is the most common twinning mode, but all four twinning modes participate over the range 77 to 600 K. Twinning at 77 K is complex and appears to account for nearly all of the initial plastic deformation (Fig. 12.21). Figure 12.22 shows second-order twinning (twins within twins) and shear deformation in neighboring grains caused by the twinning displacements. The dashed lines in Fig. 12.21 seem to suggest that all of the initial plastic strain was attributable to twinning; however, at least a small amount of dislocation slip must occur because the twins did not collapse and disappear when the stress was removed after small strains. Still, it is clear that deformation occurs primarily by twinning at 77 K, with the role of dislocation slip increasing as strain progresses. Although twinning plays a major role in the plasticity of other metals [e.g., Mg (Sec. 11.4.2.2) and α-U (Sec. 24.4.3.1)], the dominant role played by twinning in Zr at 77 K is extraordinary.

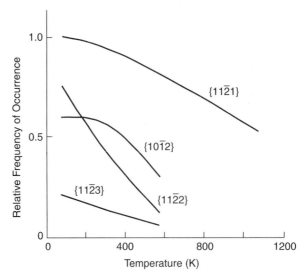

Fig. 12.20 Variation of twinning frequency with deformation temperature. Note the dominant role played by twinning on the $\{11\bar{2}1\}$ planes at all temperatures. Zr transforms to the BCC structure at 1136 K. (Redrawn from Rapperport and Hartley, 1960; with permission of AIME, *www.aimehq.org*.)

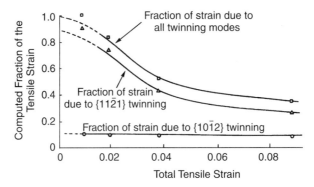

Fig. 12.21 Twinning fractions of plastic strain in Zr deformed at 77 K. Twinning accounts for nearly all the initial strain and a large fraction of deformation at higher strain levels. (Redrawn from Reed-Hill and Dahlberg, 1966; with permission of the Electrochemical Society, Inc.)

12.3.5.2 Radiation Damage in Zr Alloys. Intense radiation alters metals' mechanical properties drastically (Sec. 7.5). In the Zircaloy alloys used in reactor cores, a high fluence of high-energy neutrons more than triples yield strength while reducing elongation from 30% to 2% (Fig. 12.23). These changes result from high concentrations of self-interstitials, vacancies, and dislocation loops that accumulate as neutrons knock the metal atoms out of their normal lattice positions. Radiation and cold work have similar effects on the mechanical properties of metals. Cold-worked metal recovers and recrystallizes by self-diffusion during annealing (Sec. 3.3), and the damage induced by radiation can also be removed by annealing. In the case of Zircaloy-2 and Zircaloy-4, the damage caused by radiation anneals away almost as rapidly as it forms if the irradiation occurs above about 380°C. In Fig. 12.23 the Zircaloy was irradiated at 290°C, just below the recovery temperature. For most commercial pressurized water reactors, the water coolant is maintained at about 315°C. (Water remains liquid at 315°C under

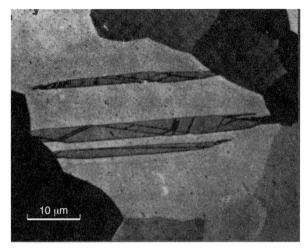

Fig. 12.22 Optical micrograph of twins in Zr showing second-order twinning (twins within twins) and deformation across grain boundaries in neighboring grains caused by the twinning shear displacements. (From Reed-Hill, 1962.)

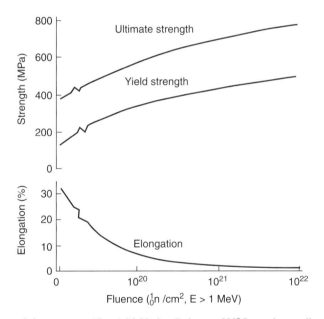

Fig. 12.23 Effect of fast neutron ($E > 1$ MeV) irradiation at 290°C on the tensile test behavior of Zircaloy-2. The metal was fully annealed when irradiation began. Fluence is the total number of neutrons that have passed through a 1-cm^2 plane within the specimen. (Redrawn from Schemel, 1977.)

the 160-atm pressure maintained in the core.) The temperature of the fuel inside the Zircaloy cladding is higher, and a temperature gradient exists across the 0.5 to 1.0 mm thickness of the cladding. This results in most of the cladding radiation damage recovering almost as rapidly as it occurs during normal reactor operation. A second challenge to Zr use in nuclear reactors is posed by H embrittlement effects (Sidebar 12.4), but these problems have been largely solved in modern commercial reactors. Only about one cladding failure occurs per 100,000 fuel rods for cladding subjected to a fluence of 8×10^{21} neutrons/cm^2.

12.4 HAFNIUM

Hf metal is metallurgically and chemically similar to its two congeners but has a higher melting temperature (2222°C), a higher HCP → BCC transformation temperature (1760°C), and much greater density (13.28 g/cm^3). Hf's coefficient of thermal expansion is low at 5.9×10^{-6}°C^{-1}, which is similar to that of Zr. Its electrical and thermal conductivities are also low, 35.1 μΩ·cm and 22.3 W/m·K, respectively. The Young's modulus of Hf is 137 GPa. Hf has a high-thermal neutron absorption cross section (Table 12.1), and it has been used as a control rod material in naval shipboard nuclear reactors and a few commercial reactors. Its adherent, protective oxide layer allows it to be used in reactors with no additional cladding. Pure Hf is moderately strong and has good ductility ($\sigma_y = 230$ MPa, $\sigma_{UTS} = 450$ MPa, 23% elongation). As with Ti and Zr (Sec. 12.2.2.1), the reactivity of Hf and its proclivity for interstitial embrittlement impose special constraints and costs on all hot-working, joining, machining, and descaling operations. The only alloying addition used in reactor applications is O, which increases strength while reducing ductility. The cost of Hf control rods is higher than that of competing materials, and Ag–Cd–In control rods are generally used in preference to Hf in commercial nuclear powerplants. Hf metal has been used as a grain refining addition to Ni superalloys and as a nozzle material for plasma arc cutting; HfO$_2$ has limited use in high-temperature ceramics. Hf is a by-product of the production of nuclear-grade Zr, an arrangement that results in growing stockpiles of excess Hf. Less than 100 tons of metal is produced per year. The quoted price of $180 per kilogram for Hf sponge is somewhat arbitrary and reflects the cost of reduction to the metallic form of a reactive, little-used metal.

ADDITIONAL INFORMATION

References, Appendixes, Problem Sets, and Metal Production Figures are available at
ftp://ftp.wiley.com/public/sci_tech_med/nonferrous

13 V, Nb, and Ta

13.1 OVERVIEW

Vanadium, niobium, and tantalum comprise group 5 of the periodic table. The outer electron configuration is $d^3 + s^2$, which can hybridize to $d^4 + s^1$. The five bonding electrons make V, Nb, and Ta refractory metals, and each of these elements is among the highest-melting metals of its period. Group 5 elements are quite electropositive, which makes them reactive and difficult to reduce to the metallic state. The metals have good to excellent corrosion resistance in aqueous solutions but poor elevated temperature oxidation resistance. V and Nb are used mostly as alloying additions to steel, Ti, and Ni superalloys. Ta's dominant use is in electronic components, although it has other uses in chemical processing, biomedical devices, and superalloys.

13.2 HISTORY AND PROPERTIES

V was the first group 5 element to be used in the metallic state. In the late nineteenth century, it was discovered that adding only 0.05 to 0.2% V to steel substantially increases yield strength and fracture toughness. Pure V metal was difficult to produce with the technology available a century ago, but Fe–V alloys (ferrovanadium) were readily produced by reduction of V_2O_5 with Fe and Al. Use of V steel grew rapidly in the early twentieth century for armor, tool steel, and automobile components. Ferrovanadium and other V master alloys for steel production have dominated V use for more than a century and comprise 80% of today's V market. Although pure V is higher melting and lighter than steel (Table 13.1), it is somewhat costly, toxic, and has a nonprotective oxide that precludes high-temperature use.

Almost 90% of all Nb used is added to steel as ferroniobium, which strengthens steel and improves the weldability of stainless steel. Most of the remaining 10% of Nb production is used in Ni and Zr alloys, corrosion-resistant materials, high-performance superconductors, and cemented carbide cutting and grinding tools. Nb is less abundant (20 ppm in Earth's crust) than V (136 ppm). Ta is the least abundant (1.7 ppm) of the group 5 elements and the only one used extensively in the pure state. Although Ta is rather costly, its oxide layer is such an excellent dielectric that Ta dominates the market for high-performance capacitors. Ta's oxide layer has outstanding corrosion resistance in both strongly oxidizing and strongly reducing aqueous solutions, making it useful in difficult chemical processing applications for valves, heat exchangers, and tubing.

13.3 VANADIUM

> Always a bridesmaid, never a bride.
> —Colloquial expression

Structure–Property Relations in Nonferrous Metals, by Alan M. Russell and Kok Loong Lee
Copyright © 2005 John Wiley & Sons, Inc.

TABLE 13.1 Selected Room-Temperature Properties of Group 5 Metals with a Comparison to Cu

Property	V	Nb	Ta	Cu
Valence	+5	+5	+5	+1,+2
Crystal structure at 20°C	BCC	BCC	BCC	FCC
Density (g/cm^3)	6.11	8.57	16.65	8.92
Melting temperature (°C)	1915	2468	2996	1085
Thermal conductivity (W/m·K)	31	52.7	54.4	399
Elastic (Young's) modulus (GPa)	128	103	186	130
Coefficient of thermal expansion (10^{-6} m/m·°C)	8.4	7.3	6.3	16.5
Electrical resistivity (μΩ·cm)	24.8	12.5	12.4	1.7
Cost ($/kg), large quantities	14[a]	60[b]	200[c]	2

[a] Price based on V content of Fe–V alloy (ferrovanadium); pure V is several times more expensive and is very thinly traded.
[b] Price for pure Nb ingot. In Fe–Nb alloy (ferroniobium), the Nb basis cost is $15 per kilogram.
[c] Price for vacuum-melted Ta ingot. Prices for electronic-grade Ta are higher ($300 to 600 per kilogram).

Although pure V has almost no engineering use, tens of thousands of tons of V are used each year as alloying additions to steel and Ti. Ferrovanadium additions to steel precipitate V carbonitride particles that serve as dislocation barriers. V also refines the grain size of steel by slowing the austenite → ferrite transformation and initiating greater numbers of ferrite grain embryos along the advancing transformation interfaces. This prevents the grains that nucleate first from growing rapidly to large size. The result is fine-grained steel with higher yield strength, higher fracture toughness, and better weldability. One ton of steel containing 1 kg of V has the strength to replace 1.4 tons of ordinary steel. V is a β-phase stabilizer in Ti, and the "workhorse" Ti alloy, Ti–6% Al–4% V, commands nearly half the entire Ti market.

13.3.1 Physical Properties of V

V is a ductile BCC metal with a high melting temperature, moderate density, and a low coefficient of thermal expansion. It has fairly good corrosion resistance in seawater and mild acids, but its elevated temperature oxidation resistance is poor because its most stable oxide melts at 678°C. V is a superconductor below 5.1 K, and some V intermetallic compounds have excellent properties for superconducting electromagnets (Sec. 13.3.2.5). V and its compounds are toxic and require careful handling to avoid inhalation of V-containing powders.

13.3.2 Applications of V

At first inspection, V appears to offer several appealing attributes for use as a structural metal. Compared to Fe, V has a higher melting temperature, better ductility, a lower coefficient of thermal expansion, lower density, and better aqueous corrosion resistance. However, V poses several difficulties that render it uncompetitive with other structural metals. Among these are its toxicity (Sec. 13.3.3), low-melting oxide, and relatively high cost. Consequently, pure V has no structural applications (see, however, Sec. 13.3.2.6).

13.3.2.1 Tool Steels. Tool steels containing 0.5 to 5% V have been used for more than a century. V is a strong carbide former in steel, and V carbides dissolve only partially in steel's austenite phase. The undissolved V carbide pins austenite grain boundaries to minimize grain growth during austenitizing prior to quenching. The V that does dissolve in austenite increases the hardenability of the steel during quenching, and during tempering it forms carbide and nitride particles only a few nanometers in diameter that raise hardness. V carbides are stable in ferrite at elevated temperature, reducing the hardness loss that occurs when hardened-steel cutting tools heat during high-speed machining (Fig. 12.5).

13.3.2.2 HSLA Steels. In the 1960s a series of high-strength low-alloy (HSLA) steels (sometimes called *microalloy steels*) were developed to provide better strength and weldability relative to plain carbon steels without adding large amounts of costly alloying additions or providing a full quench-temper heat treatment. HSLA steels achieve yield strengths of 550 MPa with surprisingly lean alloy compositions (e.g., 1.35% Mn, 0.45% Si, 0.12% C, 0.12% V, 0.04% P, and 0.02% Nb, N, and S). This is achieved by ferrite grain size refinement (i.e., Hall–Petch hardening, Sec. 3.2), suppression of grain growth in the austenite phase by V and Nb carbides and nitrides, and precipitation of extremely fine V(C,N) particles during air cooling after hot rolling. Nb additions retard austenite recrystallization and provide nucleation sites for ferrite grain formation during cooling. HSLA steels have higher yield strength, higher fracture toughness, better ductility, and easier weldability than those of plain carbon steel. The superior welding performance results from the low C content of these steels (which avoids martensite formation in the weld zone), the ability of V carbide particles to retard grain growth in the heat-affected zone, and the tendency of V to retard bainite formation. HSLA steel has been particularly useful in oil and gas pipelines (Sidebar 13.1); offshore oil drilling; automobile brackets, supports, and braces; rail steel for railroads; and truck and train car/trailer body panels.

13.3.2.3 V Additions to Forging Steels and Cast Iron. V additions also improve performance in forging steels, which typically involve thick sections of metal (e.g., steam turbine rotors, driveshafts to connect ship engines and propellers) that are hot forged to achieve their near-final shape. In these steels, V inhibits grain growth during hot working. As the steel cools, V serves as a ferrite grain refiner; fine V(C,N) precipitates strengthen the metal. Such large components are too massive to quench, making heat treatment to produce tempered martensite infeasible; V raises strength without the need for such heat treatments. High-strength rebar for steel-reinforced concrete construction is frequently made from V forging steel, and V provides similar strengthening and wear-resistance improvements to cast iron.

13.3.2.4 V Additions to Ti Alloys. Nearly 10% of V production is used as alloying additions to Ti alloys. V stabilizes the high-temperature BCC phase of Ti. V is used in $\alpha-\beta$ Ti alloys to strengthen the metal (Sec. 12.2.3.4) by forming HCP + BCC microstructures. It is also used in β-Ti alloys (Sec. 12.2.3.3), where the high-temperature BCC phase is retained as a metastable phase at room temperature for improved formability and better high-temperature strength.

SIDEBAR 13.1: V STEEL IN THE TRANS-ALASKA OIL PIPELINE

The 1968 discovery of large petroleum reserves in the United States in northern Alaska presented an opportunity to market a major new source of petroleum. But this remote location posed problems in delivering the crude oil to refineries and users. Northern Alaska's Arctic Ocean coastline is clogged with ice floes most of the year, making ocean tanker transport infeasible. The only reasonable option was transporting the oil south by pipeline. Much of the overland route to the ice-free ports in southern Alaska is a wilderness of tundra, mountains, rivers, and forest, so construction of this pipeline was one of the more challenging engineering projects of the later-twentieth century.

A key factor in the successful construction of the Alyeska pipeline from Prudhoe Bay to Valdez, Alaska was the availability of HSLA steel for the pipe material. The steel in the pipeline contains a total of 650 tons of V. The good weldability of V steel allowed reliable joining of pipe sections under difficult field conditions. HSLA steel achieves high strength without high C content; this improves weldability and keeps the steel's ductile–brittle transition temperature low. This strength and fracture toughness permit use of higher pumping pressures to move the oil more rapidly. To prevent the hot oil from melting the permafrost underground, much of the pipe is suspended above ground (Fig. 13.1). The strength of HSLA steel was an important design parameter to bear the weight of the steel and oil suspended between supports. Since operation began in 1977, the pipeline has carried more than 13 billion barrels of crude oil to the ice-free harbor in Valdez, Alaska.

Fig. 13.1 Alyeska trans-Alaska oil pipeline. Supports suspend the HSLA steel above the frozen ground so that the hot petroleum will not melt the permafrost in the ground below. V and Nb in the HSLA pipe steel allowed reliable in-field welding and give the pipe sufficient strength to span the 18-m distance between supports. The pipe is 1.22 m in diameter and 1280 km long. Thermal expansion and contraction of the pipe is accommodated by lateral movement of the sections on the supports; the empty pipeline is straight at $-60°C$ and assumes a slight zigzag shape when filled with warm petroleum. (Photo provided courtesy of *en.wikipedia.org/upload/b/b7/Alaska_Pipeline.jpg*, 2004.)

13.3.2.5 V Superconductors. Pure V is a superconductor below 5.1 K, and V intermetallic compounds [V_3Ga, V_3Si, and $V_2(Zr,Hf)$] have high critical currents in the presence of strong magnetic fields (Fig. 13.2). The Nb_3Sn intermetallic superconductor enjoys wider application than that of V intermetallics, but V-based superconductors are chosen for use in certain combinations of magnetic field, current, and mechanical stress.

13.3.2.6 V in Fusion Reactor First-Wall Construction. The neutronic and physical properties of V are attractive to scientists and engineers designing future nuclear fusion power reactors (Sidebar 10.2). When high-energy neutrons bombard an element, some of its nuclei absorb one or more neutrons and are transmuted into other isotopes and other elements. Many of those transmutation products are radioactive, and some emit intense, penetrating γ rays that pose severe safety hazards to workers who will maintain and eventually disassemble the power plant. High-energy neutrons are unavoidable in nuclear fusion, but V forms relatively benign transmutation products. The two natural V isotopes, $^{50}_{23}V$ and $^{51}_{23}V$, transmute by neutron capture into $^{52}_{23}V$, which decays within minutes to $^{52}_{24}Cr$. Subsequent neutron capture in $^{52}_{24}Cr$ leads to a long sequence of stable or short-lived Cr, Mn, and Fe isotopes that emit minimal radiation. The radiation that does exist decreases rapidly after the reactor is shut down. These favorable nuclear properties combine with V's good physical properties (good ductility, high melting temperature, resistance to attack by liquid Li or Na metal coolants, low density, and acceptable thermal conductivity) to make it a good candidate for construction of the first wall of future fusion reactors.

13.3.2.7 Use of V Compounds. V has a range of valences, from +2 to +5, that make its compounds versatile catalysts. V_2O_5 accelerates reactions such as the oxidation of SO_2 to SO_3, which is then dissolved in water to form sulfuric acid (H_2SO_4). Ammonium metavanadate is extensively used in the manufacture of polymers and ceramic pigments, and $YVO_4:Eu^{3+}$ produces the red color in some television and computer display monitors (Sidebar 23.1).

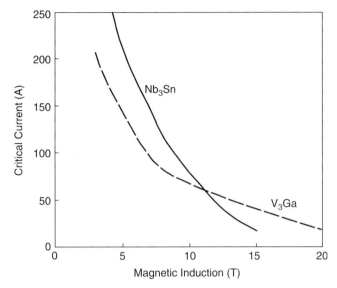

Fig. 13.2 Critical current versus magnetic induction for two superconducting materials tested as 0.4-mm-diameter wire at 4.2 K. Although Nb$_3$Sn outperforms V$_3$Ga at lower magnetic field strengths, V$_3$Ga has a higher critical current in strong magnetic fields. NbTi alloy (not plotted here) has lower superconductive performance than that of V$_3$Ga and Nb$_3$Sn, but NbTi is the most frequently used of these three materials because its ductility makes it much easier to form into wire.

13.3.3 Toxicity of V

V$_2$O$_5$, V metal, ferrovanadium, V chloride, and V–Al alloys are all toxic. Inhalation of V-bearing powders causes pulmonary edema, cough, and chest pain. V inhalation can be fatal in high doses. V is also toxic if swallowed or injected into the bloodstream, but those methods of contamination are more easily prevented. Patients receiving a survivable dose typically recover with no apparent long-term damage. Some petroleum and coal deposits contain V, and their combustion poses risks of V poisoning to those in close proximity to the exhaust gases, fly ash, and residue in oil-fired boilers. No danger of long-term, low-dosage exposure has been demonstrated conclusively, but this is a topic of continuing research.

13.3.4 Sources of V and Refining Methods

Essentially all V metal is used in alloying addition to steels and other metals, so most V production is geared toward producing Fe–V and Al–V master alloys to make specialty steels and Ti–6% Al–4% V (Sec. 12.2.3.4). V ore is widely distributed around the world but generally contains low concentrations of V (0.1 to 1.7% V$_2$O$_5$). Consequently, it is often produced as a co-product with other metals (frequently, Fe) to hold down costs. In fact, the merit of V as a steel additive was originally discovered because certain ores were a natural mixture of Fe and V oxides that produced V steel "naturally." V is also present in significant quantities in various coal and petroleum deposits (particularly from Venezuela and Mexico), and it can be recovered economically from combustion flue dust. The principal intermediate product of V mining and extraction is V$_2$O$_5$, most often obtained by a variety of processes based on leaching, roasting, and calcining. Ferrovanadium (an alloy of roughly 80% V–20% Fe) is then produced by reacting V$_2$O$_5$, Al metal, and Fe scrap to form Al oxide and ferrovanadium. Production of ferrovanadium is considerably easier and less costly than production of pure V metal, which requires Ca metal reduction of V$_2$O$_5$ followed by electron-beam melting and/or fused salt electrorefining to achieve 99.9% purity.

13.4 NIOBIUM

> You can spot old American metallurgists from a thousand miles away; their e-mails refer to niobium as "columbium."
>
> —Professor Francis X. Kayser, Iowa State University, 1996

13.4.1 Physical Properties of Nb

Element 41 was often called *columbium* until the international chemists' union agreed in 1950 to use the name *niobium*; a few troglodytic metallurgists still hold resolutely to the old name, adjuring their children against use of the "n-word." When its interstitial impurity content is low, Nb has an unusually low work-hardening rate and almost unlimited room-temperature ductility, tolerating wire drawing or rolling to well over 90% reduction without annealing. Only five metals have melting points higher than Nb's (W, Re, Ta, Os, and Mo), and all of those are much denser than Nb. Nb's Young's modulus is surprisingly low for such a refractory metal. Nb is a low-temperature superconductor ($T_C = 9.46$ K), and several of its alloys and intermetallic compounds are widely used in superconducting electromagnets (Sec. 13.4.2.5). Nb's oxide layer provides excellent protection against attack by acid and other corrosive aqueous solutions, but hot Nb ($>200°C$) oxidizes rapidly in air. Unlike its toxic congener element V, Nb is nontoxic and is so inert to body fluids that it can be used in biomedical implants.

13.4.2 Applications of Nb

Nb's main uses are as alloying additions to steel, stainless steel, Ni/Co superalloys, and Zr alloys. Smaller quantities are used in chemical processing components, superconducting electromagnets, cemented carbide cutting tools, and acoustic and electrooptical components.

13.4.2.1 Nb in Steel and Stainless Steel. Nb behaves similarly to V in steel (Secs. 13.3.2.1 to 13.3.2.3); it is a ferrite stabilizer and forms carbides that resist dissolving, even at austenitizing temperatures. Nb is usually present in tool steels and HSLA steels, and it is sometimes added to stainless steel and cast iron. Nb carbonitride precipitates that form in high-temperature steel nucleate preferentially on austenite grain and subgrain boundaries, which pins the boundaries and minimizes grain growth in austenite. If Nb carbonitride forms during the $\gamma \rightarrow \alpha$ transformation, sheet-type precipitates form along the advancing γ–α interface. Precipitates forming after the $\gamma \rightarrow \alpha$ transformation ends are smaller and more uniformly distributed throughout the grains. These precipitates and their accompanying strain fields are potent hardeners in steel. HSLA steel (Sec. 13.3.2.2) typically contains between 0.012 and 0.06% Nb and is widely used in applications where weldability, good toughness at low temperature, and a high strength/weight ratio are sufficiently important to justify HSLA steel's somewhat higher cost (Sidebar 13.1).

Nb improves stainless steel's weldability. Stainless steel contains high Cr levels ($>11\%$) to form a protective Cr_2O_3 surface layer. However, if stainless steel is held even briefly (several seconds) between about 500 and 900°C, the Cr reacts with the C in stainless steel to form $Cr_{23}C_6$ precipitates on grain boundaries. This removes Cr from solid solution near grain boundaries, leaving it vulnerable to rust, a phenomenon called *sensitization* or *weld decay*. At the mill, stainless steel is quenched rapidly to avoid sensitization, but welding sensitizes the heat-affected zone near the weld. Adding Nb to stainless steel mitigates this problem, because Nb is a stronger carbide former than Cr. Sufficient Nb can be added to "stabilize" the stainless steel by reacting with most of the C to form Nb carbides, leaving little C to react with the Cr. This maintains the protective Cr oxide surface layer, preventing corrosion near the weld. Both Nb and Ti can be used for stabilizing stainless steel. Ti costs less, but Nb is preferred because Ti reacts more strongly with S and O, forming Ti oxide and sulfide particles that make some of the Ti unavailable for carbide formation and produce surface defects. Nb carbides have the additional benefit of increasing creep strength in stainless steel.

13.4.2.2 Nb in Superalloys. Nb contents of 1 to 6% are common in many Ni-, Co-, and Fe-based superalloys (Sec. 16.3.3.6). Nb contributes to all of the strengthening mechanisms used in superalloys to maintain useful strength to temperatures as high as 1100°C. Nb atoms are larger than Ni, Co, and Fe atoms, so Nb acts as a solid-solution hardener. Nb has limited solubility in the FCC γ phase of Ni and precipitates preferentially to the coherent γ' phase; these are ordered intermetallic phases that strain the γ matrix and strengthen the superalloy at high temperatures. Nb increases the antiphase boundary energy (Sec. 25.6.1) of γ', inhibiting superdislocation motion through the γ' precipitates. Finally, Nb is a strong carbide former; Nb carbides precipitate on grain boundaries and minimize creep caused by grain boundary sliding.

13.4.2.3 Nb in Zr Reactor Alloys. Nb has a fairly low thermal neutron capture cross section (1.1 barns) and good solubility in Zr, which makes it one of the rather limited number of acceptable alloying additions to Zr for nuclear fission reactor core structures, pipes, and fuel cladding (Sec. 12.3.3.2). Alloys of Zr containing 0.5 to 2.5% Nb were initially used primarily in Canadian and Soviet reactors, but they have more recently taken a larger share of the market from the Zircaloys (Zr–Sn alloys) in commercial power reactors worldwide. Nb stabilizes the BCC β phase of Zr, and its presence (sometimes with 0.5% Cu) allows solutionizing, quenching, and deliberate overaging heat treatments to produce two-phase microstructures that are stronger than Zircaloy and are stable after irradiation to high fluences.

13.4.2.4 Nb in the Chemical Processing Industry. The undisputed champion of aqueous corrosion resistance is Ta (Sec. 13.5.1); however, Ta is expensive. Processing equipment can often deliver satisfactory service with less costly materials, such as stainless steel, Ti, Zr, polymers, Nb, or Nb–Ta alloys. Although Nb is not quite as corrosion resistant as Ta, its excellent fabricability, lower cost, and lower density make it useful in corrosive media that do not require Ta. Alloys of Ta containing up to 50% Nb have almost the same corrosion resistance as pure Ta, but are lighter and less expensive.

13.4.2.5 Superconducting Nb Alloys and Intermetallic Compounds. Ceramic compounds (e.g., $YB_2Cu_3O_{7-x}$ where B = Ba or Sr) have the highest superconducting transition temperatures, but metallic and intermetallic superconductors are more useful in many superconducting devices (e.g., high-field magnets) because they maintain their superconductivity in stronger magnetic fields while carrying larger currents. Nb is a key constituent of most commercial metallic and intermetallic superconductors. Some of the best-performing superconductors are BCC Nb–Ti and Nb–Ti–Ta alloys, which are generally used as multifilamentary wires in a Cu matrix. Electric current travels in an extremely shallow layer near the surface of a superconductor, so performance is enhanced by bundling many small filaments into the wire (Fig. 13.3) rather than using a monolithic wire of superconducting material. Cu is not a superconductor, so the Cu matrix separates the individual superconducting filaments (e.g., NbTi alloy), acting almost as an insulator between them. A thin V layer is generally used to prevent undesirable Cu diffusion into the superconducting material (one of the very few uses for pure V). If the superconductivity fails, the Cu carries the current briefly before it heats excessively, allowing time to shut down the device before the high current destroys the wire in its nonsuperconducting state. Although this microstructure requires elaborate, multistep drawing and heat-treating processes, its enhanced performance justifies the expensive production procedure.

The superconductors with the greatest ability to carry high currents in the presence of strong magnetic fields (Fig. 13.2) are the A15 (cP8) intermetallic compounds, such as Nb_3Sn, Nb_3Al, Nb_3Ge, or $Nb_3(Al,Ge)$. Nb_3Sn is most frequently used for superconducting magnets in medical magnetic resonance imaging, magnetic levitating trains, and high-energy particle accelerators. These intermetallics are brittle (Sec. 25.6) and cannot be drawn into wire. However, Nb_3Sn wire can be produced by drawing dendritic cast Cu–Nb alloy (Fig. 18.13), bundles of Cu and Nb rods, or "jelly rolls" of Cu and Nb to produce microstructures similar to that shown in Fig. 13.3. That two-phase Cu–Nb wire is then immersed in molten Sn, so Sn can diffuse into the Cu and Nb to form Nb_3Sn filaments in a bronze matrix, the final form of the superconducting wire.

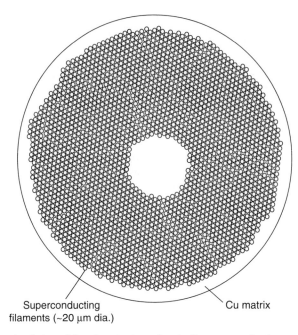

Fig. 13.3 Cross-sectional view of the microstructure of a wire in a superconducting magnet. Current travels in a shallow layer near a superconductor's surface, so performance is enhanced by bundling many small filaments rather than using one large wire of superconductor. Cu is not a superconductor, so the Cu matrix separates the individual superconducting filaments (e.g., NbTi alloy or Nb_3Sn), acting like an insulator between them.

13.4.2.6 Nb Carbide in Cutting Tools. WC is widely used as a hard, wear-resistant material for rock and masonry drills, metal cutting, and hard-face coatings on wear-prone surfaces (Sec. 14.4.3.5). WC reacts with ferrous workpieces at high cutting temperatures (Fig. 12.5); it dissolves rapidly into hot Fe near 1000°C, causing rapid tool wear. Ta and Nb carbides are much less soluble in hot Fe, and they improve hot hardness and hot wear resistance when added to WC cutting tools used on steel and cast iron.

13.4.2.7 Nb Compounds in Optical Components and Transducers. Nb compounds are versatile optical and electronic materials. Many additives will increase glass's refractive index, but Nb_2O_5 does so without greatly increasing the density of glass, making it popular for weight-sensitive products (e.g., eyeglasses, binoculars, and camera lenses). Nb_2O_5 is used in multilayer barium titanate ($BaTiO_3$) capacitors to control grain size and modify the temperature coefficient of the dielectric constant. Lithium niobate is used as an electrooptic modulator; its refractive index can be changed by applying an electric field to the crystal, which allows laser beams to be modulated, deflected, or encoded in laser optic communications devices. Nb's electronic and optical applications account for only a tiny fraction of global Nb consumption, but they are high-value products.

13.4.2.8 Use of Nb Alloys at High Temperature. Nb has several desirable qualities for use as a high-temperature structural material. Nb has a high melting temperature, good room-temperature formability and fracture toughness, and the lowest density of all refractory metals. Numerous alloying projects have sought to make Nb-based alloys the "next-generation" metals to replace Ni- and Co-based superalloys (Secs. 16.2.3.1 and 16.3.3.6); however, these efforts have led to only a few applications. The poor high-temperature oxidation resistance of pure Nb can be greatly improved by alloying with Al and Cr, but those alloys have near-zero ductility at room temperature and are difficult to fabricate. The hot strength and creep resistance of Nb–W alloys are quite impressive, offering useful strength to 1400°C. For example, at 1315°C, Nb–10% W–10% Hf–0.15% Y has $\sigma_{UTS} = 180$ MPa and a 10-hour rupture stress of 103 MPa. But these alloys suffer catastrophic oxidation damage if used at high temperature in air. Coated Nb alloys have seen limited use as nozzle flaps for jet engines, rocket combustion chambers, nose cones, and wing leading-edge structures for hypersonic aircraft. A variety of coatings has been tried with varying degrees of success to protect Nb alloys at high temperature in air. Nb silicide coatings containing Cr and Fe have performed reasonably well in service, but all coatings are vulnerable to problems from differing coefficients of thermal expansion between the metal and the coating, which can cause cracking and spalling during thermal cycling. In some systems, intermediate coatings have been applied between the Nb and the protective coating to form a liquid at the planned operating temperature that will accommodate COTE mismatches, but none of these systems has been sufficiently effective to allay completely the fears of safety-conscious designers. The fundamental problem remains that a coating failure at high temperature leads to rapid destruction of the underlying metal.

13.4.3 Sources of Nb and Refining Methods

Brazil supplies most of the world's Nb; Canada is the only other producer of any size. Numerous Nb-bearing ore bodies exist in other nations, but those deposits are not economically competitive with the rich Brazilian and Canadian deposits. Most Nb ores also contain some Ta, and vice versa. The two metals are separated by solvent extraction. Since Nb is used primarily as an alloying addition to other metals, most Nb is produced as ferroniobium and Ni–Nb master alloys for addition to steels and Ni superalloys (Sec. 16.3.3.6). Ferroniobium is produced by Al reduction of Nb_2O_5 with Fe and NaCl additions, similar to the method used for producing ferrovanadium (Sec. 13.3.4). Although ferroniobium dominates the Nb market, smaller amounts of Ni–Nb master alloy are produced for use in Ni superalloys. Pure Nb is used in chemical process equipment, fabrication of superconducting wire, medallions, jewelry, and biomedical implants. Pure Nb can be produced by several methods, but the three most widely used approaches are Al reduction of Nb_2O_5, Mg reduction of $NbCl_5$, and Na reduction of K_2NbF_7; all of these are followed by electron-beam melting or vacuum arc melting for purification.

13.4.4 Structure–Property Relations in Nb

Nb has a remarkable capacity to absorb O, H, N, and C atoms as interstitial impurities, and these impurity atoms alter Nb's physical properties. Property changes caused by interstitial impurity atoms are common in many transition metals (e.g., Fig. 12.3), but they have been especially well studied in Nb. Figure 13.4 shows the solubility of these four elements in Nb as a function of temperature. The solubilities of the three larger interstitial atoms in Nb increase with increasing temperature, but the solubility of H decreases at higher temperatures. Nb can absorb less H at higher temperatures because H dissolution in Nb is an exothermic process; the entropy of a Nb–H solution is lower than the entropy of pure Nb, thus ΔS is negative in $\Delta G = \Delta H - T \Delta S$. H behaves similarly in V, Ta, and some of the other metals in groups 4, 5, and 6. This behavior makes it relatively easy to remove H impurities from the metal simply by heating it in an actively pumped vacuum chamber.

O, N, and C dissolve into the octahedral voids of Nb's BCC lattice (Fig. 13.5a), which seems odd because BCC octahedral voids are smaller than tetrahedral voids. However, an oversized impurity atom can occupy the octahedral void by displacing only the two Nb atoms in the body-centered positions of the neighboring unit cells (Fig. 13.5b), whereas all four neighboring atoms must move to accommodate an oversized atom in a tetrahedral void. Ionized H atoms (i.e., protons) dissolve preferentially into the tetrahedral interstices. In V, Nb, and Ta, the crystal potential felt by a proton has its minimum value at the larger tetrahedral site. (This is not

214 V, Nb, AND Ta

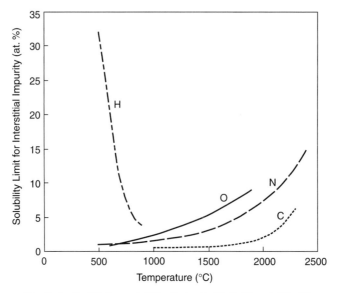

Fig. 13.4 Equilibrium solubility limits for H, O, N, and C in Nb.

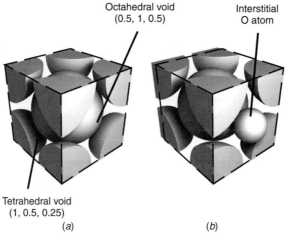

Fig. 13.5 (*a*) Location of the octahedral and tetrahedral voids in the Nb BCC unit cell. (*b*) Distortion of the unit cell resulting from an interstitial O atom in the octahedral void.

always the preferred site for H; for example, in FCC Pd, the minimum energy site for H is the octahedral void, which is larger than the tetrahedral void in the FCC lattice.) The distortions in the lattice caused by interstitial impurity atoms make dislocation motion more difficult, which in turn strengthens the metal and reduces ductility (Fig. 13.6). Ultrapure Nb is so ductile that it can be cold-worked to extreme deformations, but as interstitial content increases, the metal becomes progressively less ductile.

O and N produce a yield point drop (Sec. 6.2) in Nb at around 300°C. The stress–strain plot in Fig. 13.7 shows the yield point drop and strain-aging serrations in Nb containing 3200 at. ppm O. Impurity atoms diffuse to interstitial sites near dislocations and bind to them. The strained lattice around a dislocation produces distorted interstitial sites that can more easily

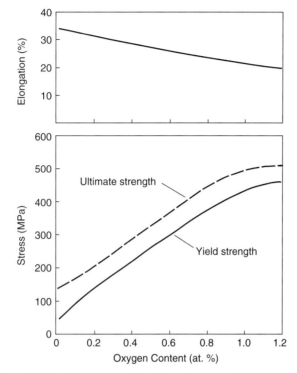

Fig. 13.6 Strength and ductility changes caused by O impurity content in Nb. (Redrawn from Neubauer, 1962, p. 31; with permission.)

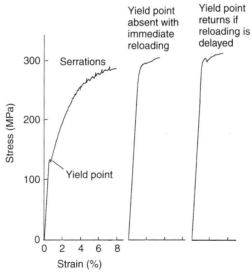

Fig. 13.7 Tensile test stress–strain plot for Nb containing 3200 at ppm O (tested at 250°C). The yield point drop and serrations in the left stress–strain curve result from O interstitial impurities binding to dislocations. If the test is interrupted and then resumed immediately (middle curve), no yield point appears; however, when the test is interrupted and then resumed after sufficient delay to allow O to diffuse to the dislocations (right curve), a new yield point appears. Strain at final fracture was 28.5%. (Redrawn from Neubauer, 1962, p. 41; with permission.)

accommodate interstitial atoms. Once a dislocation accumulates an "atmosphere" of interstitial atoms, additional stress is required to move the dislocation, because the O and N atoms bound to the dislocation line must be forced into normal (smaller) interstitial sites when the dislocation leaves. O has a binding energy to Nb dislocations estimated at 0.32 eV, and N's dislocation binding energy is about 0.15 eV. If the strain rate and temperature of the tensile test are suitable, strain-aging serrations appear in the stress–strain plot caused by interstitial atoms diffusing to dislocations that are temporarily halted by obstacles.

13.5 TANTALUM

> Tantalus was the son of Zeus and ... uniquely favored among mortals since he was invited to share the food of the gods. However, he abused the guest–host relationship and was punished by being ... immersed up to his neck in water, but when he bent to drink, it all drained away; luscious fruit hung on trees above him, but when he reached for it the winds blew the branches beyond his reach.
>
> —James Hunter, "Tantalus," *Encyclopedia Mythica*, http://www.pantheon.org/articles/t/tantalus.html, November 2003

For some reason, studying the fantasies of the ancient Greeks was considered an essential component of a high-quality nineteenth-century education. In the years when chemists were struggling to dissolve Ta oxide and to separate Ta from Nb, the difficulties of these chemical processes reminded the chemists of Tantalus's suffering in Greek mythology. When Ta was finally produced in a reasonably pure form, it was named for the tantalizing difficulty of its processing. Today, Ta is a valuable industrial material, able to perform certain tasks in electronics, chemical processing, and as an alloying addition better than any other material.

13.5.1 Physical Properties of Ta

Ta is a ductile BCC metal. Although its critical resolved shear stress rises rapidly below room temperature in the normal manner for BCC metals (Fig. 10.2*b*), it is one of the few BCC metals with no ductile–brittle transition. Ta tensile specimens tested at 4.2 K have more than an 80% reduction in area at the fracture surface. Ta has the fourth-highest melting temperature of all metals; only W, Re, and Os are higher melting. Ta's greatest attribute is its tenaciously adherent oxide coating. Ta_2O_5 is so corrosion resistant that Ta metal is often said to have the same corrosion resistance as glass. This oxide layer is an integral part of most of Ta's uses in electronics, the chemical process industry, and surgical implants. The oxide layer does not, however, protect it from oxidation in air at elevated temperature. Ta exposed to air above about 250°C oxidizes rapidly. One of the authors (A.R.) recalls once having left a Ta crucible in a 600°C air furnace over the noon hour. Upon opening the furnace after lunch, nothing remained of the crucible but a white circle of Ta oxide powder lying on the furnace floor. Ta is similar to Nb in possessing a low work-hardening rate; high-purity Ta can be cold-worked up to 97% without requiring stress-relief anneals. However, interstitial impurities sharply reduce ductility. Ta has a low coefficient of thermal expansion and is a superconductor below 4.47 K.

13.5.2 Applications of Ta

13.5.2.1 Ta Capacitors. Ta_2O_5 has outstanding dielectric strength. In contrast to ceramic and Al capacitors, the capacitance of Ta_2O_5 changes little over a broad temperature range. The oxide performs well up to about 170°C, high enough to meet almost all circuit design performance requirements. In capacitor design (Fig. 13.8), both the anode and the cathode are made of Ta. The anode (− lead) is a lightly sintered mass of Ta powder particles. The goal of this sintering operation is to assure that all the particles are in electrical contact with one another while maintaining the highest possible surface area. A deliberately thickened Ta_2O_5 layer is formed

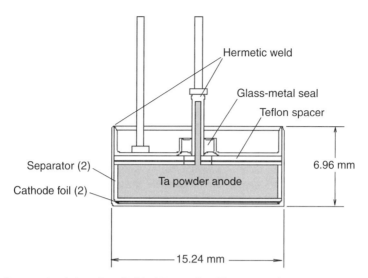

Fig. 13.8 Cross-sectional view of a cylindrical Ta capacitor. The case and lid are made from welded Ta. The anode is pressed, sintered, and anodized Ta powder at 5 to 8 g/cm^3 density (about one-third to one-half the density of Ta). The cathode is two sheets of Ta foil connected electrically; one sheet lies above the anode and the other lies below. Teflon spacers and glass seals keep the anode and cathode from shorting out. (Courtesy of Evans Capacitor Co., redrawn from *http://www.evanscap.com/THQA2%20life%20test.pdf*, 2003.)

by anodization. The cathode (+ lead) is a similarly anodized Ta wire or foil. The oxides formed on metal surfaces are often amorphous films. In the case of Ta capacitors, an amorphous oxide layer forms on Ta during anodization that has much better dielectric strength than crystalline Ta$_2$O$_5$. Small regions near impurity defects in the Ta$_2$O$_5$ layer are vulnerable to devitrification in the presence of large voltage gradients, and once the oxide crystallizes, its dielectric performance at that spot breaks down. To minimize capacitor leakage from breakdown, every effort is made to minimize impurity content in the Ta metal. High-purity Ta will typically contain about 10^4 such flaws per square centimeter; in less pure Ta, this figure can be as high as 10^{10} cm^{-2}, which substantially degrades capacitor performance. The market growth of personal computers and cell phones has greatly expanded Ta capacitor use in recent years, although this has not been fully matched by growth in demand for Ta metal, due to improvements in capacitor efficiency and miniaturization. Ta capacitors have enjoyed a strong position in the electronics industry for several decades, but competition between Ta and ceramic capacitors will undoubtedly continue into the foreseeable future, and some analysts predict that ceramic capacitors may eventually cut heavily into Ta capacitor use.

13.5.2.2 Ta in the Chemical Process Industry and Biomedical Implants. Although Ta capacitor manufacture accounts for more than 60% of global Ta demand, substantial quantities of Ta are used in the chemical process industry. The amorphous Ta$_2$O$_5$ layer on Ta metal has an ideal density to form an adherent, protective surface layer with few defects. It is thin enough to deform elastically if the underlying metal deforms, and it re-forms if the original layer is penetrated. Ta resists corrosive attack over an unusually wide pH range, resisting attack by nearly all solutions except fluorides (including HF acid), oleum, free SO$_3$, and strong caustics. Ta reaction vessels have been exposed to boiling HCl or HNO$_3$ acids continuously for as long as 30 years with no appreciable loss of Ta. Even when Ta does corrode, corrosion is usually by uniform reduction in thickness, avoiding the pitting or crevice attack that can rapidly ruin other metals. Ta is expensive, and its high density means that a kilogram is only 60 cm^3 of metal. For this reason, it is used only in applications where less costly materials fail or where stringent purity control of the product (e.g., pharmaceutical and cosmetic products) permits no contamination from process vessels. Even when Ta is chosen, the mass of Ta used is minimized by lining steel or Cu pipes with a thin inner sleeve of Ta, coating Ta onto surfaces, and alloying Ta with Nb (Sec. 13.4.2.4). Higher strength can be achieved in Ta–2.5% W and Ta–10% W alloy. The 480-MPa room-temperature yield strength of Ta–10% W (σ_y for pure Ta is only 170 MPa) permits use of thinner sections, lowering costs, and increasing heat transfer rates. Ta–10% W alloy has corrosion resistance equal to that of pure Ta in all environments. Ta crucibles are frequently chosen for single-crystal growth crucibles, rare earth–metal processing, and research applications where a ductile, refractory, easily welded container is required to resist reaction with hot metals and compounds.

13.5.2.3 Ta as an Alloying Addition. In Ni and Co superalloys (Sec. 16.3.3.6), Ta serves as a solid-solution strengthener, a carbide former, and a solute that partitions preferentially into intermetallic phases to strengthen them. Ta is particularly valuable in stabilizing the γ' phase in Ni superalloys against coarsening, maintaining fine precipitate size during service at 1100°C. TaC may be the highest-melting material known ($T_{m.pt.} \simeq 4000°C$); although HfC's melting point has also been reported to approach 4000°C. TaC resists dissolving into the Ni or Co superalloy matrix, even after prolonged exposure to high temperature.

13.5.2.4 TaC in Cutting Tools. TaC finds considerable use in specialty carbide cutting tools. About 0.3 wt% TaC is added routinely to inhibit grain growth in cemented carbide tools (Sec. 16.2.3.3). TaC has low solubility in Fe, so larger additions of NbC and TaC reduce wear in carbide tools intended to cut ferrous metals at high speed (Sec. 13.4.2.6).

13.5.2.5 Ta Alloys for Elevated Temperature Use and Munitions. Ta has useful strength at 2000°C but requires an oxidation-resistant coating if used in air. Ta–10% W alloy has an ultimate tensile strength of 345 MPa at 1315°C and a 10-hour rupture stress of 17.2 MPa at that temperature. Ta alloys have seen limited use in spacecraft and propulsion engines, but the reliability of protective coatings (Sec. 13.4.2.8) remains a major concern for such applications. One specialty use of Ta is in shaped-charge armor-piercing weapons (Sidebar 13.2).

SIDEBAR 13.2: Ta ANTIARMOR WEAPONS

Ta penetrators are used in both shaped charge and explosively formed penetrator antiarmor weapons. In these devices an explosive charge focuses exceptionally high energy on a small region of the target (Fig. 13.9). In the 1960s, shaped-charge liners were made of Cu, but in the 1970s Ta was found to produce a more effective penetrator because its density is almost twice as high, and it has excellent ductility at extremely high strain rates. The hypersonic penetrator slug or jet plunges through the armor, shattering it. When the Ta approaches the interior wall of the armor, it accelerates pieces of the disintegrating armor into the interior of the target, generating a lethal spray of high-velocity armor fragments.

The physical properties of Ta are well suited to this grim task. High density allows greater force to be concentrated at the impact point. Other high-density metals (Fig. 1.13) are either too expensive, radioactive/toxic, or lack the ductility to deform at extremely high strain rates. Ta is ductile at strain rates as high as 8000 s^{-1} (much greater than the 10^{-4} to 10^{-2} s^{-1} strain rates of normal tensile tests). As Fig. 13.10 shows, pure Ta is ductile at the high strain rate in the Hopkinson bar compression test. Since the shape of the penetrator jet or slug varies with the strength of the Ta, production of these weapons requires stringent control of purity, grain size, and grain shape to achieve reliable performance.

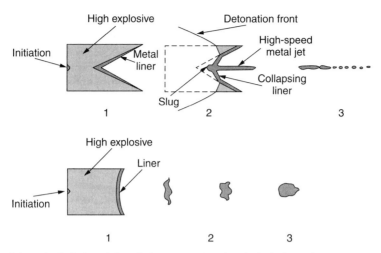

Fig. 13.9 Schematic depiction of shaped charge (upper) and explosively formed penetrator (lower) antiarmor warheads. Image 1 is the initial configuration prior to detonation of the high explosive. Image 2 shows the slug or metal jet in a partially formed state after detonation of the explosive. Image 3 is the final slug/jet shape in hypersonic free flight just before it strikes the armor surface. (Redrawn from Muller, 1996, p. 301; originally from Carleone, 1993.)

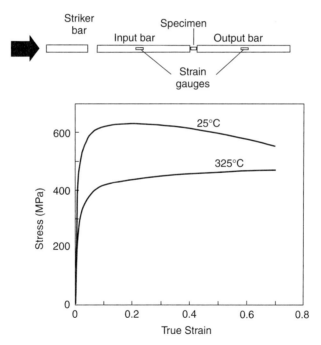

Fig. 13.10 Stress–strain data from a Hopkinson bar test performed on pure Ta at 25 and 325°C with a strain rate of 5000 s^{-1}. The striker bar is fired with an explosive charge to transfer force through the incident bar to the specimen. Strain gauges on the incident and output bars determine sample stress and strain. This high strain rate produces an ultimate tensile strength (σ_{UTS}) substantially higher than that of "normal" strain rates (e.g., about 10^{-3} s^{-1} σ_{UTS} = 450 MPa at 25°C.)

13.5.3 Sources of Ta and Refining Methods

Australia dominates world Ta production, with Brazil, Rwanda, Canada, and Congo supplying most of the remaining market demand. Although Ta production tonnages are small, its high cost makes it an industry of moderate economic importance. Most Ta ores are mixed oxides of Ta, Nb, Fe, Mn, Sn, and other transition metals. Some ores contain rare earths or actinides. Traditional beneficiation methods separate the Ta and Nb oxides from other minerals, and solvent extraction methods separate the Nb and Ta. Ta metal can be produced by several methods, the most important of which is Na reduction of K_2TaF_7 to produce Ta metal powder. Powder is the most frequently sold commercial product because Ta's high melting temperature necessitates powder metallurgy fabrication. Since most Ta is used for electronics (primarily in capacitors), high purity is desired, and vacuum arc melting or electron-beam melting is used to remove impurities.

ADDITIONAL INFORMATION

References, Appendixes, Problem Sets, and Metal Production Figures are available at
ftp://ftp.wiley.com/public/sci_tech_med/nonferrous

14 Cr, Mo, and W

14.1 OVERVIEW

The high melting temperatures of the midtransition metals reach their apex in the group 6 elements: chromium, molybdenum, and tungsten. Their nominal electron configurations (s^2d^4) hybridize easily to s^1d^5, which provides six bonding electrons. Mo and W are the highest melting elements in periods 5 and 6, and Cr has the second-highest melting point of the period 4 elements. Cr, Mo, and W are less electropositive than the early transition metals, and all three elements were first isolated as metals in the late eighteenth century by simple C reduction reactions. Their engineering value lies principally in their ability to improve steel's corrosion resistance and hardenability, as refractory metals with good high-temperature strength, and as wear-resistant carbides. Cr is a moderately abundant element (122 ppm in Earth's crust); Mo and W are two orders of magnitude scarcer (1.2 ppm each).

Cr metal forms a strongly adherent oxide that protects the pure metal and many Cr-containing alloys from corrosion. Commercial-purity Cr is brittle at room temperature and requires special alloying additions to achieve room-temperature ductility. Cr's principal use is as an alloying addition to steel and other metals and for protective coatings. Continuing success in the ongoing development of Cr-based alloys with better room-temperature ductility may expand Cr's usefulness in the near future. Mo's largest use is also as an alloying addition to steel and other metals. Pure Mo has sufficient room-temperature ductility to allow some cold work to be performed on the metal (usually after initial hot work). W has the highest melting point of all metals, and this property makes it useful in applications involving extreme temperatures (e.g., incandescent light bulb filaments). Pure W is brittle at room temperature, but the metal can be made ductile by appropriate hot-work treatments. Both Mo and W have volatile oxides that cause rapid weight loss at elevated temperature, so the metals require protective coatings for long-term high-temperature use in air. WC's high hardness (24 GPa) makes it useful for cutting tools and wear-resistant coatings. Half of all W is used as WC.

14.2 CHROMIUM

> The first thing to hit you is the hammer of thunder, the distant throaty rumble and roar of a midsummer storm that eats miles of countryside as it swells to an earsplitting howl. From behind the deafening blare come flashes of chrome like lightning, a dark tempest of denim and black leather encasing a tattooed squall, a turbulent gale of speed, power, and beautiful machinery.
>
> —Michael Jamison, "Hell's Angels," *The Missoulian*,
> http://www.missoulian.com/specials/hellsanngels/ha01.html, December 5, 2003

14.2.1 History of Cr

Throughout the nineteenth and twentieth centuries, Cr's development was closely linked to steelmaking. When about 1% Cr is added to steel, it improves hardenability by retarding the

---

Structure–Property Relations in Nonferrous Metals, by Alan M. Russell and Kok Loong Lee
Copyright © 2005 John Wiley & Sons, Inc.

austenite → martensite transformation. When Cr content exceeds 11% in steel, an adherent oxide layer forms on the surface that protects the metal from aqueous corrosion in many solutions and also greatly improves high-temperature oxidation resistance. Most Cr is produced as ferrochrome, an Fe–Cr alloy containing 45 to 95% Cr and 0.01 to 10% C used as an alloying addition to steel. Cr consumption by the steel industry (85%) dwarfs Cr's other uses: in Cr_2O_3 refractory brick, in the chemical industry, in Ni and Co alloys, and for Cr plating and coating. Annual world Cr demand rose about 4% throughout the last three decades of the twentieth century, driven primarily by increasing use in stainless steel production (about 20 million tons/yr in 2000).

14.2.2 Physical Properties of Cr

The physical properties of Cr are summarized in Fig. 14.1.

14.2.2.1 Corrosion. In room-temperature air, Cr immediately forms a protective oxide layer a few nanometers thick. If damaged, this chromia layer regenerates if even low O concentrations are present. The oxide is strongly adherent and protects the metal from oxidizing acids (e.g., HNO_3) which passivate the surface. However, other acids (e.g., HCl acid) will attack Cr. The oxide layer is an effective barrier to high-temperature oxidation, making Cr and Cr alloys useful in air at temperatures as high as 800 to 900°C. For many years, Cr plating served as an attractive and corrosion-resistant coating on mild steel parts (e.g., automobile bumpers and

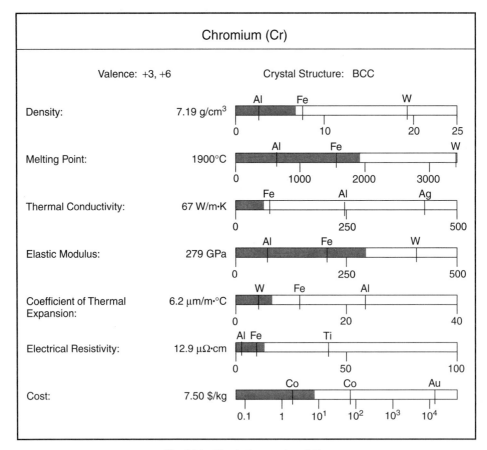

Fig. 14.1 Physical properties of Cr.

trim). However, stricter regulation of Cr use in electroplating operations and problems with waste solution disposal (Sec. 14.2.4) has sharply reduced use of Cr plating.

14.2.2.2 Elasticity and Plasticity. Cr has several desirable attributes, including the highest elastic modulus of all the period 4 metals; an adherent, protective oxide layer; a low coefficient of thermal expansion; density 10% less than that of Fe; and moderate cost. These factors would make Cr a useful structural metal, particularly at elevated temperature, were it not for the metal's room-temperature brittleness. Single crystals of Cr are highly ductile at 20°C in certain circumstances [e.g., in Cu–Cr DMMCs (Sec. 18.2.3.10)], which suggests that polycrystalline Cr's brittle behavior may be due to an inability to accommodate strain incompatibilities between neighboring Cr grains. Indeed, Cr-rich alloys can be ductile at room temperature if steps are taken to sequester N impurities and provide dispersoids to limit the extent of dislocation pile-ups at grain boundaries (Sec. 14.2.6).

14.2.3 Alloys of Cr and Cr Electroplating

Cr is used primarily as an alloying addition to other metals and as a protective coating. Cr's incompletely solved room-temperature brittleness problem limits use of Cr-rich alloys.

14.2.3.1 Ferrochrome. Approximately 5 million tons/yr of ferrochromium (ferrochrome) are used for alloying additions to steel. Adding 0.20 to 1.5 wt% Cr to steel retards C diffusion during austenite decomposition; this delays pearlite formation and permits more martensite to form during cooling, increasing hardenability. To produce an effective passivating layer of Cr_2O_3 on steel surfaces (i.e., stainless steel) Cr contents of 11 wt% or more are needed. Stainless steel resists rust in a wide range of aqueous solutions that would aggressively attack mild steel. Stainless steel also has good high-temperature oxidation resistance. Cast iron also benefits from Cr additions, and Cr contents as high as 30% suppress graphite formation in cast iron and also improve hardenability, toughness, wear resistance, and corrosion resistance.

14.2.3.2 Cr Additions to Nonferrous Alloys. Cr is a common alloying addition to Ni, Al, Cu, and Ti alloys. In small amounts Cr serves as a grain refiner, and in larger amounts it improves oxidation resistance. The Cr-containing alloys most familiar to nonmetallurgists may be the *nichrome* family of alloys. A typical nichrome alloy is Ni with additions of 21 to 25 wt% Fe, 14 to 18 wt% Cr, 1 wt% Mn and Si, and 0.15 wt% C. Nichrome's key attributes are oxidation resistance and high electrical resistivity, making nichrome ideal for heating elements in hair dryers, toasters, hair straighteners, electric ovens, and stovetop "burners." Nichrome is ductile in the annealed condition with tensile strengths of about 700 MPa (annealed) and 1350 MPa (fully work hardened). A mixed Ni–Cr oxide protects the underlying metal from oxidation during long-term operation in air at temperatures as high as 1150°C. A wide array of other Ni and Co alloys, a few of which contain as much as 60 wt% Cr, are used in severe environments in the chemical process industry (Secs. 16.3.3.4 and 16.3.3.5).

In Cu, 0.3 to 1.2 wt% Cr improves strength and wear resistance while reducing conductivity only modestly; these alloys are useful in spot-welding electrodes and rotating electrical contacts. Cr is an effective grain refiner in Al–Mg, Al–Mg–Si, and Al–Mg–Zn alloys. Cr diffuses slowly in Al, and Cr additions of 0.1 to 0.35% in Al alloys form fine dispersed phases that inhibit recrystallization and grain growth (Sec. 20.2.8.2). In Ti, Cr has limited solubility in β-Ti and forms intermetallic compounds by eutectoid decomposition of the β phase. However, this decomposition reaction is so slow that intermetallics do not form during normal fabrication and heat treatment or during service. Consequently, Cr acts as a β stabilizer in Ti (Sec. 12.2.3.3).

14.2.3.3 Cr Plating and Coating. CrO_3 dissolved in dilute H_2SO_4 or other sulfate solutions is used to electroplate Cr onto steel, Al, polymers, and other materials to improve their appearance, protect them from corrosion, harden them against wear, and lower their surfaces' coefficients of friction. Electroplated Cr layers are typically 1 to 25 μm thick (although much thicker layers can be used to rebuild worn parts and to improve wear resistance). A complex sequence of base coats (e.g., Zn, Cu, and/or Ni) typically underlies Cr electroplate (Sec. 16.3.3.2). Cr coatings can also be applied by *pack coating* or *gas coating*, processes that involve immersing the workpiece in a mixture of Cr metal powder, inert materials such as kaolin or alumina, and ammonium or halide salts. When heated in an inert atmosphere at 900 to 1300°C, this mixture deposits an adherent, protective Cr layer on the workpiece; however, the surface finish produced by these methods lacks the shiny, mirrorlike appearance of electroplated Cr.

14.2.3.4 Cr Compounds. Cr_2O_3 is a major constituent in refractory bricks and mortar. It is higher melting and stronger than many silicate-based refractories and resists corrosive attack by both acidic and basic materials at high temperature. The best chromite ore for refractories use has a high alumina content and a low silicate and Fe oxide content. Cr compounds are also widely used in pigments for paints and dyes. Smaller amounts of Cr compounds are used in leather tanning, drilling mud, corrosion inhibitors, and catalysts.

14.2.4 Toxicity of Cr

Most $3d$ transition metals are a necessary micronutrient at low doses and a potential toxin at high doses (Sec. 18.2.4). Cr^{3+} is an essential nutrient for metabolic processes involving sugars, proteins, and fats; most vitamin supplements contain microgram quantities of Cr^{3+} salts. However, hexavalent Cr (Cr^{6+}) is a dangerous toxin. Cr^{6+} compounds are used in Cr electroplating operations, which makes these processes potentially hazardous. Workers must be protected from inhalation, ingestion, or skin contact with Cr^{6+} solutions, and careless release of spent Cr^{6+} electrolytes to streams or groundwater endangers residents outside electroplating facilities. Inhaling Cr^{6+} compounds causes nasal irritation and nosebleeds. Long-term exposure to airborne Cr^{6+} compounds ulcerates and perforates the nasal septum and increases lung cancer risk. Ingested Cr^{6+} compounds cause stomach upsets and ulcers, convulsions, kidney and liver damage, and even death. Skin contact with certain Cr^{6+} compounds can cause skin ulcers. Some people suffer allergic reactions from skin contact with even trace amounts of Cr^{3+} or Cr^{6+}. Birth defects have been observed in laboratory animals exposed to Cr^{6+}.

14.2.5 Sources of Cr and Refining Methods

Chromite ($FeCr_2O_4$) is the principal ore for both ferrochrome and pure Cr production. South Africa's reserves are the world's largest, and roughly half the world's chromite is produced there. Neighboring Zimbabwe has substantial chromite reserves, but its change in political leadership in the 1980s brought egregious mismanagement of Zimbabwe's mining and agricultural resources, and its future chromite production levels are uncertain. Smaller chromite ore bodies are worked in several other locations worldwide.

High-carbon ferrochrome can be produced in a submerged arc furnace by reduction of concentrated chromite ore with C (coke). This process typically yields material that is 60 to 70% Cr, 4 to 6% C, and most of the balance, Fe, often with a few percent Si. The high C content in submerged arc furnace ferrochrome results from Cr's high affinity for C. High-carbon ferrochrome is inexpensive to produce and is used in the largest quantities in the steel industry. Medium- and low-carbon ferrochromes are produced by reducing chromite with ferrosilicon or ferrosilicochromium. This is a more expensive process and produces material with 0.5 to 4.0% C (medium-carbon ferrochrome) or 0.01 to 0.5% C (low-carbon ferrochrome).

Pure Cr metal is produced for nonferrous alloys and specialty steels by three different methods. (1) Aqueous Cr^{3+} solutions can be electrolyzed to plate pure Cr on the electrodes. (2) Bomb reduction of Cr_2O_3 by Al powder is a highly exothermic reaction (Sidebar 11.5) that heats the charge above 2000°C, assuring complete melting of the Cr and good separation of metal and slag. Both electrolytic and aluminothermic Cr metal is more than 99% purity. (3) Intimately mixed C and Cr_2O_3 powders can be held in a furnace at 1400°C for long periods (about 4 days) to produce 98% pure Cr with a combined O + C content of about 1.5%. About 70,000 tons of pure Cr is produced annually. For electronic applications, higher purities are achieved by iodide refining, zone melting, or fluxing with alkaline metals.

14.2.6 Structure–Property Relations in Cr

14.2.6.1 Cr Ductility Problem. Low-temperature brittleness is a problem in most BCC metals. Their yield strengths rise rapidly as temperature decreases, eventually becoming so high that cleavage fracture requires less energy than plastic deformation. In commercial-purity Cr, the ductile–brittle transition temperature (DBTT) is about 300°C. This is higher than most other BCC metals' DBTTs (Fig. 14.2). This gives Cr low fracture toughness and makes cold work impossible. Cr's room-temperature brittleness results from grain boundary impurity precipitates that cause slip-induced cracks leading to transgranular cleavage fracture. In most BCC metals, interstitial impurity atoms in the bulk of the grains also reduce screw dislocation mobility, which lowers ductility. N is especially troublesome in Cr. N's high solubility at elevated temperatures causes hot Cr to absorb N from the air. As the metal cools, N solubility drops, and fine, acicular precipitates form, often at grain boundaries. Production of superpurity Cr (N content <15 ppm) is prohibitively costly, and such high purity is quite difficult to maintain during subsequent hot work, welding, machining, and service use.

Research in the 1960s produced room-temperature tensile ductility in commercial-purity Cr by adding MgO dispersoids and Ti to the metal. Room-temperature tensile elongations of 10 to 20% were achieved in Cr–2 to 6 wt% MgO–0.5 wt% Ti alloys made by sintering and extruding blended powders of the three constituents. These alloys form $MgCr_2O_4$ spinel by reaction of the MgO additions and the O in commercial purity Cr. Ti "getters" N, O, and C interstitial impurities. More recent research has shown that impurities, particularly N and S, precipitate at the $MgO–MgCr_2O_4–Cr$ matrix interfaces (Fig. 14.3). Since these impurities are trapped on the oxide phase surfaces, they are effectively removed from the Cr grain boundaries, which greatly improves ductility. The $MgO–MgCr_2O_4$ dispersoids located throughout the bulk of the grains reduce the mean free-travel distances of dislocations, causing many small dislocation pile-ups

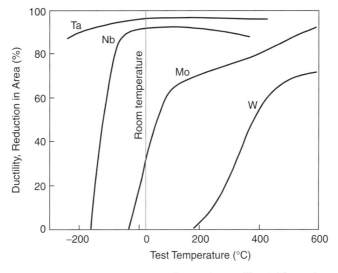

Fig. 14.2 Ductile–brittle transition in four recrystallized polycrystalline BCC metals as determined by percent reduction in area of the fracture surface in tensile tests. (Adapted from Walters, 1988, p. 422.)

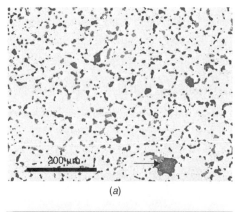

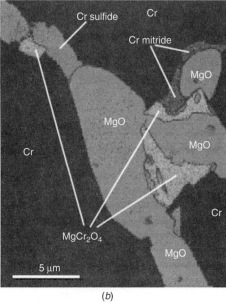

Fig. 14.3 Microstructure of ductile, hot-pressed Cr–6% MgO containing a N content typical of commercial purity Cr (about 350 wt ppm). (*a*) SEM cross-section micrograph: $MgCr_2O_4$ spinel (light particles), MgO (dark particles) in Cr matrix. Occasional large particle agglomerations (arrow) are present. (*b*) Low-voltage energy-dispersive x-ray spectrum image phase map of a $MgO/MgCr_2O_4$ particle dispersion area. N and S are trapped at the $MgO–MgCr_2O_4$ interfaces, which sequesters them from the grain boundaries. (From Brady et al., 2003; with permission.)

rather than fewer, larger pile-ups at grain boundaries. This lowers the stress at slip-induced cracks at grain boundaries, delaying crack propagation.

MgO–Ti additions are not a "cure-all" for Cr's ductility limitations. Cr–MgO–Ti alloys have high strain rate sensitivity; raising the tensile strain rate from 3.3×10^{-4} to 3.3×10^{-3} s^{-1} lowers ductility from 18% to 9%. In addition, the alloys are somewhat notch-sensitive; if 180-grit rather than 600-grit abrasive is used to polish tensile specimen gauge lengths, the 9% tensile ductility at 3.3×10^{-3} s^{-1} drops to about 2%. Nevertheless, continuing research on Cr ductility problems may make high-Cr alloys an affordable, oxidation-resistant, tough, high-temperature material for use in gas turbine engines, heat-treatment furnace fixtures, and chemical process equipment.

14.3 MOLYBDENUM

"The Hellhole Near the Sky," now with 1,400 employees producing 12,000 tons of moly ore per day, held the worst safety record of any major U.S. mine. In 1938 alone, the hospital, fortuitously located only 100 yards from the Phillipson Tunnel portal, tended to eight fatalities, forty-six "serious" accidents (involving broken bones or amputations), and countless lesser injuries.

—Steve Voynick, "The Rise of Climax Molybdenum, Mining History," *Colorado Central Magazine*, 9(16), 1994, http://www.cozine.com/archive/cc1994/00090162.htm

14.3.1 History of Mo

Mo development has been a boom and bust business (Sidebar 14.1). The metal was little used until World War I simultaneously cut off W supplies and drove up demand for the tough, high-strength steel that can be produced with W or Mo alloying additions. During the mid-twentieth century, steelmaking demand for Mo surged and slumped in synchrony with cycles of war and peace, but Mo has now expanded into many fields unrelated to steelmaking. Production from several primary Mo mines is now augmented with by-product Mo from Cu mines, and although Mo's use as an alloying addition to steel remains its largest use, pure Mo, Mo alloys, and Mo compounds serve as inexpensive components for high-temperature applications and as catalysts, lubricants, pigments, and flame retardants.

14.3.2 Physical Properties of Mo

Mo is a BCC metal with good electrical and thermal conductivity. Mo is sometimes used as a conductor at temperatures too high for Cu. Mo's low coefficient of thermal expansion combined with its high thermal conductivity and ductility (Fig. 14.4) allow it to undergo rapid temperature

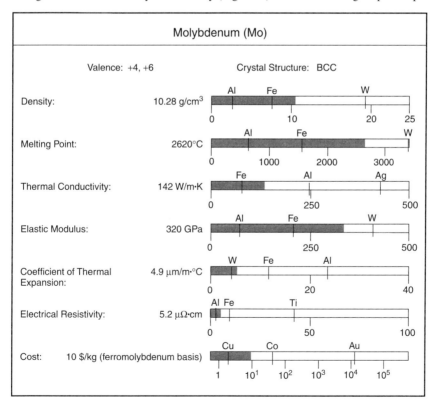

Fig. 14.4 Physical properties of Mo.

SIDEBAR 14.1: COLORADO CLIMAX Mo MINE

The rise and fall of the famous Climax, Colorado, Mo mine in the United States typifies the rough and tumble, boom and bust business of mining. Mo was little used during the nineteenth century, but World War I cut off W supplies needed for high-speed steel and armor, and Mo was the obvious substitute. This triggered a sudden demand for Mo, and in 1917, miners and speculators rushed to Bartlett Mountain, Colorado, where Mo ore had been found decades earlier. Three companies competed to produce from Bartlett Mountain in a frenzy of claim jumping, fights, equipment thefts, and lawsuits. The Climax Molybdenum Company prevailed in this contest, but the World War I Armistice brought a near-collapse of the Mo industry. Climax president Brainerd Phillipson kept the company intact until it could resume small-scale operation in 1924, and all the while he evangelized among automakers and steel companies about Mo's usefulness. His sales efforts met with growing success, but the Climax mine's finances were often near bankruptcy during the 1920s. In 1929, Phillipson made a bold "make-or-break" gamble by committing all the company's modest resources to a huge expansion effort that would make Climax the largest mine tunnel ever built. Simultaneously, he dropped the price of Mo to stimulate demand. This initiative was launched just as the Great Depression descended upon America, and to make matters worse, Phillipson died shortly after the effort began. But Climax made good on the risky expansion by augmenting U.S. sales with exports to Germany, Japan, and the Soviet Union. By the time the "New Tunnel" was complete in 1933, Climax was booming.

Climax stock had traded for as little as 10 cents per share in 1928, but in 1935 it soared to $42, and the company's net earnings exceeded 50% of gross sales. Climax was the best-performing U.S. stock during the Great Depression; Phillipson's widow and several local store and saloon keepers became wealthy. The gritty world of Mo mine workers, however, remained miserable and dangerous. High altitude, cold weather, wretched living conditions, and poor safety procedures made the miners' existence a grim montage of injuries, deaths, "rocked-up lung" syndrome, and alcoholism. Worker turnover averaged 300% per year, and the mine became known by the workers' sardonic appellation, "The Hellhole near the Sky." Using ever-expanding profits, the company gradually improved worker living quarters and mine safety, and World War II drove Mo production to record highs. When the war ended in 1945, employment at the Climax mine dropped by 75%, but the Korean War renewed Mo demand. During the 1950s Climax was the world's largest and safest underground mine. Climax produced three-fourths of the world's Mo, and it had sufficient reserves for 40 years of continuing production. In 1958, Climax merged with American Metals Company, and the next quarter century was a sad succession of labor strife, failed price manipulation strategies, and a money-losing experiment with Mo oxide processing. In 1982, the Climax mine closed.

---

changes with minimal concerns about internal stress, distortion, or cracking. Of all the metals with melting points above 2000°C, Mo and W are the least expensive. Since Mo has lower density and easier fabrication than W, Mo is often preferred for many high-temperature uses.

14.3.2.1 Oxidation and Corrosion. At room temperature, Mo resists oxidation in air and can be polished to a lustrous finish that persists indefinitely. However, above about 400°C, Mo forms a bluish-black mixed oxide patina that slowly vaporizes, causing gradual metal loss. Above 600°C, both O and Mo diffuse rapidly through the oxide layer, and the oxide vaporizes more quickly. Oxidizing Mo forms MoO_2 near the metal–oxide interface and MoO_3 near the surface. Mo oxides have low melting temperatures, and a eutectic exists at 778°C between MoO_2 and MoO_3, so the poorly protective solid oxide becomes an even less protective liquid film as temperature rises. At 1000°C, the liquid oxide evaporates so rapidly that no appreciable liquid layer persists. Mo metal exposed to air at 1000°C loses 1.4 to 2.0 $g/cm^2 \cdot h$ to this process.

Thus, the metal requires a protective coating for long-term, high-temperature service in air (Sec. 14.3.3.4). Mo can, however, be hot worked in air since the time at temperature is brief. Mo has low solubility for O and N and can be exposed to air at 1400°C without embrittling the metal. No grain boundary weakening or internal oxidation occurs during such exposure. Billets are heated in inert or reducing atmosphere furnaces, and some loss of metal inevitably occurs from oxide volatility when the metal is removed from the furnace for deformation. Such metal loss is less costly than the alternative of hot working the metal in inert atmospheres. Liquid-glass coatings can be applied to Mo surfaces for hot rolling or hot forging Mo to slow oxidation. Despite considerable research effort, no alloying additions to Mo have been found that solve its high-temperature oxidation problems while maintaining desirable mechanical properties.

Mo has good resistance to attack by many acids. Mo metal resists attack by HCl, HF, and H_2SO_4 acids at 20°C, although these acids slowly dissolve the metal at 80 to 100°C. Both dilute and concentrated HNO_3 acid corrode Mo, as do various mixtures of common mineral acids. Aqueous alkaline solutions have little effect on Mo at room temperature, but molten alkali salts attack the metal rapidly. Mo resists attack by many molten metals, including Bi, Na, and Hg. It performs reasonably well in molten Zn but is attacked by molten Al, Ni, Fe, and Co. Mo has good compatibility with many compositions of liquid glass and refractory oxides.

14.3.2.2 Mechanical Properties. Mo is a strong metal with moderate room-temperature ductility. Its high melting point allows useful strength up to 1600°C. Mo has one of the highest Young's moduli and lowest coefficients of thermal expansion of all the common commercial metals. The Young's modulus of Mo at 1000°C is considerably higher than that of steel at room temperature (Fig. 14.5). The ductile–brittle transition in Mo occurs near 20°C (Fig. 14.2); the amount of cold work Mo will tolerate depends on its microstructure and prior deformation history.

Nearly all massive Mo is produced by powder metallurgy (P/M), but arc melting is used in a few instances. In either case, the mechanical properties of the metal are improved by hot working. P/M metal usually has several percent porosity after sintering, and although this is acceptable in some applications, collapsing the porosity (usually by rolling or forging) improves performance. Arc-melted metal typically has large columnar grains, so hot breakdown deformation (usually by extrusion) improves mechanical properties. Hot work is more demanding for Mo than for steel for several reasons: (1) Mo's high melting temperature requires that hot work be performed at 1200 to 1400°C, so short contact times are essential to avoid chilling the workpiece excessively. Even with short contact times, such high temperatures cause rapid heat transfer to dies, shortening their service lifetimes. (2) The high radiative heat loss at these temperatures (emitted energy varies as T^4) combined with Mo's high thermal conductivity and low specific heat cool the metal quickly once it leaves the furnace. Deformation must be performed quickly, so hammer forging is preferred to press forging. (3) Mo's high strength, even at such high temperatures, requires more powerful presses and hammers to deform the metal. Once hot breakdown has been achieved, light deformation steps are usually performed at progressively lower temperatures so that deformation finishes below the recrystallization temperature (about 1100°C). Mo in the completely recrystallized condition has lower strength and poorer bend and impact performance, so some degree of warm work below the recrystallization temperature is usually desired.

Pure Mo is relatively strong. Cold-rolled Mo has $\sigma_y = 540$ MPa and $\sigma_{UTS} = 700$ MPa with 40% elongation when tensile tested at 27°C. If the same material is tested at 760°C, both strength figures are halved. Metal recrystallized at 1150°C for 1 hour is softer; $\sigma_y = 390$ MPa and $\sigma_{UTS} = 470$ MPa with 42% elongation when tensile tested at 27°C. Mo displays a distinct fatigue limit (Sec. 5.3 and Fig. 14.6), and its creep performance is excellent.

14.3.3 Applications of Mo

14.3.3.1 Mo as an Alloying Addition to Other Metals. Mo's largest use is as an alloying addition to steel. Mo can be added to steel as mixed MoO_3/C briquettes. The MoO_3 is reduced

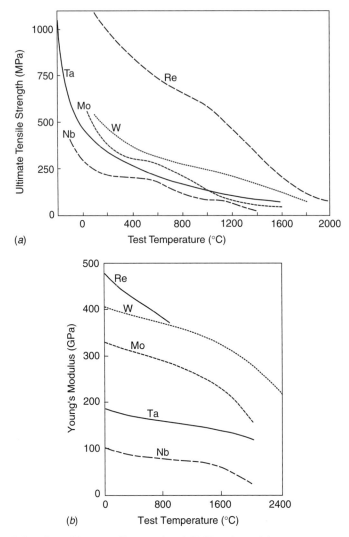

Fig. 14.5 Variation of (a) ultimate tensile strength and (b) Young's modulus versus temperature for several refractory metals. (Redrawn from ASM, 1990, pp. 557, 559; with permission.)

by the C in the liquid steel, and the evolved CO and CO_2 gases aid the turbulent mixing of the Mo metal into the liquid steel. This method is inexpensive, but it raises the steel's C content substantially; ferromolybdenum (Sec. 14.3.4) additions add less C to the metal. Most ferromolybdenum contains 55 to 75% Mo, 1.5 to 4% C, and the balance Fe. Ferromolybdenum is less expensive than pure Mo and more easily dissolved in liquid steel, since the Fe–Mo system has a peritectic at 1550°C. Mo increases steel's hardenability, corrosion resistance, toughness, and strength. The presence of Mo (usually in concentrations of 0.13 to 0.25 wt%) improves hardenability by retarding C diffusion during austenite decomposition and by slowing coalescence of Fe_3C precipitates during tempering. Mo also reduces temper embrittlement in steel by slowing diffusion of impurities (e.g., Sb, Sn, P, As) to grain boundaries. Mo additions are also common in tool and die steels, high-speed steels, gray cast iron, and high-strength low-alloy steels. In stainless steel, Mo additions from 2 to 4% inhibit crevice corrosion and raise elevated temperature strength.

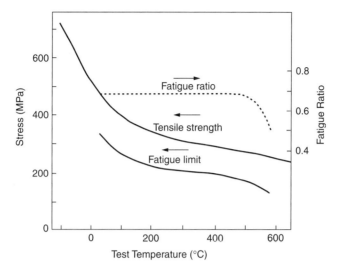

Fig. 14.6 Fatigue behavior and tensile strength of pure polycrystalline Mo versus temperature. (From Gupta, 1992, p. 23; with permission.)

14.3.3.2 Mo in Ni Alloys. Mo is frequently added to Ni-based alloys. Mo contents of 6 to 28 wt% in Ni provide good corrosion performance in HCl and H_2SO_4 at temperatures and concentrations that attack stainless steel. These Ni–Mo alloys cost more than stainless steel, but they can sometimes solve chemical process industry corrosion problems at lower cost than Zr or Ta use. Ni superalloys (Sec. 16.3.3.6) often contain several percent Mo, where it segregates preferentially to the Ni γ phase and acts as a solid-solution hardener. Mo also reduces stacking fault energy in γ-Ni, which makes cross-slip more difficult.

14.3.3.3 Mo and Mo Alloys. Pure Mo is used for such applications as mandrels for forming W filaments; supports in incandescent light bulbs to minimize W filament sagging; high-temperature parts in electronic tubes for radar devices and x-ray generation; base plates and heat sinks for power transistors and Si rectifiers; resistive heating elements in furnaces operating up to 2000°C; switch contacts resistant to arc damage; electrodes, stirrers, pumps, and molds for handling liquid glass; glass–metal seals; laboratory crucibles; grips for high-temperature tensile tests; thermal spray coatings on piston rings and other wear-prone parts that operate at elevated temperatures.

Mo alloys are used in especially demanding circumstances such as rocket nose cones, winglets, and turbine engines. The best known of these is TZM (Mo–0.40 to 0.55 wt% Ti, 0.06 to 0.12 wt% Zr, and 0.01 to 0.04 wt% C). Ti and Zr raise TZM's recrystallization temperature to 1400°C, 300°C higher than in pure Mo. This allows the beneficial effects of work hardening to be retained to higher operating temperatures. Ti and Zr form carbide precipitates that increase creep strength and raise tensile strength to 600 MPa at 1000°C. Newer variants of TZM use higher C contents plus 1.2 to 2.0 wt% Hf in addition to or in place of the Ti or Zr; these composition changes raise tensile strength to 800 MPa at 1000°C and boost the recrystallization temperature to 1550°C. Mo doped with K_4SiO_4 can be extensively deformed to develop oriented grains that are particularly effective at enhancing creep strength by orienting most grain boundaries parallel to the loading direction, thereby minimizing grain boundary sliding. Useful creep strength at temperatures as high as 1800°C has been achieved in this material.

Mo–W alloys are used to resist attack in liquid Zn for impellers and other components in Zn refining and galvanizing operations. Mo–Re alloys (Sec. 15.5.2.2) provide an excellent combination of high-temperature strength and room-temperature ductility for applications that warrant their higher cost.

14.3.3.4 Coatings for Mo and Mo Alloys. Mo and Mo alloys offer high-temperature strength and good fabricability at moderate cost, but Mo parts must operate in vacuum or inert atmosphere (as is often the case in electronic parts) to avoid ruinous oxidation losses. In certain applications (e.g., rocket components), the metal loss associated with exposing hot Mo to the atmosphere is tolerable because the component's operating lifetime is only a few minutes. However, long-term, high-temperature service in air requires a protective coating. Coatings based on aluminide intermetallic compounds, silicides, oxides (ZrO_2, Al_2O_3), and metals (Cr, PGMs) have been studied, and some are in use today. These can be applied by electroplating, dip alloying, chemical vapor deposition, physical vapor deposition, plasma spraying, pack cementation, and slurry methods. The best results achieved to date have been delivered by silicide coatings, which are protective at temperatures up to 1600°C.

An example of one type of coating operation is provided by the method used to apply an alumina–silica coating to Mo. The semifinished Mo part is immersed in liquid 70% Al–30% Si alloy at 1100°C for 30 seconds. This produces a 300-μm-thick coating that is deliberately oxidized by heating in air at 1000°C. The part is next coated with a paste of 95% Si–5% Al_2O_3 suspended in CCl_4. Finally, the part is heated to 1600°C, where the CCl_4 evaporates, and a silica–alumina glaze forms. Coatings greatly elevate Mo's maximum use temperature in air, but they remain an imperfect solution to the problem. Their long-term performance is not sufficient for some potential applications. The coatings are usually subjected to thermal cycling, elastic flexure, and in some situations, ballistic impacts. All of these factors risk cracking and spalling the coating, particularly around edges and corners and at impact sites. Metallic coatings generally have the best adherence and impact toughness but are least able to withstand use at the highest temperatures. Ceramic and silicide coatings perform best in terms of oxidative stability but are more vulnerable to separation and cracking failures. Development of better coating technology remains one of the great challenges of materials science.

14.3.3.5 Machining and Joining. Mo is readily machinable, although its hardness and hot hardness cause rapid wear in high-speed steel tools. Sintered carbide tools give longer tool life. Mo is readily welded by tungsten inert gas or resistance welding; preheating the workpieces is recommended for optimal results. Exclusion of O from the fusion zone is essential to avoid incorporating MoO_3 into the metal. For especially critical welds, electron-beam welding produces outstanding Mo weldments with fine grain sizes and low H pickup. Brazing has also been used successfully to join Mo. Ag-based brazing alloys work well for lower-temperature service, and Cu–Ni or Mo–Ni compositions can be used at higher temperatures.

14.3.3.6 Mo Compounds. Mo compounds have a broad spectrum of uses. MoS_2 is a versatile lubricant that maintains a low coefficient of friction from -150 to 400°C. It is particularly useful at temperatures and pressures that break down organic lubricants. In vacuum or protected atmospheres, MoS_2 can lubricate up to 1100°C. It has excellent adherence to metal surfaces, packing into microscopic valleys and scratches and adhering to asperities. MoS_2 is used in wire drawing, cold forming, thread rolling, and pressing operations to extend die life and improve workpiece surface finish. MoS_2 in chassis lubricants, packed bearings, and other hard-to-reach components delivers long service life with little or no maintenance. The chemical process industry uses Mo compounds for pigments and dyes, corrosion inhibitors, and fire/smoke suppression. Mo is an essential nutrient in plants and animals and is usually present in vitamin and mineral supplements and fertilizer. Mo metal, MoS_2 and MoO_3 are versatile catalysts to accelerate organic synthesis, ammonia production, and cracking reactions in the petrochemical and polymer industries.

14.3.4 Sources of Mo and Refining Methods

The principal Mo source is MoS_2, produced from both primary Mo mines and as a by-product of Cu mining. A few sources produce Mo as a by-product of W mining. Geological concentration of Mo tends to be low: 0.05 to 0.25% Mo content is typical for primary mines and 0.01 to 0.05%

in Cu ores. These low concentrations require comminution and flotation processing to produce industrial-grade MoS$_2$ (90 to 95% purity). MoS$_2$ is converted to MoO$_3$ by roasting. Mo metal powder is produced from the oxide by reduction in H$_2$ gas at about 1050°C. The high melting point of Mo makes casting difficult, so nearly all Mo parts are fabricated by powder metallurgy press and sinter methods. Sintering requires temperatures of 1600 to 2200°C and times of 3 to 30 hours in an H$_2$ atmosphere to achieve 90 to 95% density. Small quantities of Pt, Pd, or Ni can be added as sintering aids, which lowers the sintering temperature to 1300°C but also reduces postsintering ductility. Vacuum arc furnaces or electron-beam furnaces are used instead of P/M methods for about 5% of Mo products.

Low-C ferromolybdenum (1% C or less) is produced by reduction of mixed Fe and Mo oxides with Al and ferrosilicon metal (the thermit process). The reaction is highly exothermic and produces liquid Mo–Fe alloy that settles to the bottom of the reaction vessel; Al$_2$O$_3$/SiO$_2$ slag floats to the top. Higher C ferromolybdenum (1.5 to 4% C) is less expensive to produce and is made in a submerged arc furnace from mixed Fe, Mo oxides (or other Mo compounds), coke, and limestone.

14.4 TUNGSTEN

> At the center of the constellation Scorpius lies his red heart, the star Antares. Antares is a dying giant, old and immense, bloated to vast size. It appears bright to us because of its size, but its surface is 3400 Kelvins, cooler than our sun. A tungsten probe could skim its surface like a scissorbill and escape unmelted to bring back a wispy sample of Antarean star stuff.
>
> —Barbara Shelton, Westfield Academy, personal communication, 2004

14.4.1 History of W

The first W metal was produced in the late eighteenth century, but the metal went largely unused until the early twentieth century, when it burst onto the engineering scene with a rapid succession of discoveries that still define its major uses today. In 1900, high-speed steel containing W was introduced, followed by the first W incandescent light filament (1903), ductile W wire (1909), and WC–Co cermets for cutting and abrasive use (1922). The physical properties of W are exceptional: the highest melting temperature of all metals, the lowest vapor pressure of all metals, the lowest compressibility of all metals, high density, high Young's modulus, and a low coefficient of thermal expansion. Because of this remarkable combination of properties, W is considered a strategic metal, and it has several uses for which no other metal can adequately substitute.

14.4.2 Physical Properties of W

W has exceptionally strong bonding. Only 8 eV of energy is required to hybridize its $5d^46s^2$ outer electron configuration into a $5d^56s^1$ arrangement that provides six bonding electrons per atom. These strong bonds are the cause of W's exceptionally high melting point, high strength, high stiffness, and 1500°C recrystallization temperature (Fig. 14.7).

14.4.2.1 Oxidation and Corrosion. Massive W is stable in air at room temperature, and it shows no obvious reaction until heated to about 350°C, where a thin layer of bluish W oxide appears on the surface. This oxide layer is partially protective. But if the metal is heated to still higher temperatures, the oxide layer begins to crack, which accelerates oxidation of the underlying metal. Above 800°C, WO$_3$ sublimes quickly, and metal loss becomes substantial. Above 1100°C, the oxide sublimes as rapidly as it forms. The loss of metal to oxidation in the temperature range 700 to 1300°C in units of mg/cm^2 · h can be calculated using

$$-\frac{d(m/A)}{dt} = 5.89 \times 10^6 \exp\left(\frac{-12,170}{T}\right) P^{0.5}$$

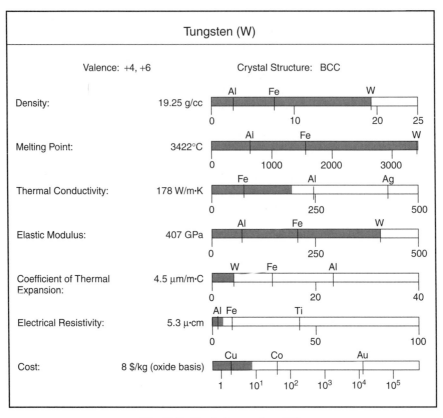

Fig. 14.7 Physical properties of W.

where m/A is the weight change per unit area (mg/cm^2), T the absolute temperature, and P the pressure in atmospheres. Massive pieces of hot W can be exposed to air for short periods without ruinous mass loss. O and N have maximum solubility of only 100 ppb at room temperature and 0.01 to 0.1% at their eutectic temperatures. This behavior contrasts sharply with the group 4 and 5 metals (Sec. 13.4.4), which can dissolve large amounts of O and N. Consequently, W is not susceptible to interstitial impurity embrittlement by hot exposure to air. W will, however, form oxides, nitrides, carbides, and silicides if high-porosity W or W powder is exposed to these elements at elevated temperatures. W is unaffected by H_2 gas, and a H_2 furnace atmosphere will protect the metal from oxidation during heating. At extreme temperatures (above 2000°C), W begins to sinter its oxide to the surface metal, which slows the metal loss rate somewhat; however, this phenomenon is not protective, and the metal quickly disappears at such temperatures. W metal reacts only slightly with several acids, including HCl, HF, HNO$_3$, and H$_2$SO$_4$. It also resists attack by molten metals, tolerating oxygen-free Li, Na, and K up to 900°C and resisting liquid or vapor Bi, Cs, Sn, Hg, and Pb up to 1000°C. W is unaffected by glass and silica up to about 1400°C.

14.4.2.2 Mechanical Properties. W metal is typically brittle at room temperature (Fig. 14.2). Polycrystalline W's DBTT occurs between 100 and 500°C, with the exact temperature dependent on purity, grain size and shape, and prior heat treatment history. For single-crystal metal, the DBTT ranges from −30°C (for high-purity W) to 80°C (for metal containing 1000 at. ppm C). It appears that three factors contribute to the brittleness of W. One factor is the interaction of dislocations with carbide precipitates, which raise the yield stress by pinning dislocations. A

SIDEBAR 14.2: EFFECT OF HYDROSTATIC PRESSURE ON DUCTILITY

A superimposed hydrostatic pressure improves the ductility of most materials. The room-temperature ductility of W has been particularly well studied, and the results of several research projects are presented in Fig. 14.8. Tensile testing was performed on W immersed in a high-pressure fluid, so the pressure is applied equally on the specimen from all directions. This pressure does not greatly affect the ability of slip to occur by dislocation motion, but it suppresses cleavage fracture by compressing any incipient cracks. To open, a crack must move the material on each side of the crack outward against the hydrostatic pressure. This makes it more difficult for the crack to expand, and slip becomes the favored deformation mode. Hydrostatic pressure can improve fabricability of marginally ductile materials. Extrusion (Sec. 8.4.4) of metals with low room-temperature ductility (e.g., Be, W, or Mg) is normally done at elevated temperature, but they can be extruded hydrostatically at 20°C (Fig. 14.9).

second factor is segregation of O and P at grain boundaries, which lowers grain boundary cohesive strength and promotes intergranular fracture. Since nearly all commercial W components are produced by P/M, the presence of some impurities at grain boundaries is unavoidable. The final factor is the fact that room temperature is only 8% of the homologous melting temperature of W, and many BCC metals are intrinsically brittle at that homologous temperature. Producing ultrapure W by arc melting, electron-beam melting, and zone refining lowers the DBTT in polycrystalline metal, but only moderately. This suggests that impurity content is not the sole cause of brittleness. The group 6 metals have much lower room-temperature ductility than the group 5 metals (Appendix D, *ftp://ftp.wiley.com/public/sci_tech_med/nonferrous*), even though all six elements are BCC. Low values of the ratio of bulk elastic modulus (K) to shear modulus (μ) are an indicator of intrinsic brittleness (i.e., the Pugh criterion). At room temperature, the group 5 metals have K/μ values between 3.12 and 4.03; by contrast, the group 6 metals K/μ values lie between 1.22 and 2.02. High pressure improves W's ductility (Sidebar 14.2).

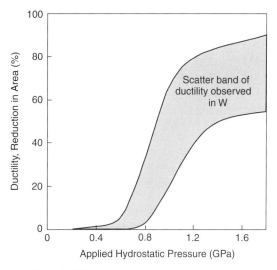

Fig. 14.8 Effect of externally applied hydrostatic pressure on the room-temperature tensile ductility of unrecrystallized P/M W. High pressure suppresses cleavage fracture, enhancing the ability of slip to plastically deform the metal. The upper edge of the scatter band is for higher-purity material; the lower edge is for less-pure metal.

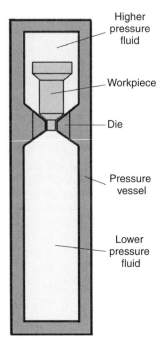

Fig. 14.9 In an hydrostatic extrusion press, high-pressure fluid in the upper chamber forces the workpiece through the die into the lower chamber. The lower tapered step on the workpiece seals the lower end against the die to prevent fluid from leaking from the upper chamber to the lower chamber. The upper tapered step on the workpiece stops the extrusion and prevents the tail and high-pressure fluid from "bursting" into the lower chamber at the end of the process.

At cryogenic temperatures W slip occurs only on the $\{110\}\langle111\rangle$ slip system. At room temperature, the $\{112\}\langle111\rangle$ system also becomes active, and between 1370 and 2760°C a third system, the $\{123\}\langle111\rangle$, is active. Below $0.1T_{m.pt.}$, W's screw dislocation mobility is much lower than its edge dislocation mobility. Consequently, dislocation loops tend to expand greatly in the direction parallel to the Burgers vector, producing long, unmoving screw segments (Fig. 14.10). This effect is thought to contribute to the rapid increase in yield strength seen in W as the temperature decreases (Fig. 14.11). W's stacking fault energy is apparently high; dissociated dislocations are not seen in W except in material quenched from 2500°C.

W sheet and wire can be ductile at room temperature if they are given extensive elevated-temperature deformation. This deformation improves ductility by two mechanisms: (1) it redistributes the grain boundary impurities over a greatly expanded intergranular area, and (2) a fibrous, fine-grained structure forms that makes it difficult for intergranular fracture to occur perpendicular to the fiber direction. To achieve this condition, pure W metal is hot worked, then deformed at progressively lower temperatures until highly elongated grains are present. The much larger total grain boundary area "dilutes" the concentration of impurities per unit area of boundary, which reduces their embrittling effect. This process for producing ductile W wire was a true breakthrough, making W the preferred filament material for incandescent lights when it was perfected in 1909.

14.4.3 Applications of W

14.4.3.1 W as an Alloying Addition to Other Metals. The first commercial use for W was as an alloying addition to steel, and this use still comprises about one-fourth of W consumption.

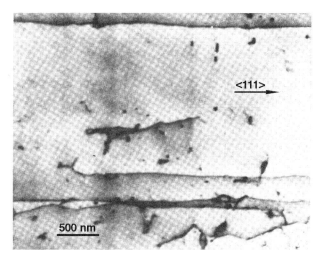

Fig. 14.10 Transmission electron micrograph of dislocation structure in single-crystal W deformed below $0.1T_{m.pt.}$ (150 K). Since the mobility of screw dislocations is much lower than the mobility of edge dislocations in W, dislocation loops tend to expand parallel to the Burgers vector, producing long, unmoving screw segments. (From Yih and Wang, 1979, p. 298; with permission.)

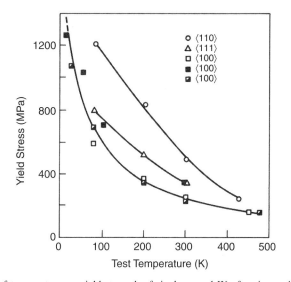

Fig. 14.11 Effect of temperature on yield strength of single-crystal W of various orientations. The crystallographic orientation of the tensile axis is indicated in the key. (From Yih and Wang, 1979, p. 293; with permission.)

W reacts with the C in steel to form WC, which is harder and more wear resistant than Fe_3C. Thus, 0.5 to 18 wt% W is present in most high-speed steels for cutting tools to increase hardness. W also increases the hardenability of steel, and the WC formed during tempering is more stable against coarsening at elevated temperature than Fe_3C precipitates, so W also improves the hot strength of steel. W can be added to steel in the form of bars or sheets of sintered pure W and pure Fe powders. This avoids the losses associated with trying to add loose W powder to molten steel, and the sintered mass is less dense than solid W, so it does not sink to the bottom of the crucible as rapidly, giving it more time to dissolve completely. Ferrotungsten is also used; ferrotungsten is produced in the same manner as ferromolybdenum (Sec. 14.3.4).

The carbide-forming ability of W is exploited in many Ni and Co superalloys. Carbides play a major role in pinning grain boundaries in these alloys to retard creep by grain boundary sliding (Sec. 16.2.3.1). The large number of bonding electrons per W atom also makes W an effective solid solution hardener in Ni superalloys. W additions, typically 2.5 to 10%, strengthen Ta without reducing corrosion resistance (Sec. 13.5.2.2).

W powder is used to make composites with Ag or Cu that reduce arc damage and wear in electrical contacts (Sec. 18.3.3.1 and Fig. 18.23). These contacts are made by sintering W powder to produce a body with interconnected porosity, then immersing the sintered W in pure liquid Ag or Cu to fill the interconnected porosity of the W body. The solubility of W in Ag and Cu is nil, so the electrical and thermal conductivity of the matrix is the same as that of pure Ag or Cu, while the surface has the resistance to arc damage and mechanical wear of pure W.

14.4.3.2 W and W Alloys. Pure W is used for incandescent light bulb filaments (Sec. 14.4.5) and furnace heating elements. W is also used in electrical switch contacts and sputtered metallizing coatings, where W's low coefficient of thermal expansion provides a good match to many semiconductor, ceramic, and glass substrates. W makes good high-temperature crucibles, rocket nozzles, thermocouples, welding electrodes, and structural components for high-temperature environments. No other material combines the high density, moderate cost, and nontoxic properties of W, so it finds numerous applications in counterweights and radiation shielding. W's high atomic number and high melting point make it a good target in x-ray tubes to produce short wavelength x-rays at high intensities. W can be plasma-sprayed onto other metal parts to increase their wear resistance. W fibers make refractory reinforcing filaments that bond well with matrix metals in metal–metal composites for high-temperature use.

Alloying additions to W expand its usefulness. W–Re alloys have good room-temperature ductility and are much easier to weld and machine than pure W (Sec. 15.5.3.2). W–Re alloys have better resistance to thermal shock than pure W, so they can be used in high-intensity rotating anodes for medical x-ray sources at higher current levels than pure W could tolerate. Two-phase *tungsten heavy metal alloys* consist of Ni–Fe or Ni–Cu binder phases that are sintered with W powder at temperatures that melt the binder phase. This produces a coarse distribution of spherical W particles that have dissolved a few tenths of a percent Fe and Ni held together by the binder phase, which dissolves about 4% W. The resulting product has high density (17.0 to 18.5 g/cm^3), high conductivity, good machinability, several percent tensile elongation, and a Young's modulus of 320 to 380 GPa. These make excellent "stand-ins" for pure W in applications that do not require elevated temperature strength. Several W alloys have been developed that achieve remarkably high strength at both room temperature and high temperature (Sidebar 14.3) and excellent creep strength.

14.4.3.3 Coatings for W and W Alloys. Coatings can protect W and W alloys from high-temperature oxidation; however, the same difficulties described in Sec. 14.3.3.4 for Mo coatings apply to W coatings as well. Ni–Cr coatings work well between 1000 and 1400°C. At higher temperatures, Ni–Cr coatings melt, but Pt can provide protection to 1650°C. Pt coatings interdiffuse rapidly with the W substrate, which gradually shifts the coating composition to a nonprotective Pt–W alloy; however, intermediate barrier coatings of Ru, Ir, or Re can delay this breakdown. At still higher temperatures, metal oxides are more stable than metal coatings; however, the coefficients of thermal expansion of most oxides are higher than that of W (ZrO_2 being a notable exception), which can lead to cracking and spalling during thermal cycling. It is also difficult to achieve good adherence of oxide coatings on W metal. Many oxides diffuse O so rapidly above about 2000°C that their protection times become very brief. Silicides can also be used as protective coatings; they oxidize to form silica outer layers that operate reasonably well up to about 1800°C, although they sometimes suffer from "pest" problems (Sec. 25.4) at intermediate temperatures. Coatings allow W to be used in air at higher temperatures and for

SIDEBAR 14.3: HIGH-STRENGTH ALLOYS

"What is the strongest metal available for engineering structural use?" Ask that seemingly simple question in a meeting of mechanical metallurgists, and the debate may rage for hours. Those interested in strength/weight ratio will put forward β-Ti alloys, which have a 1550-MPa yield strength (σ_y) and ultimate strength (σ_{UTS}) of 1700 MPa. Al reinforced with B-coated SiC filaments can boast $\sigma_{UTS} = 1500$ MPa, but that strength is highly anisotropic, and purists will disqualify composites containing nonmetallic phases. The ferrous metals group can nominate their maraging steels, some of which reach yield strengths of 2400 MPa. One of the highest-strength metals is ordinary cold-drawn hypereutectoid pearlitic steel wire, which has $\sigma_{UTS} = 4800$ MPa parallel to the wire axis. However, strength perpendicular to the wire axis is considerably lower. Very fine pure W wire has been reported to have $\sigma_{UTS} = 4500$ MPa at extremely high drawing ratios while maintaining a measure of ductility. A strength of $\sigma_{UTS} = 4700$ MPa has been attained in fully work-hardened W–5% Re alloy containing 100-nm HfC precipitates. Amorphous $Co_{43}Fe_{20}Ta_{5.5}B_{31.5}$ is isotropic with $\sigma_{UTS} = 5185$ MPa, but little ductility. A truly extraordinary $\sigma_{UTS} \approx 8000$ to 9000 MPa has been reported in Mo–Re alloys cold-rolled more than 99%. Even this exceptionally high value probably does not represent the highest possible strength for metals if lower temperatures are considered. *Ab initio* calculations of the ideal shear strengths of W for dislocation slip on {110}, {112}, and {123} planes indicate that slip on all three of these planes would require about 18,000 MPa at 0 K. This is 11% of the shear modulus of W. However, calculations also indicate that the theoretical {100} cleavage strength of W at 0 K would be about 13,600 MPa, suggesting that brittle fracture would precede slip.

If the question is changed to, "What metal has the highest strength at high temperature?", the debate quickly narrows to a "W versus Re" argument. Commercial W–25% Re solid solution alloy has $\sigma_{UTS} = 560$ MPa at 1000°C and is quite ductile at room temperature as well. A W–20% Ta–12% Mo solid-solution alloy has a yield strength of 700 MPa at 1400°C (near the melting point of steel), and its yield strength is still 190 MPa at 2000°C. W with 2% ThO_2 added for dispersion hardening can deliver $\sigma_{UTS} = 100$ MPa at 2400°C. Precipitation-hardening W alloys have still better strengths above 2000°C; W–12% Nb–0.29% V–0.12% Zr–0.07% C has a yield strength of 170 MPa at 2200°C. At still higher temperatures, the difference between W and W alloys fades, and strengths at 3000°C drop to less than 10 MPa. Although Re alloys have been developed less extensively than W alloys, the hot strength of Re is roughly comparable to W despite Re's lower melting point (Sec. 15.5.5).

---

longer times than would be possible for bare metal, but the limitations of coatings still impose frustrating constraints on design engineers seeking to exploit W's excellent high-temperature mechanical properties.

14.4.3.4 Machining and Joining. W is difficult to machine due to its hardness and limited ductility. Nearly optimal conditions (extremely sharp tools and rigidly fixed tools and workpiece) must be maintained to cut W successfully. Electrical discharge machining is sometimes used in preference to traditional machining methods. Porous W (W sintered with 15 to 35% porosity that is subsequently infiltrated with Cu) is much more easily machined, and the Cu can be removed at high temperature after machining. It is possible to weld W, but the metal tends to form large grains in the fusion zone that make the weld weaker than the surrounding metal. Electron-beam welding and laser welding (Sec. 8.6.2) produce better results but are more difficult and expensive. Diffusion welding typically gives better results than fusion welding, and this method can be used to join W to W as well as to other metals. Temperatures of 1300 to 2000°C and clamping pressures of 2 to 20 MPa are sufficient to join the metals in vacuum or H_2 atmosphere. A thin-foil intermediate layer (e.g., Pd, Ni, Rh, or Ru) at the joint interface accelerates diffusion welding. Brazing with Rh, Pd, or Cu–Ni alloys in vacuum or inert atmosphere provides another joining option for W.

14.4.3.5 WC (Hard Metal). Half of all W is used to produce WC for cutting and abrasive applications. WC's hardness (24 GPa) is lower than that of ultrahard materials such as diamond (70 to 100 GPa), cubic BN (45 to 50 GPa), or various boron compounds (32 to 45 GPa), but WC is less expensive and performs well on a wide range of materials. WC is often called *hard metal*, although it is not truly a metal. It is produced by reacting W with C, and it is usually used as a sintered WC + Co composite (Sec. 16.2.3.3). The ductile binder metal makes the WC + Co composite tougher because pure WC is brittle. WC–Co cermets are used in saws, drills, tools for lathes and mills, mining and petroleum drilling tools, abrasive wheels and belts, and wear-resistant coatings.

14.4.3.6 W Compounds. One particularly intriguing family of W compounds are the tungsten bronzes, which have the formula M_xWO_3, where M is a metal, H^+, or $(NH_4)^+$ and $0 < x < 1$. When x approaches 1, the compounds take on a metallic or semiconducting character. These compounds have a range of uses as catalysts and intensely colored pigments. When M is an alkali metal or H, the compounds become photochromic and can be added to glass to produce eyeglasses that darken in strong light and become transparent in dim light. W compounds are generally nontoxic, but WF_6 (used in semiconductor fabrication for low-pressure or plasma-enhanced CVD of tungsten and tungsten silicides) is hazardous because it reacts with water and water vapor to produce HF acid.

14.4.4 Sources of W and Refining Methods

The two commercially significant W ores are $CaWO_4$ and $(Fe,Mn)WO_4$. The typical W content of workable ore deposits is only about 0.5%; deposits richer than 2% are rare. China has the world's largest W reserves and accounts for 80% of world production. W is often shipped and sold as the oxide WO_3 (sometimes called *tungstic acid* when hydrated, a mixture of $H_2O \cdot WO_3$ and $2H_2O \cdot WO_3$). The water of hydration is easily removed by heating, and the oxide is reduced in H_2 to produce metal powder. W's exceptionally high melting temperature makes melt processing and casting difficult, so P/M is used to produce almost all bulk W. P/M production of a W part begins with compaction at about 200 MPa, followed by sintering in flowing H_2 for several hours at 2800°C to achieve 90 to 97% density. The sintering also purifies the metal; many impurities vaporize at the sintering temperature and are carried away with the flowing H_2. If fully dense W is required, the sintered metal is hot worked, starting at 1500 to 1700°C and continuing at progressively lower temperatures (the recrystallization temperature drops as deformation progresses), with the final deformation performed below the recrystallization temperature.

14.4.5 Structure–Property Relations in W

14.4.5.1 Nonsag W Filaments. W can operate at 3000°C as filaments in incandescent light bulbs and electronic tubes (Sidebar 14.4). High filament temperatures produce more lumens of visible light per watt, but near its melting point, W wire undergoes grain growth and grain boundary creep (Fig. 14.12) that lead to premature filament burnout. In most metals, grain boundaries can be pinned to prevent grain growth by precipitating second phases at the grain boundaries or introducing impurity solute atoms. However, these methods fail in the special case of a W filament operating at 3000°C. Adding soluble alloying elements to pin the grain boundaries lowers the melting temperature of W and raises electrical resistivity. All elements with low solubility in W melt below 3000°C, forming a liquid phase that leads to rapid failure. C additions form refractory carbide precipitates at lower temperatures, but the W–C phase diagram has multiple eutectic points near 2700°C. The problem was solved by adding $KAlSi_3O_8$ prior to sintering the W powder into bars. Most of this dopant is washed out or vaporized before sintering

SIDEBAR 14.4: W FILAMENTS IN FORENSIC ACCIDENT ANALYSIS

In lawsuits and insurance adjustments following automobile accidents, a key issue is often, "Were the headlights, turn signal lights, and brake lights in use at the time of the crash?" Drivers' memories and bystanders' recall are often erroneous or conflicting, but the W ductile–brittle transition often indicates whether a light was operating at the moment of impact. A cold W filament will fracture in a severe impact. Microscopic examination of the broken filament reveals a brittle fracture surface if the light was cold. A hot W filament is ductile, and a crash will usually deform the filament, sometimes even bringing it into contact with the glass of the light bulb and fusing part of the glass where it touched the hot metal. If the crash breaks the light bulb glass while the filament is hot, part of the W metal flashes into WO_3, leaving a characteristic yellowish powder on cooler surfaces located nearby. The case of a turn signal lamp might seem an unreliable source of evidence since it flashes on and off in use, and the impact could occur during the "off" phase. However, the W filament does not cool sufficiently during the "off" cycle of a turn signal indicator to become brittle, so a lamp that was on a moment before impact will still be ductile.

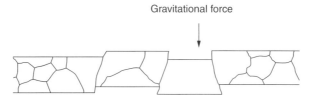

Fig. 14.12 Offset in an undoped W filament caused by prolonged operation at high temperature. Grain growth followed by grain boundary sliding leads to premature burnout of the filament.

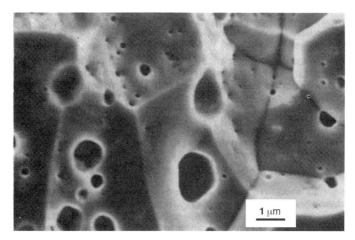

Fig. 14.13 Scanning electron micrograph of the fracture surface of a K-doped W ingot after initial sintering but before the metal is drawn into fine wire. The larger voids (about 1 μm) are ordinary sintering pores; these are empty and will collapse during swaging and wire drawing. The smaller defects (about 100 nm) are K bubbles that formed during initial sintering. These smaller bubbles will also collapse during subsequent cold work, but they contain minute amounts of solid K that will elongate during the swaging and wire drawing. When the light bulb filament is first turned on, these needle-shaped K phases vaporize, forming a string of 10-nm-diameter bubbles that pin grain boundaries and prevent filament sag. (From Horacsek, 1989, p. 180; with permission.)

is complete, but about 100 wt ppm K remains trapped within the filament. These compounds decompose at 2300°C, so as the W powder is initially sintered into rods, K metal is liberated. This K forms bubbles about 100 nm in diameter (Fig. 14.13). The Al, Si, and O diffuse into the W, but K atoms are too large to diffuse any appreciable distance. As the W rod is swaged and drawn into fine wire, the latter portions of the deformation are performed at room temperature, where the K is a solid. This elongates the K metal into needle-shaped regions that become rows of still smaller K bubbles (about 10 nm in diameter) when the filament is heated in the light bulb. K is completely insoluble in W, even above 3000°C, so the bubbles remain small and immobile during the operating life of the filament. When grain boundaries encounter these bubbles during grain growth, the bubbles pin the grain boundaries, which prevents the offset sag depicted in Fig. 14.12.

ADDITIONAL INFORMATION

References, Appendixes, Problem Sets, and Metal Production Figures are available at
ftp://ftp.wiley.com/public/sci_tech_med/nonferrous

15 Mn, Tc, and Re

15.1 OVERVIEW

The group 7 elements, manganese, technetium, and rhenium, lie at the midpoint of the transition metals. Their d subshells are nominally half-filled, giving them an outer electron configuration of $d^5 + s^2$, which may hybridize to $d^6 + s^1$. All three elements are exceptional in several regards (Table 15.1). Mn has the most complex crystal structure of all the transition metals. It ranks fifth among metals in annual production tonnage (after Fe, Al, Cu, and Zn). Tc has no stable isotopes. Trace amounts of Tc occur in U ore from spontaneous fission events, but ^{235}U fission in nuclear reactors is the only substantial Tc source. Re was the last nonradioactive element to be discovered. It is a rare, dense metal with a high melting temperature, high elastic modulus, and high strength.

15.2 HISTORY AND PROPERTIES

Although Mn was not produced in high-purity metallic form until the twentieth century, it was one of the first elements used by humankind. MnO_2 pigments were used to make 17,000-year-old Paleolithic cave paintings, and Mn has been found in the steel of ancient Spartan weapons and in glass from ancient Egypt and Rome. Mn is essential in deoxidizing and desulfurizing steel. Mn has been present in almost every ton of steel produced since the Bessemer process was adopted in the 1860s. Mn is also an alloying addition to certain Al, Cu, and Mg alloys, and Mn metal powder is a constituent of welding electrodes and fluxes. Mn compounds find a remarkably wide range of engineering applications in dry-cell batteries, as oxidizing agents in chemical processing, ceramic colorants, food additives, and water and wastewater treatment.

Mendeleev predicted the existence of element 43 in the 1860s, but searches for it in minerals were fruitless because it has no stable isotopes. All Tc produced by the stellar fusion events that generated Earth's other heavy elements (Sidebar 17.1) decayed long ago. Tc was finally obtained by artificial synthesis in 1937. Tc constitutes about 6% of ^{235}U fission products, so hundreds of tons of Tc have accumulated as a by-product from reprocessed, spent fission reactor fuel. Tc is used in nuclear medicine, but its high cost and radioactivity discourage other use.

Re was predicted by Mendeleev long before its actual discovery. Re is one of the rarest elements (Fig. 17.1), and it was not isolated and purified until 1925. Modern Re production is only about 35 tons/yr, obtained from condensation of volatile Re oxides in flue dusts and concentrates from Cu and Mo mining. Re's strength, stiffness, formability, and high melting temperature are all valuable metallurgical attributes, but its scarcity and high cost restrict its use to specialty applications such as alloying additions to superalloys, rocket nozzles (Sidebar 15.1), catalysts, and thermocouples. One of natural Re's two isotopes is slightly radioactive ($^{187}_{75}$Re), but its half-life is so long (4.3×10^{10} years) and the energy of the emitted β particles so low (0.003 MeV) that it poses no hazard.

Structure–Property Relations in Nonferrous Metals, by Alan M. Russell and Kok Loong Lee
Copyright © 2005 John Wiley & Sons, Inc.

244 Mn, Tc, AND Re

TABLE 15.1 Selected Room-Temperature Properties of Group 7 Metals with a Comparison to Cu

Property	Mn	Tc	Re	Cu
Valence	+2, +4, +7	+4, +7	+4, +7	+1, +2
Crystal structure at 20°C	Complex cubic (cI58)[a]	HCP	HCP	FCC
Density (g/cm^3)	7.43	11.5	21.04	8.92
Melting temperature (°C)	1244	2157	3180	1085
Thermal conductivity (W/m·K)	7.8	50.6	39.6	399
Elastic (Young's) modulus (GPa)	198	407	469	130
Coefficient of thermal expansion (10^{-6} m/m·°C)	22.3	8.2	6.7	16.5
Electrical resistivity (μΩ·cm)	185.0	14.9	19.3	1.7
Cost ($/kg), large quantities	0.55[b]	60,000[c]	900	2

[a] Mn has four allotropes: α-Mn (cI58) is stable to 727°C; β-Mn (cP20) from 727 to 1100°C; γ-Mn (cF4) from 1100 to 1138°C; and δ-Mn (cI2) from 1138°C to the melting temperature (1244°C).
[b] Price based on Mn content of Mn–Fe alloy (ferromanganese); pure electrolytic Mn is several times more expensive.
[c] ^{99}Tc available from the Oak Ridge National Laboratory to licensed buyers only.

15.3 MANGANESE

> The modern Age of Steel began in earnest with the large tonnage usage of manganese ore.
> —Weiss (1977)

Pure Mn is too brittle to have any structural use, but millions of tons of Mn are used in steelmaking and smaller amounts are used as alloying additions to other metals. Mn compounds find wide-ranging uses in products as diverse as batteries, food additives, wastewater treatment processes, and chemical processing.

15.3.1 Physical Properties of Mn

Mn is located in the periodic table (Fig. 1.1) amid transition elements with simple crystal structures (HCP, BCC, or FCC). The complex crystal structure of α-Mn and its odd physical properties (Fig. 15.2) seem strangely out of place. The primary determining factor for α-Mn's complex

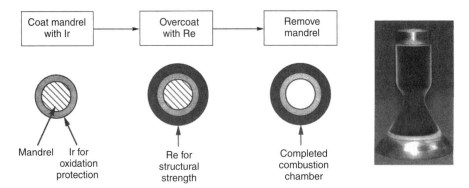

Fig. 15.1 Re/Ir rocket combustion chamber fabrication. A Mo or graphite mandrel is CVD coated, first with Ir, then with Re (layer thicknesses are not drawn to scale). The mandrel is then removed chemically, leaving a hollow Re chamber with an internal Ir coating. A completed combustion chamber is shown at the right. The creep strength of Re allows the combustion chamber to operate at 2000°C, generating 445 N of thrust for several hours. The Ir protects the Re from oxidizing as the rocket fires. (From Fortini, 2000; with permission.)

SIDEBAR 15.1: Re ROCKET COMBUSTION CHAMBERS

Rocket engine efficiency increases with increasing engine operating temperature according to Carnot's law:

$$E = \frac{T_1 - T_2}{T_1}$$

where E is Carnot efficiency, T_1 the conversion temperature (hot), and T_2 the rejection temperature (cooler). The cost to orbit payloads can be as high as \$22,000 per kilogram, but the cost could be reduced if E were raised by using materials that can withstand higher combustion chamber temperatures. The most important commercial space missions launch communications satellites into geosynchronous orbits 36,000 km high. The boost from low-Earth orbit to geosynchronous orbit requires that a small, low-thrust engine fire continuously for four to six hours. This places severe demands on combustion chamber materials. In earlier missions, the boost engine combustion chamber was fabricated from C103 Nb alloy (Nb–10% Hf–1% Ti) coated with disilicide to prevent oxidation. However, C103 engines suffer coating breakdown and catastrophic oxidation failure if operated above 1400°C.

A combustion chamber using Re instead of Nb withstands 2000°C combustion chamber temperatures. Re has good strength at 2000°C, but its oxidation resistance is hopelessly inadequate to withstand the 14,000-second burn times for geosynchronous boost. To prevent Re oxidation, an internal coating of pure Ir is used. Ir melts at 2443°C and is weak at 2000°C. However, its oxidation performance is sufficient to withstand exposure to the oxidizing gases in the combustion chamber. Re provides the structural support, and Ir protects the Re from oxidation.

The Ir layer must be free from cracks or pinholes that would expose the Re to oxidizing gases, so chemical vapor deposition (CVD) is used to assure a defect-free coating. The combustion chambers are produced (Fig. 15.1) by coating a Mo or graphite mandrel with a 50- to 75-μm-thick CVD Ir layer, coating the Ir with a 0.75-mm-thick CVD Re layer, and by dissolving the mandrel chemically. As the engine fires, Ir is gradually lost to oxidation, but test firings show that sufficient Ir remains to protect the Re during more than 21,000 seconds of rocket operation. Some Re is also lost while the engine fires due to diffusion along Ir grain boundaries that carries Re to the combustion chamber inner surface, where it oxidizes to Re_2O_7 and is vaporized into the rocket exhaust. Thermal shock is a serious design concern in the combustion chamber because the rocket engine may start and stop several times during the mission, but the nearly ideal match in coefficients of thermal expansion (Re: 6.7×10^{-6}°C^{-1} and Ir: 6.4×10^{-6}°C^{-1}) minimizes interlaminar stresses between the two metals. The first Re–Ir rocket combustion chamber was used in a 1999 communications satellite launch. The hotter burn raises efficiency, allowing a 17% increase in payload weight, which saves over \$100 million per launch.

Still higher temperatures may be possible in future launches using an experimental combustion chamber that adds a layer of HfO_2 over the Ir. The HfO_2 acts as a thermal barrier, lowering Ir's temperature and oxidation rate. This permits combustion chamber temperatures to exceed 2500°C without melting the Ir. In one ground-based test firing of a HfO_2/Ir/Re rocket, the engine fired continuously for 1.5 hours at a temperature of 2550°C, at which point the fuel injector melted. The combustion chamber wall was unharmed when the test ended.

crystal structure appears to be its half-filled $3d$ subshell, which gives each Mn atom a large magnetic moment. The α-Mn and β-Mn crystal structures (Table 15.2) align the atoms' magnetic moments antiferromagnetically (i.e., half the moments are positioned to be antiparallel to the other half, resulting in no net magnetic moment for the crystal). Below 727°C the minimum energy arrangement to accomplish this is the complex cubic α-Mn structure (cI58) (Fig. 15.3).

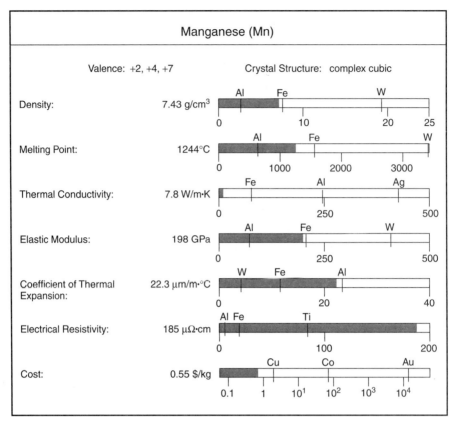

Fig. 15.2 Physical properties of Mn.

TABLE 15.2 Crystal Structure Data for the Four Cubic Allotropes of Mn

	Lattice Parameter (nm)	Pearson Symbol
α-Mn	0.8894	cI58
β-Mn	0.6290	cP20
γ-Mn	0.3774	cF4
δ-Mn	0.3080	cI2

Four different Mn atom coordination numbers exist in α-Mn, and the effective radius of the Mn atom is different in each of the four states. In this regard, α-Mn resembles a quaternary intermetallic compound with the four Mn atom types bonding as if they were different elements. The β-Mn structure (stable from 727 to 1100°C) is also complex (cP20); and it contains two different Mn coordination numbers, each with different radii. At higher temperatures, greater lattice vibrations nullify magnetic effects, and Mn transforms to "normal" metallic crystal structures, FCC (1100 to 1138°C) and BCC (1138°C to melting at 1244°C). Mn's β → α transformation is sluggish; cast Mn can be entirely β phase at room temperature if the casting cools rapidly, and it can take many days for α-Mn to replace β-Mn at 20°C. Even the γ phase can be partially retained at room temperature by severe quenching from above 1100°C. The volume changes accompanying these transformations often crack Mn.

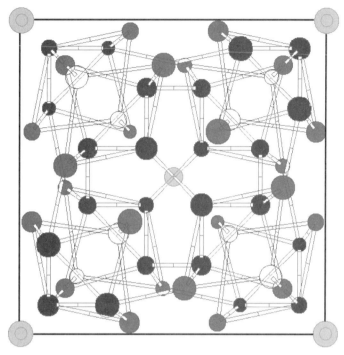

Fig. 15.3 Perspective view of atom positions in the unit cell of α-Mn (cI58) as seen from a [001] viewing direction. The Mn atoms in α-Mn have four different coordination numbers and four different effective atomic radii; these are distinguished by the white, light gray, dark gray, and black atoms. Atoms are shown disproportionately small for clarity. (Courtesy of the Center for Computational Materials Science, Materials Science and Technology Division, Naval Research Laboratory.)

Mn's complex α and β crystal structures make its physical properties markedly different from other transition metals. The α and β allotropes are hard and brittle with high electrical resistivity (185 $\mu\Omega \cdot$ cm for α-Mn and 90 $\mu\Omega \cdot$ cm for β-Mn) and low thermal conductivity. Dislocation motion is nearly impossible in these large unit cells. Mn can be plastically deformed only when the metastable FCC γ phase is present at room temperature (Sec. 15.3.2.2). Mn is more electropositive than any of its neighbors in the periodic table. It forms a superficial oxide layer in dry room-temperature air, but discolors in humid air, and tarnishes rapidly if heated. Mn reacts with all mineral acids, and it even slowly dissociates pure water, liberating H_2 gas as it forms Mn hydroxide. Finely divided Mn metal is pyrophoric, burning in air by reacting with both O_2 and N_2.

15.3.2 Applications of Mn

15.3.2.1 Mn Alloying Addition to Steels and Cast Iron. The Bessemer process was developed in the 1850s to convert pig iron into steel, and it dominated steelmaking until it was replaced by the basic oxygen process (BOP) in the mid-twentieth century. The Bessemer process was one of the great inventions of human history because it allowed steel to replace cast iron as the primary industrial metal. Although they are no longer used, Bessemer converters pumped large volumes of air through molten pig iron (about 4% C, 3% P, 2% S) to oxidize most of these impurities and remove them from the metal, converting brittle pig iron into ductile steel. However, the process left significant amounts of O in the metal, causing undesired casting defects. It also left excess S in the steel that formed liquid phase at grain boundaries in hot steel from a low-melting

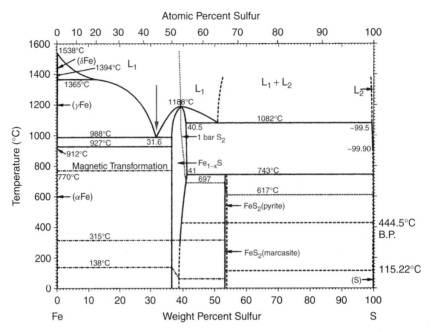

Fig. 15.4 Fe-rich portion of the Fe–S equilibrium phase diagram. Note the eutectic at 988°C to the left of FeS. This eutectic can cause liquid formation at austenite grain boundaries in steel, leading to hot shortness at high temperature and embrittlement at lower temperatures. Mn reacts preferentially with S, avoiding FeS formation. (From Massalski, 1986, p. 1103; with permission.)

eutectic (988°C) with FeS (Fig. 15.4). This causes hot shortness and room-temperature brittleness. The modern BOP also leaves excess O and S in the steel. When Mn is added to steel, it reacts preferentially with O, preventing gas porosity in the steel. Mn also forms refractory MnS precipitates that are distributed harmlessly throughout the steel grains, avoiding FeS formation on grain boundaries. In addition, Mn increases steel's hardenability, provides solid-solution strengthening, lowers the ductile–brittle transition temperature, and controls carbide distribution. Steel and cast iron usually contain from 0.3 to 1% Mn, and since ferrous metals production is so great (900 million tons/year), several million tons of Mn are consumed each year for alloying additions to cast iron and steel.

Mn is usually added to steel in one of two forms: ferromanganese (80% Mn–20% Fe) or silicomanganese (55% Mn–25% Si–20% Fe) master alloys (Sec.15.3.4). Some specialty steels contain more Mn. Steel with 1.75% Mn has better pearlite refinement and enhanced hardenability. Hadfield steel is a high-C, high-Mn steel (Fe–12% Mn–1.1% C) that can be rapidly quenched to produce a nearly 100% austenite microstructure at room temperature. Its strength (σ_{UTS} = 950 MPa), high work-hardening rate, and toughness make it useful for severe service components such as excavators and dredges, Soldier's helmets were fabricated from Hadfield steel during World Wars I and II.

15.3.2.2 Electrolytic Mn. Most Mn is used as an alloying addition to steel, and pure Mn is unnecessarily costly for that use, but other applications require pure Mn. Electrolysis of aqueous $MnSO_4$ solutions can produce 99.7 to 99.9% purity Mn. The electrolysis product can be metastable, ductile, FCC γ-Mn (which transforms slowly at 20°C to α-Mn over a period of weeks). Alternatively, if electrolyte chemistry is adjusted suitably, α-Mn can be produced directly. More than 100,000 tons of electrolytic Mn is produced each year for use as an alloying additive in Al (Sec. 20.2.4.2), Cu (Sec. 18.2.3.5–6), stainless steel, and Mg (Sec. 11.4.2.1) and for use in welding electrodes. In austenitic stainless steel, Ni can be replaced by Mn if N is also added to the metal; this lowers costs since Mn is considerably less expensive than Ni. Electrolytic Mn is sometimes converted to MnO_2 and other compounds when high purity is required.

15.3.2.3 Mn–Cu–Ni Ternary Alloys. Due to Mn's odd properties and reactivity (Sec. 15.3.1), only a few Mn-rich alloys have any engineering importance. Mn can be ductilized by quenching Mn–Cu alloys from the high-temperature FCC γ-phase field. When these alloys have compositions near 80% Mn–20% Cu, they have excellent vibration damping capability, even at high stress levels. Their ability to damp mechanical vibrations results from a thermoelastic martensitic transformation (Sec. 15.3.5). However, these alloys are vulnerable to embrittlement by γ → γ + α reactions upon heating to temperatures only modestly above ambient. Ductile and more thermally stable ternary alloys exist for large ranges of composition in the Cu–Mn–Ni system (Fig. 15.5). These alloys can be hardened by either the ordering reaction (forming tP4 and cP2 structures) of MnNi intermetallic (region A on Fig. 15.5) or by γ → γ + α precipitation hardening (region B on Fig. 15.5). When solutionized at 650°C and aged at 350 to 450°C, 60% Cu–20% Mn–20% Ni alloys achieve a 0.1% offset yield strength of 1130 MPa and $\sigma_{UTS} = 1400$ MPa with 2% elongation; these properties rival those of Cu–Be alloys (Sec. 18.2.3.8).

Most metals experience a decrease in electrical resistivity as pressure increases, but in an 84% Cu–12% Mn–4% Ni alloy called *manganin*, resistivity increases with pressure. The resistivity of manganin is about 75 μΩ · cm at room temperature, and the resistivity changes by only 0.001% per °C near room temperature. These properties make manganin useful for pressure measurement devices. Each 100-MPa increase in pressure raises the resistivity of manganin by 0.27% in a linear manner. Manganin can be used at pressures as high as 40 GPa, and its resistivity change is rapid (a few nanoseconds), allowing it to measure accurately the fast pressure changes in explosion shock fronts.

15.3.2.4 MnO_2 Batteries. MnO_2 acts as a depolarizer in dry-cell C–Zn batteries, allowing them to continue operating without accumulation of H_2 gas on the electrode. Although such batteries are often called "dry cells," they would be more accurately named "moist cells," because the presence of small amounts of water is essential in their pastelike electrolyte solution. MnO_2

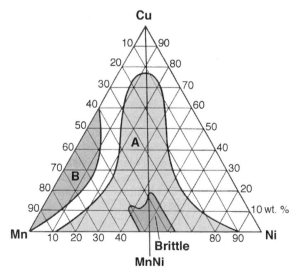

Fig. 15.5 Hardenable regions in Cu–Mn–Ni ternary alloys. Region A can be hardened by ordering reactions in the NiMn intermetallic phase; region B can be hardened by the γ → γ + α reaction. (Adapted from Dean and Anderson, 1941.)

prevents accumulation of H_2 gas around the electrode by providing O atoms, so H_2O forms on the electrode rather than H_2 gas. Without the MnO_2, a layer of H_2 gas would jacket the electrode, sharply reducing current flow from the battery. In alkaline batteries, Zn powder is packed around a brass current collector to act as the anode in the center of the can. The electrolyte is a Zn/KOH gel. The MnO_2 cathode is contained between the steel can wall and a separator membrane that prevents direct contact between the cathode and anode. Li batteries have a Li foil anode, a MnO_2 cathode, and a Li-based electrolyte.

15.3.2.5 Other Mn Compounds. Mn compounds are used in a remarkably wide array of products. $KMnO_4$ is an effective oxidizing agent in several processes, including drinking water purification, sewage treatment plants, and various wastewater systems in the chemical and paper industries. $KMnO_4$ also acts to remove odors from air in building HVAC systems. MnO_2, ferromanganese, and sometimes Mn metal powder are present in welding fluxes to form protective fluxes over the molten metal and to scavenge O and S from the liquid metal. MnO_2 is a major constituent in groundcoat frit to coat steel with glass (enameling); the MnO_2 layer adheres strongly to steel but has an undesirable black appearance. A second enamel coat is applied over the groundcoat with the desired final color. (When an enameled appliance such as a washing machine is chipped or gouged, the black MnO_2 groundcoat often becomes visible.) MnO_2 can be mixed with other metal oxides to form pink and reddish-brown pigments for ceramic glazes and in brick and tile colorants. Mn oxides and carbonates are often used in magnetic ferrites. Other Mn compounds are added to fertilizers and food additives, and MnO_2 is used as an oxidizing agent in numerous processes to make artificial dyes, flavors, and fragrances.

15.3.3 Toxicity of Mn

Like most $3d$ transition metals (Sec. 18.2.4), Mn is a necessary nutrient for both plants and animals; however, excessive intake can be toxic. Mn is present in some enzymes and acts as an activator for others; consequently, it is necessary for proper growth and health. Mn is vital in plants for chlorophyll production, and Mn sulfate is added to fertilizers for use in Mn-deficient soils. In animals and humans, excess Mn causes neurologic problems similar to those of Parkinson's disease (e.g., tremors, spasms, and lack of motor coordination). Inhalation of Mn dusts poses a greater risk of toxicity than ingestion into the gastrointestinal tract because inhaled Mn can be absorbed directly into the blood without being metabolized by the liver.

15.3.4 Sources of Mn and Refining Methods

High-grade Mn ore (35% Mn content or higher) is widely distributed around the world, and several nations are major producers. The most common ores are hydrated or anhydrous Mn oxides in sedimentary deposits. Most ore is used for ferromanganese production, but ore with low Fe content and minimal amounts of elements electronegative to Zn (i.e., Cu, Ni, Co, As) commands a higher price for battery use. Similarly, particular ores are preferred for production of electrolytic Mn metal. Although huge quantities of Mn-rich accretion nodules exist on much of the world's deep ocean floors (Sidebar 15.2), no economical method of exploiting this resource has yet been devised.

Mn ore is concentrated by the usual methods: crushing, screening, washing, flotation, and heavy media and magnetic separation. Roasting is frequently used to convert more complex Mn minerals to MnO_2 for ferromanganese production. Ferromanganese is made by mixing concentrated Fe and Mn ores (which normally contain considerable Fe) and reducing this mixture in a submerged arc furnace with coke and limestone. Although practically all the Fe ore is converted to metal by this process, only about 80% of the Mn ore is recovered, due to the greater volatility of Mn. For these reasons a Mn ore/Fe ore ratio as high as 9 or 10 is charged into the furnace to yield ferromanganese with 80% Mn.

MANGANESE

> **SIDEBAR 15.2: Mn NODULES ON THE OCEAN FLOOR**
>
> Large regions of the deep ocean floor are littered with enormous numbers of Mn nodules, each roughly the size and shape of a potato. The nodules are concretions of the colloidal suspensions of Mn, Fe, and other metal oxides and hydroxides washed into the sea from eroding minerals on the land. Compositions vary widely, but a typical nodule contains about 30% Mn, 24% Fe, 1% Ni, 0.5% Cu, and 0.5% Co. Many nodules contain an obvious nucleation site at the center, such as a shark tooth or piece of bone. Over 10^{12} tons of them exist, most at depths of 4000 to 6000 m, which is too great to permit economical recovery. Their age and their mechanism(s) of accretion are the subject of continuing research, but they are most abundant in regions with low sediment and strong bottom currents. Nodules apparently grew fastest about 50 million years ago; agglomeration continues slowly today (10^7 tons per annum).

15.3.5 Structure–Property Relations in Mn

Mn–Cu and Mn–Cu–Ni alloys are useful alloys for warship propellers because their high damping capability reduces noise and makes it easier for the ship to avoid sonar detection. A 73%

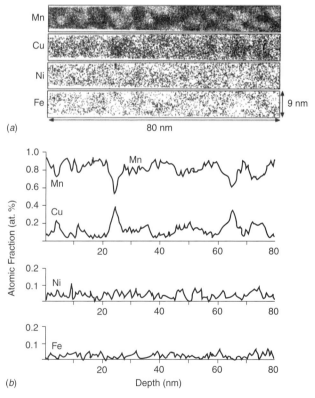

Fig. 15.6 Three-dimensional atom probe composition maps of Mn, Cu, Ni, and Fe (*a*) in a slice volume of 9 nm × 2 nm × 80 nm; (*b*) corresponding composition profiles of each element in a selected volume of 2 nm × 2 nm × 80 nm from part (*a*). Each black dot in part (*a*) represents a single atom. Note how Mn-lean and Cu-rich regions coincide in regions 5 to 10 nm wide. These are zones where the FCC γ phase decomposed into Mn-enriched face-centered tetragonal regions during cooling from 1173 K to room temperature. (From Wu et al., 2002, p. 319; with permission.)

Mn–20% Cu–5% Ni–2% Fe (at%) alloy is one member of the family of high-vibration damping Mn alloys. During slow cooling (0.025°C/s) from the high-temperature FCC γ-phase field, this alloy remains nominally FCC when it reaches room temperature. However, two changes occur in the alloy during cooling: (1) Although the alloy is not magnetically ordered at high temperature, it becomes antiferromagnetically ordered during cooling, and (2) small regions (about 5 to 10 nm wide) of Mn–Cu segregation occur (Fig. 15.6). The Mn-rich regions shown are a slight tetragonal distortion of the FCC γ phase. The interfaces between the face-centered tetragonal Mn-rich regions and the FCC Cu-rich regions are coherent, but the 0.8% difference in the c dimension of the unit cells causes elastic strain near the interface. The damping capacity is believed to result from {110} twin boundaries formed in conjunction with the strain caused by this tetragonal distortion. As the metal vibrates, the spacing between the atoms oscillates, and this causes both the twin boundaries and the magnetic domain boundaries to shift slightly back and forth in the microstructure, dissipating vibrational energy.

15.4 TECHNETIUM

> It would be a great loss to the world to have any significant quantity of this valuable metal be buried in deep holes in the ground, as per the current plan set up by those unaware of the potential use for technetium.
>
> —Edwin Sayre, "Technetium, a Manmade Sister Element and Backup Alloying Agent for Rhenium," presented at the 1997 TMS Meeting

15.4.1 Physical Properties of Tc

Tc is a strong, ductile HCP metal. It is a type 2 superconductor, and it has the highest ambient pressure superconducting transition temperature (11.4 K) of any metallic element. All Tc is radioactive, but three of the isotopes are relatively long-lived (Table 15.3), and their radiation consists of low-energy β rays and very weak γ rays that pose only a moderate safety threat. Hundreds of tons of Tc have been produced in nuclear reactors, so Tc is more readily available than most other radioactive elements, and Tc's major physical properties are well established. Tc binary phase diagrams have been partially or totally determined with about a dozen metallic elements. Tc improves the room-temperature ductility of Mo and W just as Re does in Re–Mo and Re–W alloys (Sec. 15.5.2.2).

Tc forms a thin, stable oxide in dry air, but it tarnishes noticeably in humid air. At elevated temperatures, the oxide is volatile and nonprotective. Tc resists attack by hydrochloric and hydrofluoric acids, but it is corroded by nitric and hot sulfuric acids.

15.4.2 Applications of Tc

Most Tc is handled as nuclear waste and stored with other radioactive fission products. Tc has few engineering uses due to safety concerns related to its radioactivity and because it is difficult and costly to separate from other fission products. The radiation from bulk Tc is neither intense nor penetrating, but it poses an inhalation hazard when powdered or vaporized, and since its

TABLE 15.3 Decay Behavior of Selected Tc Isotopes

Isotope	Decay Half-Life	Radiation	Decay Product
$^{97}_{43}$Tc	2.6×10^6 yr	e^- capture	$^{97}_{42}$Mo (stable)
$^{98}_{43}$Tc	4.2×10^6 yr	β^-	$^{98}_{44}$Ru (stable)
$^{99}_{43}$Tc	2.1×10^5 yr	$\beta^- + \gamma$	$^{99}_{44}$Ru (stable)
$^{99m}_{43}$Tc	6.0 h	Internal transition + γ	$^{99}_{43}$Tc

oxide is volatile at high temperatures it must be handled with caution. To date, the principal use for Tc has been in nuclear medicine, but potential future applications are both numerous and appealing.

15.4.2.1 Tc in Nuclear Medicine. One of the most widely used radioisotopes in medical practice is ^{99m}Tc. This metastable variant ($t_{1/2} = 6$ hours) of ^{99}Tc emits penetrating γ rays that allow it to be "imaged" by an x-ray camera or scanner after it has been absorbed preferentially by the body organ of interest. The short half-life requires that it be produced at the hospital just before its injection into the patient, and this is done by weekly delivery to the hospital of a $^{99}_{42}$Mo radioisotope ($t_{1/2} = 66$ hours). The $^{99}_{42}$Mo is a ^{235}U fission product produced by irradiating U at a central laboratory. The $^{99}_{42}$Mo produces fresh ^{99m}Tc continuously as its decay product, and this can be eluted from the Mo source at the hospital, reacted with the appropriate carrier chemical to "tune" it to concentrate in a particular organ, and injected into the patient. The short half-life allows easy imaging of the organ, but the radioactivity diminishes so quickly after diagnosis that the patient receives a minimal radiation dose. Tc is a prospective alloying addition to stents (Sec. 12.2.6.1) used to hold open arteries that had been occluded by plaque buildup. Research has shown that a continuous light β-radiation dosage can prevent plaque from redepositing on the stent, and the radiation from ^{99}Tc could serve that purpose.

15.4.2.2 Possible Future Uses of Tc. Tc's ability to improve room-temperature ductility in Mo and W may allow it to replace Re in these alloys, conserving supplies of scarce Re. Re is added to many Ni superalloys to improve fatigue strength and to permit operation at higher temperatures (Sec. 15.5.3.4). The similarity of Tc and Re may allow Tc to be substituted for Re in this application, which would lower the alloy's weight and provide a "built-in" radiation tracer that would permit monitoring of oxidation or wear of combustion zone turbine blades during engine operation. The pertechnetate ion (TcO$_4^-$) suppresses corrosion of steel. Concentrations of only 5 to 50 ppm are needed in water to arrest corrosion of mild steel at temperatures up to 250°C. Reduction in stress corrosion cracking is seen when Tc is added to stainless steel at the 100 to 1000 ppm level.

15.4.3 Production of Tc

The natural abundance of Tc is vanishingly small, so all Tc is produced in nuclear reactors. The fission of ^{235}U produces a wide variety of elements, and Tc (particularly ^{99}Tc) is produced in large quantities during routine reactor operation. It is estimated that the total world inventory of Tc is more than 250 tons, with an additional 4.5 tons added to the total annually. Most Tc is stored in nuclear fuel reprocessing waste solutions or is present in unprocessed spent nuclear fuel assemblies. The chemical processing technology exists to separate the many elements present in spent reactor fuel. Some nations (e.g., Russia, France, England, and Japan) use this technology to reprocess used fuel, but they do not presently utilize its Tc content in significant quantities. Other nations, most notably the United States, simply store used fuel and plan to bury it in long-term secure storage facilities to prevent environmental damage from the fuel's more intensely radioactive constituents. Whether the world's inventory of Tc will be utilized or discarded remains to be determined, and the decision may hinge more on political factors than on engineering or economic factors.

15.5 RHENIUM

> Even at 2,800°C, about 380° below its melting point, rhenium has a useable and reproducible level of strength.
>
> —Bernd Fischer, D. Freund, and D. Lupton, *Rhenium and Rhenium Alloys*, Minerals, Mining, and Materials Society, Warrendale, PA, 1997

15.5.1 Overview of Re

Re is a metal of extremes. It is rare, expensive, and dense. Re is the strongest pure metal possessing room-temperature ductility, and it is ductile over a wider temperature range (0 to 3453 K) than any other solid. It has the second-highest melting point of all metals (after W). Os and Ir are the only other metals with higher Young's moduli. These properties make Re useful in solving engineering problems, but its high cost and scarcity restrict it from wider use.

15.5.2 Physical Properties of Re

15.5.2.1 Corrosion.
Re tarnishes only slightly at room temperature in humid air but oxidizes rapidly in air at temperatures above about 350°C. Re_2O_7 boils at 363°C and is visible as a white smoke emanating from the surface of hot Re metal exposed to air. Thus, Re use at high temperatures requires a vacuum or inert atmosphere or protection by an oxidation-resistant coating. Re has good resistance to sulfuric, hydrofluoric, and hydrochloric acids but is attacked rapidly by nitric acid and alkalis.

15.5.2.2 Mechanical Properties.
Re is HCP from 0 K to its melting point. It is among the strongest pure metallic elements, and it has the highest work-hardening rate of any metal. Annealed pure Re has $\sigma_{y,0.2\%} = 290$ MPa and work hardens rapidly as the tensile test proceeds to achieve an ultimate strength of 1070 MPa. In fully work-hardened Re, $\sigma_{UTS} = 2300$ MPa. This high strength is accompanied by good ductility at room temperature; Re has been cold-rolled to 60% reduction at room temperature, which is unusually high for an HCP metal. Room-temperature tensile elongation is 15 to 20%, but ductility declines to only a few percent elongation above 800°C (Table 15.4). Re's grain boundary strength is inadequate to permit extensive hot deformation; this hot shortness makes it more amenable to cold work than to hot work. Re recrystallizes at 1400 to 1600°C. After recrystallization, Re appears not to experience grain growth at 2400°C, even when annealed for extended times. Re's high work-hardening rate makes it essentially unmachinable. Despite its ductility, electrodischarge machining or diamond grinding is necessary to shape pure Re at room temperature. Re is readily weldable, but the high melting temperature and oxidation behavior make electron-beam welding or inert-gas welding the preferred techniques. For these reasons, powder metallurgy with hot isostatic pressing is the dominant Re fabrication technique; both flake and spherical powders are commercially available. Wrought products can be produced from pure Re, but extensive dimensional change requires costly cycles of cold work followed by vacuum annealing. Re alloys with W and Mo

TABLE 15.4 Comparison of Pure Re, 52.5% Mo–47.5% Re, and 75% W–25% Re Properties

Property	Re	52.5% Mo–47.5% Re	75% W–25% Re
Density at 20°C	21.04	13.70	19.70
Young's modulus (GPa)	469	367	431
σ_{UTS} (MPa) at 20°C	1070	1030	1310
σ_{UTS} (MPa) at 800°C	620	480	1030
σ_{UTS} (MPa) at 1600°C	210	100	230
Tensile elongation (%) at 20°C	>15	19	>15
Tensile elongation (%) at 800°C	5	18	—
Tensile elongation (%) at 1600°C	2	17	—
Ductile–brittle transition temperature (°C)	Ductile at −273	−273 to −170	−25 to −100
Thermal expansion coefficient (°C^{-1}) at 500°C	6.12×10^{-6}	5.72×10^{-6}	4.48×10^{-6}

Source: Carlen (1997); with permission.

(Sec. 15.5.3.2) have lower costs and fewer limitations on fabrication; consequently, they are preferred for several applications.

Re single crystals slip on two slip systems: $\{10\bar{1}0\}\langle11\bar{2}0\rangle$ (τ_{CRSS} = 21 MPa) and (0001) $\langle11\bar{2}0\rangle$ (τ_{CRSS} = 14 MPa), and slip is accompanied by deformation-induced $\{11\bar{2}1\}$ twinning. Some evidence of slip on the $\{10\bar{1}1\}$ plane has been obtained from deformed polycrystalline Re. Rapid work hardening occurs due to the metal's high elastic constants. Twinning is thought to play a key role in Re's ductility, reorienting grains to allow further slip to occur as deformation progresses. The work-hardening rate is highly anisotropic, and this causes the intergranular fracture often observed in Re, particularly at elevated temperatures. The addition of Mo and W to Re reduces work-hardening anisotropy and mitigates polycrystalline Re's hot shortness problem.

15.5.3 Applications of Re

15.5.3.1 Pure Re. Re delivers excellent performance for applications requiring strength at 2000°C, room-temperature ductility, low sublimation losses for heating elements, or minimal erosion in high-voltage switch contacts. Pure Re's high cost and the difficulties in forming it (Sec. 15.5.2) limit its use to applications where the special properties of pure Re are essential. These include rocket nozzles (Sidebar 15.1), instrument filaments and heating elements, high-energy electron tubes, wear- and arc-resistant switch contacts, targets for x-ray generating tubes, and Re–W thermocouples for use up to 2200°C.

15.5.3.2 Re Alloys. Both W and Mo have extensive solubility in Re (Fig. 15.7), and they are often alloyed with Re to make it lighter, less expensive, and more easily fabricated. The most widely used high-Re alloys are 52% Mo–48% Re and 75% W–25% Re. As Table 15.4 shows, the alloys have many of pure Re's outstanding physical properties, including room-temperature ductility and good elevated-temperature strength. In addition, the alloys can be machined (with difficulty) using conventional methods. The best properties for these single-phase, BCC, solid-solution alloys occur near the solubility limit of Re; unfortunately, the phase regions shown on Fig. 15.7 are not clearly established because diffusion in these materials is quite slow. For alloys intended for prolonged service at high temperature, somewhat lower Re content is sometimes used to avoid possible formation of the brittle σ and χ phases.

15.5.3.3 Re Catalysts. Pt–Re and Pt–Re–In catalysts supported on Al_2O_3 are used to promote hydrocarbon re-forming reactions that improve the yield of high-octane constituents in gasoline during petroleum refining (Fig. 17.5). In addition to gasoline production, Re catalysts also improve yields for several pharmaceutical and chemical synthesis processes. These catalysts contain a variety of Re compounds (e.g., CH_3ReO_3), and they have better resistance to poisoning by N, S, and P than other catalysts. Their use rose dramatically when a U.S. government regulation in the 1970s mandated that only Pb-free gasoline be sold. Since catalysis consumes more than a third of world Re production, the price of Re experienced wide swings ($660 to $6600 per kilogram for Re metal) during the last decades of the twentieth century as demand for Re catalysts rose and fell. Competing catalytic compositions (e.g., Pt–Sn) have eroded Re's market share in recent years, but Re use is still substantial.

15.5.3.4 Re as a Superalloy Addition. Modern Ni-based single-crystal superalloys are strengthened by ordered intermetallic γ′ precipitates that are coherent with the solid-solution γ-phase matrix (Sec. 16.3.3.6). Mechanical properties of superalloys are improved by adding such refractory elements as Ta, Re, W, and Mo. Re is particularly effective at improving creep strength at 1000 to 1150°C, and all modern single-crystal superalloys contain 1 to 2 at% Re. Re partitions primarily into the matrix phase, where it slows γ′ coarsening. Re also forms clusters about 1 nm in diameter spaced about 20 nm apart within the γ matrix that are more effective strengtheners than other elements in solid solution.

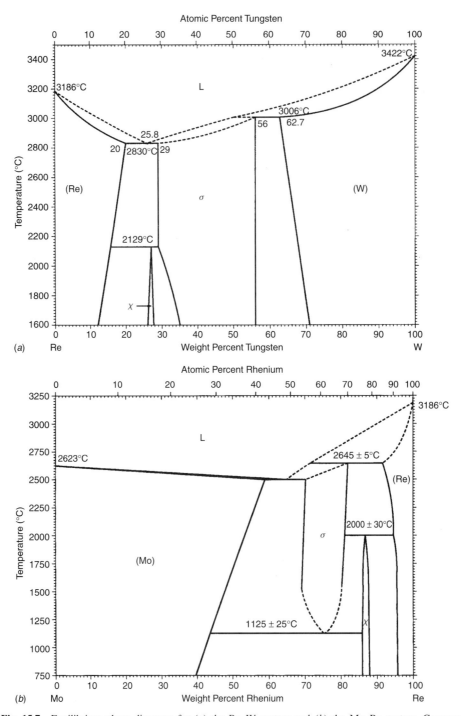

Fig. 15.7 Equilibrium phase diagrams for (*a*) the Re–W system and (*b*) the Mo–Re system. Commercial alloys have compositions in the W- and Mo-rich solid solution phase fields. The χ phases (α-Mn, cI58 structure) and the σ phases (tP30 structure) are brittle. (From Massalski, 1986, p. 1923, 1968; with permission.)

SIDEBAR 15.3: Re EMISSIONS FROM RUSSIA'S KUDRYAVY VOLCANO

Russia's Kuril Islands lie in the northwestern Pacific. Long known for fog, storms, timber, fur, and fish, this volcanic archipelago now has the potential to add the distinction of possessing the world's first Re mine. In 1992 volcanologists discovered that the Kudryavy Volcano was emitting about 20 tons/yr of Re in the form of Re sulfide vapor. Experimental trials performed intermittently between 1996 and 2003 have shown that a Zeolite-packed dome built over the vent could extract the Re sulfide from the 800°C stream of vapors in sufficient quantity to be profitable. Heretofore, no concentrated Re deposits had ever been found, and all existing Re recovery operations produce Re as a by-product of Cu and Mo mining. There are considerable logistical and safety challenges associated with operating a "filter mine" in the crater of an active volcano, but Re's market value of nearly $1 million per ton provides a strong incentive to exploit the resource.

15.5.4 Sources of Re and Refining Methods

There are no Re mines (see, however, Sidebar 15.3). Re is produced as a minor by-product of Cu and Mo mining. Re and Mo are often found together because they respond similarly to geological segregation processes. Re is extracted with Mo from Cu in the form of sulfurous sludge, which is roasted to release volatile Mo and Re oxides. The Re oxide collected is usually converted to ammonium perrhenate (NH_4ReO_4), which is traded internationally. NH_4ReO_4 is reduced by H_2 at high temperatures to produce Re metal powder for catalytic or metallurgical use. A significant fraction of annual Re demand is satisfied by recycling Re from spent catalysts.

15.5.5 Structure–Property Relations in Re

Only a few metals have the capability to deliver useful strength above 2000°C, and creep becomes the dominant failure mechanism in this temperature regime for even the most refractory metals. In creep tests between 2100 and 3000°C (Table 15.5), pure Re displays mixed ductile and intergranular fractures, with the latter predominating at the higher temperatures (Fig. 15.8). One group of specimens was prepared by powder metallurgy (P/M) and sintered at 2700°C to 90.5% of theoretical density; the other specimens were prepared by the same initial P/M processing, but these were subjected to wire drawing deformation to achieve 98.6% of theoretical density. The 98.6% dense specimens showed superior strength from 2100 to 2500°C, presumably because

TABLE 15.5 Stress Rupture Strengths of Pure Re Specimens for a 10-Hour Time to Failure

Temperature (°C)	10-h Stress Rupture Stress for 90.5% Dense P/M Specimens (MPa)	10-h Stress Rupture Stress for 98.6% Dense Wrought Specimens (MPa)
2100	12.5	24.0
2500	6.0	18.0
2700	a	9.1
2800	5.0	4.2
3000	3.4	b

Source: Fischer et al. (1997); with permission.

[a] No tests were performed at this temperature on P/M specimens.
[b] Early failures prevented reliable determination of 10-hour stress for these specimens.

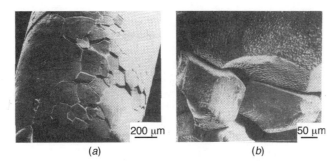

Fig. 15.8 Intergranular fracture in Re creep specimen tested at 2800°C showing (*a*) specimen surface with several grain boundary separations visible, and (*b*) the intergranular fracture surface. (From Fischer et al., 1997, p. 319; with permission.)

they have fewer and smaller pores to serve as crack initiation sites for intergranular fracture. However, at the highest temperatures tested, the 90.5% dense specimens had better creep strength than the denser material. The denser material experienced rapid grain growth at 2700°C and above during the test, while the higher porosity of the lower-density metal delayed grain growth until the porosity was finally reduced by further sintering during the test. The difference in grain size combined with the presence of preexisting surface flaws on the drawn metal weakened it at the highest temperatures. The ability of Re to hold a 3.4-MPa load for 10 hours at 3000°C is rivaled by only one other metal, W, the highest melting of all metals (Sec. 14.4).

ADDITIONAL INFORMATION

References, Appendixes, Problem Sets, and Metal Production Figures are available at
ftp://ftp.wiley.com/public/sci_tech_med/nonferrous

16 Co and Ni

16.1 OVERVIEW

In the late transition metals, the d subshell is more than half-filled, and d electrons have begun to withdraw into the atoms' inert electron cores. Consequently, Co ($3d^7 4s^2$) and Ni ($3d^8 4s^2$) have fewer bonding electrons, lower maximum valences, lower melting and boiling points, and lower reactivity than the earlier $3d$ transition metals. Co and Ni are ductile, tough, and strong at ambient and elevated temperatures. Their high electronegativities and adherent, diffusion-resistant oxide layers give them the oxidation resistance that defines much of their engineering use. Both metals are used in heat-resistant superalloys, and Ni and Co are often electroplated onto baser metals to protect them from corrosion. Ni is the FCC stabilizer in austenitic stainless steels, and Co is the most commonly used binder metal in carbide cermets for cutting and grinding. Both Co and Ni are ferromagnetic and used in a variety of magnetic materials.

Co is one of the least abundant $3d$ transition elements, comprising 29 ppm of Earth's crust; only Sc (25 ppm) is scarcer. Ni is somewhat more abundant (99 ppm) and is produced in far larger quantities than Co, but both Co and Ni are costlier than their neighboring $3d$ transition metals Mn, Fe, Cu, and Zn. Consequently, their use is limited to especially challenging applications where other metals cannot be substituted.

16.2 COBALT

> One can appreciate why Co users were anxious about the reliability of Co supplies. Most production came from two central African nations that were in an almost continuous state of civil war and political unrest. One of these was so unstable that the names of the country itself and its major cities kept changing every few years.
>
> —David C. Fisher, Agilent Technologies, Inc., Palo Alto, CA, private communication, 2004

16.2.1 History of Co

Co minerals have been used as blue pigments for thousands of years. Glass and pottery glazes from ancient Egypt, Greece, Persia, and China contain Co oxide fused with potash and silica. Reduction of the oxide to produce Co metal did not occur until 1735, and Co metal saw little use until the early twentieth century, when a series of wear-resistant, high-temperature Co–Cr alloys (*stellites*) were introduced. In the 1920s, Co was adopted for the metallic binder phase in WC cermet cutting tools and abrasives. Interest in Co expanded again in the 1930s when Al–Ni–Co alloys were found to have high magnetic-energy products. The 1940s saw the introduction of the first superalloys, Co- and Ni-based alloys used for jet engine combustion zone turbine blades. In the later twentieth century, stronger permanent magnets made of RCo_5 (where R = a rare earth element) were introduced, and magnetic cobalt-alloy thin films became the dominant material for thin-film magnetic storage technology. In the mid-twentieth century, the primary sources of Co were Congo (Zaire) and Zambia, both of which were politically unstable. The importance of

---

Structure–Property Relations in Nonferrous Metals, by Alan M. Russell and Kok Loong Lee
Copyright © 2005 John Wiley & Sons, Inc.

Co for jet engines, carbide tools, and high-performance magnets led several nations to stockpile Co in strategic reserves, but those reserves became less necessary in the 1990s as Co production increased from more stable suppliers in Russia and Australia.

16.2.2 Physical Properties of Co

The physical properties of Co are summarized in Figure 16.1.

16.2.2.1 Co Crystal Structures. As Co cools, its high-temperature FCC α phase transforms at 421°C to an HCP structure ($\varepsilon, c/a = 1.6233$). This occurs in a martensitic phase change by dislocation movement on the FCC octahedral planes; Co's density increases 0.15% during the FCC → HCP transformation. Since the α → ε transformation energy is only about 360 J/mol, the transformation is sluggish, and pure Co and Co alloys often have mixed HCP and FCC microstructures at 20°C (Fig. 16.2). Since FCC and HCP differ only in the stacking sequence of their densest-packed planes, α and ε regions are separated by stacking faults where the FCC {111} planes' stacking sequence (*ABCABCABC*...) changes to the HCP (0001) planes' sequence (*ABABABA*...). The FCC ⟨212⟩ direction lies parallel to the HCP ⟨10$\bar{1}$0⟩ direction at these stacking faults. Pure Co's stacking fault energies are low: 31 mJ/m^2 in ε-Co at 20°C and 18.5 mJ/m^2 in α-Co at 710°C; these energies are roughly comparable to twin boundary energy in Cu (25 mJ/m^2).

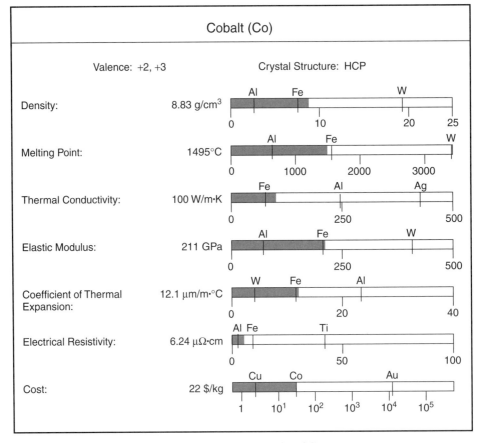

Fig. 16.1 Physical properties of Co.

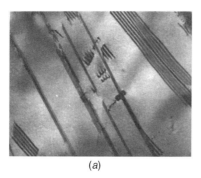 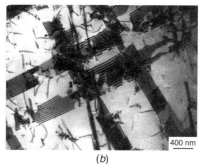

(a) (b)

Fig. 16.2 Transmission electron micrographs of stacking faults in FCC Co superalloy (Co–25 wt% Cr–11 wt% Ni–7.5 wt% W–0.8 wt% Al–0.5 wt% C–0.25 wt% Ti–0.2 wt% Mo–0.15 wt% Zr–0.14 wt% Ta–0.05 wt% B) taken (a) before and (b) after deformation. Electron wave interference effects make each stacking fault appear as a group of closely spaced parallel lines. (From Lu et al., 1999; with permission.)

The stability of Co's two allotropes is sensitive to the purity and grain size of the metal. Fine grain structure and certain impurities (e.g., Fe) favor the presence of metastable FCC at room temperature. The FCC phase usually is the more common structure at 20°C in fine powders, fibers, and thin films. Since α and ε are often present in varying ratios at room temperature, the physical properties of pure Co are somewhat variable. For example, the coefficient of thermal expansion (COTE) is higher for α than for ε, so the COTE of a given sample of Co varies with the α/ε ratio.

16.2.2.2 Corrosion and Gas Solubilities. Co forms a protective oxide layer at room temperature that resists corrosive attack with roughly the same effectiveness as Ni (Sec. 16.3.2.1). However, Co is attacked rapidly by HCl, HNO_3, and H_2SO_4 acids; HF acid reacts slowly with Co. Even pure H_2O corrodes Co very slowly (about 5 μm/year). At elevated temperatures, Co's oxide layer thickens at a parabolic rate. Between 400 and 900°C, the oxide has two stoichiometries, Co_3O_4 near the oxide–air interface and CoO near the oxide–metal interface. Above 900°C, the oxide becomes entirely CoO. Although pure Co is somewhat oxidation-resistant at elevated temperature, it oxidizes about an order of magnitude faster than Ni. The oxidation resistance of many Co alloys is markedly superior to that of pure Co. Co suffers more rapid attack in air containing S compounds due to the presence of a eutectic between Co and Co_4S_3 at 877°C. Pure Co is rarely used in engineering practice, so corrosion and oxidation performance comparisons between pure Ni and pure Co are of limited importance. O, N, and H have low solubilities in solid Co, but their solubilities are higher in liquid Co (Table 16.1).

16.2.2.3 Elasticity and Plasticity. Pure Co is hard and tough with relatively high ductility for an HCP metal. The primary slip plane in ε-Co is the basal plane (0001); secondary slip has been

TABLE 16.1 Solubility (wt%) of O, N, and H in Co[a]

Impurity Element	600°C	1200°C	1500°C	1750°C
O	0.006	0.013	0.125	0.4
N	nil	nil	0.004	0.006
H	<0.0001	—	0.002	0.0026

[a]The melting point of Co is 1495°C; the data for 1500 and 1750°C are for liquid Co.

observed on $\{10\bar{1}1\}$ and $\{11\bar{2}2\}$. Co twins on two planes: $\{10\bar{1}2\}$ and $\{11\bar{2}1\}$. Room-temperature tensile yield strength of 99.9% purity Co ranges between 200 MPa (annealed, cast metal) and 300 MPa (P/M metal). Co's σ_{UTS} is typically 700 to 875 MPa, with the usual variations associated with purity and thermal history. Tensile elongation is typically 15 to 30%. An unusual aspect of work hardening in Co and Co alloys (which are usually FCC) is the contribution made by stacking-fault pile-ups. The number and density of stacking faults increase as plastic deformation progresses (Fig. 16.2b), with HCP zones and twins forming at stacking-fault intersections.

16.2.2.4 Magnetic Behavior. Ferromagnetic materials can have a large and permanent magnetic moment, even in the absence of an external magnetic field. Ferromagnetic behavior results from the combined effects of the spin of each electron (the major factor determining the net magnetic moment) and the orbital motion of each electron (a minor factor). Fe, Co, and Ni are ferromagnetic at room temperature. In a ferromagnetic crystal lattice, the atomic magnetic moments interact and align parallel to one another. The quantum mechanics (Heisenberg) model of ferromagnetism describes the parallel alignment of magnetic moments in terms of an exchange interaction between neighboring atoms moments.

An isolated atom of Fe, Co, or Ni has two $4s$ electrons in the unexcited ground state. However, in a crystal of Fe, Co, or Ni metal, the average density of electrons in the $4s$ level drops from 2 to about 0.6. In such crystals, each atom has five of the $3d$ electrons in the spin "up" mode and a smaller, fractional number of $3d$ electrons in the spin "down" mode. Fe, for example, has five spin-up and 2.4 spin-down electrons, yielding a net magnetic moment of 2.6 ($5 - 2.4 = 2.6$). Co has one more electron in the $3d$ subshell, and it has 3.4 electrons in the spin-down mode, yielding a net magnetic moment of 1.6. Ni has yet another $3d$ electron, so its net magnetic moment is only 0.6. The actual saturation magnetizations deviate somewhat from the predictions of this simple model (Fig. 16.3). Fe is the most strongly ferromagnetic pure element, followed by Co and then Ni.

A ferromagnetic material's magnetic domains can be aligned by a strong external magnetic field, and much of this alignment remains after the external field is removed (point R on Fig. 16.4). Heavy plastic deformation or high temperature destroy this remanent magnetization. Although Co does not have the largest saturation magnetization of the pure elements, it does have a large coercivity (point C on Fig. 16.4) and it retains its ferromagnetism to unusually

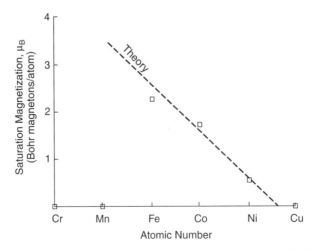

Fig. 16.3 Theoretical (dashed line) and actual (open squares) saturation magnetizations of the late $3d$ transition elements.

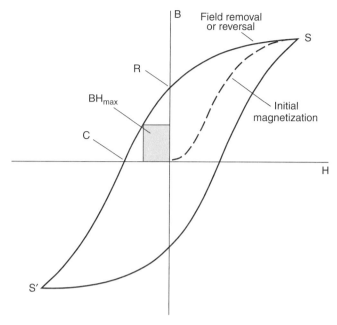

Fig. 16.4 Relation between the external magnetic field strength (H) and the magnetization (B) of a ferromagnetic material. An unmagnetized ferromagnet has magnetic domains oriented in many different directions, and the overall net effect is for these domains to cancel one another out, yielding no net magnetization in the whole body, the condition at the origin of this plot. When the external field is strong enough to align all the domains, saturation is achieved (S). When the external field is removed, the material retains some of the domain alignment (R). The reversed external magnetic field strength needed to bring the material's magnetization back to zero is the coercivity (C). The strength of a permanent magnet can be expressed by the area of the largest rectangle (shaded area) that can fit inside the hysteresis loop's second quadrant, labeled BH_{max}. (From Callister, 2000, p. 690; with permission.)

high temperature. The Curie point is the temperature at which thermal vibration overwhelms the magnetization effect and the metal loses its ferromagnetic behavior. The Curie point of Co is 1121°C, which is higher than the Curie points of Fe (770°C) and Ni (358°C). Co's ease of magnetization is highly anisotropic; at room temperature the $\langle 0001 \rangle$ direction (the c axis) can be saturated by a relatively weak external field, while saturation magnetization in the perpendicular $\langle 10\bar{1}0 \rangle$ direction requires a much stronger field (Fig. 16.5). This crystalline anisotropy in HCP Co is about 10 times greater than in Fe and 100 times greater than in Ni. However, Co's magnetic aniosotropy declines as the temperature rises above ambient, and at 250°C a single crystal of Co is magnetically isotropic. Above 250°C, the c axis is more easily magnetized than the $\langle 10\bar{1}0 \rangle$. Above the $\varepsilon \rightarrow \alpha$ transition temperature, the anisotropy is lessened in the cubic structure.

16.2.3 Applications of Co

Pure Co is seldom used, but many Co-based alloys are valued for their high-temperature performance and magnetic behavior. Co is also frequently added to other alloys.

16.2.3.1 High-Temperature Co Alloys. The first commercial Co alloy was a Co–Cr heat-resistant alloy introduced in 1907. These alloys have been improved continuously to achieve the modern Co *superalloys* in use today. Although several compositions of Co-based high-temperature alloys are commercially available, most contain about 35 to 65% Co–20 to 25% Cr–5 to 15% W–0 to 30% Ni–0 to 10% Ta, and 0.25 to 1% C. Small amounts of Fe, Si, Zr, Nb,

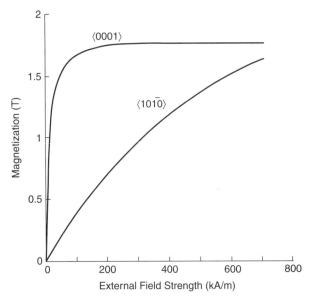

Fig. 16.5 Room-temperature magnetization response in two different crystallographic directions in single-crystal HCP Co. The $\langle 0001 \rangle$ direction is the "easy" direction at room temperature; magnetization is more difficult in the $\langle 10\bar{1}0 \rangle$ direction. This plot is equivalent to the first quadrant of Figure 16.4.

Mo, Mn, rare earth metals, and Ti are frequently present. High Cr content is essential in these alloys to form an oxide layer that is sufficiently protective for use at 1100°C. Cast alloys tend to have low Ni content, and wrought alloys have more Ni to stabilize the FCC phase and facilitate ductility. W is a particularly effective solid solution strengthener in Co; most Co alloys contain several percent W. The presence of a few ppm concentrations of tramp elements (e.g., P, S, Pb, Bi, Te, Se, and Ag) is assiduously avoided in Co superalloys, because these elements can concentrate at grain boundaries and produce lower-melting phases. Mg and Ce can scavenge some tramp elements, forming harmless precipitates and reducing grain boundary accumulation. The alloys are predominately FCC, although stacking faults (Fig. 16.2) often introduce some HCP regions. The strength of Co superalloys is largely attributable to solid-solution hardening and carbide precipitates. Precipitation hardening is used only occasionally in Co alloys, although some alloys derive part of their strength from Ni_3Ti or Co_3W precipitates.

Co forms no binary carbides, but carbide formers in Co alloys (Cr, W, Mo, Ta, Nb, Zr, Ti) produce carbides on grain boundaries, retarding creep strain from grain boundary sliding. Some carbides also precipitate on stacking faults during high-temperature service, which strengthens the metal but lowers ductility. Four types of carbides are common in Co alloys: MC carbides, where M = Ti, Zr, Nb, or Ta; $M_{23}C_6$ and M_7C_3 carbides, where M = Cr (and sometimes W or Mo); and M_6C, where M = W or Mo. These alloys can be heat-treated to optimize carbide distribution for creep resistance. The cast metal is annealed at the highest temperature possible (up to 1250°C) without risking partial melting. This anneal solutionizes much of the C and metallic alloying elements; although complete solutionizing cannot be achieved in the solid state. The solutionizing anneal is followed by aging at 700 to 925°C to form a fine distribution of semicoherent carbides and $M_{23}C_6$ (Fig. 16.6) as a Widmanstätten structure on {111} planes. The best properties result from high concentrations of carbides on grain boundaries (to slow grain boundary sliding) while avoiding a continuous carbide film at grain boundaries that can permit low-energy intergranular crack propagation. A fine dispersion of carbides in the bulk of the grains also aids strength, although the carbides tend to be larger than the ideal dislocation blocking size.

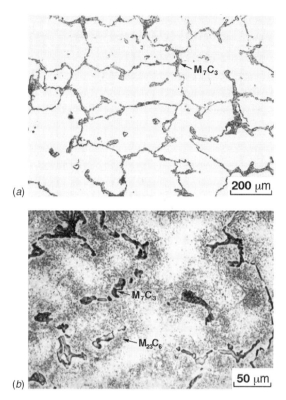

Fig. 16.6 Co K40S superalloy (Co, 25.5 wt% Cr, 10.5 wt% Ni, 7.5 wt% W, 0.8 wt% Si, 0.8 wt% Mn, 0.5 wt% C, and 0.004 wt% B) in (*a*) the as-cast condition, showing coarse carbides primarily on grain boundaries, and (*b*) after annealing at 700°C, which produces discontinuous carbides on the grain boundaries and a finer carbide dispersion in the bulk of the grains. (From Yang et al., 2003; with permission.)

Co superalloys have lower stress rupture values than Ni superalloys (Sec. 16.3.3.6), but Co alloys have good structural stability and slightly better oxidation resistance than do Ni-based alloys at high temperatures. Co alloys are often chosen for components not subjected to high stress, such as nonrotating vanes and nozzles. The relative ease with which Co alloys can be repaired by welding also makes them attractive competitors with Ni superalloys for certain components in turbine engines, burners, and furnace hardware. Large swings in Co prices in recent decades (Sec. 16.2.5) prompted some turbine engine manufacturers to specify less Co in their engines. Several thousand tons of high-temperature Co alloys are still produced each year, but this is far smaller than Ni superalloy demand.

16.2.3.2 Specialty Alloys Containing Co. Various Co-based alloys are used in special applications. Vitallium (Co–30% Cr–5% Mo–0.5% C–0.5% Si) is used in surgical bone replacement plates and in dental prostheses. A related alloy (Co–12% Cr–9% Ni–8% Ti–1% Mo) has the same coefficient of thermal expansion as dental porcelain, allowing dental crown restorations to be fired to fuse the metal to the porcelain without cracking the porcelain during cooling. In low-expansion Co alloys, the normal thermal expansion is offset by a contraction associated with a decline in ferromagnetism as temperature increases. These offsetting effects give the alloys unusually small net changes in length as temperature changes. A 54% Co–37% Fe–9% Cr alloy has an expansion coefficient near zero between −60 and 60°C, which is useful for maintaining exact tolerances in assemblies used outdoors in diverse climates. Alloys of similar composition display near-zero Young's modulus change with temperature.

266 Co AND Ni

16.2.3.3 Co Binder Phase in Carbide Cermets. WC (Sec. 14.4.3.5) has excellent hardness (about 24 GPa) and is less costly than ultrahard materials, such as diamond and cubic BN. However, carbides are brittle, and they chip and fracture easily during cutting and grinding operations. Consequently, they perform better as composites of WC particles in a ductile metal matrix (Fig. 16.7). Many metals can serve as the matrix in these composites, but Co has several properties that make it the best binder phase for carbide tooling. Co forms a eutectic with WC at 1275°C that accelerates sintering and bonds the WC and Co surfaces. W and C dissolve into the Co at high temperature (Fig. 16.8) but precipitate as the cermet cools to room temperature, stabilizing the FCC phase of Co and strengthening it by solid-solution hardening and formation of sub-micrometer-sized carbide and Co–W intermetallic particles. The strong bond formed at the Co–WC interface, combined with the high fracture toughness and good hot strength of Co, make the cermet a much better cutting material than pure WC. The fracture toughness of WC–Co cermets quadruples as Co content in the cermet rises from 6% to 25%; however,

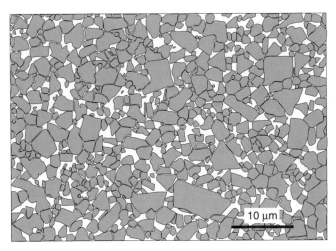

Fig. 16.7 Microstructure of a WC–Co cermet containing 9.5 wt% Co. Light areas are Co; dark areas are WC. (Redrawn from Betteridge, 1982, p. 117.)

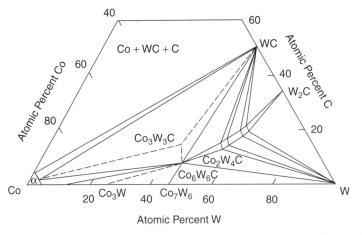

Fig. 16.8 Equilibrium ternary phase diagram for the C–Co–W system at 1000°C. (Redrawn from Betteridge, 1982, p. 54.)

hardness and wear resistance decrease with increasing Co content, so Co content is varied to optimize performance for different uses. These cermets are often produced as mixtures of WC and Ti, Ta, and Nb carbides to improve hot strength and minimize the tendency of the carbides to weld to workpiece metals.

16.2.3.4 Co in Magnetic Alloys. Although Fe, Co, and Ni are ferromagnetic, the magnetic performance of alloys and intermetallic compounds is much superior to that of the pure elements. A strong permanent magnet must resist demagnetization by an external magnetic field that is antiparallel to the remanent magnetization vector; this resistance to demagnetization increases BH_{max} (often called the magnet's work function) (Fig. 16.4). Co additions to permanent-magnet alloys and intermetallic compounds often increase coercivity and enlarge BH_{max}. In addition, coercivity is affected by the metal's microstructure. Lattice elastic strain can increase coercivity, as can small, elongated particles or grains, which make domain reversal more energetically costly.

The twentieth century witnessed a steady progression of ever-stronger permanent magnet materials. The earliest permanent magnets were quench-hardened steels that had better coercivity than ordinary Fe from the residual strain in the lattice caused by the martensitic transformation. These alloy steels have BH_{max} values of 2 kJ/m^3. In 1917 it was shown that adding 35% Co to such steels could increase their BH_{max} to 7 kJ/m^3; Co introduces magnetostrictive strain in the lattice that raises coercivity. Fe–Ni–Co–Al–Cu–Ti alloys [often collectively called *Alnico magnets* (Sec. 16.3.3.7)] were developed in the 1930s with BH_{max} values of 10 to 16 kJ/m^3. These alloys could be heat-treated to achieve high-coercivity microstructures. Cooling them in a strong external magnetic field further boosts BH_{max} to 45 kJ/m^3. Directional solidification forms columnar grains with their long dimensions parallel to the $\langle 100 \rangle$ crystallographic direction, which further raised energy products to 55 to 75 kJ/m^3 in Alnico permanent magnets. In the late 1960s, RCo$_5$ magnets were introduced (where R represents a rare earth element, Sec. 23.4.2.2). When R is Pr or Sm, the RCo$_5$ magnets achieve energy products as high as 175 kJ/m^3. The RCo$_5$ magnets are brittle intermetallics, and they are expensive because Co and rare earth metals (Table 23.1) are both costly. The 1980s saw introduction of a still higher BH_{max} magnet material, Nd$_2$Fe$_{14}$B, which attains energy products of 250 kJ/m^3 with lower rare earth content and no Co content (Sec. 23.4.2.2). Although still brittle, this material is less expensive than the RCo$_5$ magnets, and Nd$_2$Fe$_{14}$B are now widely used in applications such as speakers, headphones, and small electric motors. Nd$_2$Fe$_{14}$B magnets have a relatively low Curie point (310°C), and Co is sometimes partially substituted for Fe in these magnets to raise the Curie point, although Co additions decrease BH_{max}.

Co is a constituent in amorphous alloys that have ultralow coercivities and zero magnetostrictive response. These metallic glasses (Sec. 7.3) are strong materials whose lack of crystal structure makes their magnetic response completely isotropic. They have low eddy current losses and are used in magnetic shielding, amplifiers, and transformers. The most recent Co magnetic alloy to achieve widespread engineering use is the family of Co–Pt alloys used in thin-film magnetic storage on computer hard disks and related devices (Sidebar 16.1).

16.2.3.5 Co in Protective Coatings. Co alloys can be electrodeposited from solutions of Co salts plus alloying element(s) salts. These coatings improve wear resistance and corrosion resistance of base metal surfaces. Co–Ni coatings are frequently used (Vickers hardness of about 500), finding applications in glass molding dies, internal combustion engines, and turbine engines. Co–W and Co–Mo electrodeposited coatings have still greater hardness (up to 700 Vickers), and they retain their hardness well at elevated temperatures, making them useful wear-resistant coatings on hot forging dies. Still higher wear resistance can be achieved by entraining SiC, Al$_2$O$_3$, or other hard ceramic particles in the coatings. Co–Cr–Al–Y and other Co-based alloys can be applied by thermal spray techniques, and these may contain WC or other hard particles to form cermet coatings on a wide range of surfaces, including graphite–epoxy composites.

SIDEBAR 16.1: THIN-FILM Co ALLOYS FOR MAGNETIC DATA STORAGE

Early computer hard disk drives used Fe_2O_3 particles dispersed in an epoxy phenolic resin as the magnetic recording material. The needle-shaped particles were aligned roughly parallel to the motion of the reading/writing head, and data were stored as binary 0 or 1 bits by magnetizing all the particles in one small area to either north or south. These disks worked well for moderate data storage densities (e.g., 150,000 bits/mm^2), but higher storage density (10^7 bits/mm^2) requires magnetic thin films rather than particles, and these replaced Fe_2O_3 particle hard disks in the late 1990s. In thin-film media, each grain in the metal film acts as a single magnetic domain, and the grains are typically 12 to 14 nm in thickness and width (Fig. 16.9). Ferromagnetic materials with large, square hysteresis loops (Fig. 16.4) deliver the best magnetic data storage. A "square" loop has enough coercivity to resist easy demagnetization (making the film more robust against accidental data erasure), and it also assures that the magnetization of the grain can be reversed by a narrow range of applied field strengths. About 1000 grains are needed for each stored bit to minimize "noise" and assure reliable data storage.

A thin film is deposited (usually by sputtering) a ferromagnetic material as a crystalline layer with extremely small grain size. Several Co-based thin-film compositions are used, including Co–Pt–B, Co–Pt–Cr–B, and Co–Cr–Ta. The Co–Pt intermetallic compound is a disordered FCC phase at high temperature, but orders as it cools to a FCT (tP4) structure, which has a suitable hysteresis loop for magnetic data storage. B is sometimes added to these materials as a grain refining agent, reducing average grain size. Grains smaller than about 10 nm begin to spontaneously "flip" from north to south (or vice versa) from thermal activation at room temperature. Since this compromises stored data accuracy, using still smaller grains to increase storage density would require that the disk be refrigerated continuously. Other approaches are under study to achieve still higher storage density. These include orienting the easy magnetization direction perpendicular to the disk surface rather than parallel, thus allowing much of each grain's volume to be vertical so that a grain of a given volume would occupy less surface area. Alternative materials are also under study, such as Co–Fe–O alloys that resist spontaneous room temperature polarity flipping at smaller grain sizes.

16.2.3.6 Co Additions to Steel and Ni Superalloys. Co is not used widely as an alloying addition to steel. Co is used in certain specialty steel alloys, such as the maraging steels (typical composition: Fe–18% Ni–10% Co–5% Mo plus small amounts of Al, Ti, and C). Maraging steels can be hardened by aging to form fine precipitates in Fe–Ni martensites. The high Ni content allows a relatively soft martensite to form by simple air cooling. As cooled, these alloys have tensile strengths of 650 to 825 MPa. During aging, fine Ni_3Mo, Ni_3Ti, Fe_xMo_y, and Fe_xTi precipitates form that raise tensile strengths to 1400 to 2400 MPa. Although Co is not present in the precipitates, Co appears to promote precipitation hardening by reducing Mo's solubility in the martensite matrix. Co may also cause short-range ordering of the solid-solution matrix of maraging steels, which further improves strength. Co is also used in some tool steels to raise the ferrite → austenite transition temperature, allowing the metal to maintain hardness at higher temperatures without having its carbides redissolve in the austenite phase.

Many Ni superalloys (Sec. 16.3.3.6) contain Co in amounts ranging from 1 or 2% up to as much as 18%. Co makes only a small contribution to solid-solution hardening in Ni because the Ni and Co atoms have similar sizes and electron structures. However, Co reduces the stacking fault energy in Ni, which allows partial dislocations to separate more widely; this improves high-temperature strength by making cross-slip more difficult (Sec. 4.4–5). Co also substitutes for Ni in the γ' (A_3B) precipitate phase of Ni superalloys, where A is Ni, Co, or Fe and B is Al, Ti, or Nb.

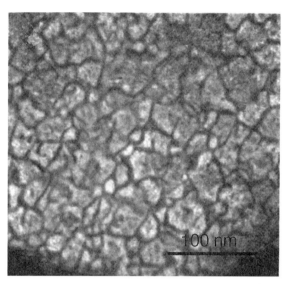

Fig. 16.9 Transmission electron micrograph of grain structure in 84% Co–12% Cr–4% Ta intermetallic compound thin film used for magnetic data storage. (From Wittig et al., 1999, p. 12; with permission.)

16.2.3.7 ^{60}Co *Radioisotope.* Co has only one stable isotope, $^{59}_{27}$Co. When ^{59}Co captures a neutron, radioactive ^{60}Co is produced that decays ($t_{1/2} = 5.3$ years) to stable ^{60}Ni by emitting β rays and penetrating γ rays (1.17 and 1.33 MeV). ^{60}Co is produced in fission reactors and is used for food sterilization, weld radiography, and medical radiotherapy. A nuclear weapon casing lined with ^{59}Co forms large amounts of ^{60}Co in the explosion's fallout debris. Immediately after the explosion, γ radiation from fission products would be much more intense than the ^{60}Co radiation, but fission product radioactivity drops off rapidly. One year after the explosion, radiation from the ^{60}Co in the fallout would be eight times more intense than that from the bomb's fission products, and after five years, the ^{60}Co radiation would be 150 times more intense. No government has ever acknowledged producing "Co-salted" atomic bombs.

16.2.3.8 Co Compounds. About 40% of world Co production is used to produce Co compounds for decolorizing and colorizing glass and pottery, catalysis, paint and ink additives that accelerate drying, and additives to rechargeable Li-ion batteries. The yellowish tint in glass and ceramics caused by Fe oxide impurity can be countered by adding 1 to 2 ppm of Co oxide, rendering the material colorless. Larger Co oxide additions (0.1 to 0.5 wt%) tint glass and ceramics a deep blue. Glassy enamels containing mixed Al and Co oxides have been used to protect plain carbon steel from corrosive atmospheres at 750°C. A number of Co compounds are used to produce blue and violet tints in artists' paints, pigments for polymers, dyes in Al anodizing solutions, textile dyes, and printing inks. Co's strong complex-forming tendencies are exploited in catalysis. Mixed Co–Mo oxides supported on alumina catalyze hydrogen desulfurization of petroleum. Other Co compounds promote hydrogenation, dehydrogenation, polymerization, and ammonia synthesis reactions. Organic salts of Co serve as drying agents in oil-based paints and inks. The organic component of the molecule dissolves in the oil carrier of the paint or ink, and the Co metal ion catalyzes the oil's oxidation, rapidly drying the applied film.

16.2.4 Toxicity of Co

Co is an essential trace element in human nutrition and is vital to prevent anemia; the recommended daily dietary allowance of 2.4 mg of vitamin B_{12} contains 40 ng of Co. However, prolonged exposure to much higher doses of Co and Co compounds can cause allergic sensitization of the skin, chronic bronchitis, and impaired thyroid function. Co is a low-grade toxin; large doses or prolonged exposure are required to produce problems. The principal threats to health are posed by inhalation of powders, ingestion of Co compounds, and skin reactions from handling Co metal and compounds. Increasingly stringent regulations on Co handling, dust levels, and release to the environment have led to more careful monitoring of airborne levels of Co dust and to tighter control of Co in industrial waste streams.

16.2.5 Sources of Co and Refining Methods

Co is present in hundreds of minerals, but only a few of these allow economical Co recovery. The important ores are sulfides, arsenides, and oxides, usually associated with Ni and Cu ores, and to a lesser extent in Au, Ag, Pb, and Zn ores. The amount of Co present in these ore bodies is generally much less than the amount of Ni and Cu, so Co is typically a by-product or co-product of Ni or Cu mining. Co is usually concentrated from these ores by gravity separation or froth flotation, followed by roasting to convert the sulfides and arsenides to oxide. Co oxide can be reduced to metal by heating in the presence of lime (CaO) and coke (C). In some ore bodies, water-soluble Co salts are produced that are purified by solvent extraction methods and electrolyzed to plate 99.5+% purity Co metal onto the cathode.

Over the past few decades, Co has experienced sharp price changes attributable to its status as a by-product metal whose production does not respond directly to market forces. This situation was exacerbated by a series of strikes, mine cave-ins, political unrest, and strategic reserve sales that caused Co prices to oscillate between $13 and $66 per kilogram during the period 1992–2002. Enormous reserves of Co are present in the world's seafloor Mn nodules (Sidebar 15.2), which contain an estimated 6 billion tons of Co, along with substantial amounts of Cu, Ni, Mn, and Fe. However, these nodules cannot be recovered profitably at current metal prices.

16.3 NICKEL

> Although in the overall composition of the earth, nickel is believed to be a most abundant element, making up 3 per cent of the whole and being exceeded only by iron, oxygen, silicon, and magnesium, it is much less prominent in the earth's crust—less than 0.01 per cent of the whole.
>
> —Betteridge (1984)

16.3.1 History of Ni

Ni is a marvelously useful metal, but most of Earth's Ni inventory lies in the planet's core. This unfortunate segregation occurred because dense elements such as Ni sank when the Earth was young and hot, while lighter elements (e.g., Al, Si, O) formed the crust. Those Ni ores that are accessible in Earth's crust are often mixed with Cu ore. Early European miners considered NiAs ores a disappointment because they looked similar to Cu ore but yielded toxic As_2O_3 vapor when smelted. In fact, the name *nickel* originates from the German *kupfernickel* (false copper). Small amounts of Cu–Ni metal have been traced to central Asia as early as 200 B.C., but their use was apparently limited. In the Middle Ages, Chinese metallurgists concurrently reduced mixed Cu, Ni, and Zn ores to produce *pai thung*, an alloy of roughly 40% Cu–32% Ni–25% Zn–3% Fe, resembling Ag in appearance and corrosion resistance. In the nineteenth century, European metallurgists caught up with their Chinese counterparts by producing their own Cu–Ni–Zn alloys called *German silver* or *nickel–silver* (although no Ag is present). Ni was not produced in a reasonably pure form until 1751, and production of pure Ni did not attain 1000 tons/yr until the late nineteenth century. Ni's usefulness in ferrous alloys combined with expanded use of Ni-based alloys that exploit this element's excellent physical properties (Fig. 16.10) have driven the annual world Ni production to well over 1 million tons today.

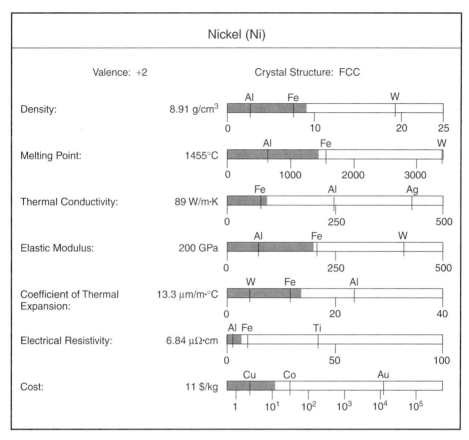

Fig. 16.10 Physical properties of Ni.

16.3.2 Physical Properties of Ni

16.3.2.1 Corrosion. Ni forms a protective, adherent oxide layer at room temperature, and Ni is often electroplated onto steel for corrosion protection. When used indoors, Ni remains bright almost indefinitely, but exposure to weather forms a darker patina, usually a sulfate layer from trace amounts of atmospheric SO_2 or H_2S. Ni performs well in flowing seawater, typically corroding less than 25 µm/yr; however, stagnant seawater can cause pitting, even though the overall loss rate is quite low. HCl and HF acids attack Ni slowly, sulfuric acid somewhat more aggressively. Dilute HNO_3 reacts rapidly with Ni, but concentrated HNO_3 acid can passivate the surface. Ni has outstanding corrosion resistance to organic materials and most caustic solutions (NH_4OH being the prime exception), which allows its use in food-processing equipment, in the manufacture and handling of caustic soda, and for electrodes in alkaline batteries. Ni exposed to boiling 50% caustic soda has a corrosion rate of only about 25 µm/yr.

Ni does not react with atmospheric N_2, and its oxidation is retarded by formation of a protective NiO layer that slows oxidation, even at high temperature (Fig. 16.11). Diffusion is the rate-limiting factor on the growth of Ni's oxide layer, following the classic parabolic relation $x^2 = at$, where x is the oxide layer thickness, a is a constant at a fixed temperature, and t the O_2 exposure time. For high-purity Ni, a can be determined over the temperature range 750 to 1240°C from

$$a = 3.2 \, \exp\left(\frac{-45,000}{RT}\right) \quad kg^2/m^4 \cdot s$$

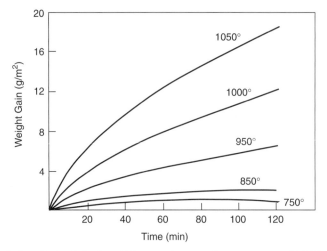

Fig. 16.11 Parabolic oxidation behavior of Ni exposed to pure O_2 at a pressure of 10 kPa (about 0.1 atm) at the temperatures indicated. (Redrawn from Betteridge, 1984, p. 23.)

where R is the gas constant (1.987 cal/mol·K) and T is the absolute temperature (K). The oxidation rate of Ni is sensitive to impurities in both the metal and the atmosphere. Impurities in Ni often increase the diffusion rate of Ni and O atoms through the oxide layer. For example, Cr impurity in Ni causes a to increase when the Cr concentration is less than 7%, but a decreases when more than 7% Cr is present. The most important atmospheric impurities affecting Ni oxidation are S-containing gases such as H_2S, because a eutectic exists in the Ni–S system at 643°C between Ni and Ni_3S_2, causing intergranular penetration and embrittlement that sharply increase the oxidation rate. Traces of S are common in fuels, and they can degrade Ni performance in components exposed to combustion gases (e.g., turbine blades and furnace fixtures).

H_2 diffuses readily through Ni. In the familiar expression for the diffusion coefficient:

$$D = D_0 \exp\left(\frac{Q_d}{RT}\right)$$

Q_d is 42 kJ/mol and D_0 is 1.07×10^2 m²/s, where R is the gas constant and T is the absolute temperature (K). At 1 atm pressure, Ni will dissolve 2 ppm H at 300°C, 8 ppm at 900°C, and 37 ppm in liquid Ni at the 1455°C melting temperature. Although figures expressed in ppm sound small, the 37 ppm value of H solubility in liquid Ni corresponds to more than 3 cm³ of H_2 gas (at normal temperature and pressure) for every 1 cm³ of metal.

16.3.2.2 Elasticity and Plasticity. Ni is a tough, ductile, moderately strong metal that deforms by the usual $\{111\}\langle\bar{1}10\rangle$ FCC slip system. In hot-rolled 99.0% purity Ni, $\sigma_y = 170$ MPa and $\sigma_{UTS} = 490$ MPa with 50% tensile elongation. Ni retains strength well at higher temperatures; at 600°C, $\sigma_y = 110$ MPa and $\sigma_{UTS} = 250$ MPa with 60% tensile elongation. Ni retains its toughness and ductility even at cryogenic temperatures. Ni forms stacking faults on $\{111\}$ planes with relatively high energy (about 250 mJ/m²).

16.3.2.3 Magnetic Behavior. Ni is one of three metallic elements (Fe, Co, Ni) that are ferromagnetic at 20°C. Ni's ferromagnetism is the weakest of the three (Fig. 16.3), and it has the lowest Curie temperature (Sec. 16.2.2.4), but Ni has the largest magnetostrictive response. When Ni reaches saturation magnetization in an external magnetic field, it contracts by a factor of 38×10^{-6} from its magnetostrictive response, which is several times greater than the response of Fe. Some intermetallic compounds have a much larger magnetostrictive response than pure Ni

(Sidebar 23.3), and those materials have largely replaced Ni in high-performance systems. However, they are brittle and much more expensive than Ni, which leaves some uses for Ni and Ni alloys in low-end transducers for ultrasound generators, strain gauges, and fish-locating systems.

16.3.3 Applications of Ni

Ni stabilizes Fe's FCC austenitic phase and retards C diffusion to delay pearlite formation in steel. These attributes make it a heavily used alloying addition to austenitic stainless steel and high-hardenability alloy steels. Ni's good hot strength and protective oxide layer allow Ni-based alloys to operate in air at temperatures as high as 1100°C, and Ni alloys are used in jet engines and other high-temperature systems and structures. Ni has excellent resistance to aqueous corrosion in alkali solutions, and Ni is frequently used to manufacture and process caustics. Ni's ferromagnetism also finds application in some magnetic materials.

16.3.3.1 Ni Additions to Stainless Steels and Alloy Steels. Most austenitic stainless steels use Ni to retain FCC γ-Fe phase as a metastable room-temperature phase. Ni concentrations are usually 7 to 9% in these steels, although Ni concentrations up to 20% are used in some alloys. Two-thirds of all stainless steel production is austenitic stainless steel, and about half of all world Ni production is consumed in stainless steel production. Austenitic stainless steels have the best overall aqueous and elevated-temperature corrosion resistance of the stainless steels. They have excellent ductility and formability, and they can be hardened by (1) cold work, which induces some of the austenite to transform to martensite; (2) heat treatment to form martensite, carbide, NiAl and $NiAl_3$ precipitates; and (3) formation of a dual ferritic–austenitic microstructure (duplex steels). These alloys are used in an enormous range of applications, including food and chemical processing, architectural trim, structural components in transportation systems, heat exchangers, tableware, exhaust gas manifolds, and furnace parts.

A smaller but still substantial quantity of Ni is used to make high-hardenability alloy steels. In these steels, Ni concentrations range from 0.3 to 2%, often in combination with elements such as Cr and Mo, to retard the diffusion of C in steel. These alloying additions slow the austenite $\rightarrow$ pearlite transformation as the steel cools, allowing martensite to form rather than pearlite. This improved hardenability permits tempered martensite to be formed in thicker sections and at slower cooling rates, delivering much higher strengths than pearlitic steel. Ni lowers the ductile–brittle transition temperature of steel, and steel containing 3.5 to 9% Ni is often specified for arctic service and cryogenic storage tanks. Ni is the key constituent that produces the high strength in maraging steels (Sec. 16.2.3.6).

16.3.3.2 Pure Ni and Ni Coatings. Commercial-purity Ni is typically about 99.0% Ni, 0.5% Co, with smaller amounts of Mn, Fe, Si, Cu, and C impurities in solid solution. Although not particularly strong (Sec. 16.3.2.2), pure Ni has good ductility and formability, and it can be strengthened by cold work to boost σ_y to as high as 650 MPa (although ductility is low at this level of cold work). The metal is readily weldable if C content is held below 0.02% to avoid graphite formation in the heat-affected zone. Pure Ni's resistance to corrosion in caustic solutions makes it useful in food- and chemical-processing equipment, heat exchangers, and electrodes. Ni is often roll-bonded to steel substrates to afford corrosion protection at a cost lower than that of a solid Ni component. Ni's corrosion resistance and easily formability make it useful for coinage and spark plug electrodes. Ni powder can be sintered into porous structures for high-temperature filtration and electrodes with high surface area. Gaseous diffusion separation of U isotopes (Sec. 24.4.2.1) utilized porous Ni metal tubing as diffusion membranes.

Large amounts of Ni are plated onto other metals (particularly steel, Cu, Al, and Zn alloys) for corrosion protection. Deposition can be done either by electroplating or by electroless deposition, in which a $NiCl_2$ solution is catalyzed by NaH_2PO_2 to form a uniform, hard Ni coating containing 5 to 10% P. Ni–Co alloys can be applied from mixed Ni–Co electrolytes (Sec. 16.2.3.5) to produce especially hard coatings. In many cases, small steel components need only a 0.25- to

0.5-μm-thick Ni coating to provide corrosion protection for items such as paper clips, scissors, keys, and buckles. Items used in more corrosive service environments (e.g., process equipment in the paper, pharmaceutical, and food-processing industries) may receive Ni coatings up to 25 μm thick or more. "Chromium-plated" components are usually multilayer coatings of Cr over Ni over the base metal. Ni has better adherence to the base metal than Cr does but is less reflective than Cr. The Ni actually provides better corrosion protection than the Cr. Cr is brittle and has a lower coefficient of thermal expansion than all the commonly plated base metals, so Cr platings almost invariably contain microscopic cracks. The presence of Ni beneath the cracks in the Cr layer assures corrosion protection. Heavier layers of Ni (up to several millimeters thick) can be electrodeposited onto worn parts, which are then remachined to bring their dimensions back to the level required for continued service. The difference between a new engine and one that is too worn to run is only about 200 g of metal lost to wear. In many cases, rebuilding worn components with Ni or Ni–Co plating can extend the service life of valuable machinery at much lower cost than replacement.

16.3.3.3 Ni–Cu Alloys (Monel). Ni–Cu alloys were first used over 2000 years ago. Ni and Cu ores are sometimes found together, and both metals can be reduced from their sulfides by simple reduction methods. Ni–Cu alloys form a complete series of solid solutions. The Ni-rich alloys are harder than the pure metals, but still ductile. Since Ni is considerably more expensive than Cu, alloys are usually chosen with the lowest Ni content that can meet the application's corrosion and strength requirements. The most popular of the higher Ni alloys contain about 65% Ni, with the balance Cu plus small amounts of Fe, Mn, and Si; these alloys are often called by the trade name *Monel*. Monel possesses better corrosion resistance than pure Ni in reducing environments and can be used in some dilute mineral acids (including HF acid) and in high-velocity seawater (e.g., propellers and pumps). Monel is immune to chloride-induced stress corrosion cracking, which can be a problem with many stainless steels. Annealed monel has $\sigma_y = 170$ to 310 MPa, $\sigma_{UTS} = 550$ MPa with 35 to 60% tensile elongation (Fig. 18.7). The less costly Cu–Ni alloys with lower Ni content (typically, 10 to 30% Ni) are not as strong but have adequate corrosion performance for use in low-velocity seawater and are commonly used in marine heat exchangers and desalination plants (Sec. 18.2.3.5). Ni–Cu alloys are easily welded and work harden rapidly. Monels with Al and Ti additions are precipitation-hardenable, forming $Ni_3(Al,Ti)$ precipitates that boost strength to $\sigma_y = 650$ MPa, $\sigma_{UTS} = 1000$ MPa with 30% tensile elongation. Monel maintains good strength up to about 350°C.

16.3.3.4 Ni–Cr Alloys (Nichrome). Ni can dissolve up to 30% Cr (Fig. 16.12) to form solid-solution alloys that are single-phase at 20°C. Ni–Cr alloys have excellent high-temperature oxidation resistance (Sec. 14.2.3.2). Lower concentrations of Cr in Ni (0 to 12%) actually *increase* the oxidation rate vis-á-vis pure Ni by increasing diffusion rates through the oxide layer. However, concentrations greater than about 15% Cr sharply decrease the parabolic rate constant (a) for oxidation (Sec. 16.3.2.1). In alloys with 25 to 35% Cr, a is four times lower than in pure Ni. Ductility and electrical conductivity both decrease as Cr content rises, and the most popular alloys contain about 20% Cr, since that composition provides good fracture toughness and sufficient ductility for fabrication, yet still provides much better oxidation resistance than pure Ni. These alloys (often called *nichrome*) are widely used as heating elements in industrial furnaces. The oxide layer on these alloys is a mixture of NiO and Cr_2O_3 with the spinel structure, $NiCr_2O_4$. For applications with many heating and cooling cycles, the adherence of the oxide is improved by small additions of more reactive metals, such as Si, Ca, Zr, and Ce. The σ_{UTS} of nichrome is 750 MPa at room temperature with 45% elongation; this declines to 70 MPa at 1000°C with 70% ductility. These alloys have high resistivity (about 100 $\mu\Omega \cdot cm$), which changes only about 4% over the temperature range 20 to 1000°C. Nichrome heating elements can operate as hot as 1150°C and still deliver good oxidation resistance. Less costly Ni–Cr alloys containing up to 25% Fe perform well up to 1000°C (Sec. 16.3.3.5) for items such as kitchen stove heating elements. For applications involving S-bearing fuel ash, extra-high

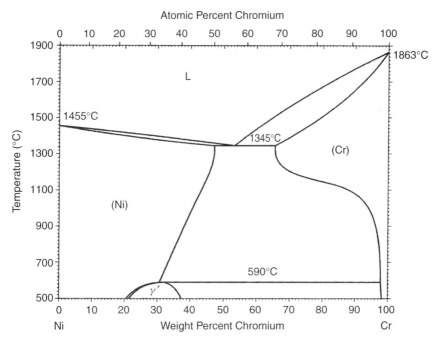

Fig. 16.12 Ni–Cr equilibrium phase diagram. Note the extensive solubility of Cr in Ni. Alloys up to 30% Cr are usually single phase; the ordered γ′ region indicated is slow to form. (From Massalski, 1986, p. 842; with permission.)

Cr content (up to 60%) can resist the corrosive effects of liquid Ni_2S_3 (Secs. 16.2.2.2 and 16.3.2.1). Cr_2S_3 ($T_{m.pt.}$ = 1550°C) forms preferentially to Ni_2S_3, which delays formation of the low-melting S–Ni_2S_3 eutectic. Cr is less effective in protecting Ni–Cr alloys from formation of V_2O_5 ($T_{m.pt.}$ = 678°C) in V-bearing fuel ash; V_2O_5 has a fluxing action on the $NiCr_2O_4$ layer, greatly accelerating attack. The tensile ductility of high Cr-content Ni–Cr alloys is only 1 to 2% at 20°C, since much of the microstructure is brittle α-Cr (Sec. 14.2.2.2). Ni–Cr alloys are also used as wear-resistant materials. Ni alloys with compositions such as Cr (8 to 12%), C (0.3 to 1.0%), Si (3 to 4%), and B (1.5 to 2.5%) form carbide and boride phases that combine with the good oxidation resistance of the Ni–Cr matrix to make a hard, corrosion-resistant coating that can be plasma-sprayed onto substrates subject to hot, abrasive environments.

16.3.3.5 Ni–Cr–Fe Alloys (Inconel). Ferrochrome is much less expensive than pure Cr (Sec. 14.2.3.1), and early alloy developers sought to lower Ni–Cr alloy costs by adding ferrochrome rather than pure Cr to Ni. The result was a lower-cost Ni–Cr–Fe alloy with good hot strength and much of the corrosion resistance of binary Ni–Cr alloys. The early alloys were called Inconel (Inco is a major Canadian Ni producer), and they have found wide application as a sort of "super stainless steel." The term *inconel* gradually became a generic descriptor for Ni–Cr–Fe alloys, and Inco Ltd. adopted a new trade name, Incoloy, for these alloys. Most Inconels have a simple solid-solution microstructure (Fig. 16.13), containing scattered carbide and/or nitride particles that deliver good strength at elevated temperature (Table 16.2). Inconel 600 is similar to the first Ni–Cr–Fe alloys. Newer Inconels, such as 625, often contain Mo and Nb additions to improve high-temperature solid-solution hardening and weldability. Nb reacts preferentially with residual C in the metal, avoiding intergranular corrosion (sensitization) caused by Cr carbide precipitation on grain boundaries (Sec. 13.4.2.1). Some Inconels have small Al additions (about 1.5%) to improve the oxide's diffusion resistance. The 800 series Inconels

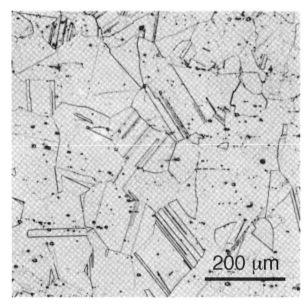

Fig. 16.13 Microstructure of Inconel 625 (61% Ni–22% Cr–9% Mo–4% Nb–3% Fe–0.5% Co–0.25% Mn–0.25% Si) cold-drawn and annealed at 1150°C. Carbide and nitride particles are visible as gray and black dots distributed throughout the solid-solution matrix. Annealing twins are visible in many of the grains (Sec. 16.3.6.1). (From ASM, 1972, p. 316; with permission.)

TABLE 16.2 Yield Strengths and Compositions of Three Ni–Cr–Fe Alloys[a]

	σ_y at 20°C (MPa)	σ_y at 600°C (MPa)	σ_y at 800°C (MPa)	Approx. Composition (wt%)
Inconel 600	185–300	190	95	75Ni–16Cr–8Fe–0.5Mn–0.2Si
Inconel 625	600	490	400	62Ni–22Cr–9Mo–3.6Nb–3Fe–0.2Mn–0.2Si
Inconel 825	300	200	135	46Ni–24Fe–22Cr–3Mo–2.2Cu–1Mn–0.9Ti–0.5Si–0.2Al

[a] Tensile elongations are high (40%+) for all alloys at all temperatures shown.

contain more Fe, which lowers cost, and these alloys are usually selected for demanding aqueous or elevated temperature corrosion challenges rather than for their mechanical properties. Inconel is used in furnaces, internal combustion engines, jet engine afterburners and thrust reversers, and equipment for handling corrosive aqueous solutions. Inconels are costlier and slightly heavier than stainless steel, but they have better high-temperature oxidation resistance, resistance to formation of the brittle σ intermetallic during prolonged high-temperature service, and resistance to chloride-induced pitting and stress corrosion cracking.

16.3.3.6 Ni Superalloys. If engineers ever convene to choose alloys for a "metallurgical hall of fame," the Ni superalloys will be a unanimous selection. These remarkable alloys can withstand long-term operation at 1100°C under tensile loads in oxidizing environments. They are essential for efficient air transportation and play a major role in electric power production. A small army of scientists and engineers has labored for the past 60 years to improve them, and each increment of additional performance saves billions of dollars in fuel costs. Ni superalloys are best known for use in the combustion zone of aircraft turbines (Figs. 16.14 and 16.15), but they are also

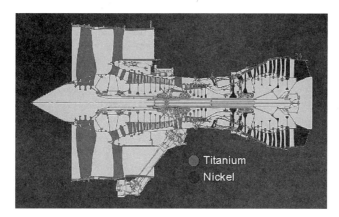

Fig. 16.14 Cutaway view of an aircraft turbine engine labeled to indicate the locations where Ti and Ni alloys are used. The air flows from left to right through the engine. The lower density of Ti alloys serves well in the large compressor blades in the forward portion of the engine, but the high temperatures in and aft of the combustion zone require Ni superalloy blades. (Courtesy of M. Cervenka, Rolls-Royce PLC; with permission.)

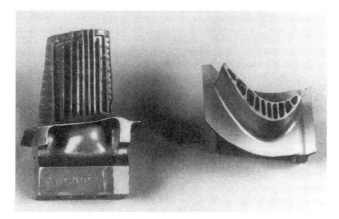

Fig. 16.15 Cutaway view of a single-crystal Ni superalloy combustion zone turbine blade. Note the internal channels, used to pump cooling air through the blade's interior. The absence of grain boundaries in this component reduces creep strain. (From Durand-Charre, 1997, p. 4; with permission.)

used in marine turbines, ground-based turbines for power generation, rockets, furnaces, and other applications where high-temperature strength and oxidation resistance are required.

Ni superalloy development began in the 1940s with efforts to improve the high-temperature creep performance of Ni–Cr alloys. These simple, early alloys have now evolved into sophisticated materials containing an unusually large number of alloying additions. They are often given multistep heat treatments to optimize their properties, and the Ni superalloys turbine blades are routinely produced with a single-crystal matrix to avoid grain boundary sliding creep. They achieve their high strength and creep resistance by three strengthening mechanisms: (1) solid-solution hardening, (2) hardening with coherent precipitates, and (3) precipitation of incoherent carbides on grain boundaries (to retard grain boundary sliding) and in the Ni matrix (to inhibit dislocation motion). Superalloys are known by several proprietary names, including Astroloy, Incoloy, Nimonic, Rene, Udimet, and Waspaloy; compositions vary considerably but usually fall within the ranges shown in Table 16.3.

TABLE 16.3 Typical Alloying Additions to Ni Superalloys

Alloying Addition	Content (%)	Function
Cr	15–20	Improves oxidation resistance; forms Cr carbide precipitates; solid-solution hardening
Co	1–20	Inhibits cross-slip by lowering stacking fault energy in Ni; element A in A_3B precipitates; minor solid-solution hardening effect
Mo, Ta, Re, W	0–10	Strong solid-solution hardeners; diffusion inhibitors
Al	0.8–6	Element B in A_3B precipitates; solid-solution hardener
Ti, Nb	2–5	Element B in A_3B precipitates; solid-solution hardeners
C	0.04–0.2	Carbide formation for grain boundary pinning and carbide precipitation in γ phase
B	0.006–0.03	CrB_2 precipitation at grain boundaries
Zr, Hf	0.02–2	Grain boundary strengtheners
Y_2O_3	<1	Dispersion hardener

Ni superalloys are strengthened by solid-solution solutes with varying degrees of misfit of atomic radius and electronic structure with the Ni matrix. Their strengthening effects vary with the degree of misfit in these two parameters, ranging from slight (Co) to moderate (Cr, Ti, Al, Fe) to strong (Re, Mo, W, Ta, Nb). The latter group of elements have a greater solid-solution hardening effect because they have more bonding electrons, which slows diffusion in their vicinity. However, these heavy elements raise alloy density and can form undesirable topologically close-packed (TCP) intermetallic phases. Thus, there are limits on the amount of solid-solution hardening elements that can be added (Sidebar 16.2). On a fine scale, the γ matrix of a Ni superalloy is inhomogeneous. Atom probe microanalysis shows small regions (1 to 4 nm) of short-range order and solute enrichment [particularly in alloys containing Mo, W, and Re (Sec. 15.5.3.4)] that are thought to provide additional strengthening vis-á-vis a perfectly homogeneous solid solution.

A coherent γ' precipitate also plays a major role in strengthening Ni superalloys. Ni superalloys' γ' volume fraction often exceeds 60%. The γ' phase is nominally Ni_3Al, although it is more accurately described as A_3B, where A is a mix of the more electronegative elements Ni, Co, and Fe, and B is a mix of electropositive elements such as Al, Ti, and Nb. The γ' phase

TABLE 16.4 Common Topologically Close-Packed (TCP) Phases in Ni Superalloys

TCP Phase	Example Compositions	Comments
σ	$Cr_{46}Fe_{54}$, $Ni_8(Cr,Mo)_4(Cr,Mo,Ni)_{18}$	Tetragonal phase similar to the σ hexagonal phase in the Cr–Fe system, forms Widmanstatten structures that lower ductility and facilitate crack propagation
Laves (χ)	A_2B, where A = Co, Fe, or Ni, B = Nb, Mo, Ta, or Ti	Hexagonal phase, more common in Fe-containing alloys, usually appears as irregular or platelet-shaped particles after long exposure to high temperatures
μ	$(Co,Fe,Ni)_7(Mo,W,Cr)_6$	Rhombohedral phase; precipitates on grain boundaries, promoting intergranular fracture; sometimes contains C, forming an $M_{12}C$ phase

SIDEBAR 16.2: TCP PHASES AND PHACOMP

Topologically close-packed (TCP) phases are undesirable in Ni superalloys. TCP phases tend to form specific orientation relationships with the Ni matrix (γ), often as thin plates parallel to the $\{111\}_\gamma$ planes. TCP phases (Table 16.4) form during heat treatment or during prolonged service at high temperature (Fig. 16.16) and degrade alloy performance by (1) removing solid-solution strengthening atoms from the Ni matrix, (2) forming thin plates of brittle phases in the matrix that sharply lower ductility, and (3) altering the coherent strain formed by the γ' precipitation-hardening phases. For these reasons, solid-solution hardening elements need to be kept below concentration levels that will form TCP phases. In the early days of superalloy development, this was done more or less by trial and error; however, an alloy design calculation precept called *phase computation* (PHACOMP) has been developed to predict the maximum alloying element additions that can be used without forming TCP phases. PHACOMP uses the effective number of unpaired d electrons (N_v) to indicate optimal amounts of alloying additions. The following equations predict that TCP phase formation will be suppressed in the matrix if the effective number of unpaired d electrons, N_v, is held below 2.45 to 2.50:

$$N_v = 4.66\text{Cr} + 2.66\text{Fe} + 1.71\text{Co} + 0.66\text{Ni} + 3.66\text{Mn} + 5.66\text{V} + 6.66\text{Si} + 9.66(\text{W} + \text{Mo})$$
$$+ 7.66\text{Al} + 6.66\text{Ti} + 5.66(\text{Ta} + \text{Nb}) + 9.66\text{Re}$$

where the element symbol represents the atomic fraction of that element present in the alloy. More recently, refined versions of PHACOMP have been presented that also account for the atomic-radius-size mismatch, $M\,dt$:

$$M\,dt = 0.717\text{Ni} + 1.142\text{Cr} + 1.90\text{Al} + 1.655\text{ W} + 1.55\text{Mo} + 0.777\text{Co} + 2.271\text{Ti} + 2.224\text{Ta}$$
$$+ 2.11\text{Nb} + 1.267\text{Re}$$

This expression predicts σ-phase suppression when $M\,dt \leq 0.991$.

A major difficulty in applying PHACOMP is that one needs to know the amount of each alloying element present in the γ matrix after the precipitation of γ' phase and carbides. Thus, simply knowing the nominal composition of the overall alloy is not sufficient. Assumptions are required to determine what portion of the alloying additions are consumed by formation of the carbide and γ' phases, and these assumptions are inherently imprecise. Nevertheless, PHACOMP provides a valuable "starting point" for formulating new superalloys.

---

is an ordered intermetallic with the cP4 structure (AuCu$_3$-type, Sec. 25.6.2 and Fig. 25.6) that nucleates homogeneously. The cP4 structure has atoms in the same locations as ordinary FCC, but the face-centered atoms are A atoms and the corner atoms are B atoms. The mismatch at the γ–γ' interface is typically only about 0.1%, so even comparatively large precipitates remain coherent with the γ matrix. The lowest-energy interface occurs on {100} planes, so in many superalloy compositions the precipitates are cuboidal (Fig. 16.17) to minimize surface energy. Multistage heat treatments can produce a range of γ' precipitate sizes. Dislocations originating in the γ matrix can shear through coherent γ' precipitates, but dislocations moving in the ordered γ' structure create antiphase boundaries (APBs) (Sec. 25.6.1), and the energy required to produce APBs makes dislocation motion more difficult.

The ability of γ' to resist shearing varies with composition, and this is a key factor in determining alloy compositions. Dislocations can also bow past γ' particles (Fig. 16.18) in the narrow channels of γ phase between the γ' particles. Due to locking of dislocations by cross-slip

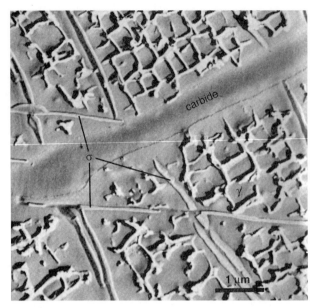

Fig. 16.16 Replica electron micrograph of σ-phase plates emanating from carbide in Ni superalloy. Precipitated γ′ is visible in much of the field. The σ phase lowers ductility and fracture toughness. (From ASM, 1972, p. 190; with permission.)

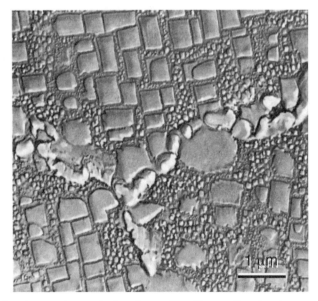

Fig. 16.17 Replica electron micrograph of a polycrystalline Ni superalloy given heat treatments at progressively lower temperatures to form γ′ precipitates of varying sizes. The large cuboid γ′ particles formed first at 1080°C; the finer γ′ formed at 840 and 760°C. Note the carbide particles precipitated on the Y-shaped grain boundary intersection running through the center of the micrograph. (From ASM, 1972, p. 171; with permission.)

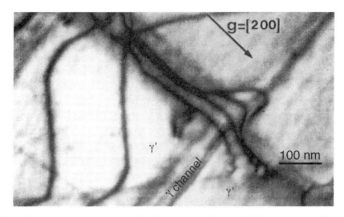

Fig. 16.18 Transmission electron micrograph of a pile-up of bowed dislocations in a 50-nm-wide γ channel between γ′ precipitates in Ni superalloy. (From Benyoucef et al., 1995; with permission of Taylor & Francis Co.)

from {111} to {100} planes, γ′ precipitates have the unusual and highly desirable property of increasing in strength as the temperature rises from ambient to 750°C. Since γ′ strength peaks at 750°C, most superalloys have fairly constant σ_y, from room temperature to 800°C (typically, 1100 to 1200 MPa); σ_y then falls to about 600 MPa at 950°C and to 200 MPa at 1150°C.

As in Co superalloys (Sec. 16.2.3.1), Ni superalloys carbide precipitates reduce grain boundary sliding and strengthen the γ matrix. The most common carbide stoichiometries are MC, $M_{23}C_6$, and M_6C. The M element in MC is usually an electropositive element such as Ti, Ta, Nb, and W. MC particles form almost immediately after solidification; they are quite stable and will not dissolve completely during solutionizing heat treatments and prolonged high-temperature service. Their persistence at high temperature limits grain growth. In the $M_{23}C_6$ carbides, M is predominately Cr with some Fe, Mo, Co, and W on M sites. When Mo and/or W content exceeds 6 to 8%, M_6C carbides tend to form. Both the $M_{23}C_6$ and M_6C carbides form preferentially on grain boundaries (Fig. 16.19) and at lower temperatures (between 760 and 1000°C) than MC carbides. These more complex carbides form by decomposition of MC or by reaction between dissolved C and alloying elements. Their ability to minimize grain boundary sliding enhances creep strength in polycrystalline superalloys, and the ideal structure is a heavy but discontinuous carbide population at grain boundaries (Fig. 16.17).

A steady progression of superalloy improvements (Fig. 16.20) has allowed higher turbine blade operating temperatures, which substantially boost engine performance. A major improvement was achieved by replacing equiaxed grain turbine blades with directionally solidified blades possessing columnar grains lying parallel to the tensile stress direction. Still greater improvement was realized by replacing those columnar grained blades with single-crystal blades (Sec. 16.3.6.2) and by adding Re as a solid-solution hardener. Typical creep rupture performances of Ni and Co superalloys are plotted in Fig. 16.21. Improvements in superalloy mechanical properties have been accompanied by improved oxidation performance. Increasing the volume fraction of γ′ to 60% or sometimes even 70% of the microstructure has raised strength, but such high γ′ fractions are achieved by lowering Cr content and raising Al content. This puts more Al_2O_3 and less Cr_2O_3 into the oxide layer. Al_2O_3 is a more stable, diffusion-resistant material than Cr_2O_3, but trace S contamination of fuel and of salt spray intake in the engine airstream can produce a fluxing action that attacks the oxide aggressively (Fig. 16.22). Mitigating these problems requires high Cr content.

At today's high operating temperatures, the lower Cr/higher Al contents of single-crystal turbine blades cannot fully protect the alloys from oxidation and sulfidation. Protective Cr_2O_3, SiO_2, and Al_2O_3 coatings are applied to form a more effective barrier. Cr_2O_3-rich coatings are

282 Co AND Ni

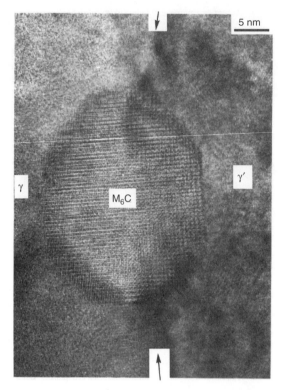

Fig. 16.19 High-resolution transmission electron micrograph of a small carbide particle at a γ–γ' interface (marked by arrows) in MC2 Ni superalloy. Note that all three phases have coherent interfaces with their neighboring phases. (From Durand-Charre, 1997, p. 82; with permission.)

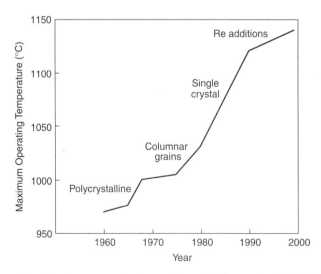

Fig. 16.20 Progression of maximum operating temperatures for Ni superalloy turbine blades. Even small increases in operating temperature significantly improve engine power and efficiency.

NICKEL **283**

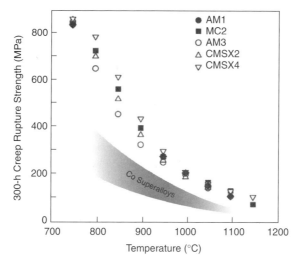

Fig. 16.21 Ni and Co superalloy 300-hour creep rupture strengths. The five alloys named in the key are all Ni single-crystal superalloys; the shaded region indicates the relative performance of Co superalloys. (From Durand-Charre, 1997, p. 106; with permission of Taylor & Francis Co.)

Fig. 16.22 Sulfidation hot corrosion damage to a Nimonic 100 superalloy turbine blade. (From Donachie and Donachie, 2002, p. 300; with permission.)

most effective when hot salt corrosion is the key problem (e.g., terrestrial and marine turbines), but aluminide coatings are generally preferred for aviation turbines to limit diffusion mixing with the superalloy and minimize brittle phase formation. Chromia-rich coatings are vulnerable to TCP phase formation at the coating–alloy interface, and alloy composition adjustments are sometimes made to minimize TCP formation near the coating. Coatings can be applied as MCrAlY (pronounced "em'-craw-lee") alloys, where M is Ni and/or Co (Fig. 16.23). Small amounts of Y in these alloys improve oxide adhesion. After coating, a high-temperature anneal diffusion bonds the MCrAlY layer to the superalloy. Small additions (about 100 ppm) of Hf, Si, or heavy rare earth metals (Er, Y) can improve the oxidation resistance of single-crystal superalloys by improving oxide adherence and tying up S to avoid reaction with Ni (Sec. 16.3.2.1). Coatings also insulate the underlying metal, reducing heat transfer from the hot combustion gases to the superalloy.

16.3.3.7 Ni in Magnetic Alloys and Low Expansion Alloys. Alnico permanent magnets contain about 20% Ni (Sec. 16.2.3.4) and achieve their high BH_{max} values by formation of a BCC matrix containing a fine precipitate of ordered BCC second phase based on Fe_2NiAl. The lattice parameters of the two phases are nearly equal, which strains the metal and improves coercivity. Ni–Fe alloys with 72 to 83% Ni are soft magnetic materials with high permeability. Their hysteresis loops can be optimized by appropriate heat treatments to achieve low remanence and low coercivity for transformers and magnetic shielding. Some Fe–Ni–Co–Si–B magnetic alloys can be made amorphous by rapid quenching from the melt (Sec. 7.3).

Metastable FCC single-phase Fe–36% Ni alloys have unusually low coefficients of thermal expansion between -100 and $100°C$. In these alloys the decline in ferromagnetism upon heating partially offsets the normal thermal expansion effect, producing alloys with coefficients of expansion (COTEs) similar to those of many glasses. Consequently, these alloys (often called

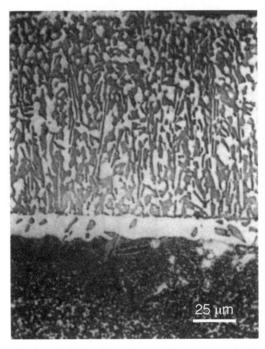

Fig. 16.23 Cross-sectional view of NiCrAlY overlay coating (upper region) on Ni superalloy (lower region). Within the coating, NiAl comprises the gray regions dispersed within white Ni solid-solution alloy. (From Donachie and Donachie, 2002, p. 317; with permission.)

invar) make good glass–metal seals; the Ni content can be adjusted from 32 to 53% to match the COTE of particular glass compositions. By adding Co to these alloys, the COTE can be reduced to near zero (Sec. 16.2.3.2).

16.3.3.8 Ni in Batteries. Ni–Cd batteries are compact, high-power, rechargeable batteries that use Ni(OH)$_2$ and Cd electrodes with a KOH electrolyte (Sec. 19.3.3.1). However, Ni–Cd batteries have two drawbacks: Cd's toxicity (Sec. 19.3.4) and the "memory effect," which limits battery power after repeated "shallow" charge–discharge cycles that don't discharge the battery fully. Newer Ni metal hydride batteries avoid both of these problems and provide 20 to 40% higher energy density than Ni–Cd batteries, making them popular in cell phones and other portable electronic devices. Ni metal hydride batteries have KOH electrolyte solutions with small additions of other compounds. They are charged by the following reactions (discharge reactions proceed with the reaction arrows pointing leftward):

$$LaNi_5 + H_2O + e^- \rightarrow LaNi_5 \text{ (with absorbed H)} + OH^-$$

charging reaction of negative electrode

$$Ni(OH)_2 + OH^- \rightarrow NiOOH + H_2O + e^-$$

charging reaction of positive electrode

The LaNi$_5$ intermetallic crystal lattice can absorb large quantities of H at 20°C. The La in the LaNi$_5$ intermetallic compound can be replaced with less costly mixed rare earth metals (Sec. 23.4.1), although this reduces battery performance.

16.3.3.9 Ni Compounds. Only about 2% of Ni is used in nonmetallic form. Ni(CH$_3$COO)$_2$, NiCO$_3$, Ni(CO)$_4$, and NiO are all used as catalysts for processes such as petroleum refining and hydrogenating oils (e.g., vegetable oils). NiCO$_3$ serves as a pigment in coloring ceramics and glass. Ni(CH$_3$COO)$_2$ is used as a mordant in the textile industry to make dyes colorfast. NiO is used in fuel cell electrodes and in coloring and decolorizing glass and in semiconducting resistance thermometers or thermistors.

16.3.4 Toxicity of Ni

Ni is an essential micronutrient in humans and higher animals, but excessive doses can be harmful or even fatal. Within cells, Ni changes membrane properties and influences oxidation–reduction systems. Ni deficiency causes abnormal bone growth, low blood glucose levels, poor absorption of Fe^{3+}, and altered metabolism of Ca, vitamin B$_{12}$, and nutrients. The human requirement for Ni is estimated to be 100 μg/day; Ni is present in so many foods that Ni deficiency is rare. Excess Ni damages chromosomes and other cell components, alters hormone and enzyme activities, accelerates or retards movement of ions through membranes, and impairs immune function. The most serious health effects from excess Ni occur from inhalation. Different chemical forms vary greatly in toxicity. Workers exposed to Ni dust or Ni carbonyl have increased risk of asthma and nasal and sinus problems. Some cases of nasal and lung cancers have been linked to prolonged Ni exposure, particularly NiO dust. At lower doses, about 10% of women and 2% of men are highly sensitive to Ni and develop skin rashes from direct contact with Ni metal or compounds. Environmental sources of Ni include tobacco, dental or orthopedic implants, stainless steel tools and utensils, and stainless steel jewelry. Long-term overexposure can lower growth rate, reduce bone development, and lower resistance to infection. High Ni doses given to lab animals caused weight loss, low-birthweight infants, and stillbirth; very high doses can be fatal.

16.3.5 Sources of Ni and Refining Methods

The principal Ni ores are complex oxide silicates (termed *lateritic*) and mixed Fe–Ni sulfides. Geological concentration of Ni tends to be low; ores containing 1 to 1.5% Ni are considered rich

SIDEBAR 16.3: METALS FROM HEAVEN

One of Earth's richest ore bodies lies in southern Ontario, Canada. The impact of an asteroid about 10 km in diameter (Fig. 16.24) traveling 30 km/s caused an upwelling of material from deep beneath Earth's crust that mixed with the asteroidal material to produce high concentrations of Cu, Ni, Co, Fe, Au, Ag, Te, Se, and Pt group metals. These now form the present-day Sudbury Basin, a 30 km × 60 km depression ringed by mines. Discovered during excavation for the Canadian Pacific Railway project in 1883, mining of the Sudbury deposit began in the 1890s and continues today, generating $3 billion in annual revenue. Early smelting operations released SO_2 evolved during roasting of the Cu and Ni sulfide ores, and the surrounding forests were severely damaged by acid rain (Sidebar 18.3). In recent decades, tight SO_2 emission controls and a major reforestation effort have largely restored the area's flora.

ore bodies. One of the world's largest Ni deposits resulted from an Earth–asteroid collision that occurred nearly 2 billion years ago in what is now Ontario, Canada (Sidebar 16.3). Sulfide ores are easier to process, and they produce the majority of the world's Ni, even though they comprise only about 20% of Ni ores. The sulfide ores are concentrated by grinding, froth flotation, and in some cases magnetic separation. Years ago, concentrated sulfides were roasted in open air to produce NiO, but stricter SO_2 emission standards prohibit this practice in most, but not all (Sidebar 18.3) smelters. SO_2 is usually captured and converted into sulfuric acid. Many Ni deposits contain Co, which is somewhat difficult to separate; a substantial fraction of total Co

Fig. 16.24 Artist's rendering of an Earth–asteroid collision. Such an impact occurred 1.85 billion years ago at the site of the Sudbury Basin ore body in Canada. The asteroid is estimated to have been 10 km in diameter, and its 30-km/s impact caused a massive upwelling of material from beneath Earth's crust. The minerals left near the surface after this impact are rich in Cu, Ni, Co, and Pt group metals. (Courtesy of Don Davis, NASA.)

content is abandoned in slags or retained in the final refined Ni as a harmless impurity (usually, well below 1%). NiO is easily reduced by H_2 or by coke and limestone to produce Ni metal. Higher-purity Ni is produced electrolytically or by Ni's unusual ability to react with CO. At 60°C, Ni will form Ni carbonyl, $Ni(CO)_4$, and the carbonyl will decompose at 180°C to pure Ni and CO gas. These reactions allow Ni to be purified, since other metal carbonyls (including Co carbonyl) require higher temperatures and/or pressures to undergo the same reactions.

About half of all Ni production is used to make austenitic stainless steel. An additional 12% is used in high-hardenability alloy steel. Ni is added to these ferrous metals as NiO (which can be reduced to Ni metal concurrently with the Fe oxide) or as ferronickel (20 to 50% Ni, 1.5 to 1.8% C, and most of the balance Fe). The lateritic ores contain both Fe and Ni, and these are often used to produce ferronickel. After initial pyrometallurgical and hydrometallurgical concentration, laterites are calcined to drive off H_2O and smelted in submerged arc furnaces to produce ferronickel.

16.3.6 Structure–Property Relations in Ni

16.3.6.1 Annealing Twins in Ni and Austenitic Stainless Steel. A crystal is twinned when the positions of atoms on the far side of a planar boundary can be generated by a reflection of the atoms on the near side of the boundary. The crystal *structure* is the same on both sides of the twin boundary; only the crystal *orientation* differs (Sec. 2.4.2). In an FCC metal such as Ni, the twin boundary is parallel to {111} planes with a stacking sequence across the boundary of ... *ABCABCABACBACBA* ... (the underlined B layer is the twin boundary). Twins form coherent, low-energy boundaries between the twin and the matrix that are usually flat.

Annealing twins are simply "growth accidents" that occur during recrystallization of cold-worked FCC metals such as Ni, γ-Fe, and Cu. They are seen in metal that has been partially (Fig. 16.25) or completely recrystallized. Annealing twins differ from mechanical twins, which form in response to high stress to minimize strain energy. Annealing twins are straight, because

Fig. 16.25 The straight boundaries of annealing twins are visible in several places in this TEM micrograph of cold-worked 302 austenitic stainless steel (Fe–18% Cr–9% Ni) that has partially recrystallized by annealing one hour at 700°C. The unrecrystallized metal has a high dislocation density, but the recrystallized regions containing the twins have almost no dislocations. The diffraction contrast caused by the different orientations on opposite sides of the twin boundaries makes some areas appear dark and others light. Note that the ends of the twinned regions are blunt; they do not come to a sharp point as do the mechanical twins in Figure 12.22. (From Sourmail et al., *http//www.msm.cam.ac.uk/phase-trans/abstracts/annealing.twin.html#note*, 2004; with permission.)

minimization of interfacial energy is the only factor determining their shape. Mechanical twins (Fig. 12.22) are often lenticular (lens shaped), because that shape minimizes elastic strain.

Annealing twin formation is favored by high grain growth velocity during recrystallization and by low stacking fault energy. High growth velocity results in more growth accidents, and when twin boundary energy is low, the driving force to "correct" the stacking sequence error is also low. A simple equation relates twin density (ρ) to the grain size (d) and a constant (b) that is proportional to the inverse of the stacking fault energy:

$$\rho = \frac{b}{d} \log \frac{d}{d_0}$$

where d_0 is the grain size at which $\rho = 0$.

16.3.6.2 Single-Crystal Ni Superalloy Turbine Blades. Diffusion-driven grain boundary sliding (Sec. 6.4.3.2) is a major factor in most creep deformation. In 1960, Ni superalloy turbine blades were ordinary polycrystalline castings, and the factor most directly limiting their maximum operating temperature and service lifetime was creep by grain boundary sliding. When these blades were replaced by directionally solidified turbine blades in the 1970s (Sec. 16.3.3.6), operating temperatures could be raised, because the stress in the blade now lay approximately parallel to the grain boundaries, giving low resolved forces for grain boundary sliding. Although columnar-grained turbine blades performed better, they still suffered some grain boundary creep

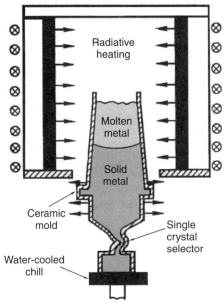

Fig. 16.26 Production of single-crystal Ni superalloy turbine blades. The mold is filled with molten metal and gradually lowered, withdrawing it from the hottest part of the furnace. Solidification begins in the chilled lower portion of the mold, nucleating many crystals. As the solid–liquid interface moves upward, it encounters the pigtail-shaped single-crystal selector, a helical tube through which only one of the many grains can advance into the liquid in the upper portion of the mold. Above the selector this single crystal grows through the remaining liquid metal, and by the time the mold has been completely withdrawn from the furnace, only a single γ-matrix orientation exists. The solidified turbine blade is not a true single crystal because it contains second-phase precipitates, but the matrix γ is aligned with its $\langle 100 \rangle$ direction parallel to the long dimension of the blade. (From Durand-Charre, 1997, p. 57; with permission of Taylor & Francis Co.)

because stress is not purely uniaxial where the blade's root attaches to the hub and because the grain boundaries were imperfectly aligned with the radial direction. Performance was further improved in the 1980s by producing single-crystal turbine blades, grown in the manner depicted in Fig. 16.26.

Single-crystal turbine blades not only eliminate the problem of creep by grain boundary sliding effects but also permit alloy composition to be adjusted in several advantageous ways. In blades with grain boundaries, C, Zr, Hf, and B are needed to produce separated carbides to pin grain boundaries. With the elimination of grain boundaries, these elements become unnecessary. Their removal raises the solidus temperature by 80 to 100°C, permitting higher solutionizing temperatures to dissolve coarse-grained γ' formed during the slow cooling of castings. With more complete solutionizing, finer γ' precipitates can be formed at lower aging temperatures, which improves strength. Oxidation resistance is also improved by the absence of grain boundaries, which provide pathways for more rapid diffusion. Single-crystal turbine blades can also have higher Al and Ti contents, which raise the γ' fraction and improve oxidation resistance. It also becomes possible to replace some W with Ta, which further improves oxidation resistance. Re additions slow diffusion in the γ phase, stabilizing the γ' precipitate structure at the higher temperatures possible in single-crystal blades.

ADDITIONAL INFORMATION

References, Appendixes, Problem Sets, and Metal Production Figures are available at
ftp://ftp.wiley.com/public/sci_tech_med/nonferrous

17 Ru, Rh, Pd, Os, Ir, and Pt

17.1 OVERVIEW

Ruthenium, rhodium, palladium, osmium, iridium, and platinum (Ru, Rh, Pd, Os, Ir, and Pt) are called the *platinum group metals* (PGMs) for their similarity to Pt, the most prominent group member. PGMs are dense, refractory, and oxidation resistant. They are among the least abundant elements in Earth's crust (Appendix C, *ftp://ftp.wiley.com/public/sci_tech_med/nonferrous*; and Sidebar 17.1), which makes them expensive. PGMs all have densest-packing crystal structures (FCC or HCP), relatively good thermal and electrical conductivities, and good oxidation and corrosion resistance. Pt and Pd are ductile and easily fabricated; Rh and Ir have limited ductility, and Ru and Os are hard, brittle metals with exceptionally high elastic moduli. PGMs solve a number of engineering problems, and they are used in catalysis, jewelry, glass fabrication, and electronics.

17.2 HISTORY, PROPERTIES, AND APPLICATIONS

> For a long time, platinum was considered a harmful impurity in silver, and a decree by the Queen of Spain Isabella prescribed the officials to... dump it into the Amazon River, so as to prevent abuses in the manufacture of gold and silver.
>
> —E. Savitsky et al. (1975)

17.2.1 History

Rare samples of mixed PGMs were found by ancient peoples in native form, but they were almost unknown until the European discovery of the Americas. Pt was first isolated in the sixteenth century from ores shipped to Europe from South American colonies; at the time it was deemed an adulterant in Ag and Au. Its high melting temperature posed problems for early metallurgists, and it was little used. The remaining five PGMs were not separated and identified until the early nineteenth century. Today, PGMs are more widely utilized, but their scarcity and high cost limit their use to applications where no other materials can be substituted. The total value of all six PGMs annual production is only about one-third the value of world Au production.

17.2.2 Physical Properties

17.2.2.1 Density. The PGMs' most striking physical property may be their high densities (Table 17.1 and Figs. 17.2 and 17.3). Few people have the opportunity to handle a 1-kg ingot of Ir or Pt, but those who have done so find it remarkable that something the size of a small chocolate bar can be so startlingly heavy. Ru, Rh, and Pd are the densest metals of period 5, and Os, Ir, and Pt are the densest of all metals (Fig. 1.13). The densest metals lie near the middle of each period of transition elements because those elements have large numbers of unpaired d

---

Structure–Property Relations in Nonferrous Metals, by Alan M. Russell and Kok Loong Lee
Copyright © 2005 John Wiley & Sons, Inc.

SIDEBAR 17.1: SCARCITY OF THE PT GROUP METALS

Figure 17.1 shows several trends in the terrestrial abundances of the elements: (1) Low-atomic-number (Z) elements are generally far more abundant than high-Z elements (see, however, Sec. 11.3.1); (2) even-Z elements have greater nuclear stability and are usually more common than odd-Z elements (hence the sawtooth pattern of the plotted line); and (3) elements with a $Z = 44$ to 46, 52, 75 to 79, and 83 are especially rare. All six PGMs are among the 10 rarest elements.

Light elements are more abundant because nuclear fusion in stars is exothermic for $Z \leq 26$ and endothermic for $Z > 26$. Thus, formation of light elements is the natural product of the primary fusion reactions occurring in stars. In contrast, elements heavier than Fe are formed only in stellar "side reactions" that produce free neutrons, such as

$$^{13}_{6}C + ^{4}_{2}He \rightarrow ^{16}_{8}O + ^{1}_{0}n$$

$$^{21}_{10}Ne + ^{4}_{2}He \rightarrow ^{24}_{12}Mg + ^{1}_{0}n$$

These neutrons can be captured by another nuclide to form elements endothermically that are heavier than Fe. However, the final abundances of these heavier elements will depend on the neutron capture cross sections of elements with $Z > 26$. Elements with low-neutron-capture cross sections, such as Sr, Zr, Ba, Ce, and Pb, are especially stable and resist transmutation into heavier nuclides. Thus, these elements tend to accumulate in high proportions in the cores of stars. By the same logic, elements with high-neutron-capture cross sections tend to be scarce, because they are likely to transmute rapidly to still heavier elements. This slow neutron capture process occurs over periods of millions to billions of years and is thought to be the primary mechanism for synthesizing elements from $_{29}$Cu to $_{83}$Bi. When the matter from old stars re-forms into new stars and planets (e.g., Earth), those new bodies "inherit" the elemental abundances of the old stars.

Slow neutron capture works poorly to produce elements heavier than Bi, because elements with Z between 84 and 89 have short half-lives; they decay to lighter elements faster than slow neutron capture can promote them to the relatively long-lived elements with $Z = 90$ and 92. However, there is also a rapid neutron capture process that occurs during a period of about 0.01 to 100 seconds during supernovae explosions of dying massive stars. In these immense explosions, an intense torrent of neutrons and protons is liberated. An Fe nucleus can capture dozens or even hundreds of neutrons and protons in quick succession. This allows the rapidly growing nuclei to "race" through the ($84 \leq Z \leq 89$) region of short decay half-lives, forming actinide and trans-actinide elements that fission spontaneously or decay to long-lived Th and U or stable Pb. These stellar element-building processes produce relatively few PGM atoms. PGMs have high-neutron-capture cross sections that disfavor formation by the slow and fast neutron processes, so they are rare. Another factor exacerbates the rarity of PGMs (and also Re); they are heavy and they are efficiently dissolved by molten Fe. When Earth was young and hot, most of the small quantities of these metals initially present were dissolved in the Fe that settled toward the planet's core. They remain there today, leaving only a small fraction of Earth's initially meager PGM and Re concentrations in the mantle.

electrons available to participate in bonding. Ru, Os, Rh, and Ir have high elastic moduli and short interatomic spacings. Elements farther to the right in the periodic table (e.g., Pd, Pt, Ag and Au) have greater atomic mass, but with fewer unpaired d electrons available for bonding, their atoms are larger and less strongly bonded. PGM weight-basis costs are high, but their volume-basis costs are made especially high by their high densities; for example, 1 liter of Pt costs $500,000.

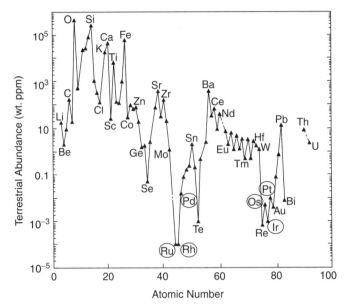

Fig. 17.1 Relative terrestrial abundances of the elements, excluding H, the inert gases, and elements lacking long-lived isotopes.

17.2.2.2 Mechanical Properties. Pt and Pd are ductile, relatively soft FCC metals (Table 17.2). The other two FCC PGMs, Rh and Ir, are considerably harder and less ductile. As-cast Rh has low ductility at room temperature, but it can be hot-worked to produce a fine, equiaxed grain structure that will tolerate a moderate degree of cold work. Ir is the least ductile of all FCC metallic elements. Ir's low ductility is only partially attributable to impurity effects; even

TABLE 17.1 Properties of the Platinum Group Metals with a Comparison to Cu

Property	Ru	Rh	Pd	Os	Ir	Pt	Cu
Valence	−2, +1, +2, +5, +6, +7, +8	−1, +1, +2, +4, +5, +6	+2, +4	−2, +1, +2, +3, +5, +6, +7, +8	−1, +1, +2, +5, +6	+5, +6	+1, +2
Crystal structure at 20°C	HCP	FCC	FCC	HCP	FCC	FCC	FCC
Density (g/cm^3)	12.45	12.41	12.02	22.61[a]	22.65[a]	21.45	8.92
Melting temperature (°C)	2310	1960	1554	3050	2443	1769	1085
Thermal conductivity (W/m·K)	105	150	76	87	148	73	399
Elastic (Young's) modulus (GPa)	447	275	121	552	528	168	130
Coefficient of thermal expansion (10^{-6} m/m·°C)	6.4	8.0	11.67	5.1	6.4	8.9	16.5
Electrical resistivity (μΩ·cm)	6.71	4.33	9.33	8.12	4.71	9.85	1.7
Cost ($/kg)[b]	1080	15,500	6400	11,800	2700	23,500	2

[a] Os and Ir are the densest of all metals, and their densities are nearly equal. Physical measurements and calculations from x-ray data give slightly different values; it is uncertain which element is denser.
[b] PGM prices vary widely from year to year (see Sec. 17.3).

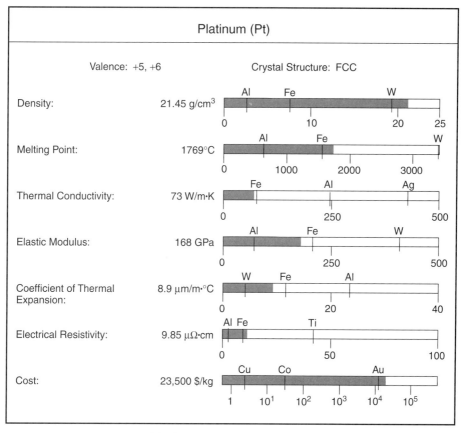

Fig. 17.2 Physical properties of Pt.

high-purity Ir has inherently low ductility (Sec. 17.5.1). HCP Ru is a hard metal with low room-temperature ductility. Ru can be hot worked at 1500°C, but even at this temperature it is hard and vulnerable to edge cracking. In addition, if hot-worked in air, Ru loses mass as its volatile, toxic oxide (RuO_4) vaporizes (Sec. 17.3). HCP Os is unworkably hard and brittle, even at high temperature. Os's hardness at 1200°C is 140 VHN, a remarkably high value for that temperature. The Young's moduli of Os, Ru, and Ir are among the highest of all metals.

PGM mechanical properties can be improved by alloying. For example, Pt's strength can be greatly increased by Ir additions. A 70% Pt–30% Ir alloy has $\sigma_{UTS} = 1100$ MPa in the annealed condition and 1380 MPa when cold-worked. This high-strength alloy has limited ductility, and although still higher strengths are attainable with Ir content $\geq 35\%$, the alloys become so difficult to form that they are seldom used.

17.2.2.3 Corrosion. PGMs' high electronegativities (Fig. 1.14) give them exceptional resistance to oxidation and chemical attack in corrosive aqueous solutions. They are intrinsically resistant to attack and do not rely on a protective oxide layer. Some PGM oxides are stable within certain temperature ranges; for example, Pd forms an oxide layer between 400 and 850°C that decomposes to Pd metal and O_2 gas at higher temperature. This oxide layer is metastable at room temperature, so Pd must be cooled rapidly through the range 850 to 400°C if a bright finish is desired. The aqueous corrosion resistance of PGMs is similarly outstanding. Ir has been called the most corrosion resistant of all metals; it is unaffected by hot mineral acids and even resists aqua regia.

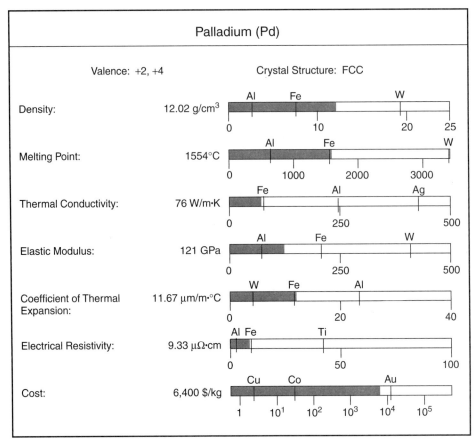

Fig. 17.3 Physical properties of Pd.

TABLE 17.2 Mechanical Properties of Commercial-Purity Annealed Polycrystalline PGMs

	Ru	Rh	Pd	Os	Ir	Pt
Vickers hardness number	350	120	40	400	220	37
Yield strength ($\sigma_{0.2\%offset}$, MPa)	380	70	50	NA[a]	90	50
Ultimate strength (σ_{UTS}, MPa)	500	420	190	NA[a]	500	120
Tensile elongation (%)	3	9	30	0	3	40
Reduction in cross-sectional area at fracture (%)	2	20	80	0	10	90

[a]No tensile strength data exist for Os, due to its extreme brittleness.

Pt and some Pt alloys are the only true metals that can tolerate prolonged service in air above 1100°C without a protective coating. In Ni and Co superalloys, both strength and oxidation resistance decline sharply above 1100°C. Au and Ag have complete oxidation resistance but melt below 1100°C. Although numerous intermetallic compounds (e.g., Mo silicides) have sufficient oxidation resistance to provide reliable service above 1100°C, their room-temperature fracture toughness and ductility are poor. Among the PGMs, Ru and Os have the poorest resistance to oxidation in high-temperature air; they form volatile oxides and lose mass in a linear manner. Ir and Pd perform better at high temperature, but they gradually lose mass from loss of volatile

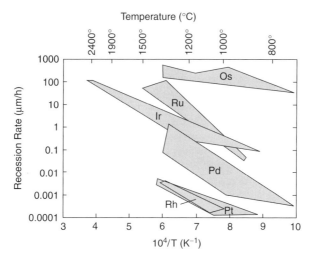

Fig. 17.4 Loss rate from flat PGM surfaces by oxidation and vaporization in air. Recession rates vary with purity, stress levels, and atmospheric humidity, so a range of loss rates is shown for each metal. (Adapted from Fortini et al., 2000, p. 220.)

oxides and from sublimation (Fig. 17.4). Pt and Rh deliver the best high-temperature oxidation resistance. Above 1300°C, Pt outperforms Rh; Pt's mass loss rate is low for both volatilization of oxides and sublimation of the metal, and its melting temperature is high enough to provide useful strength at 1500°C when alloyed in solid solutions with other PGMs. Addition of 10 wt% Rh to Pt forms a solid-solution alloy that melts at 1850°C and has 26 MPa tensile strength at 1500°C. The 90% Pt–10% Rh alloy forms a thin bluish oxide between 750 and 1150°C, but this decomposes to metal and O_2 at higher temperatures. This alloy is used for high-temperature thermocouples and resists attack by molten glass, making it useful in nozzles and spinerettes for producing glass fibers for insulation and fiberglass.

17.2.2.4 Conductivity. Although PGMs have large numbers of unpaired electrons in the conduction band, their conduction electrons experience electronic interactions with the core electron orbitals that lower conductivity. As a result, PGMs are only mediocre conductors. Ag and Cu have fewer charge carriers, but their filled d shells minimize interactions between the moving electrons and the atoms, giving them better room-temperature conductivities than the PGMs.

17.2.2.5 Catalysis. The surfaces of PGMs provide active sites that strongly adsorb many gases and organic chemicals. Compounds adsorbed onto the PGM surface react with one another much more rapidly than they would without the PGM surface (Fig. 17.5). When the reaction is complete, the PGM has not been altered, and this catalyzing process can continue indefinitely without consuming the PGM. Many materials act as catalysts, but PGMs are generally superior to base metal catalysts because PGMs can withstand more corrosive chemical species, and they often catalyze reactions at lower temperatures. PGMs are particularly useful in catalyzing reactions involving oxidation, hydrogenation/dehydrogenation, and organic synthesis. PGM catalysts are frequently supported on oxide substrates that stabilize fine PGM particles and prevent them from sintering to one another and losing their effectiveness. Catalysts account for more than half of all PGM use.

17.2.3 Applications

PGMs' high costs restrict their use, but their exceptional catalytic abilities and corrosion resistance make them irreplaceable materials for several applications. Ingenious methods have been

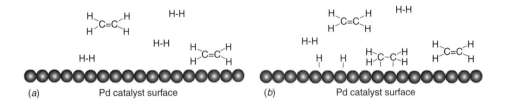
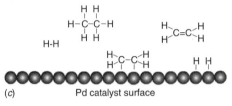

Fig. 17.5 Catalyst action of Pd in hydrogenation of ethylene gas; (*a*) ethylene and H$_2$ gaseous reactants approach the catalyst surface; (*b*) reactants adsorb to surface, breaking the double C bond in ethylene molecules; (*c*) reactants combine on the Pd surface to form ethane. The Pd surface atoms are not consumed in the reaction and remain available to catalyze additional reactions.

devised to minimize the quantity of a PGM needed to accomplish a desired task (Sidebar 17.2), and these methods, combined with careful attention to recycling, have made PGMs cost-effective materials for a number of uses.

17.2.3.1 Ruthenium. Ru is the least costly PGM. Its low ductility and comparatively high reactivity with air make it less desirable than other PGMs for structural and elevated-temperature service. Ru's largest application is in the electronics industry, where it is used in resistors and hybrid integrated circuits. A thin coating of Ru on computer hard disks improves data storage densities, and some Ta capacitors use thin Ru coatings to enhance performance. Ru is used extensively as a catalyst in the petroleum refining and chemical process industry, particularly for acetic acid production. Ru has also been used as a catalyst in fuel cell proton exchange membranes. The corrosion resistance of Ti pipes is markedly improved by Ru additions of only 0.1 wt%. Ti–Ru pipe is used in sour gas oil wells, geothermal conversion systems, and in chemical process equipment (Sidebar 12.3).

17.2.3.2 Rhodium. By far the largest demand for Rh is in catalytic converters for cars and trucks; Rh is particularly effective in reducing NO and NO$_2$ to N$_2$ and O$_2$ in exhaust gases (Sidebar 17.2). Most catalytic converters contain about five times as much Pd/Pt as Rh. Smaller quantities of Rh are used in other catalytic applications in the chemical process industry and as an alloying additive to Pt for handling molten glass. Although Rh has some ductility, it is less easily formed than Pd and Pt, and it is seldom used in the pure state for structural applications. Rh and Rh alloys provide wear-resistant, oxidation-resistant switch contacts for electrical circuits and are used occasionally in jewelry.

17.2.3.3 Palladium. Pt and Pd are the two major PGMs; they are considerably more abundant than the other PGMs in Earth's crust, and they are the PGMs in greatest demand because of their superior catalytic performance, high ductility, and resistance to high-temperature oxidation. Catalytic converters for cars and trucks (Sidebar 17.2) consume more than half of world Pd production. Pd catalysts are also employed for a wide range of tasks in the chemical process and petroleum refining industries (Fig. 17.5). Many polymer precursor materials are formed by Pd-catalyzed reactions. Pd raises the liquidus temperature of Au dental alloys (Sec. 18.4.3.3),

> **SIDEBAR 17.2: PGM CATALYSTS IN AUTOMOBILE EXHAUST SYSTEMS**
>
> The exhaust gases of internal combustion engines contain unburned hydrocarbons, CO, and NO_x ($x = 1$ or 2). In areas with high traffic density, these pollutants pose a health hazard, and many nations regulate the maximum permissible exhaust emissions of motor vehicles. The most effective way to minimize internal combustion engine exhaust pollutants is to pass the exhaust gases through a catalyst bed that converts them to less harmful compounds. Catalytic converters contain mixtures of Pt and Pd to oxidize the unburned hydrocarbons and CO in exhaust gases into H_2O and CO_2. Converters also contain Rh to reduce NO_x gases to N_2 and O_2. For best performance, the converter must present a high effective surface area of PGMs to the flowing exhaust gas. This is done by forming refractory oxides or stainless steel into porous structures containing long, narrow channels (Fig. 17.6). These structures contain as many as 140 channels/cm^2 and have wall thicknesses as small as 0.06 mm; they are "washed" with a ceramic slurry containing CeO to produce a rough, irregular surface with an effective area of about 10^4 m^2, and this is followed by a second wash of PGM salts to deposit the catalyst on the surface seen by the flowing exhaust gases. Ceria stores O to maintain sufficient concentrations for oxidation reactions during rapid changes in fuel–air ratios. A typical converter contains 3 to 4 g of PGMs in a thin layer that serves as the active catalyst on the interior surfaces of the converter.
>
> Catalysis does not begin until the PGMs reach 250 to 275°C, and optimal operation occurs at 400 to 600°C. The converter is heated by the exhaust gas, and it takes time for the converter to reach these temperatures. Consequently, the converter is ineffective during the first few minutes the engine runs. The converter requires O_2 in the exhaust gas to oxidize the hydrocarbons and CO. A Pt O_2sensor measures exhaust gas O_2 content and uses that information to regulate the fuel injectors to assure adequate O_2 is present in the exhaust gas. At optimal operating temperature, PGMs convert about 90% of the pollutants to H_2O, N_2, and O_2.

allowing ceramic veneers to be fired onto metal supports at temperatures that would melt most other Au alloys. Dental alloys account for about 15% of Pd use, although Pd use in dentistry fluctuates with the market price of Pd because Pd competes with other less costly (and lower-performing) dental alloys. Pd is also used in electronic components such as multilayer ceramic capacitors and conductive pastes for hybrid integrated-circuit assembly. Smaller amounts of Pd are used in jewelry to strengthen Pt alloys and as an additive to white Au alloys.

17.2.3.4 Osmium. Os is the least used PGM. Os is seldom used in pure form, but it can be alloyed with other PGMs to increase hardness and elastic modulus. Os has the highest Young's modulus of all metallic elements and the third-highest melting temperature; only W and Re are higher melting. Os is brittle, even at 1200°C, and it forms a volatile, toxic oxide (OsO_4, Sec. 17.3), making it poorly suited for elevated temperature service. It can be added to other metals to increase their hardness and stiffness for applications such as premium-quality fountain pen tips, instrument pivots, and electrical contacts. However, even these relatively few uses are losing market share to ceramic materials. The tetroxide is used as a staining agent to detect fingerprints and for optical microscope examination of fatty tissues.

17.2.3.5 Iridium. Ir has the second-highest elastic modulus of all metallic elements, but it has minimal tensile ductility (Sec. 17.5.1). Ir has outstanding corrosion resistance in many aqueous solutions, and its high-temperature oxidation resistance is relatively good. These attributes make Ir an effective barrier coating to prevent aqueous corrosion or high-temperature oxidation of base metal substrates. Chemical vapor deposition methods are used to deposit Ir coatings (Sidebar 15.1). The largest user of Ir is the semiconductor production segment of the electronics

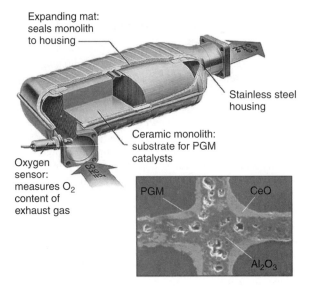

Fig. 17.6 Internal structure of a PGM catalytic converter for automobile exhaust systems. Exhaust gas from the engine enters at the lower left. A thin coating of Pt, Pd, and/or Rh on the interior surfaces of the honeycomb-structured ceramic monolith catalyzes the oxidation of unburned hydrocarbons and CO into CO_2 and H_2O and the reduction of NO_x gases into N_2 and O_2. Inset shows SEM micrograph of a ceramic monolith with a CeO wash coat overlain with PGMs. (Courtesy of J. Buchdahl, Manchester Metropolitan University, and of the U.S. Department of Energy.)

industry, where Ir crucibles are used for growing single crystals of Si and related semiconductors. Ir–Ru alloys are used as catalysts in acetic acid production and to coat electrodes for processing caustic soda and chlorine. Small amounts of Ir are used in erosion-resistant, high-performance spark plugs.

17.2.3.6 Platinum. The jewelry industry and manufacturers of catalytic converters for car and truck engines each consume about a third of world Pt production. The ability of Pt jewelry to serve the dual roles of adornment and investment drives the Pt jewelry market much as it does the larger Au jewelry market. Pt jewelry is particularly popular with Asian consumers; its silvery luster, great weight, and higher value vis-á-vis Au generate $2 billion sales of Pt to jewelry producers. The retail value of Pt jewelry sales is considerably higher, since jewelry markups are large. The corrosion resistance and easy formability of Pt allow intricate shapes to be crafted; however, the higher melting temperatures of Pt alloys compared to Au make Pt jewelry somewhat more challenging to cast, braze, and solder. Industrial use consumes about a fifth of world Pt production for chemical process catalysts in manufacture of silicones and other polymer precursors, thermocouples, and spinerettes and nozzles for glass fiber and LED glass component production. Smaller amounts of Pt are used in Pt–Co magnetic alloys for computer hard disks (Sec. 16.2.6), dental alloys, biomedical implants and surgical devices, and turbine blade coatings. Pt also finds use in niche markets, such as precious metal investment, coins and medallions, and spark plug coatings.

17.3 TOXICITY

Although most PGMs are so chemically inert that they can be used for surgical implants, Ru and Os are toxic. Both metals form malodorous, volatile oxides. OsO_4 forms slowly on bare

TABLE 17.3 Production of Platinum Group Metals, 2002 (Tons)

	Pt	Pd	Rh	Ru	Ir	Os
South Africa	138	67	15	—	—	—
Russia	30	60	3	—	—	—
North America[a]	12	31	1	—	—	—
Total of all producing nations	186	163	19	13	2.4	0.3[b]

[a] Canada plus U.S. production.
[b] Estimated.

Os metal at room temperature and then vaporizes, but the rate of formation on massive metal at room temperature is low. The vapor pressure of OsO_4 is 0.08 atm at room temperature, and it boils at 130°C. OsO_4 is a powerful oxidizing agent that causes irreversible burnlike injury to eyes, skin, mucous membranes, and the respiratory tract. Airborne Os concentrations as low as 0.1 $\mu g/m^3$ are injurious; direct skin contact is especially dangerous. High doses are fatal. Heating Os metal, particularly finely divided metal, in air releases dangerous amounts of toxic vapor. RuO_4 is similar to OsO_4 in both volatility and toxicity, but it is even more reactive with organic materials and is also explosive.

17.4 SOURCES

The largest producers of PGMs are South Africa and Russia (Table 17.3); smaller amounts are produced in Canada and the United States. All six PGMs are present in most ore bodies, but nearly all sources are richest in either Pt or Pd, and the output of that metal determines the amounts of other PGMs from a given source. PGMs are by-products of Ni mining in Sudbury, Canada (Sidebar 16.3) and Norilsk, Russia (Sidebar 18.3). Consequently, PGMs, particularly Ru, Rh, Os, and Ir, are produced in quantities somewhat unrelated to their demand. Large price swings are common because normal supply-and-demand precepts are not in play. For example, the spot price of Rh reached the absurd price of $217,000 per kilogram for one day in 1990, then dropped 97%, to $6220 per kilogram in 1997. Over the past few decades, most PGMs have price maxima/minima ratios greater than 10. Investors with strong nerves and a high risk tolerance have made (and lost) fortunes in PGM commodities trading.

PGMs are most often found as microscopically fine particles of native metal mixed with other minerals or as complex, mixed sulfides of Ni, Cu, Sb, and/or Bi. Annual world production totals for Pt and Pd (Table 17.3) are usually roughly equal; together, their market values total several billion dollars. Rh production has an annual market value of about $250 million; total sales of the other PGMs are much less. The annual world production of Os, about 13 liters of metal, could be stacked atop a (strong) desk with room to spare for a computer and a cup of coffee; however, this modest volume of Os has a market value of about $3.5 million.

17.5 STRUCTURE–PROPERTY RELATIONS

17.5.1 Low Ductility of Polycrystalline Ir

FCC metals are known for excellent ductility and high fracture toughness, even at cryogenic temperatures. However, polycrystalline FCC Ir of ordinary purity elongates only about 3% in tension before fracture, and even high-purity Ir elongates only about 10%. Tensile specimen fracture surfaces show mixed intergranular and transgranular cleavage. This behavior is surprising because single-crystal tensile and compression tests, TEM study, and *ab initio* calculations all indicate that Ir deforms by the same $1/2\langle 110\rangle\{111\}$ slip system common to all FCC metals.

As in other FCC metals, these dislocations split in Ir into two partial ⟨112⟩ dislocations with an intervening stacking fault (Sec. 4.5).

Ir's poor ductility may be caused by grain boundary weakness that is either inherent or induced by trace impurities. This is consistent with three observations: (1) Single-crystal Ir is highly ductile, elongating up to 80% before failure; (2) small amounts of P and Si reduce fracture toughness and cause intergranular fracture in Ir alloys; and (3) Ir alloys are made more ductile by adding a few ppm of certain elements (e.g., Th and Ce) that segregate to grain boundaries.

Stress concentrations from dislocation pile-ups at grain boundaries can cause intergranular fracture in metals and intermetallics, and this effect can be quantified by the expression of Liu and Inouye relating fracture strain (ε_f) to grain boundary cohesive strength (σ_c):

$$\varepsilon_f = \frac{\sigma_0 - \sigma_i}{k} + \frac{\sigma_c}{k}\left(\frac{s}{d}\right)^{1/2}$$

where σ_0 is the friction stress resisting dislocation motion on slip planes, k and σ_i are material constants, s is the distance to the tip of the pile-up and d is the grain diameter. According to this relation, a plot of fracture strain versus $d^{-1/2}$ should produce a straight line with slope equal to σ_c. Two such plots are shown in Fig. 17.7, one for an Ir–0.3 wt% W alloy and a second plot for that alloy with Th dopant additions. [W is often added to Ir as a solid-solution hardener. W also raises Ir's recrystallization temperature, permitting deformation at higher temperatures (where Ir is more ductile) without causing recrystallization (which lowers strength).] In Fig. 17.7 the slope of the line is greater for the Th-doped alloy, indicating a higher grain boundary cohesive energy with Th doping. Th is nearly insoluble in Ir (<5 ppm), and Auger analysis shows that adding as little as 5 ppm Th raises the grain boundary concentration to several percent Th because nearly all the Th segregates to the grain boundaries. Figure 17.7 also shows that adding larger amounts of Th does not produce a higher σ_c value than small Th additions; specimens with varying Th contents all fall near the same line. Adding more than 5 ppm Th serves only to form Ir_5Th intermetallic precipitates, which do not affect σ_c. The ductilizing effects of Th and Ce in

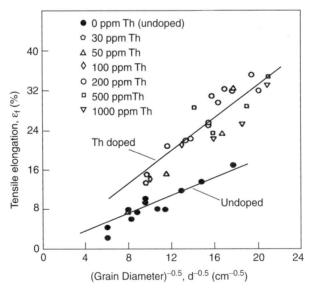

Fig. 17.7 Effect of Th doping on grain boundary strength in Ir–0.3% W alloy. The slopes of the lines are proportional to the cohesive strengths of the grain boundaries. Th increases the grain boundary strength, which improves ductility and fracture toughness by making intergranular fracture more difficult. (Redrawn from George et al., 2000.)

STRUCTURE–PROPERTY RELATIONS

TABLE 17.4 Comparison of Low-Ductility FCC Metals (Rh, Ir) with High-Ductility FCC Metals (Pt, Pd, Ni, Au, Al)

	Melting Temperature (K)	Shear Modulus,[a] $\mu_{\{111\}}$ (GPa)	Free Surface Energy,[b] γ_s (J/m^2)	Rice–Thomson Criterion,[c] $\mu_{\{111\}}b/\gamma_s$
Ir	2716	198	2.5	24.5
Rh	2233	133	2.5	14.6
Pt	2042	57.5	2.3	7.1
Pd	1827	35.9	2.0	5.0
Ni	1726	68.5	2.3	7.6
Au	1338	23.7	1.3	6.1
Al	933	24.8	1.2	5.9

[a] Shear modulus for dislocation motion in the {111} plane.
[b] Calculated energy to form a free surface (by crack propagation).
[c] According to the Rice–Thomson criterion, $\mu_{\{111\}}b/\gamma_s$ values greater than 10 predict brittle cleavage failure in preference to blunting of crack tips by dislocation motion. b, length of the Burgers vector.

Ir alloys suggest that raising grain boundary cohesion is the key to better Ir ductility, a finding that offers promise of expanding uses for this strong, corrosion-resistant element.

The fundamental question remains: Why is Ir's grain boundary cohesive strength low? A partial answer is provided by comparison of the values in Table 17.4. The shear moduli of Ir and Rh are much higher than those of more ductile FCC metals. Rice and Thomson developed a criterion to predict whether an incipient crack will grow or experience blunting by dislocation propagation at the tip; by this criterion values of (shear modulus/free surface energy) greater than 7.5 to 10 predict that cleavage requires less energy than crack blunting. This ratio is high for Ir and somewhat high for Rh, which is consistent with the macroscopic behavior of the polycrystalline metals (i.e., low ductility for Ir, low to moderate ductility for Rh). Another metric for ductility (Appendix D, *ftp://ftp.wiley.com/public/sci_tech_med/nonferrous*) also indicates Ir to be less ductile than other FCC metals. Simply put, the dislocation energies ($\sim \mu b^2$) in Ir and Rh are high, and the free surface energies need to be concomitantly high to satisfy the Rice–Thomson criterion; in Ir and Rh they are not. The high shear moduli of Ir and Rh are thought to result from extensive hybridization of the d orbitals; this makes the bonding between the atoms stronger and more directional than it is in other FCC metals, which have lower shear moduli.

ADDITIONAL INFORMATION

References, Appendixes, Problem Sets, and Metal Production Figures are available at
ftp://ftp.wiley.com/public/sci_tech_med/nonferrous

18 Cu, Ag, and Au

18.1 OVERVIEW

Copper, silver, and gold (Cu, Ag, and Au) are among the few elements existing in the metallic state in Earth's crust. For this reason, these are thought to be the first metals discovered and used by humankind. Traces of hammered metallic Au artifacts have been found in Spanish caves used by Stone Age families. The "electrum" of ancient Mediterranean civilizations was a naturally occurring alloy of Au and Ag. Masses of metallic Cu have been found in mines in Michigan's Keweenaw Peninsula, some too large to be lifted to the surface, forcing mine shafts to detour around them.

The outer electron configuration of Cu, Ag, and Au is $d^{10}s^1$. Following Hund's rule, one of the two s electrons has moved to the d subshell to achieve the lower energy of a completed d subshell. This electron structure suggests that Cu, Ag, and Au might have properties similar to the heavy alkali metals (K, Rb, and Cs), because they also have a single s electron in their p^6s^1 outer electron configuration. However, their physical properties and chemical compounds are drastically different. The filled p shell of an alkali metal shields the nucleus from the outer s electron much more effectively than do the filled d shells of Cu, Ag, and Au; this makes the alkali atom larger, more easily ionized, and more weakly bonded to neighboring atoms. Consequently, the alkali metals are less dense, much more reactive, and have lower melting and boiling temperatures than those of the group 11 metals. The high electronegativities of the group 11 metals (Sec. 1.6) give them inherently good corrosion resistance. The filled d subshell and free s electron of Cu, Ag, and Au contribute to their high electrical and thermal conductivity. Transition metals to the left of group 11 experience complex interactions between s electrons and the partially filled d subshell that lower electron mobility. In group 11 metals, the d subshell is filled, thereby improving the mobility of the s electrons and making the group 11 metals excellent electrical and thermal conductors. The conductivity, oxidation resistance, ductility, and color of these metals make them useful for a wide range of applications, including wire, plumbing pipe, jewelry, photographic film, and batteries.

The terrestrial abundances of the group 11 metals are low: Cu = 68 ppm, Ag = 0.08 ppm, and Au 0.004 ppm. Even Cu, the most abundant of these elements, is 1000 times scarcer than such common metals as Fe or Al. Fortunately, geologic processes have concentrated these elements in certain formations so they can be recovered without the extreme expense their relative abundances might suggest. Still, many Cu mines must process ore that is less than 1% Cu. Most Ag ore is too lean to make recovery profitable, so Ag is produced primarily as a by-product of other mining operations. Some Au mines operate with ore as lean as 1 g of Au per ton of rock (i.e., 1 ppm Au).

18.2 COPPER

I am a copper wire slung in the air,
Slim against the sun I make not even a clear line of shadow.

Structure–Property Relations in Nonferrous Metals, by Alan M. Russell and Kok Loong Lee
Copyright © 2005 John Wiley & Sons, Inc.

Night and day I keep singing—humming and thrumming;
It is love and war and money; it is the fighting and the tears, the work and want,
Death and laughter of men and women passing through me, carrier of your speech,
In the rain and the wet dripping, in the dawn and the shine drying,
A copper wire.
—Carl Sandburg, "Under a Telephone Pole," *Chicago Poems*,
Henry Holt and Company, New York, 1916

18.2.1 History of Cu

Cu holds a special place in metallurgical history. The earliest archeological evidence of human use of Cu metal is a trinket fabricated in Iran about 9000 B.C. Cu knives and other tools were left in Egyptian graves in 5000 B.C. Both metallic (native) and combined states of Cu exist in nature. Early Cu use was limited to native metal, but Cu ores are relatively easy to reduce, and smelting operations began in Cyprus and the Sinai in about 3800 B.C. These crude early Cu smelting operations were the birth of metallurgy, and they eventually led to deliberate alloying of Cu with Sn to form bronze. Harder and stronger than Cu, bronze was also easier to cast, and it became the dominant metal for weapons and implements during the next 2000 years. The rich Cu deposits of Cyprus made that island a crucial strategic battleground for a succession of Mediterranean cultures, and uncounted lives were lost battling over the Cu supplies underpinning the Bronze Age.

18.2.2 Physical Properties of Cu

Bronze was supplanted long ago by iron as the preferred metal for weapons and structures, but Cu's excellent electrical and thermal conductivity (Fig. 18.1) combine with its easy formability and good corrosion resistance to make it a key industrial metal. Cu has the second-highest electrical conductivity of any material at 20°C; Ag's conductivity is slightly greater. Nonelectrical applications for Cu exploit its corrosion resistance, easy formability, and unique color.

18.2.2.1 Color. Most metals are good reflectors of visible light. When a photon of light strikes metal, its energy promotes a free electron in the metal to a higher, previously unoccupied energy level. This excited state is unstable, and the electron decays immediately to a lower energy level, emitting a photon of light of the same wavelength as the incident photon. For most metals this reflection process is more than 90% efficient at all visible wavelengths, so reflected light has the same color as the incident light. This same effect occurs in Cu for red light, which is reflected at 97.5% efficiency. However, green and blue photons promote the electron to higher energy levels from which decay sometimes occurs in two steps rather than one (Fig. 18.2). Each of these two decay steps emits an infrared photon, so the eye detects fewer green and blue photons in the reflected light than were present in the incident light. Only about two-thirds of the green light and half of the blue light is reflected at the green and blue wavelengths; the rest reflects as infrared light. The brain interprets the paucity of green and blue in the reflection as the familiar reddish hue of Cu. A similar process accounts for the color of Au and brass, a Cu–Zn alloy.

18.2.2.2 Corrosion. Cu is relatively resistant to corrosive attack in many environments and is frequently used in applications where corrosion resistance is important. Cu oxidizes slowly at room temperature in dry air. When heated in air at temperatures below 200°C, Cu forms a strongly adherent layer of reddish cuprous oxide (Cu_2O), which is partially protective against further oxidation. Cu is sometimes heated to 150°C to deliberately form this protective oxide layer. At temperatures above 200°C, scaly, nonprotective black cupric oxide (CuO) forms. In polluted industrial atmospheres, particularly those containing sulfides, a black film of Cu sulfide forms on Cu. After prolonged exposure, a complex mixture of carbonates and sulfates forms with colors ranging from blue to red–green to brown to the classic green patina frequently seen

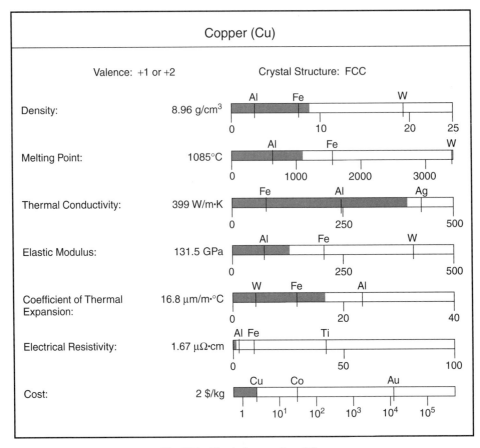

Fig. 18.1 Physical properties of Cu.

on Cu exposed to the atmosphere for many years. This patina is protective and aesthetically pleasing; indeed, artificial means have been developed to accelerate patina formation.

18.2.2.3 Plasticity. Like all FCC metals, Cu deforms by $\{111\}\langle 110\rangle$ slip (Sec. 4.6). Cu's critical resolved shear stress (τ_{CRSS}) on this slip system is 0.63 MPa (99.9999% purity) or 0.94 MPa (99.98% purity). Cu has excellent ductility and tolerates heavy deformation at room temperature without cracking. Cu can be fabricated by all standard methods (e.g., rolling, swaging, drawing, stamping); rolling reductions as large as 90% in a single pass can be used on pure Cu. Most Cu alloys are somewhat less ductile. Cu work-hardens substantially at room temperature. Cold work can double the tensile strength and quintuple the ultimate strength of annealed Cu. Ductility is even greater at elevated temperatures since dynamic recovery and recrystallization occur above about 150°C (varying with purity). Hot work minimizes texture formation and refines the grain size of cast metal. Cold-worked Cu machines reasonably well; but fully annealed Cu is somewhat difficult to machine due to built-up edge problems (chip material adhering to the cutting tool edge). Cu also deforms by twinning across $\{111\}$ planes, shearing in the $\langle 112\rangle$ direction.

18.2.2.4 Conductivity. Most Cu is used for electrical applications, and since conductivity and ductility are best for pure Cu, nearly all electrical Cu is used in the form of unalloyed metal. Impurities raise resistivity, although some elements are much more deleterious than others (Fig. 18.3). As one might expect, efforts to purify Cu for electrical use focus primarily

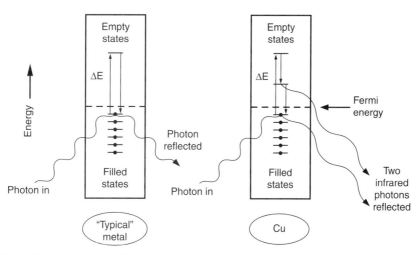

Fig. 18.2 When a photon of visible light strikes a typical metal, it excites one of the metal's electrons to a higher, unstable energy level. That electron's energy collapses back to its rest energy almost instantaneously, emitting one photon with energy equal to that of the incident photon. Red light is reflected by Cu in this manner. When a blue photon is absorbed in Cu, it also excites an electron to a higher, unstable energy level above the Fermi level. However, when this excited state collapses, the electron reverts to its former energy level either by emitting one photon of blue light or two photons of infrared light. Since infrared photons are unseen, much of the blue light is missing from the reflection, which gives Cu its reddish hue.

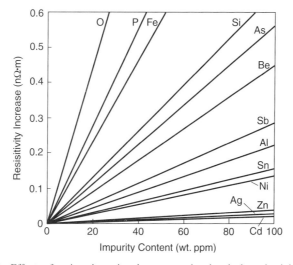

Fig. 18.3 Effects of various impurity elements on the electrical conductivity of Cu.

on removing elements (e.g., Fe, Si, As, and O) that cause considerable loss of conductivity. Electrolytic refining can achieve 99.99% purity; such material has the best combination of electrical properties and ductility and has become standard industry practice for preparing electrical Cu. Small amounts of Ag or Cd can be added to Cu to minimize softening at elevated temperatures (e.g., during soldering or polymer packaging of microelectronic circuits) without lowering conductivity substantially (Sidebar 18.1).

> **SIDEBAR 18.1: NARloy-Z IN THE *SPACE SHUTTLE* MAIN COMBUSTION CHAMBER LINER**
>
> The inner liner of a rocket engine combustion chamber is an extraordinarily severe service environment (Fig. 18.4). The combustion chamber material is exposed to the fierce heat of burning H_2 and O_2, it is vulnerable to oxidation and hydrogen embrittlement, and it must withstand the high stresses associated with acceleration to orbital velocity. For the *Space Shuttle* main combustion chamber, the engineering challenge is greater still because the engine must be reusable; design goals call for 300 firings of the rocket engine before the liner is replaced.
>
> The combustion chamber wall is actively cooled by the liquid H_2 fuel being pumped toward the fuel injectors. Without cooling, no metal could withstand the heat of combustion, and since cooling effectiveness is critically dependent on high thermal conductivity, Cu is the preferred material for this application. NARloy-Z (Cu–3% Ag–0.5% Zr+ ~50 ppm O) is precipitation hardened to form a Cu–Ag–Zr intermetallic precipitate with ZrO_2 for dispersion strengthening. Sheets of NARloy-Z 0.50 to 0.75 mm thick are bonded to an Inconel outer chamber wall, and liquid H_2 is pumped between the Ni and Cu layers to hold the NARloy-Z's maximum operating temperature below 540°C.

Pure Cu's high thermal conductivity makes welding somewhat difficult; heat is conducted away from the intended fusion zone so rapidly that high power inputs are essential. Oxyacetylene welding can be used on Cu, but traces of H in the flame cause uneven weld quality due to internal steam bubble formation (Sec. 18.2.3.1). Most Cu welding is done by tungsten–inert gas or metal–inert gas welding (Sec. 8.6.1). Electric resistance welding (spot welding) can join Cu sheet, but this process demands high power input, and electrode sticking can become a problem. Cu is often joined by soldering, since the common low-melting solders such as Pb–Sn and Pb–free solders wet Cu surfaces well.

18.2.3 Applications of Cu

18.2.3.1 Electrical Use. Electrical use consumes more than 60% of total world Cu production. Electrical Cu is usually sold as wire or cable, although strip and sheet are sometimes used. Since this is Cu's dominant use, most Cu refining processes are selected to optimize electrical conductivity. O impurity poses particular problems in electrical Cu. Liquid Cu can dissolve several percent O, but the equilibrium solubility of O in solid Cu is only 0.03 at% at the melting point and near zero at 20°C. During freezing, dissolved O is redistributed as a small amount of O in solid solution plus Cu_2O precipitate particles. Both dissolved O and Cu_2O precipitates cause problems. Even small amounts of dissolved O degrade electrical conductivity (Fig. 18.3). Cu_2O precipitates do not appreciably affect conductivity, but they reduce ductility for both hot and cold work. Moreover, Cu_2O precipitates can react with H to form steam bubbles inside the metal by the reaction $Cu_2O + 2H \rightarrow 2Cu + H_2O$. Since H_2O molecules are too large to diffuse into the metal, small steam bubbles accumulate, building to internal pressures high enough to crack the metal. H can enter hot Cu from atmospheric humidity or from the combustion products of furnace fuel used to melt the metal. H can also be introduced into Cu by entrapped electrolyte in Cu cathodes produced by electrolytic refining methods.

Cu and Cu alloys can be deoxidized by additions of Li, Be, B, C, Na, Mg, Al, Si, P, or Ca. If P is added to the melt as a deoxidizer, any excess P will severely degrade conductivity (Fig. 18.3). Consequently, P is not used to deoxidize electrical Cu unless CaB_6 is also added to purge excess P by forming Ca_3P_2 slag. H can be removed from molten Cu by adding Li (which also deoxidizes the metal) or by entrainment with dry Ar, N_2, or CO_2 bubbled through the molten metal. Specially refined oxygen-free high-conductivity (OFHC) Cu is available for

Fig. 18.4 Test firing of the *Space Shuttle's* main rocket engine. The outer wall of the combustion chamber is Ni superalloy (Sec. 16.3.3.6); the inner wall is a Cu alloy selected for its high thermal conductivity. (Courtesy of NASA.)

high-temperature service in H-rich environments, but it is substantially more costly than ordinary Cu (Sec. 18.2.5). For electrical applications where maximum conductivity is not essential, "tough pitch" Cu (Cu containing O) can be used to lower costs. Small amounts of dissolved O actually improve conductivity in lower-purity Cu by forming oxides with Fe and other metal impurity atoms. Metal oxide particles raise resistivity much less than do solid-solution atoms.

18.2.3.2 Plumbing and Heat Exchanger Use. Pure Cu is widely used for tubing in plumbing and heat exchangers. Purity requirements are sometimes less stringent for these applications than for electrical Cu. However, high ductility is helpful for forming tubing, and since high-purity Cu has the best ductility and the best corrosion performance, electrolytically refined Cu is commonly used for Cu tubing. Tubing products consume 12% of annual Cu production. Cu gives excellent service with neutral or mildly basic water, but acidic solutions (pH < 6.5) can dissolve dangerously high levels of Cu ions from the tube's inner wall. This introduces potentially toxic contamination to beverages or foods made with such water. Even when toxicity is not a concern, acidic water eventually perforates Cu tubing.

18.2.3.3 Architectural Use. Pure Cu roofing, flashing, exterior cladding, and trim pieces provide an attractive finish to buildings and offer long service life. Cu roofs typically last 80 years

308   Cu, Ag, AND Au

or more. Although Cu is more expensive to install than competing roofing materials, Cu's long-term cost is often lower, because repair and replacement are so infrequent. Architectural Cu exposed to the weather forms a protective patina of sulfides and carbonates that vary in color with the composition of local atmospheric pollutants and with climate factors.

18.2.3.4 Brass.
Zn has extensive solid solubility in Cu and is added in amounts ranging from 5 to 45% to make brass. Commercial brasses often contain small amounts of Al, Mn, Ni, Pb, Si, and Sn to enhance specific properties for particular applications. Single-phase solid-solution brass (Fig. 18.5) has excellent cold formability, higher strength and ductility, and lower conductivity than pure Cu (Fig. 18.6). Brass's attractive yellow color can be varied by adjusting Zn content. Brasses with as much as 35% Zn will yield a single-phase solid solution from normal casting and annealing operations.

For Zn contents between 35 and 40%, a two-phase $\alpha + \beta$ or $\alpha + \beta'$ microstructure forms. The β phase is BCC, and β' phase is an ordered CsCl-type (cI2) structure. Since dislocation motion is much easier in the high-temperature disordered β phase than in the low-temperature β' phase (Sec. 25.6.1), the $\alpha + \beta$ brasses are usually deformed exclusively by hot work. Pure β' has good high temperature strength, but its room-temperature brittleness makes it is unsuitable for engineering use unless it is quenched to retain a metastable β phase at room temperature. No commercial brasses are produced with Zn content $\geq 50\%$, since all such alloys are brittle.

The Cu–Zn phase diagram illustrates the *electron phases* (also called *Hume–Rothery phases*). The four phase fields labeled (Cu), β, γ, and ε contain progressively larger amounts of Zn. Since Cu has one s electron and Zn has two, the bonding electron concentration rises with increasing Zn content. As the electron concentration per atom approaches 1.5, the FCC (cF4) crystal structure of α-brass becomes less stable than the BCC structure (cI2) of β-brass. Further increase in Zn content shifts the electron concentration per atom to 1.6, forming the complex γ (cI52)

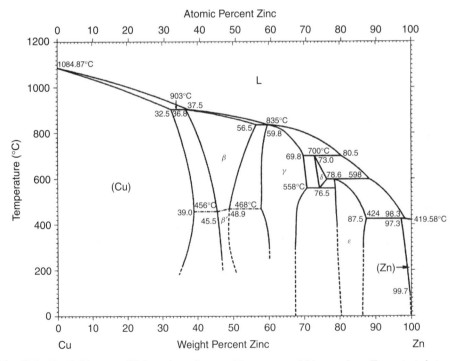

Fig. 18.5 Cu–Zn binary equilibrium phase diagram. Most commercial brasses have Zn contents between 5 and 40%. (From Massalski, 1986, p. 981; with permission.)

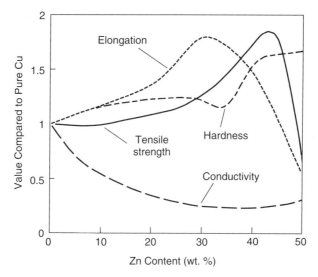

Fig. 18.6 Effect of varying Zn content on conductivity and mechanical properties in brass.

intermetallic compound with nominal composition Cu_5Zn_8 ($5 \times 1 + 8 \times 2 = 21$, and $21/13 \approx$ 1.6). Still higher Zn content causes yet another shift to the hexagonal ε $CuZn_3$ phase when the electron ratio exceeds 1.75. This correlation of crystal structure with electron concentrations occurs in many other group 11 elements when they are alloyed with elements from groups 12+. Electron phases also appear in the binary phase diagrams of group 8 to 10 transition metals with the group 12+ metals.

Ternary alloys comprised of 17 to 45 wt% Zn, 8 to 30 wt% Ni, and the balance Cu have properties similar to those of Cu–Zn brasses, but their corrosion resistance is improved by the Ni. These alloys are often called *nickel silvers* (although they contain no Ag). Most compositions are single-phase solid-solution alloys. The alloys lack the reddish tint common to most Cu alloys, but several other colors can be produced by suitable adjustments to composition. Low Ni compositions (10 to 12 wt% Ni) are a pale yellow; low Ni content with 30 to 45 wt% Zn confers a greenish tinge; and higher Ni contents (>20 wt%) have a slight pinkish hue. They are used for decorative trim, costume jewelry, base metal for silver plate, and a variety of fasteners, springs, and fixtures.

18.2.3.5 Copper–Nickel Alloys (Marine Alloys). Ni and Cu form single-phase solid solutions at all compositions. Cu alloys containing 10 or 30% Ni and small amounts (1% or less) of Fe and Mn provide excellent resistance to erosion corrosion by high-velocity seawater, deposit attack, and pitting corrosion. They are used for ship hull and marine structure cladding, condenser service, seawater pipe work in both ships and offshore oil and gas platforms, and for desalination plants. Marine biofouling is a common problem on underwater structures, ships, and pipes. Hundreds of biofouling organisms, such as slime algae, sea mosses, sea anemones, barnacles, and mollusks attach themselves to such objects, particularly in warm water with low flow rates. Biofouling causes excess drag on ships and restricts flow through pipes. Steel, concrete, Ti, and Al foul rapidly, but Cu-based alloys, including Cu–Ni, resist biofouling. Thus, use of Cu alloys in marine environments often saves the costs associated with painting and cleaning fouled surfaces or in chemically treating water to avoid biofouling problems. The reasons for Cu alloys' biofouling resistance are incompletely understood, but studies suggest that the Cu_2O film acts against the attachment mechanisms of the organisms at the point of contact. Many antibiofouling marine coatings and paints contain large amounts of Cu_2O, which allows non-Cu surfaces to achieve the biofouling resistance of Cu alloys as long as the coating is present (typically, three to five years).

The addition of Ni to Cu follows the usual trends in solid-solution effects on properties (Sec. 3.4). As Ni content increases, yield, tensile, and fatigue strengths all increase, as does resistivity (Fig. 18.7). Cupronickels are less ductile than pure Cu, but they maintain reasonably good ductility at all Ni contents. An alloy with 55 wt% Cu–45 wt% Ni has nearly constant electrical resistivity over a wide temperature range, making it useful in certain electrical instruments.

18.2.3.6 Tin Bronze and Silicon Bronze. Bronze (Cu + Sn) is one of the most famous alloys in human history. Cu will dissolve a maximum of about 15 wt% Sn at 500°C (Fig. 18.8). Although the equilibrium phase diagram indicates that a two-phase structure should form during cooling, ε phase (Cu_3Sn, oC80) forms quite sluggishly. Bronze with less than about 11 wt% Sn will be a metastable, single-phase solid solution when cooled at normal cooling rates. Most commercial wrought bronzes contain from about 1.25 to 10% Sn with about 0.1% P added as a deoxidizer (Cu–Sn–P alloys are frequently called *phosphor bronzes*). Excess P forms Cu_3P, a hard second phase that strengthens the metal. Bronze with 10 to 20 wt% Sn is also used, but these alloys form intermetallic compounds (Fig. 18.9), and their low ductility restricts their use to castings. Bronze is harder than brass and somewhat more expensive. It is most often used for bearings, springs, bushings, piston rings, gears, and bells. Large amounts of Pb (10 to 30%) may be added for bearing alloys. Bronze with lower Sn contents (e.g., about 5%) have excellent ductility but work-harden rapidly. If bronze requires extensive deformation, stress-relief anneals are used to avoid excessive loads on processing equipment. Bronze has high fatigue strength, and tensile strengths of 1000 MPa can be achieved in cold-worked phosphor bronze.

Cu–Si solid-solution alloys are sometimes used as lower-cost alternatives to tin bronze. Cu dissolves more than 4 wt% Si at 600 to 800°C, and these single-phase solid solutions can be retained at room temperature. The commercial silicon bronzes contain 1 to 3 wt% Si (usually with small Fe or Mn additions). They have strengths and work-hardening rates similar to those of tin bronzes. Their corrosion behavior is generally similar to that of tin bronze, including good resistance to seawater attack; however, their impingement corrosion resistance is lower.

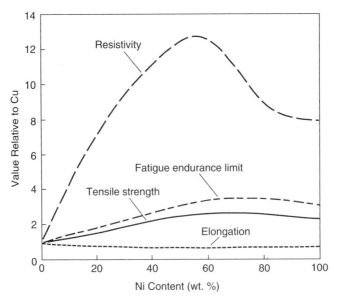

Fig. 18.7 Effect on conductivity and mechanical properties of varying Ni contents in cupronickel alloys. The fatigue endurance limit data were measured at 10^8 cycles.

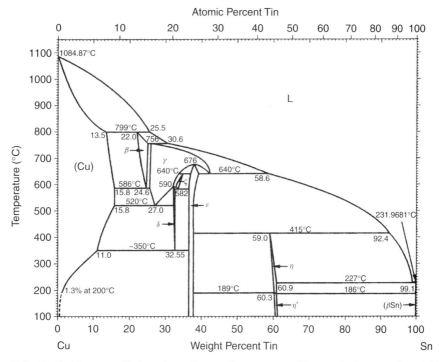

Fig. 18.8 Cu–Sn binary equilibrium phase diagram. Bronze compositions typically range from 1.25 to 15% Sn. (From Massalski, 1986, p. 965; with permission.)

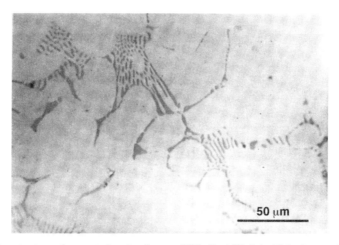

Fig. 18.9 Microstructure of as-cast phosphor bronze (90% Cu–10% Sn). Alpha bronze (light), δ phase (gray), and a few very small Cu_3P particles (dark gray) are visible. (From West, 1982; with permission.)

18.2.3.7 Aluminum Bronze. Cu alloys containing up to about 11 wt% Al are used for applications requiring a combination of high strength, high fracture toughness, and good corrosion resistance. Cu will dissolve several wt% Al at elevated temperature (Fig. 18.10), which allows formation of solid-solution-strengthened alloys possessing a mixed-Cu–Al oxide surface film that is more protective against high-temperature oxidation than are the oxides of pure Cu. At

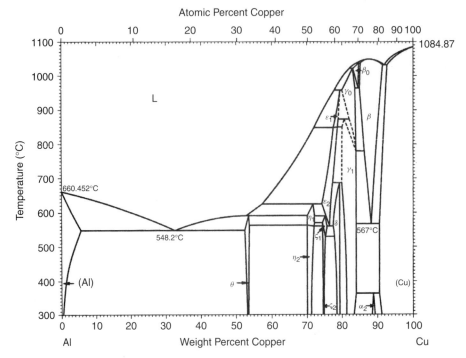

Fig. 18.10 Cu–Al binary equilibrium phase diagram. Aluminum bronze compositions typically range from 5 to 11% Al. (From Massalski, 1986, p. 106; with permission.)

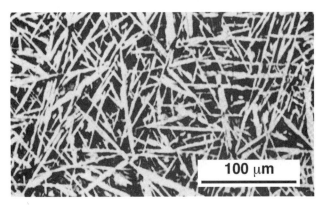

Fig. 18.11 Quenched Cu–11.8% Al. The acicular structures are metastable BCT β′ phase formed martensitically by quenching. (From ASM, 1979; with permission.)

higher Al contents, the eutectoid decomposition of β phase (BCC, cI2) at 567°C allows a variety of heat treatments analogous to those of steel. An alloy at or near the eutectoid composition (11.8 wt% Al) can be solutionized above the eutectoid temperature and water quenched to trigger a martensitic transformation to a metastable, tetragonal β′ phase. The quenched eutectoid material (Fig. 18.11) is hard and brittle, but it can be tempered at 400 to 650°C to improve toughness. Small Fe and Ni additions (about a few percent) to aluminum bronze shift the eutectoid point somewhat and increase the strength of the quenched and tempered alloy. Quenched and tempered ternary and quaternary aluminum bronzes can achieve over 600 MPa yield strength and 1000 MPa ultimate tensile strength with 10% or better tensile elongation. These alloys are sometimes used to produce nonsparking tools for use in environments with flame or explosion hazards. Their high thermal conductivity and low free energy of oxide formation makes them much less likely to throw sparks when dropped or dragged on rough surfaces.

18.2.3.8 Copper–Beryllium Alloys. Cu containing 0.5 to 2.8 wt% Be forms precipitation-hardenable alloys that are the strongest conventional Cu alloys (Sec. 18.2.3.9). Commercial Cu–Be alloys often contain 1 to 2% Co or Ni. Alloys can be solutionized at 800°C in the α-Cu phase field (Fig. 18.12), then quenched and aged at 250 to 330°C to precipitate GP zones (Secs. 3.5 and 20.2.4.4). These GP zones strain the lattice, impeding dislocation motion. The first precipitates form coherently on Cu {100} planes as plates one atom thick in the supersaturated solid solution. As aging progresses, these plates grow to 1 to 7 nm in diameter and about three atoms thick, gradually evolving into semicoherent BCT rods or plates (γ′ phase). Overaging weakens the alloy by forming δ phase, an incoherently ordered BCC structure. If the quenched material is cold-worked before aging, the additional dislocations serve as nucleation sites for GP zones, giving the aged alloys yield strengths as high as 1350 MPa and ultimate strengths of 1460 MPa. Ductility is low (1 to 3% elongation) at these maximum strength levels. Greater ductility can be obtained by aging to somewhat lower strengths. Cu–Be alloys provide outstanding strength, good corrosion resistance, and fair conductivity, but the high cost of Be (Sec. 11.1) limits their use.

18.2.3.9 Cu Matrix Deformation Processed Metal–Metal Composites. The excellent ductility of pure Cu has been exploited to form a class of metal–metal composites with extraordinarily high strength and high conductivity. These deformation-processed metal–metal composites (DMMCs, sometimes called *in-situ composites*) are produced by extensively drawing or rolling a two-phase microstructure until the second phase is only tens of nanometers thick. The best-known DMMC is the Cu–Nb composite, which typically is 80% Cu and 20% Nb. Cu and Nb are

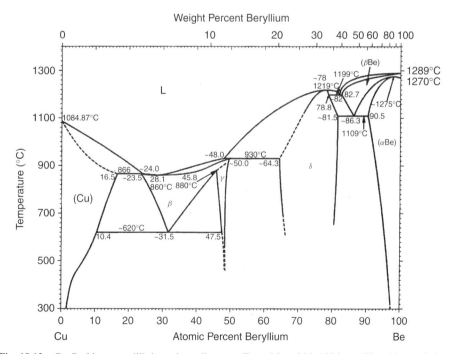

Fig. 18.12 Cu–Be binary equilibrium phase diagram. (From Massalski, 1986, p. 450; with permission.)

314 Cu, Ag, AND Au

miscible as liquids but have almost no solid solubility. When cast, the Nb forms a dendritic second phase in a Cu matrix. As the metal is cold-worked, the Nb dendrites are deformed into long filaments (by drawing) or lamellae (by rolling). True strains as high as 12 are used to deform the Nb to exceptionally small thicknesses (Fig. 18.13). Such high true strains correspond to huge dimensional changes in the ingot; a true strain of 12 converts a 100-mm-diameter cast ingot into 0.25-mm-diameter wire. These composites have ultimate tensile strengths as high as 2400 MPa while maintaining electrical conductivities 35% that of pure Cu (Sidebar 18.2). The principal disadvantage of the process is that only wire and thin sheet can be produced unless costly intermediate chopping and rebundling/restacking operations are performed partway through the processing, or equichannel angular extrusion is used.

Cu DMMC wire can be immersed in molten Sn, which allows Sn atoms to diffuse through the Cu (forming bronze) to react with the Nb to form Nb_3Sn filaments (Sec. 13.4.2.5). Nb_3Sn is a high-performance superconductor, and having thousands of ultrafine filaments of the superconductor embedded in a bronze matrix is a particularly advantageous assembly for use in the wire windings of superconducting electromagnets.

18.2.4 Cu in the Biosphere

Cu is an essential micronutrient in plants and animals. Cu is required for a wide range of metabolic processes, and metalloorganic Cu compounds exist in all tissues. The human body contains 50 to 120 mg of Cu per kilogram of body weight. As with many metals essential

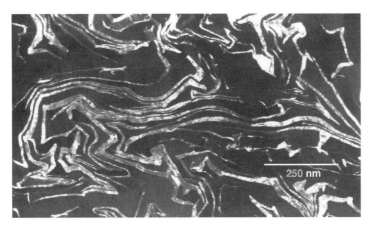

Fig. 18.13 Transmission electron micrograph of Cu–Nb deformation processed metal–metal composite in transverse cross section. The Nb filaments (light gray) are 10 to 40 nm thick; their original thickness in the cast ingot before deformation processing began was 5 to 12 μm. The filaments lie perpendicular to the plane of the photograph and are several millimeters long. (Courtesy of F. C. Laabs, Ames Laboratory, U.S. Department of Energy.)

SIDEBAR 18.2: SPACE TETHERS

Cu DMMC wire (Sec. 18.2.3.9) may provide the best material for space tethers to be extended several kilometers alongside an orbiting spacecraft (Fig. 18.14). The tether cuts Earth's magnetic field lines, inducing a current in the tether that can be converted to electromagnetic radiation and beamed back to the spacecraft from the tether's end to complete the electrical circuit to power onboard systems on the spacecraft. The tether must be both strong and conductive, and Cu DMMCs have a better combination of those two properties than any solid solution or precipitation-hardening Cu alloy.

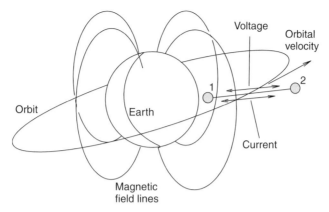

Fig. 18.14 Artist's depiction of a space tether deployed for power generation on an orbiting spacecraft. Current is induced as the tether cuts Earth's magnetic field lines. Power can be beamed back to the spacecraft from a pod at the tether's far end. The tether must have high strength and high electrical conductivity.

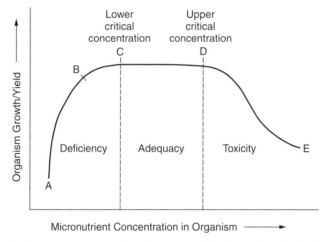

Fig. 18.15 Generic dose–response curve for micronutrients in plants and animals. Cu and many other metals display this behavior in the human body; deficiency or toxicity problems occur when the concentration falls outside the desired range (C to D). Severe deficiency results from concentrations between A and B; mild deficiency results from concentrations between B and C. Concentrations greater than D can be injurious or even lethal.

to metabolism, the human body displays a dose–response curve for Cu of the sort shown in Fig. 18.15. Cu deficiency causes slow growth, loss of hair and skin pigmentation, blood vessel damage, impaired immune function, and even sudden death by heart failure. Cu requirements are two to three times greater during pregnancy and for infants, since metabolic processes accelerate in rapidly growing tissues. Many regions have too little Cu in soils and water to maintain adequate Cu levels in crops and drinking water; Cu supplements are added routinely to animal feed and to infant formula and vitamin supplements for human consumption. A regular daily intake of 2 to 5 mg of Cu is considered optimal for human adults. The body can tolerate some excess Cu, but high doses induce nausea, vomiting, and even death. Acute Cu toxicity has been reported in cases involving water ingested from badly corroded Cu water heaters, corroded cocktail shakers, and use of Cu tubing in carbonated soft-drink vending machines. A 10- to 20-g dose of Cu sulfate can be fatal to humans.

18.2.5 Sources of Cu and Refining Methods

Although native Cu nuggets as large as 400 tons have been found, most Cu is extracted from mineral compounds. Cu has a strong affinity for S (chalcophilic), and most common ores are mixed sulfides. Cu often occurs in mineral deposits containing Ag, Au, As, Sb, Bi, Co, Ni, and Zn. In principle, Cu sulfide ores can be reduced simply by heating them in air (direct smelting), since Cu sulfide decomposes at high temperature. But most ores are so lean that direct smelting wastes a great deal of energy by heating worthless rock (gangue) along with the Cu minerals. Direct smelting in air also generates SO_2 pollution, causing serious environmental damage (Sidebar 18.3); most nations regulate such emissions.

Most Cu mining and refining operations grind ore to fine powder and use a froth flotation method in violently agitated and aerated aqueous suspensions to separate the Cu compounds from the gangue. Specially formulated flocculents attract the Cu mineral particles to air bubbles in the solution, but not to gangue particles. These bubbles float the Cu compounds to the surface, where they are skimmed away as concentrated Cu minerals. The unwanted gangue remains at the bottom of the tank and is pumped to holding ponds, where it settles and is discarded onto tailings piles. Flotation processes are custom tailored to maximize the yield of the particular Cu minerals found in each mine; with careful process control, up to 95% of the Cu minerals can be separated from the original ore.

The concentrated Cu minerals may require heating to 500 to 750°C to reduce the sulfur content. Sulfur released by this roasting process is captured and used to produce sulfuric acid. Conversion of the roasted concentrate to metallic Cu is a two-step process. First the mixture is melted with silicate fluxes, and Fe present in the concentrate is oxidized and removed as slag. Cu metal is then produced by blowing air or O_2 through the molten Cu sulfide. The metallic Cu from this reaction (called *blister Cu*) ranges in purity from 96 to 99%, and subsequent refining by further oxidation and electrolysis produces 99.9+% pure metal (cathode Cu) with good mechanical and electrical properties. Electrolysis also recovers whatever precious metals (Ag, Au, and the Pt group metals) were originally present in the ore. For certain specialty applications, OFHC Cu is made from electrolytic Cu by remelting in a vacuum or a reducing atmosphere (usually, N_2 and CO). OFHC Cu is free of the Cu_2O interdendritic eutectic precipitates normally found in cathode Cu (Sec. 18.2.3.1).

Substantial amounts of Cu are now produced from Cu oxide-bearing ores rather than Cu sulfide ores. Cu oxide ores are treated with a dilute acid solution, producing a dilute Cu sulfate solution that is electrolyzed to yield Cu metal. This *electrowon* Cu offers economic advantages

---

SIDEBAR 18.3: SMELTER POLLUTION

Cu ores are usually sulfides, and roasting the ore releases large amounts of S. In early smelters, S was released to the air as SO_2, a toxic pollutant that causes acid rain. Most modern smelters capture the SO_2 by liquefaction and convert it to sulfuric acid. Each ton of refined Cu metal produced in the United States yields 1.8 tons of by-product H_2SO_4 acid. Unfortunately, not all smelters recover their S. In Norilsk, Russia, a Cu and Ni smelting operation releases 2 million tons of SO_2 to the atmosphere each year. This toxic emission has seriously degraded the health of Norilsk's inhabitants and killed all trees downwind of the smelter in a region 120 km long and 25 km wide (Fig. 18.16). A much larger region (about 25,000 km^2) has suffered critical environmental damage. As one resident described the polluted arctic city: "It's unbearable. Winter is long here, and life is short." Located north of the 70th parallel in Siberia, Stalin established the city in the 1930s as a mining and smelting operation manned by gulag prison labor. Funding is unavailable to purchase the pollution recovery equipment normally used in Cu and Ni smelting operations.

COPPER 317

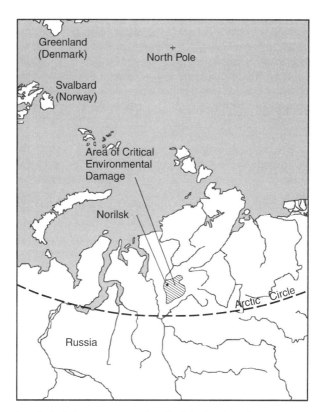

Fig. 18.16 Area of ecological damage surrounding Norilsk, Russia. All trees within a 3000-km^2 region around Norilsk have been killed by large-scale release of SO$_2$ from Cu and Ni smelting operations. Less severe but still serious environmental damage has occurred in the 25,000-km^2 crosshatched area.

over the conventional processes used with Cu sulfides and is also used to recover Cu from tailings and from recycled scrap metal.

Pure Cu and Cu alloys are designated by the letter C followed by five numerals (e.g., C80100); a handbook or Web site is needed to determine the composition. Metallic Cu comprises 94% of all Cu use. Cu metal is both durable and valuable, and a large fraction of all the Cu ever mined has been recycled, often repeatedly, and is still in use today. Scrap recycling supplies 40% of the Cu used each year. About one-third of recycled Cu comes from used objects (old scrap); the balance comes from manufacturing waste such as turnings and powders (new scrap). An active scrap Cu recycling market exists in all industrialized nations.

18.2.6 Structure–Property Relations in Cu

18.2.6.1 Stacking-Fault Energy and Mechanical Properties. In FCC metals, dislocations frequently dissociate into two partial dislocations with a stacking fault between them (Sec. 4.5). Solute atoms often lower stacking fault energy in FCC metals, particularly when the solute is of higher valence than the solvent metal. For example, Zn atoms in solid solution in Cu substantially reduce the stacking fault energy (Fig. 18.17), which alters the metal's plastic deformation behavior in two ways. First, the separation distance between the two partial dislocations is proportional to Gb/γ, where G is the shear modulus, b is the Burgers vector length, and γ is the stacking-fault energy. Lower stacking-fault energy causes partial dislocation pairs to be spaced much farther apart. This makes cross-slip more difficult because (1) extended dislocations must

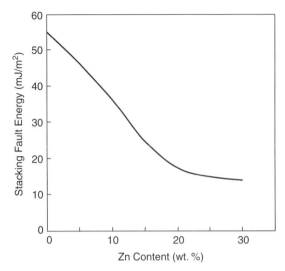

Fig. 18.17 Stacking fault energy in α-brass as a function of Zn content.

collapse their stacking faults and become perfect dislocations before they can cross-slip (this occurs by thermal fluctuations in atom positions and becomes more difficult as stacking-fault ribbons grow wider), and (2) widely extended partial dislocations have a higher probability of intersecting one another. Intersection of one partial dislocation pair/stacking fault by another causes an extended jog at the intersection point (Sidebar 4.1), and jogged extended dislocations require considerably more energy to move. For these reasons, work-hardening rates tend to be high in alloys with low stacking-fault energies. Brass, bronze, silicon bronze, and other solid-solution Cu alloys usually work-harden more rapidly than pure Cu. In addition, both stacking faults and twins in FCC metals develop by **a**/6⟨112⟩{111} shear, so low stacking-fault alloys also twin more easily. Cold-worked brass and numerous Cu alloys with low stacking-fault energies (e.g., Cu–Al, Cu–Ga, Cu–Ge) twin profusely. Not all alloying additions to Cu lower stacking-fault energy; Cu–Mn alloys, for example, have stacking-fault energies similar to those in pure Cu, and they have lower work-hardening rates and limited twin formation. Metals with different stacking-fault energies display markedly different dislocation structures. In pure Cu (and Cu–Mn), dislocation "tangles" are chaotic and tend to form cells due to extensive cross-slip (Fig. 18.18a). However, in alloys with low stacking-fault energies, dislocations often stack up in regularly spaced rows because cross-slip is more difficult for dislocations approaching obstacles (Fig. 18.18b).

18.2.6.2 Nanocrystalline Strengthening and the Inverse Hall–Petch Relationship. For more than half a century, materials engineers have used the Hall–Petch equation to describe the relation between a metal's yield strength and its average grain size [Sec. 3.2 and equation (3.1)]. The Hall–Petch relation predicts behavior accurately in metals with "ordinary" grain sizes (i.e., a few micrometers to a few hundred micrometers) (Fig. 18.19).

Nanocrystalline metals possess an average grain size smaller than about 100 nm (Sec. 7.2.2.1). Such ultrafine-grained metals often possess desirable magnetic behavior and very high strength, but research shows that the Hall–Petch relation fails to describe behavior at very small grain sizes. Much of this research has been performed on Cu (Fig. 18.20). Metals typically follow the Hall–Petch relation when the average grain size is 100 nm or larger, but Hall–Petch behavior breaks down at smaller grain sizes. Indeed, an "inverse Hall–Petch relationship" appears to exist at very small grain sizes, with the yield strength actually *decreasing* as the grain size decreases. It should be noted that many of the specimens plotted in Fig. 18.20 were produced

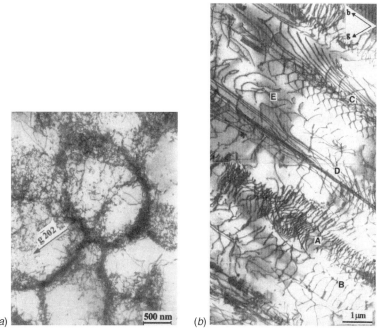

Fig. 18.18 Transmission electron micrographs of dislocation structures in (*a*) pure Cu, and (*b*) Cu–16% Al solid-solution alloy, which has a low stacking-fault energy. Note the tangled dislocation cell structures in pure Cu (where cross-slip can occur relatively easily) and the regularly spaced rows of piled-up dislocations in the Cu–16% Al alloy, where dislocations tend to remain on their original slip planes, due to the difficulty of cross-slip. Dislocation structures are marked in part (*b*) with letters to identify primary edge dislocations (A), edge dislocation trains (B), dislocation net (C), primary screw dislocations (D), and mixed character dislocations (E). Vector **b** = [111] and **g** = [0$\bar{2}$2]. [(*a*) From Li et al., 2002; (*b*) from Li et al., 2003; with permission.]

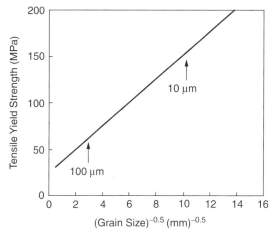

Fig. 18.19 Hall–Petch relation between grain size (*d*) and tensile yield strength in single-phase solid-solution brass (Cu–30 wt% Zn). Note that larger grain sizes are on the left side of this plot, with grain size decreasing on a nonuniform scale (as $d^{-0.5}$) from left to right.

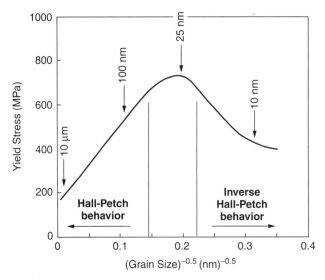

Fig. 18.20 Yield stress as a function of (grain size)$^{-0.5}$ in nanocrystalline Cu. These data were compiled from several different studies.

by condensing Cu vapor in an inert-gas chamber onto a chilled finger, then cold pressing these particles under high pressure into bulk specimens. This method introduces defects (porosity, oxide inclusions, and adsorbed and absorbed gases at grain boundaries) into the pressed specimens that can affect mechanical properties. However, "inverse Hall–Petch" behavior has been observed in specimens where such defect concentrations were low, and there is a growing body of evidence that the Hall–Petch relation does not apply to nanocrystalline metals.

Why does the Hall–Petch relation fail at very small grain sizes? One obvious answer is that for grain diameters only a few atoms wide, the Hall–Petch equation predicts yield strengths much higher than the theoretical maximum ($\sim G/30$). The limiting case of grain size equal to atomic size is, of course, just another way of describing a single crystal, and single-crystal yield strengths are only a few MPa. Hence, common sense demands that the Hall–Petch relation must fail at some point for extremely fine-grained material.

Nanocrystalline Cu specimens with 10 nm average grain size are much weaker than the $G/30$ strength of Cu (2.7 GPa). One of the premises of the Hall–Petch relation, the pile-up of dislocations at grain boundaries, becomes questionable for grains tens of nanometers wide. Dislocations of like Burgers vectors cannot be forced arbitrarily close to one another unless large amounts of energy are available to overcome their mutual repulsion. A grain only 20 nm in diameter is approximately 80 atoms wide. For such small grains, the concept of multiple dislocations piling up becomes implausible. If a dislocation pile-up is considered essential to Hall–Petch behavior, such behavior must "saturate" at very small grain sizes. Moreover, a certain minimum grain volume is needed to operate a Frank–Read source (Secs. 4.4 and 7.2); one component of the yield stress needed to operate a Frank–Read source is equal to Gb/d. This term is minor in "normal"-size grains, but it becomes large for grains only tens of nanometers wide. For example, a 20-nm-diameter Cu grain has $G = 41,000$ MPa, $b = 0.2556$ nm, and $d = 20$ nm, so this factor adds 520 MPa to the yield stress. A 10-nm-diameter grain of Cu requires that an extra 1050 MPa be added to the yield stress just to operate a Frank–Read source in such a tiny volume. High-resolution TEM study indicates that dislocation pile-ups at grain boundaries are absent in nanocrystalline metals. Grain boundaries appear to act both as sources and sinks of partial dislocations, and grain boundary sliding phenomena appear to become a major deformation mechanism for nanoscale grains.

18.3 SILVER

> These places being covered with woods, it is said that in ancient times these mountains [the Pyrenees] were set on fire by shepherds and continued burning for many days and parched the earth so that the abundance of silver ore was melted, and the metal flowed in streams of pure silver like a river.
>
> —Diodorus Sicilus, *Historical Library, Book V*, Chapter 2

18.3.1 History of Ag

Ag was known to ancient peoples because it is sometimes found in native form and is easily extracted from its common ores by heat, including, as Sicilus reports, by deliberately setting forest fires. Ag is considered a precious metal and has been used for coins and jewelry for thousands of years, but it also possesses several unusual attributes (Fig. 18.21) that make it an important industrial material.

18.3.2 Physical Properties of Ag

18.3.2.1 Reflectivity and Corrosion.
The optical reflectivity of polished Ag is 95% for visible light and 98% for infrared light. Unlike its congener elements Cu and Au, Ag's excellent reflectivity is colorless, and indeed the term *silvery* has become the generic descriptor for true-color

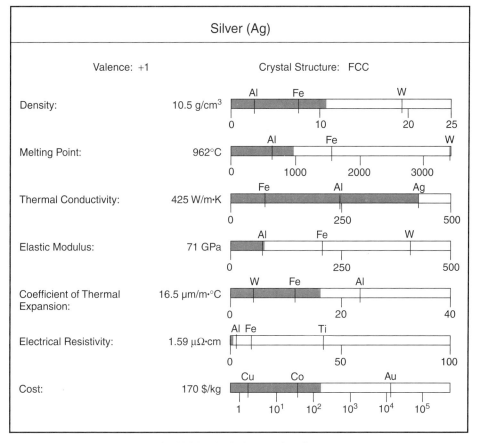

Fig. 18.21 Physical properties of Ag.

reflection. Before the widespread use of Al, Ag was the material of choice for telescope, light house, and vanity mirrors. Unlike more reactive metals such as Al and Ti, whose oxidation resistance relies on impervious oxide films, Ag is inherently oxidation resistant. AgO is only slightly more stable than Ag metal at room temperature and decomposes to Ag metal and O_2 when heated above 300°C in air. However, Ag does corrode in air containing SO_2 or H_2S, forming a protective but unsightly black Ag sulfide tarnish that is all too familiar to owners of jewelry, coin collections, and silverware. Ag is resistant to some corrosive aqueous solutions, but it is not as noble as its congener Au. Liquid Ag has an extraordinary capacity to dissolve atmospheric O. One liter of molten Ag can dissolve 20 liters of O in O at 1 atm pressure. During solidification, the dissolved O is forcefully rejected at the solid–liquid interface in a potentially hazardous eruptive "spitting" action.

18.3.2.2 Plasticity. Ag's FCC crystal structure slips on $\{111\}\langle 110\rangle$ (Sec. 4.6) with $\tau_{CRSS} = 0.37$ MPa (99.999% purity). It is the second-most ductile of all metals after Au; a 20-kg ingot of pure Ag can be drawn into a wire long enough to encircle the Earth. High-purity Ag recovers and recrystallizes dynamically at room temperature as deformation progresses, making ductility almost unlimited.

18.3.2.3 Conductivity. Ag has the highest electrical conductivity of any material at 20°C, and its thermal conductivity is the highest of all metals. However, its conductivity is only slightly higher than Cu, and Ag's much higher cost prevents it from competing with Cu for electrical wiring and heat exchanger applications. Ag is used heavily for electrical switch contacts, which benefit from its high conductivity and relative freedom from oxide layer formation. Ag also makes an excellent bearing material because its rapid heat dissipation prolongs service life.

18.3.3 Applications of Ag

Ag has long been used as a store of monetary value. Until the 1960s, Ag was widely used for coinage, but this use ended when rising Ag prices made the value of the coins' Ag content greater than their face value. Some governments and banks still maintain Ag bullion reserves, but Ag's industrial uses greatly surpass demand for monetary Ag. Ag has three categories of commercial use, each of which accounts for about one-third of annual consumption: (1) jewelry, silverware, coins, and medallions; (2) halide compounds for photography; and (3) industrial and medical use. The latter category includes such uses as switch contacts, batteries, catalysts, dental implants, bearings, and brazing alloys.

18.3.3.1 Electrical Contacts and Electronics. Billions of electrical switch contacts are manufactured each year and used in systems ranging from computer keyboards to automobiles to electric power generation and transmission equipment. The requirements of a switch seem simple; they must conduct electrical current with minimal resistance when the switch is closed and interrupt current flow when the switch is opened. These tasks are made more difficult, however, by contamination of the contact surfaces and damage from the wear and arcing that can occur during opening and closing of the switch. A perfectly smooth, clean metallic contact surface has the least resistance when a switch is closed; however, such a surface is a practical impossibility. Even the most carefully prepared switch contact surface has topographic relief a few hundred atoms high, and this surface grows rougher as the switch wears. In addition, O_2 and H_2O vapor adsorb onto contact surfaces. In air, O_2 molecules are attracted to the switch's contact surfaces by van der Waals forces, and disassociated into individual O atoms that share electrons with metal atoms at the surface. In most metals these O atoms diffuse into the surface, forming a gradually thickening metal oxide surface layer. In addition, the strongly polar H_2O molecules in air form van der Waals bonds atop the oxide layer. Such surface layers are helpful, since they prevent metal surfaces from spontaneously cold-welding to one another upon casual contact; however, they increase the resistance to current flowing through the contacts and can cause overheating, contact failures, and even fires. These problems can be minimized by selecting appropriate contact metals or by sealing the contacts in inert atmospheres. Even if the problem of insulating surface layers can be overcome, contacts carrying high power are vulnerable to damage by electric arcs that form briefly while the switch is opening or closing (Fig. 18.22). Near these arcs, metal is melted, vaporized, and even converted to plasma, transferring metal from one switch contact to the other and forming asperities on the contact surfaces.

The problems of insulating surface layers and arcing contacts can be reduced if the contact materials have high electrical and thermal conductivity and resist oxide formation. Both Au and Ag provide this combination of attributes. Ag has the highest electrical and thermal conductivity of all metals and is much less costly than Au. Ag will form thin oxide and sulfide layers, but these tend to be removed in sliding contacts or at elevated temperatures. For these reasons, pure Ag is widely used as an electrical contact material in low- and medium-current switching devices. However, pure Ag has a fairly low melting point and is damaged rapidly by arcing if used in high-power switches.

Ag membrane switches are used in the keyboards of most computers, as well as in flat-panel keypads on some telephones, microwave ovens, ATMs, and similar devices. These switches are made of conductive ink containing Ag flakes and carbon particles in a polyester binder. This film is silk-screened into an electrical circuit pattern on each of two parallel Mylar sheets. The two surface patterns of Ag face each other separated by a narrow air gap. When the key is pushed, the two sheets touch, closing the switch. Connecting paths for electronic circuit boards are often made with epoxy resin/Ag formulations. Ag-filled resins provide higher conductivity than Cu, and do not suffer from Cu's tendency to form oxide films when heated during soldering operations or when used in corrosive atmospheres.

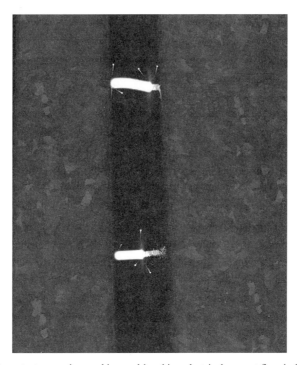

Fig. 18.22 Switch contacts arc when making and breaking electrical current flow in high-power circuits. Arcing heats the metal sufficiently to form liquid droplets, vapor, and even plasma, all of which roughen the contact surfaces and degrade switch performance.

Ordinary household light switch contacts made with metals such as Al form oxide layers that cause overheating and fire hazards. This problem is ameliorated with Ag switch contacts. A modern automobile may contain more than 40 Ag electrical switches. For more demanding applications, Ag is often alloyed with Cu, Pt, Cd, or Pd to increase its hardness. Composite materials, comprised of a pure Ag matrix containing refractory particles such as Mo, W, or WC, combine the excellent conductivity of Ag with the refractory material's resistance to arc damage. These composites perform well in high-voltage switchgear. Since Ag and these refractory materials are mutually insoluble, these composites consist of two pure phases intermixed by either melt infiltration or by powder metallurgy (Fig. 18.23). Ag also serves as the ideal matrix material to form high-temperature superconducting wire (Sidebar 18.4).

18.3.3.2 Dental Alloys. The most commonly used material for "filling" a portion of a tooth damaged by decay (caries) is an Ag amalgam. A powder of 67 to 70 wt% Ag, 25 to 28 wt% Sn, 0 to 5 wt% Cu, and 0 to 2 wt% Zn is mixed with Hg immediately before implantation in the patient's tooth. Typically, eight parts of Hg are blended with five parts of the Ag alloy, any excess Hg is squeezed out, and the mixture is pressed quickly into the prepared cavity while it is still plastic so that it can react and harden inside the tooth cavity (Sidebar 19.4). The high Ag content assures satisfactory corrosion resistance. The Ag/Sn ratio in the powder is selected to produce the stoichiometry of Ag_3Sn, which minimizes contraction of the filling as it hardens in the tooth. Ag is also a major alloying addition to many Au-based dental implants (Sec. 18.4.3.3).

18.3.3.3 Sterling Tableware and Jewelry. The corrosion resistance and beautiful luster of polished Ag make it an excellent metal for tableware and jewelry. Pure Ag has only moderate strength and wear resistance, so Ag alloys are generally used for these applications. Sterling silver, a 92.5 wt% Ag–7.5 wt% Cu binary alloy, is the usual choice for these applications. Cu is the only metal with which Ag forms a simple eutectic between two well-defined solid solutions (Fig. 18.25).

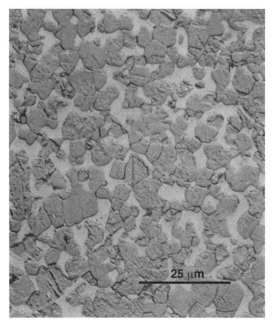

Fig. 18.23 Micrograph of a Ag–W composite for switchgear. Pure Ag (lighter phase) forms a continuous network around the W particles (darker phase) to carry electricity and dissipate heat. The W is refractory and resists arc damaged when the switch is opened and closed. (From Butts and Coxe, 1967; with permission.)

SIDEBAR 18.4: HIGH-TEMPERATURE SUPERCONDUCTING WIRE

Since their initial discovery in the 1980s, a family of ceramic superconducting materials ($YBa_2Cu_3O_7$, $HgBa_2Ca_2Cu_2O_8$, and related compounds) have been under continuing development. These materials are superconducting at liquid N_2 temperatures (although their current-carrying capacity is lower than metallic superconductors, such as Nb_3Sn). They could be used with lower losses than ordinary metallic conductors while avoiding the prohibitive expense and technical problems associated with cooling metallic superconductors to 4.2 K with liquid He to maintain their superconducting properties. However, these ceramic superconducting materials are brittle and difficult to fabricate into wire.

High-temperature superconducting (HTS) wire can be fabricated by loading the HTS material precursors into a Ag can (Fig. 18.24), welding the can shut, extruding and drawing the encapsulated powders into rod, cutting the rod into short lengths that are bundled into another Ag can, and extruding and drawing that second can into small-diameter rods. This rod is then rolled into ribbon, and the ribbon is heated in O_2 to diffuse O through the Ag sheath. This forms the ceramic oxide HTS material *in situ* within the ribbon. The sheath used for this process must be ductile and an excellent electrical conductor, and the sheath must also allow O diffusion through the sheath wall without reacting with either the O or the HTS material. Only Ag has all of these capabilities.

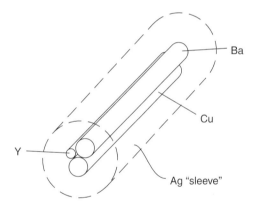

Fig. 18.24 Starting geometry used to produce a high-temperature superconducting composite wire. The Ag matrix forms a sheath around the precursor Ba, Cu, and Y metals. This assembly is extruded and drawn into fine wire, then perfused with O to form the brittle superconducting ceramic oxide ($YBa_2Cu_3O_{7-\delta}$). Ag diffuses O readily but does not react with it.

At equilibrium, sterling silver would be a homogeneous single-phase solid solution near the 780°C eutectic temperature; however, cast sterling silver cools too rapidly to attain equilibrium and forms a Ag-rich primary solid solution containing the Ag–Cu eutectic as a second phase. Since Cu's solid solubility in Ag decreases sharply as the cast part approaches room temperature, a Cu-rich precipitate forms within the primary Ag solid-solution phase. If cast sterling silver is cold worked and annealed, the microstructure changes to the equilibrium structure predicted by the phase diagram, a Ag matrix with a few percent Cu in solid solution plus a Cu-rich solid solution second phase containing some Ag. It is possible to perform a classic precipitation-hardening heat treatment on sterling silver by solutionizing at 750°C, quenching, and aging at 300°C. This is rarely done, however, because the alloy has very low strength at the solutionizing temperature. In addition, small amounts of P are often added to molten sterling silver as a

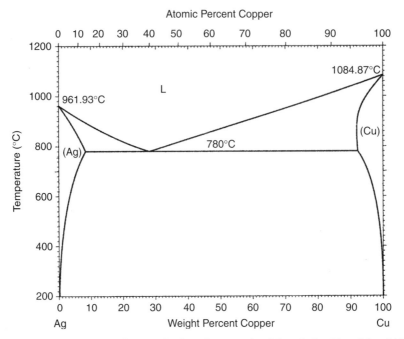

Fig. 18.25 Ag–Cu binary phase diagram. Sterling silver contains 7.5 wt% Cu. (From Massalski, 1986, p. 19; with permission.)

deoxidizer, and if excess P remains in the metal, it will be hot short and may "slump" out of shape during solutionizing.

High cost and tarnish problems have caused sterling silver tableware to lose market share to stainless steel tableware. Sulfur compounds in certain foods (e.g., eggs) and in ambient air form an unsightly black sulfide film over sterling silver, which must be removed by fine abrasive polish or electrolytic methods. Simply heating pure Ag in air will reduce both Ag_2S and Ag_2O to Ag metal, but heating tarnished sterling silver is counterproductive, since the Cu oxides and sulfides are not reduced below the eutectic temperature, and corrosion of the Cu in sterling silver worsens with prolonged high-temperature exposure. Various attempts have been made to reduce the tarnish problem with lacquers or Rh plating or by replacing the Cu in the alloy with Pd, but each of these methods diminishes the metal's luster. The recent introduction of Ag–Cu–Ge alloys solves these problems, but at a higher cost (Sidebar 21.1).

18.3.3.4 Bearings. Materials used for bearings need several physical and chemical properties for good performance:

- *Strength*: to withstand the forces exerted against the bearing
- *Lubricity*: a low coefficient of friction with other components sliding or rolling over the bearing surface
- *Fatigue resistance*: the ability to withstand cyclical applications of stress without cracking
- *High thermal conductivity*: to dissipate heat generated at the bearing's surface
- *Corrosion resistance*: to resist attack by organic acids that sometimes form in lubricating oils during operation
- *Embeddability*: the ability of the bearing to embed small dirt particles present in lubrication systems, trapping the particles in the bearing surface, where they can do no harm

- *Conformability*: the ability of the material to flow slightly to compensate for misalignments or small dimensional changes of mating parts
- *Antiscoring tendencies*: the material's ability to resist galling or seizing during the dry metal-to-metal contact that may occur during startup or lubrication failures

Although no material meets all of these requirements optimally, Ag does well in most of these areas. Ag bearings possess outstanding fatigue resistance, high thermal conductivity, and good corrosion resistance. The embeddability and conformability of Ag bearings are poor, but they can be improved greatly by adding overlay materials of softer metals, such as Pb–Sn or In. Steel bearings electroplated with high-purity Ag have excellent fatigue strength and load-carrying capacity and are used in severe service and heavy-duty applications. Such bearings are used in jet engines. The main shaft of an aircraft turbine engine rotates on steel ball bearings that roll within steel retaining rings, called *cages*. Similar bearings are required for the connecting gear boxes that drive accessories, such as hydraulic pumps and fuel pumps. Steel itself has a poor coefficient of friction, but electroplating the steel with Ag greatly reduces the friction between the steel balls and the steel cage, increasing performance and reducing maintenance costs. In addition, the lubricity of the Ag is sufficient to protect expensive turbine engines in the event of a complete lubrication failure, allowing time for the engine to be shut down before serious damage occurs.

18.3.3.5 Brazing and Soldering. The joining of two parts is facilitated by introducing a second, low-melting metal between them. This process is called *brazing* (Sec. 8.6.3) when the materials are joined above 600°C or soldering if performed at lower temperatures. The parts to be joined do not melt during brazing and soldering, which distinguishes the processes from welding. Ideally, the brazing alloy or solder not only wets the surfaces of the parts to be joined but actually diffuses into the parts to enhance joint strength. Ag brazing alloys produce strong, leaktight, corrosion-resistant joints and have melting points ranging from 143°C to over 1000°C. They are widely used in air-conditioning and refrigeration equipment, power distribution equipment, and automobile and aerospace assemblies. Ag brazes and solders possess excellent wettability on most metal and glass surfaces (and even some ceramics) and provide high tensile strength, ductility, and thermal conductivity. Sn–Ag solders and Sn–Ag–Cu solders (Sidebar 21.4) eliminate the toxicity hazard of Pb-based solders.

Ag brazing alloys are usually ternary or higher-order alloys typically classified as eutectic or noneutectic. Eutectic alloys have a discrete melting temperature above which they flow freely and are useful for joints with narrow clearances (i.e., 25 to 100 μm); indeed, they are so fluid when melted that they are difficult to retain in larger gaps. By contrast, noneutectic alloys are part liquid and part solid over a range of temperatures; in this semisolid state they are much less fluid than a molten eutectic alloy and are useful to join wide gaps or to "build up" material for fillets. The Ag–Cu–Zn ternary alloy is typical of the many commercial Ag brazing alloys. The 50% Ag–34% Cu–16% Zn composition is a typical noneutectic brazing alloy with a solidus temperature of 660°C and a liquidus temperature of 760°C (Fig. 18.26). The 72% Ag–28% Cu binary is often used where the discrete melting point (780°C) of a eutectic brazing alloy is preferred.

18.3.3.6 Batteries. Ag–Zn batteries provide high-energy output from a small light cell. Although these batteries have shorter life and higher cost than many competing batteries, they are widely used in watches, calculators, cameras, hearing aids, and various aerospace and military systems. Ag–Zn batteries provide three times the watthours of energy per gram of battery weight than that of Ni–Cd batteries and four times that of Pb acid batteries; however, the Ni–Cd batteries have 10 times greater service life than Ag–Zn batteries. More than 1 billion small Ag–Zn "button" cells are used each year, but manufacturing these batteries consumes only a few hundred tons of Ag. Other batteries are made from Ag combined with Cd, Mg, and H, but their use is limited primarily to military systems.

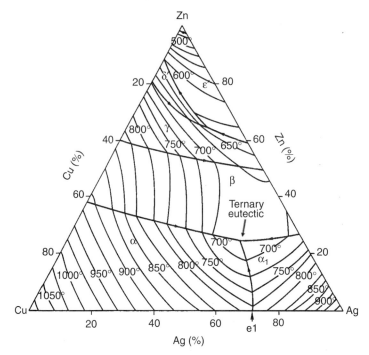

Fig. 18.26 Ag–Cu–Zn ternary equilibrium liquidus surface. Point e1 is the binary Ag–Cu eutectic point. (From Gebhardt et al., 1962; with permission.)

18.3.3.7 Catalysts. Ag's weak interaction with O makes it an excellent catalyst for production of various chemicals. For example, the catalytic action of high-purity Ag removes two H atoms from methanol (CH_3OH) to form formaldehyde (CHOH). In ethylene oxide (C_2H_4O) production, Ag adsorbs molecular O_2 from air, disassociates these O_2 molecules into monatomic adsorbed O atoms, then reacts these O atoms to ethylene gas molecules (C_2H_4) to form C_2H_4O. Several million tons of both formaldehyde and ethylene oxide are produced annually by Ag catalysis. These compounds are intermediates in various polymer production processes. A global inventory of over 700,000 kg of Ag is used in such catalytic oxidation reactions; the Ag is not consumed by the reaction and is regenerated chemically and electrolytically to remove impurities that accumulate during use.

18.3.3.8 Photography. The first photographic image was made in 1813 by reducing Ag nitrate to metallic Ag. Similar reactions are used in modern photographic films. Although electronic methods have taken some of the photography market from Ag, predictions of the rapid end of Ag-based film overlook the low cost and extraordinary resolution of fine detail in Ag films. Photographic films are Ag halide crystals suspended on a polymer substrate. These salts decompose into metallic Ag when exposed to light, and they are subsequently "developed" by reaction with mild reducing agents. Microscopic, opaque crystals of Ag metal produce the dark regions of the negative image where the film was exposed to light. Many metallic salts are photosensitive, but Ag halides are especially efficient for photographic use. Just four or five photons of light incident on a Ag halide crystal will reduce an equivalent number of Ag atoms. This small number of Ag atoms then catalyzes reduction of as many as 10^{12} Ag atoms when the film is developed. Approximately 200 photographs can be taken using 1 g of Ag. Ag halide crystals are also sensitive to x-ray wavelengths. Medical diagnoses and nondestructive examination for flaws deep inside engineering parts rely heavily on Ag halide films. X-ray inspection is often required by code for castings and welds used in pipelines, pressure vessels, and other safety-critical assemblies.

18.3.4 Sources of Ag and Refining Methods

Ag is widely dispersed in Earth's crust in both the metallic state and in compounds. Native Ag usually occurs in particles too fine to be easily seen by the unaided eye, and it is almost always alloyed with other metals, such as Pb, Zn, Cu, and Au. Coarser metallic dusts and nuggets are rare. The most common Ag-bearing compounds are sulfides and chlorides. Few ore bodies contain sufficient Ag to justify extraction for the Ag alone. Three-fourths of all Ag is produced as a by-product of Pb, Zn, Cu, and Au mining. Primary and by-product mine production meet only about two-thirds of world demand for the metal; scrap recycling and drawdown from investment inventories provide the balance.

Since most Ag is produced as a by-product, the mining, concentration, and refining techniques employed to produce Ag vary with the primary metal being processed. Some primary Ag mining occurs, and there the crushed ores containing free Ag and AgCl are mixed with Hg to form amalgams that can be separated from the rock and distilled to produce Ag metal. When Au is the primary valuable constituent of the ore, the crushed rock is leached with basic dilute sodium cyanide solution. The Ag (and Au) react with the sodium cyanide to form cyanide solutions that are then treated with metallic Zn dust to precipitate the Ag. The precipitated Ag is melted and cast into crude Ag bullion bars (Sidebar 18.5). These bars can be refined to high-purity Ag either by electrolysis or other methods. Since Ag does not oxidize, bars containing base metals can also be purified by melting in air or by injection of O_2 into the molten metal to oxidize the base metals. As the lighter oxides float to the surface of the melt, they are skimmed off, leaving purified Ag.

18.4 GOLD

> Mr. Bond, all my life I have been in love with gold. I love its colour, its brilliance, its divine heaviness. I love the texture of gold, and I love the warm tang it exudes when I melt it down into a true golden syrup. But, above all, Mr. Bond, I love the power that gold alone gives to its owner—the

---

SIDEBAR 18.5: SILVER'S ROLE IN THE EVOLUTION OF THE DOLLAR

In the time between the fall of the Western Roman Empire and the Spanish conquest of Mexico, most monetary Ag was supplied by mines and mints in central Europe. Early Ag coins were usually small, but larger transactions required greater quantities of precious metal, and demand rose for a larger standard coin. A mine at Joachimsthaler in Bohemia minted the first large Ag trade coins in 1486. These coins were initially called *Joachimsthalers*, but this was soon shortened to *thalers*, which finally evolved into *dollars*.

Shortly thereafter, Spain began minting coins from the huge Ag deposits discovered in Mexico, and this trading coin, the piece of eight (reales), came to dominate maritime trading among Atlantic and Pacific Rim nations. These coins were generally called *Spanish dollars*, and one of the designs (Fig. 18.27) had on its reverse side the royal arms of Spain, which contained two Baroque architectural columns representing the pillars of Hercules at Gibraltar. These pillars were wreathed in an S-shaped swirl of ribbon bearing a Latin inscription. Thus, the abbreviation for the dollar came to be a simplified image of the coin's design, two vertical pillars superimposed on the S-shape of the ribbon. At the time of the American revolution, the Spanish dollar was the most familiar trading coin, so the new coinage of the United States of America continued to use the name *dollar* and its well-known symbol, $.

Fig. 18.27 Spanish piece of eight. Note the vertical columns with wrapped ribbons that came to be abbreviated with the "$" symbol for the dollar. (Courtesy of Aarons Gifts, Inc., *http://www.aaronsgifts.com/mp003l.htm*.)

magic of controlling energy, exacting labor, fulfilling one's every wish and whim and, when need be, purchasing bodies, minds, even souls.

—Auric Goldfinger, in Ian Fleming's *Goldfinger*, Jonathan Cape Ltd., London, 1959

18.4.1 History of Au

The small quantities of Au in Earth's crust frequently occur as native metal, and its availability as nuggets and dust made it one of the first metals found and used by humans. Au has been excavated from the graves of Paleolithic humans buried tens of thousands of years ago, and the metal has played an integral part in the development of civilization, serving as a means of monetary exchange for several thousand years. All the Au ever mined in history could fit into a cube 18 m on a side, yet this metal was the motivation for wars in the ancient Mediterranean, Columbus's and Pizarro's brutalization of the native peoples of America, and a frenzied Gold Rush that jump-started California's population growth. Even today, humankind's fascination with Au drives legions of workers to mine it from depths nearing 3000 m, refine it, and transport it under armed guard, only to "bury" much of it again in underground bank vaults (Sidebar 18.6).

18.4.2 Physical Properties of Au

The beauty, scarcity, and monetary significance of Au often overshadow appreciation for its unusual combination of physical properties (Fig. 18.29). Au has outstanding oxidation resistance, high electrical conductivity, and an ability to form extensive series of alloys and intermetallic compounds with many other metals. It has the highest infrared reflectance of any metal, 99%

SIDEBAR 18.6: WORLD GOLD PRODUCTION

Annual world Au production is a relatively small volume of metal. Total world Au production in 2000 had a market value of $25 billion, but this quantity of Au would fit inside a cube just 5 m on a side (Fig. 18.28). By contrast, if the 900 billion kg of steel produced each year was placed on the same 5 m × 5 m "footprint," it would tower to a height of 4500 km, 10 times the orbital altitude of the International Space Station. Despite its impressive bulk, this huge mass of steel has a market value only about 40 times greater than that of the 5-m Au cube.

Fig. 18.28 The world's annual production of Au could be melted into a single cube 5 m on a side. If the world's annual production of steel were melted into one ingot with the same 5-m² "footprint," it would be over 4000 km tall.

at near-infrared wavelengths. Au has the highest electronegativity (2.4) of all metallic elements (Fig. 1.14), which makes it strongly resistant to oxidation. In fact, in compounds with the alkali metals (e.g., AuCs), Au is the anion, and these compounds are transparent, more like a salt than a metal. When vaporized, Au forms diatomic molecules like those of gaseous nitrogen and oxygen. Its yellow color originates from the same reflection phenomenon that occurs in Cu (Sec. 18.2.2.1).

Au's FCC crystal structure slips on the $\{111\}\langle 110\rangle$ (Sec. 4.6) with $\tau_{CRSS} = 0.91$ MPa (99.99% purity). It is the most ductile metal; 1 kg of Au can be beaten into a sheet only 10 nm thick with an area of 5000 m², about the size of a football field. High-purity Au recovers and recrystallizes dynamically at room temperature as cold work progresses. Au has high electrical and thermal conductivity, making it one of the few materials to combine oxidation resistance, ductility, and high conductivity.

18.4.3 Applications of Au

Au has long been used as a store of monetary value (Sidebar 18.7). Thousands of tons of Au are held as monetary reserves by governments and financial institutions, usually as 1000-troy ounce bars (32.15 troy ounces = 1 kg). The metal in monetary reserves is largely "stagnant"; it is seldom moved and is uninvolved with the annual production and consumption of Au metal.

18.4.3.1 Jewelry. Jewelry production consumes over 2000 tons of Au each year, nearly 90% of world production. Pure Au has low tensile strength and poor wear resistance, but jewelry alloys can provide improved wear resistance and strength, while maintaining the color, formability, and corrosion resistance of pure Au. In Asian markets particularly, Au jewelry serves not only as ornamentation but also as a store of value. For this reason, alloys with the highest possible Au content are favored. Jewelers designate Au content in alloys by the *karat* unit; pure Au is

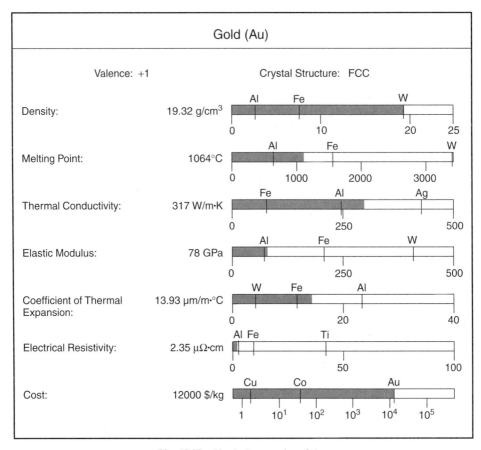

Fig. 18.29 Physical properties of Au.

SIDEBAR 18.7: CRASSUS'S THIRST FOR Au

History is replete with stories of greed and outrageous acts committed in pursuit of wealth. Marcus Licinius Crassus, proconsul of the Roman Empire during the first century B.C., was particularly obsessed with the acquisition and lavish expenditure of money. He acquired a vast fortune from slave trading and Ag mining, and he was famous for extravagant expenditures designed to endear himself to the Roman citizenry. In one of his typically grand gestures, he hosted a festival at which 10,000 tables were set for a feast. The guest list included most of the free citizens of Rome.

Crassus's lifestyle required a large and continuous stream of fresh income, and at the age of 60, he led an army to Syria to plunder whatever Au and other riches could be found. His troops rampaged through the eastern Mediterranean, raided the temple in Jerusalem, and started a war with the Parthians. In the climactic battle of this war, Crassus was defeated and beheaded. The head of Crassus was presented to Orodes, the victorious Parthian king. The king, well aware of Crassus's lust for Au, ordered his men to pour molten Au into the mouth of Crassus's severed head. As they did so, Orodes said, "There, sate thyself now with that metal of which in life thou wert so greedy." Despite his ignominious demise, Crassus lives on in language with the word *crass*, meaning overly commercialized or vulgar.

24 karat. During the 1990s, a precipitation-hardening 99.0% Au–1.0% Ti alloy was developed that is particularly popular in Asia. This alloy is 23.7 karat, which can be marketed as 24 karat by rounding to the nearest integer. This alloy can be quenched and age hardened at room temperature to a tensile strength of 1000 MPa and a hardness of nearly HV 200.

In other jewelry markets, older alloys of Au still dominate, in part because their lower Au contents are less costly. Many of these have been developed in the ternary Au–Ag–Cu system and in the quaternary Au–Ag–Cu–Zn system. Au contents in these alloys typically range between 41.7 wt% (10 karat) and 75 wt% (18 karat), with most of the rest of the alloy consisting of Ag and Cu; 3 to 10 wt% Zn is often present to improve castability. Au, Ag, and Cu will not dissolve completely at all temperatures and concentrations, which allows these alloys to be homogenized at high temperature, quenched, and age hardened in the multiphase regions of the ternary phase diagram (Fig. 18.30). Hardness values as high as 200 Brinnell can be achieved in age-hardened Au–Ag–Cu ternary alloys. Zn additions act as deoxidizers and lower liquidus temperatures, reducing the hardness of quenched metal and shifting the alloy color to a whiter tone.

The ratio of Cu and Ag in these alloys affects their color, and extensive studies have been performed to allow accurate "tuning" of alloy color by adjusting the ratios of elemental constituents. Ag additions tend to color the alloy with a slight greenish tint in low concentrations which becomes a whitish tone at higher concentrations. Increasing Cu concentrations lead to a redder hue, and Zn and Pd turn the color to white. *White gold alloys* contain high Ni and Ag contents with lower Au concentrations. These alloys were developed to provide a less-expensive alternative to Pt jewelry, but they are difficult to heat-treat because phase separations occurs easily, degrading workability, strength, and corrosion resistance. They are also vulnerable to *sulfur poisoning*, which causes hot shortness (cracking during hot deformation) from low-melting sulfides at grain boundaries.

AuCu and $AuCu_3$ (Fig. 25.6) intermetallic compounds can strengthen binary Au–Cu jewelry alloys. These ordered compounds contribute to the alloy's precipitation hardening by order strengthening (Sec. 25.6.1). Dislocation with a Burgers vector of 1/2[110] passing from the matrix into a coherent precipitate particle juxtaposes like atoms across the slip plane within the

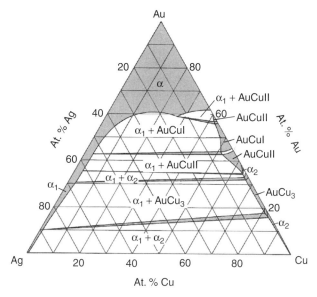

Fig. 18.30 Duplex regions at 300°C in the Au–Ag–Cu system. Shaded areas are single phase; unshaded areas are duplex regions that can be precipitation hardened. (From Prince et al., 1990, p. 33; with permission.)

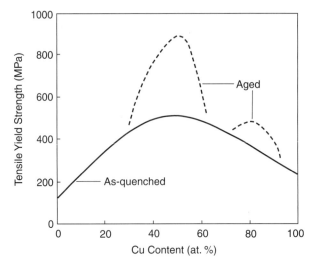

Fig. 18.31 Strength of binary Au–Cu alloys in the as-quenched (metastable disordered FCC phase) and aged (ordered FCC intermetallic precipitates) conditions. The greatest strength occurs when Au–50 at% Cu forms tetragonal AuCu intermetallic precipitates; a secondary maximum in strength occurs near the stoichiometry of the cubic $AuCu_3$ phase at 75 at% Cu.

precipitate. The resulting anti-phase boundary disrupts the order of the intermetallic compound and raises the energy of the crystal. A second dislocation of the same type moving on the same slip plane is required to "reestablish order" in the precipitate and lower the crystal energy. The effect of these anti-phase boundaries on strength is clearly evident in Fig. 18.31. The strength of aged Au–Cu alloys reaches a maximum at 50 at% Cu with a second, local maximum near 75 at% Cu.

18.4.3.2 Electronic Uses. Although most Au is used in jewelry and monetary reserves, a significant quantity is used in industrial applications. Au's unique attributes can solve a variety of engineering problems. No other metal delivers Au's combination of ductility, corrosion resistance, reflectivity, and high thermal and electrical conductivity. The electronics industry consumes more than half of industrial Au. Most metals accumulate gradually thickening oxide scales when exposed to air, and these oxides increase contact resistance. But Au forms no oxide scale; its surfaces possess the lowest contact resistance possible. For this reason Au provides the most reliable contact surfaces for a wide spectrum of electronic devices. For example, an ordinary telephone keypad contains 33 Au-plated contacts. For switches carrying low voltages and amperages, Au containing about 1% Co or Ni is most often used; but Au contacts for switchgear subject to sliding wear or the arcing and erosion associated with higher voltages and amperages are usually made of binary or ternary alloys of Au with Ag, Pd, or Cu. Au's other major role in electronic devices is interconnect wire used to connect microprocessors in electronic circuits. These wires are often as small as 20 μm in diameter. A major electronic assembly facility will typically consume 100 km of interconnect wire per day.

18.4.3.3 Dental Uses. Au has been used in dentistry for more than 3000 years, and dental use still accounts for 20% of worldwide industrial Au consumption. Au's easy formability and outstanding corrosion resistance make it the premier material for dental inlays, bridges, and crowns. Since pure Au is relatively soft, dental Au alloys typically contain 50 to 90% Au, with the balance consisting of Ag, Pd, Pt, In, Sn, and other metals. These alloying additions increase the Vickers hardness from 25 for pure Au to 140 to 170 in dental alloys. The strength of dental alloys is developed through both solid-solution hardening and precipitation-hardening mechanisms. Ternary and higher-order alloys are selected from compositions that avoid dendritic solidification to minimize the galvanic corrosion that may result from inhomogeneous element concentrations. Au dental alloys are often bonded to ceramic veneers to improve the appearance of the dental implant; these implants require Au dental alloys with melting temperatures above 1000°C to avoid melting the alloy at the ceramic's firing temperature. Additions of Pd and Pt raise the liquidus temperature of Au, and these alloys can be precipitation hardened when Pt contents exceed about 6 wt%.

Since Au dental alloys often contain five or more elements, it is difficult or impossible to rely on phase diagrams to predict melting ranges and precipitation-hardening behavior. For this reason, extensive empirical testing is required in most dental alloy formulation and product development. Au's high cost has motivated development of dental alloys with lower Au content or no Au at all. But Au's biocompatibility, formability, and corrosion resistance have yet to be equaled in other materials, and it seems likely that Au-based alloys will continue to play a major role in dentistry for the foreseeable future.

18.4.4 Sources of Au and Refining Methods

Most heavy elements are rare, but Au is especially so. Only one isotope ($Z = 197$) exists that is stable against radioactive decay, and its average abundance in Earth's crust is only 4 parts per billion. Due to its great scarcity, Au can be extracted economically only where geological forces have concentrated it. Primary Au deposits are formed by hydrothermal fluids either derived directly from magmatic bodies or produced by their heat. These are often associated with quartz veins. Secondary Au deposits (placers) occur when primary deposits erode and redistribute the

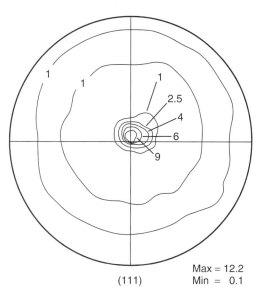

Fig. 18.32 X-ray diffraction ⟨111⟩ pole figure of highly textured polycrystalline Au wire. The wire axis lies at the center of the stereographic pole projection, and the strong x-ray diffraction intensities near the center of the pole figure indicate that most of the Au grains have a ⟨111⟩ direction nearly parallel to the wire axis. Most FCC metals display similar texture effects, although some FCC metals (e.g., Cu) have a mixed texture with some grains showing a preferred orientation of ⟨111⟩ parallel to the wire axis, while other grains in the same specimen show a ⟨100⟩ preferred orientation. (Courtesy of F. C. Laabs, Ames Laboratory of the U.S. Department of Energy.)

Au at the base of streambeds or along shorelines, where Au's high density causes it to collect from the action of flowing water or waves.

The Au in ore bodies is usually present as native metal, not chemical compounds; however, these metallic Au particles are usually too small to be seen by the unaided eye. Occasionally, Au forms larger particles called *Au dust* or *nuggets*; these were readily recognized by ancient peoples and prospectors. The largest nugget ever discovered was a 3000-troy ounce (over 90 kg!) specimen found in 1872 in Australia. Au sometimes occurs in telluride compounds. Modern Au mining requires either precipitating the dense metallic particles of Au in hydraulic recovery and settling operations or concentrating it by reacting finely crushed ore with cyanide compounds or Hg metal. The Au recovered often contains a few percent base metals such as Pb or Cu, which can be removed by oxidation, leaving Au containing a few percent Ag and small amounts of Pt group metals, which are separated electrolytically.

18.4.5 Structure–Property Relations in Au

Metals often become textured (Sec. 3.3) by plastic deformation. When Au interconnect wire is produced, a 50-mm-diameter Au ingot may be drawn to 20-μm wire with only one intermediate stress-relief anneal. This reduces the ingot's diameter by a factor of 2500, a true strain of 15.6. Such drastic plastic deformation makes the Au heavily textured (Fig. 18.32). Like many FCC metals, axisymmetrically deformed Au assumes a dual texture; both $\langle 111 \rangle$ and $\langle 100 \rangle$ fiber textures are present. This means that some of the grains are oriented with a $\langle 111 \rangle$ direction within a few degrees of the wire axis, and most of the remaining grains are oriented with a $\langle 100 \rangle$ direction within a few degrees of the wire axis. These orientations still leave three or four slip systems positioned with favorable Schmid factors (Sec. 4.6) for further slip, allowing the metal to accommodate still more plastic flow.

ADDITIONAL INFORMATION

References, Appendixes, Problem Sets, and Metal Production Figures are available at
ftp://ftp.wiley.com/public/sci_tech_med/nonferrous

19 Zn, Cd, and Hg

19.1 OVERVIEW

Group 12 is the last group of transition metals. The nominal outer electron configuration of these elements is s^2d^{10}. The d electrons are more tightly held than the s electrons, so the typical valence is +2, but all three elements have high ionization energies, making their metallic bonds relatively weak. These elements have a number of physical and chemical similarities to alkaline metals and are the lowest melting of all transition metals.

Zn is one of the world's major metals. Although it is not especially abundant (76 ppm of Earth's crust), its annual production of 8.5 million tons ranks fourth behind Fe, Al, and Cu. Zn's largest use is for protective coatings on steel to retard corrosion and to harden the surface, but it is also used to make brass (Sec. 18.2.3.4) and Zn-based pressure die-casting alloys. Cd is similar to Zn (Table 19.1) but is much less abundant (0.16 ppm) and has far smaller commercial utilization. Cd's largest use is in Ni–Cd rechargeable batteries, but it also finds application in coatings and pigments. Hg is a rare metal (0.08 ppm) with niche uses in liquid electrodes, dental amalgams, precious metals extraction, and batteries. Both Cd and Hg are toxic, and their use poses a broad range of medical and environmental problems.

19.2 ZINC

Zinc is not a glamorous metal.

—S. W. K. Morgan, Imperial Smelting Processes, Ltd.

19.2.1 History of Zn

The first use of metallic Zn was in brass made by smelting mixed Cu and Zn ores. Substantial amounts of brass were being produced by 1400 B.C., although rare examples are found thousands of years earlier. Pure Zn is more difficult to produce than brass because reduction of ZnO by C must be performed above 1000°C. At that temperature Zn is a vapor. Consequently, pure Zn was available to the ancients only in tiny amounts condensed inside smelting chimneys. The dual problems of excluding air from the reaction zone and condensing the evolving Zn vapor were first solved in India in the fourteenth century and shortly thereafter in China. Today Zn is a "workhorse metal"—widely used but unglamorous. Zn is a low-melting, low-stiffness metal with moderate strength (Fig. 19.1). Its primary virtues are its corrosion resistance, low cost, and excellent castability. Metallic Zn is little used in exotic, high-tech engineering applications, but it solves a multitude of more mundane engineering problems as a protective barrier coating on steel, an alloying addition to Cu, in die-cast parts, roofing, and batteries.

19.2.2 Physical Properties of Zn

19.2.2.1 Corrosion. Zn has little reaction with dry air at room temperature, but humid air containing CO_2 forms an adherent gray, hydrated Zn carbonate that inhibits further corrosion.

TABLE 19.1 Selected Room-Temperature Properties of Group 12 Metals with a Comparison to Cu

Property	Zn	Cd	Hg	Cu
Valence	+2	+2	+1, +2	+1, +2
Crystal structure at 20°C	HCP	HCP	Rhombohedral (hR1)	FCC
Density (g/cm^3)	7.14	8.65	13.53	8.92
Melting temperature (°C)	420	321	−39	1085
Boiling temperature[a] (°C)	907	765	357	2570
Thermal conductivity (W/m·K)	113	93.5	8.2	399
Elastic (Young's) modulus (GPa)	70[b]	55	25[c]	130
Coefficient of thermal expansion (10^{-6} m/m·°C)	39.7	30.8	61	16.5
Electrical resistivity (μΩ·cm)	5.9	6.8	95.8	1.7
Cost ($/kg), large quantities	0.80	1.30	5.20	2

[a] Temperature at which vapor pressure equals 1 atm.
[b] Pure Zn is anelastic at room temperature; the 70-GPa elastic modulus is valid only for very brief time intervals.
[c] Liquid at room temperature; bulk modulus.

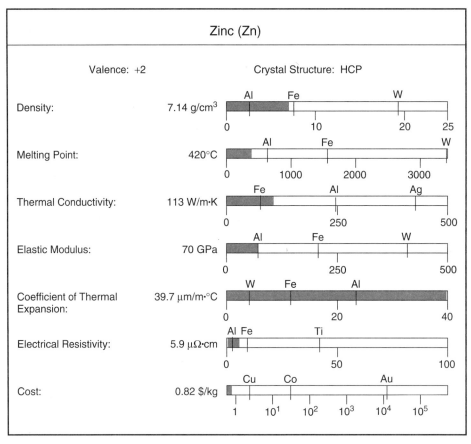

Fig. 19.1 Physical properties of Zn.

Above 200°C, however, Zn oxidizes rapidly. Zn is attacked by most mineral acids (dilute H_2SO_4 being the exception) and alkalis. Zn has a low seawater electrode potential (-1.10 V). Since Fe's electrode potential is higher at -0.68 V, Zn acts as the sacrificial anode in Zn–Fe galvanic couples, making Fe the protected cathode. As a member of the American Galvanizers Association once put it: "Zinc is like the Secret Service for steel; it takes the bullet when corrosion attacks." The largest use of Zn is as a protective coating for steel. The adherent corrosion barrier on Zn's surface coupled with its action as a sacrificial anode when that barrier is scratched or damaged makes it a more effective anticorrosion coating for steel than simple barrier coatings such as paint or nobler metals (Sec. 19.2.3.1).

19.2.2.2 Crystallography and Mechanical Properties. Zn has an HCP crystal structure with a *c/a* ratio (1.856) much higher than the ideal value (1.633). Thus, the interatomic separation in the *c* direction is considerably larger than in the *a* direction, and Zn is highly anisotropic. Bonding within the (0001) plane is much stronger than bonding along the *c* axis. At cryogenic temperatures, Zn single crystals cleave readily to expose (0001) fracture surfaces (Fig. 5.4). The coefficient of thermal expansion (COTE) is 14.3×10^{-6} K^{-1} parallel to the basal plane but 60.8×10^{-6} K^{-1} perpendicular to the basal plane.

At room temperature, high-purity Zn is ductile, but lower-purity Zn is somewhat brittle and difficult to deform without cracking. Commercial-purity Zn (about 99%) has an ultimate tensile strength of about 110 MPa. Room-temperature slip occurs almost entirely on the (0001) $\langle11\bar{2}0\rangle$ slip system, although $\{11\bar{2}2\}$ $\langle11\bar{2}3\rangle$ slip has also been observed (Sec. 19.3.6). Above room temperature, slip also occurs on the ($10\bar{1}0$) prism planes. Deformation is aided by twinning on the ($10\bar{1}2$) pyramidal planes that achieve a net reduction in thickness of 6.75% in the $\langle0001\rangle$ direction and reorient the basal planes at a 94° angle to their original position. Thus, the twinning shear action not only deforms the metal but also repositions slip planes to new positions with more favorable Schmid factors (Sec. 4.6) for continuing slip. Kink bands (Sec. 2.4.2) have also been observed in deformed Zn.

High-purity Zn's low critical resolved shear stress (0.08 MPa) results from its weak bonding between (0001) planes and the fact that room temperature is 43% of the homologous melting temperature. Diffusion is an active process at room temperature, and pure Zn creeps appreciably at ambient temperatures. Values of elastic modulus for Zn at room temperature must be used with caution by design engineers, because the metal responds anelastically to load, much as polymers do. Work hardening begins to anneal away in high-purity Zn in just a few seconds at room temperature, and the metal will recover and recrystallize completely within several hours (Sidebar 19.1). Certain alloying additions (e.g., Cu, Cd), slow recovery and recrystallization, and commercial alloys often are formulated to improve room-temperature creep strength.

19.2.2.3 Zn Alloying Behavior. There are surprisingly few elements with appreciable solid solubility in Zn. Au and Ag dissolve up to about 10 wt% in Zn; Cd, Cu, Mn, and Al have maximum solubilities of a few percent. Zn's solubility in other metals is somewhat greater, but only a few metals will dissolve large amounts of Zn. Zn has great solubility in Al (over 80 wt%), Cu and Fe can dissolve 40% Zn, and Mg and Cd can dissolve several percent Zn. In many other common metals the ability to hold Zn in solid solution is near zero, and some (e.g., Pb) even have large liquid miscibility gaps with Zn. These solubility limitations are presumably attributable to electronic–electronegativity effects since the atomic radius of Zn (0.133 nm) is rather ordinary. The short list of elements with extensive solubility in and for Zn constrains alloying options; however, several useful alloys have been developed to improve mechanical properties and enhance coating performance (Sec. 19.2.3).

19.2.3 Applications of Zn

The engineering applications of Zn are defined largely by its corrosion resistance, low melting point, and low cost. Iron and steel are the dominant metals in modern engineering practice, so

SIDEBAR 19.1: ROOM-TEMPERATURE RECRYSTALLIZATION IN Zn

Zn's low melting point makes room-temperature deformation *hot work*. Work hardening at 20°C nearly doubles the yield strength of Zn single crystals pulled 50% in tension, but after a 30-second pause, about a third of that increase has already disappeared. At temperatures slightly below room temperature, this effect is slowed, allowing the data plotted in Fig. 19.2 to be conveniently obtained. Recovery and recrystallization follow an Arrhenius-type relationship,

$$\frac{1}{\tau} = Ae^{-Q/RT}$$

where τ is the time needed to complete a given fraction of the total yield point recovery (often chosen to be one-half), A is a constant, Q the activation energy for the process, R the gas constant, and T the absolute temperature. If the reaction proceeds to the same extent at two different temperatures, then

$$\tau_1(Ae^{-Q/RT_1}) = \tau_2(Ae^{-Q/RT_2})$$

which can be rearranged to

$$\frac{\tau_1}{\tau_2} = \frac{e^{-Q/RT_2}}{e^{-Q/RT_1}} = e^{(-Q/R)(1/T_2 - 1/T_1)}$$

For the Zn recovery conditions in Fig. 19.2, Q is 20 kcal/mol (84 kJ/mol). Thus, appreciable recovery occurs in seconds at room temperature, minutes at 0°C, and days at −40°C.

Zn's ability to protect ferrous metals from rusting at minimal cost is extraordinarily important. Zn's low melting point makes it easy to apply to steel and cast iron surfaces. Zn coatings on ferrous metals consume half of the world's Zn production. Zn die-casting alloys have strength and cost intermediate between polymers and Al die castings, and this makes them the preferred material for many applications. Over 50 billion Zn die-cast components are produced each year. Zn also sees substantial use as an alloying addition in brass and 7000 series Al alloys.

19.2.3.1 Zn Coatings on Steel. The Fe–Zn phase diagram (Fig. 19.3) shows near-zero solubility of Fe in Zn and several intermetallic compounds. When steel with a well-cleaned surface is submerged in molten Zn at about 450°C (hot-dip galvanizing), the two metals react at the interface. The molten Zn quickly reaches its saturation concentration of Fe (<0.02%). Zn atoms diffuse rapidly into the steel, and Fe atoms diffuse more slowly outward into the surface layers, forming thin layers of intermetallic compounds (Fig. 19.4). These layers not only provide good adherence for the outermost layer of nearly pure Zn, they also increase surface hardness and wear resistance. The Si content of the steel strongly affects galvanizing behavior. Steel containing intermediate levels of Si (0.06 to 0.5%) reacts more rapidly with Zn, and thicker coatings result, but at high Si contents, reactivity between the steel and the molten Zn drops to low levels, producing thin coatings. Zn purity also affects galvanizing behavior. Most hot-dip galvanizing is performed with Zn of 98.5 to 99.5% purity, with the major impurity being Pb. The presence of about 1% Pb in molten Zn lowers the surface tension of the molten Zn against steel by about 40%, which improves Zn's wettability on the steel surface. Small amounts of Al (<0.005%) in the Zn improve the luster of the final product, but larger amounts of Al inhibit formation of the Fe–Zn intermetallic layers and alter the thickness and structure of the coating. Some steel sheet and wire is galvanized by a continuous process in which the steel moves at high speed (up to 3 m/s) through a Zn bath. For continuous galvanizing, higher Al content (0.1 to 0.2%) is

ZINC **341**

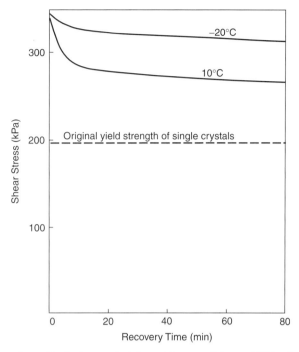

Fig. 19.2 Recovery of pure Zn single crystals deformed by easy glide at −50°C and annealed at the temperatures indicated. (Adapted from Drouard et al., 1953; used with permission of AIME, *www.aimehq.org*.)

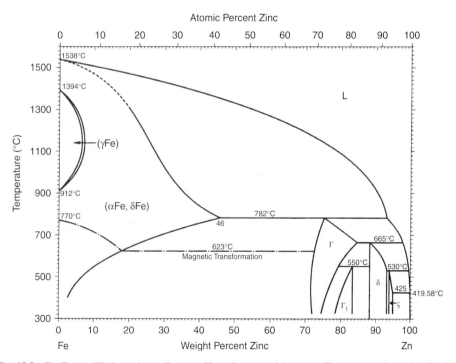

Fig. 19.3 Fe–Zn equilibrium phase diagram. Note the several intermetallic compounds in the Zn-rich regions of the diagram and the near-zero solubility of Fe in Zn. (From Massalski, 1986, p. 1128; with permission.)

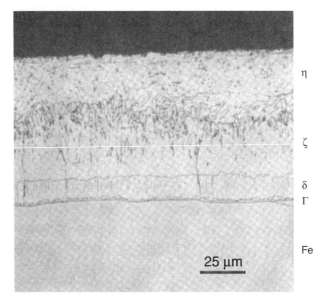

Fig. 19.4 Cross-section micrograph of galvanized, low-Si steel. Three intermetallic layers are visible in this micrograph: η is nearly pure Zn; ζ is monoclinic $FeZn_{13}$; δ is hexagonal $FeZn_{10}$; and the thin Γ layer is a complex blend of intermetallic compounds, Fe_5Zn_{21}, $FeZn_2$, Fe_3Zn_{10}, and $FeZn_3$, not all of which are equilibrium phases. (From Zhang, 1996, p. 8; with permission.)

used in the Zn bath to break up the Zn layer; this prevents the galvanized surface from cracking during later forming operations. The total immersion time for the steel workpiece is only a few minutes for batch processes and considerably less for the continuous process. Zn–55% Al and Zn–5% Al–0.1% mischmetal (Sec. 23.4.1) alloys are sometimes used instead of commercial-purity Zn to produce different coatings for special corrosion environments or to achieve more ductile coatings.

Careful steel surface preparation is crucial to remove surface contamination and assure uniform coverage with the Zn layer. Hot alkaline solutions clear oil and grease from the steel. Heavy oxide scale and adherent mold sand may require grit blasting, but pickling in HCl or H_2SO_4 acid is sufficient to remove lighter oxide films. After removal from the pickling bath, the steel is coated with a $ZnCl_2 \cdot 3NH_4Cl$ flux to remove thin FeO and ZnO surface layers from both metals and promote rapid wetting of the steel surface as the part is immersed in the molten Zn. In continuous galvanizing, the steel's surface preparation involves many of the same solutions, adapted for use "on the fly." In some continuous galvanizing processes, the steel is preheated in a furnace with a reducing atmosphere ($H_2 + N_2$) to remove the oxide layer.

Electroplating can be used in place of hot dipping to apply Zn to steel. Galvanizing steel body panels for automobiles poses special challenges because the Zn surface must be very smooth to be "paintable." Ordinary hot-dip galvanizing does not meet this requirement, so electroplating with alloys of Zn–Ni, Zn–Fe, or other Zn alloys is used. By adjusting solution pH and current density, electroplated layers of Zn containing 9 to 13% Ni or Zn containing about 18% Fe are deposited. Zn coatings can also be applied by oxyfuel spray of liquid Zn onto the steel surface. Electrolytic and thermal spray coatings usually are less effective than hot-dip processes at protecting steel, so additional protection can be imparted by treating the surface with phosphoric acid to form a crystalline layer of insoluble $Zn_2Fe(PO_4)_2$ phosphate ranging from 3 to 50 μm thick. This improves corrosion performance and paint adherence. Chromate coatings were once heavily used to enhance the corrosion protection provided by the Zn layer; however, chromate coatings use has declined due to safety concerns about Cr solutions (Sec. 14.2.4).

During the first weeks after galvanized metal is placed into service, a protective Zn carbonate surface layer forms. The hard intermetallic layers make the metal more resistant than bare steel to mechanical damage, but even if the coating is cut or scratched, the Zn surrounding the damaged site acts as a sacrificial anode, preferentially oxidizing while protecting the exposed steel from rust. Only when the Zn near the damaged site has been consumed completely does the steel begin to rust. Galvanizing can prevent rust on steel exposed to the weather for periods of 20 years or more on sites with low air pollution. However, in humid atmospheres containing SO_2, sulfurous acid (H_2SO_3) forms on the surface, attacking the Zn carbonate layer and shortening the protective life of the galvanizing layer.

Steel can also be protected from corrosion by attaching massive pieces of Zn metal to surfaces such as ships' hulls and undersea pipelines. The Zn acts as a sacrificial anode, corroding slowly while protecting the surrounding steel (Sec. 11.4.3.5). For this application, the Fe content of the Zn should be kept below 0.0015% to avoid forming impervious corrosion product layers on the Zn that reduce its effectiveness. Still another corrosion protection strategy is the use of paint containing high volume fractions (92 to 95 vol% in the dried paint) of Zn metal dust. These paints offer much of the galvanic protection of hot-dip galvanizing and are less expensive to apply to large parts, but their adherence and service lifetimes are inferior to those of hot-dip treatment, and they do not increase surface hardness because no Zn–Fe intermetallics form.

19.2.3.2 Zn as an Alloying Addition to Other Metals. The solubility of Zn in other metals is generally low, but Zn does have extensive solubility in the two most important nonferrous metals, Al and Cu. Brasses are Cu–Zn alloys with Zn content ranging from a few percent up to nearly 50%. Brass was among the first alloys used by humankind, and brass remains widely popular today for its strength, easy formability, corrosion resistance, attractive color, and moderate cost (Sec. 18.2.3.4). About 10% of the world's Zn production is consumed by the brass industry. Al has great solubility for Zn (Fig. 19.5), and Zn acts as a precipitation hardener in the 7000 series Al alloys based on the Al–Zn–Mg ternary system (Sec. 20.2.4.6). These alloys contain up to 6 wt% Zn and have the highest strengths of all the commercial Al alloys. They are widely used in the aircraft and automotive industries. Zn is also a useful strengthener in Mg-based alloys (Sec. 11.4.3.1) and in Sn, Ag, Au, and solder alloys.

19.2.3.3 Zn Die-Casting Alloys. The first attempts to make die-cast parts (Sec. 8.2.9) from Zn alloys were a miserable failure. In the early twentieth century, commercial-purity Zn was used for die casting, and although the manufacture of the products proceeded well enough, their service life was often as short as a few weeks, due to ruinous intergranular corrosion. Sn and Pb impurities are nearly insoluble in solid Zn, and their precipitation on grain boundaries in the castings caused rapid galvanic corrosion in moist environments. In the 1920s the industry began using Zn of 99.99+% purity to formulate die-casting alloys, which solved the intergranular corrosion problem. Today, Zn pressure die casting produces components that are stronger than injection-molded polymers and less expensive than Al and Mg die castings.

The most widely used pressure die-casting alloys are based on the eutectic at 381°C in the Al–Zn system (Fig. 19.5). Al reduces the rate of attack of molten Zn on steel dies and plungers. Al contents greater than 4.3 wt% are usually avoided since fracture toughness drops with Al contents near the eutectic. Cu is sometimes added (0.25 to 3.5%) to Zn–Al die-casting alloys to increase strength on aging, but Cu lowers fracture strength and causes greater dimensional change during postcasting aging. Mg (0.08% maximum) is often added to reduce the intergranular corrosion effect from tramp impurities such as Pb, Sn, In, As, Sb, Bi, and Hg.

Rapid solidification occurs when the liquid alloy is injected under pressure into the water-cooled steel mold. Since the commercial alloys are slightly hypoeutectic, the first solid to form is pro-eutectic β phase (Zn with a small amount of Al in solid solution). The remaining liquid segregates during freezing into Zn-rich β phase and Al-rich α phase (Fig. 19.6). After casting, the alloy gradually ages at room temperature, precipitating Al-rich α phase within the β regions to accommodate the 0.07 wt% maximum solubility of Al in Zn at room temperature. This

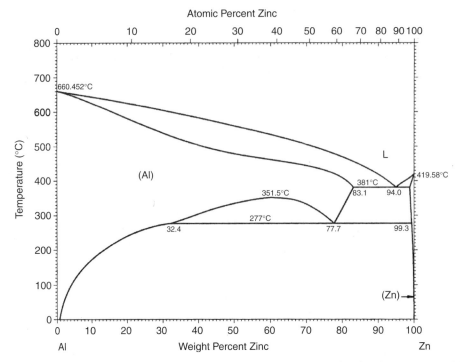

Fig. 19.5 Zn–Al equilibrium phase diagram. Note the large solid solubility of Zn in Al. Most commercial Zn die-casting alloys are hypoeutectic, containing 3.5 to 4.3% Al. (From Massalski, 1986, p. 185; with permission.)

precipitation strengthens the alloy and causes slight shrinkage (about 1 mm/m) over a period of several months after casting.

Pressure die casting with Zn–Al alloys is well suited to produce intricate shapes and large production runs. Zn die castings are often used for frames for electronic devices, complete with holes cast with internal threads. Designs that formerly contained several parts can sometimes be replaced with a single Zn die casting, simplifying assembly operations and lowering costs. Shrinkage factors are considerably smaller than in Al die castings, and the lower temperatures needed for Zn alloys reduce energy costs and prolong die life. Yield strengths of Zn–4% Al die castings are typically 220 to 240 MPa; ultimate tensile strengths range from 280 to 380 MPa, with the higher strengths corresponding to Cu-containing alloys. Tensile elongation is typically 8 to 16%. Fracture toughness is relatively good at room temperature but decreases sharply between 10°C and −10°C. Creep performance of these alloys is much better than that of high-purity Zn; however, load-bearing Zn–Al structures still creep at 25°C (Fig. 19.7). Development work shows promise for substantial improvement in creep performance for Zn casting alloys. At 140°C under 31-MPa load, experimental Zn die-casting alloys (5.5 to 7% Al, 2.4 to 3.6% Cu, 0.75 to 1% Ti, and 0.0004 to 0.004% B) have achieved 10 to 20 times lower creep rates than commercial Zn–4% Al alloys.

Although the Zn–4% Al alloys are the most heavily used, hypereutectic Zn–Al alloys with Al contents of 8, 12, and 27 wt% are becoming increasingly popular for die-cast and sand-cast components requiring higher strength and wear resistance. These high-Al alloys have ultimate tensile strengths of 350 to 440 MPa, although their ductility is somewhat lower than Zn–4% Al. Zn–Al alloys with higher-Al contents are more difficult to die cast due to their higher melting temperatures, but their densities are lower than Zn–4% Al (about 6.7 g/cm^3) and their strengths are higher.

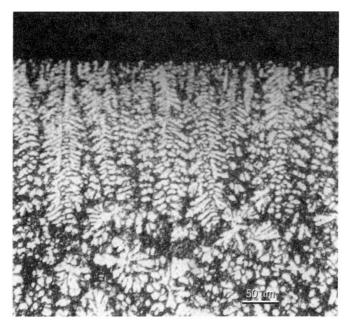

Fig. 19.6 Typical dendritic microstructure of a die-cast Zn–4% Al alloy (Mazak 3). (From Morgan, 1985, p. 181; with permission.)

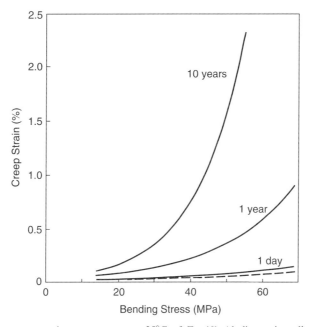

Fig. 19.7 Bending creep strain versus stress at 25°C of Zn–4% Al die casting alloy (Alloy 3) with composition 3.8 to 4.3% Al, 0.25% maximum Cu, 0.02 to 0.06% Mg, 0.10% maximum Fe. The dashed line is the instantaneous elastic response to loading. (Adapted from Kelton and Grissinger, 1946.)

19.2.3.4 Wrought Zn Products. About 10% of all Zn is fabricated by rolling into plate, sheet, rod, wire, and foil. Zn with 1% Pb and 0.05% Cd is rolled into thick sheet and formed into battery cans for C–Zn dry-cell batteries. A Zn–0.8% Cu alloy plated with 6 μm of Cu is rolled and stamped to produce the U.S. one-cent coin. Zn foil is applied by adhesive to ferrous structures too large to galvanize by hot dipping. Wire is used for Zn thermal spray. The most popular Zn alloy for roofing, gutters, and flashings (0.7 to 0.9% Cu, 0.08 to 0.14% Ti) is stronger and much more creep resistant than commercial-purity Zn. Zinc plate and sheet are formed in a continuous casting/hot rolling operation. The metal is strengthened by Cu in solid solution and by Zn–Cu ε phase (Fig. 18.5) and $TiZn_{15}$ intermetallics. The intermetallics form at grain boundaries during solidification (Fig. 19.8a) but break up during rolling to produce long strings of fine particles (Fig. 19.8b). The hot rolling induces dynamic recrystallization, distributing $TiZn_{15}$ particles on recrystallized grain boundaries; these particles are sufficiently fine and discontinuous to avoid intergranular corrosion. Zn–Cu–Ti alloy given suitable rolling and annealing treatment can sustain a load of 115 MPa at room temperature with creep strain of only 1 to 2% per year, a creep performance about two orders of magnitude better than that of pure Zn and nearly as good as commercial-purity Al.

19.2.3.5 Zn Compounds. About 10% of Zn production is used to produce Zn compounds. ZnO finds uses in such diverse fields as rubber and polymer vulcanization, ceramics, dehydrogenation catalysts, agricultural soil conditioners, cosmetics, sunscreen lotions, and paper manufacture. $ZnSO_4$ is the electrolyte in Zn plating operations and is also used to treat Zn deficiency in crops. ZnS containing trace activator elements such as Tb, Mn, and Cu is a fluorescent material that produces yellows and greens in lights and computer–television displays. $ZnCl_2$ acts as a mordant for textile dyes, and it also dissolves many metal oxides, making it useful in metallurgical fluxes. $Zn_3(PO_4)_2$ is used to reduce corrosion by passivating steel surfaces and is often added to metal primer paint. Zn salts of stearic acid ($C_{17}H_{35}COOH$) are used as plasticizers and stabilizers in the polymer industry and to accelerate rubber vulcanization.

19.2.4 Zn in the Biosphere

Zn is essential to plant and animal life. It is present in many enzymes involved with growth, tissue repair, sexual development, and proper immune system response. In humans, daily intake of 15 mg of Zn is recommended in the form of Zn oxide, carbonate, or sulfate. Zn-deficiency is a common problem in pregnant and lactating women, the elderly, alcoholics, and athletes (Zn is lost through perspiration) and also among persons with heart or liver disease, ulcers, and some cancers. The onset of puberty can be delayed for several years by Zn deficiency. Severe Zn deficiency causes deformed, discolored fingernails and stretchmarks and wrinkles

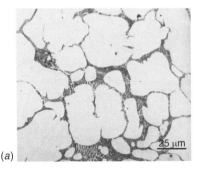
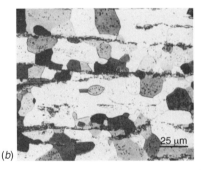

Fig. 19.8 Zn–1% Cu–0.1% Ti wrought alloy (*a*) as-cast and (*b*) after hot rolling to break up $Zn_{15}Ti$ and ε-ZnCu intermetallic phases and distribute them throughout a recrystallized grain structure. (From Morgan, 1985, p. 174; with permission.)

in the skin. Patients consuming the recommended 15 mg of daily Zn can still be Zn deficient since substances found in coffee, tea, alcohol, milk, whole grains, and vegetables bind with Zn, forming complexes that cannot be metabolized. Zn-deficiency problems also occur in livestock and crop production. Corn, beans, and fruit trees are particularly sensitive to inadequate supplies of available Zn in the soil.

Excess Zn intake has also been reported from two principal sources: (1) inhalation of Zn fumes in the workplace, and (2) contamination of drinking water, often from acidic beverages mixed in galvanized containers. Molten Zn must be used in well-ventilated areas to avoid *zinc fume fever*, a condition with symptoms similar to a severe cold or influenza. Zn overdoses are debilitating for one to two days, but cause no permanent injury. Welders working on galvanized steel in poorly ventilated workspaces are susceptible to inhaling Zn vapor. High Zn content gives beverages a metallic taste and a milky appearance, but these are sometimes masked in fruit drinks. Zn contamination in food or beverages is seldom lethal, since high concentrations induce vomiting. Zn_3P_2 is used in rodenticides that kill by forming phosphine gas (PH_3) in the animal's gut. This becomes entrained in inhaled air, asphyxiating the animal by a bizarre sort of "death by burping."

19.2.5 Sources of Zn and Refining Methods

The principal Zn ore is ZnS, which is usually found mixed with Pb and Cd minerals. Minerals containing Cu, Ag, and Mn are often associated with Zn. Zn ores are typically complex mixtures of several commercially valuable minerals, so crushing, grinding, and froth flotation are necessary to separate the various ores. ZnS is difficult to reduce, so it is roasted to convert it to ZnO. C will reduce ZnO to metallic Zn, but temperatures above Zn's boiling point are required, producing Zn vapor. Retorts reduce ZnO with C and condense the Zn vapor (Sidebar 19.2), achieving purities of about 98.5%, which is satisfactory for hot-dip galvanizing. Exclusion of air is essential to minimize reoxidation of the Zn vapor. The retort process for reducing ZnO was augmented in the early twentieth century by blast furnace, electrolytic, and electrothermic processes, the latter process involves withdrawing Zn vapor produced in an electrically heated reaction zone through a pool of liquid Zn, using the liquid metal to condense the Zn vapor. The electrolytic process gradually became the dominant method for ZnO reduction and now accounts for more than 80% of world production. ZnO is treated with H_2SO_4 to produce $ZnSO_4$, which is then electrolyzed to deposit high-purity Zn metal (99.95+% purity) on Al cathodes. One might expect that Zn metal could not be produced by aqueous electrolysis since the standard electrode

---

SIDEBAR 19.2: CONVERTING WASTE Zn DUST INTO USEFUL PRODUCTS

The condensation of Zn vapor during reduction and distillation produces a certain fraction of fine Zn dust (sometimes called *blue powder*). The dust particles are coated with ZnO, which makes it difficult to sinter or melt the powders into bulk Zn. Blue powder was once considered an undesirable waste product but has since found so many uses that some plants now modify their processes to produce larger quantities. Zn dust can also be produced by remelting, vaporizing, and condensing Zn scrap. It is the "active ingredient" in corrosion-resistant paints for steel (Sec. 19.2.3.1) and is often added to bronze castings to deoxidize them. Zn dust precipitates Au and Ag from cyanide solutions in precious metals production (Sec. 18.3.4). Blue powder also serves as the electrode in alkaline batteries (Sec. 15.3.2.4) and in Ag–Zn batteries (Sec. 18.3.3.6). Coarser Zn powders are used in explosive primers, tear gas canisters, smoke grenades, and fireworks. The petroleum industry uses Zn dust as a catalyst and condensing agent, and Zn powder makes an excellent starting material for production of several Zn compounds used in bleaching textiles and paper and in producing nitrobenzene.

potential of Zn is −0.763 V, which indicates that H_2 evolution should occur in preference to metal deposition. However, a hydrogen overvoltage is maintained at the Zn electrode, causing a voltage shift large enough to counteract the low standard electrode potential of Zn; this allows metal to form. The electrolytic process is quite sensitive to impurities in the electrolytic solution. Low concentrations must be maintained for Fe and more electropositive impurities (e.g., Cu, As, Sb, In, Te, Ge); these elements can lower the hydrogen overvoltage, causing the deposited Zn to redissolve.

About 20% of Zn requirements are met by recycling. This figure is lower than recycling percentages in the Al and Cu industries because Zn cannot be economically recovered from galvanized steel.

19.2.6 Structure–Property Relations in Zn

19.2.6.1 Superplastic Zn–Al. The eutectiod point at 77.7% Zn in the Zn–Al system (Fig. 19.5) can be exploited to produce an exceptionally fine-grained material that becomes superplastic (Sec. 6.6) between 200 and 275°C. In the superplastic condition, the alloy achieves tensile elongations of about 1000% and can easily be formed into complex shapes by the same methods as those used for thermoplastic polymers (e.g., blow molding, vacuum forming, deep drawing). Commercial superplastic alloys typically contain 22% Al, 0.4 to 0.6% Cu, and up to 0.03% Mg. The metal is prepared for superplastic forming by rapid cooling from just above the eutectoid temperature to produce a two-phase microstructure with a 1-μm average grain size (Fig. 19.9a). This grain size is an order of magnitude smaller than the grain sizes seen in ordinary metals, so the Zn–Al eutectoid metal has an exceptionally high grain–phase boundary area. In addition, the Al-rich phase is strengthened by Zn precipitates smaller than 100 nm (Fig. 19.9b). The high boundary area permits the metal to deform between 200 and 275°C by grain boundary sliding rather than by normal slip mechanisms. Superplastic metal shows equiaxed grains both before and after heavy deformation, which indicates that grain boundary sliding is occurring rather than slip, which would flatten or distort the grains. After the metal has been formed superplastically, it can be heated to 350°C and slow cooled to return it to a normal, coarse-grained state. This coarse-grained material has $\sigma_{0.2\% \, offset}$ = 385 MPa, σ_{UTS} = 440 MPa, 9% elongation, Young's modulus = 68 GPa, and the ability to creep less than 0.01% under a 55-MPa load at 25°C for 1000 hours. These properties are superior to polymer materials. The easy formability and good mechanical properties of Zn–22% Al superplastic sheets allow fabrication of such parts as ballpoint pen and lipstick cases in one-step, deep-drawing operations.

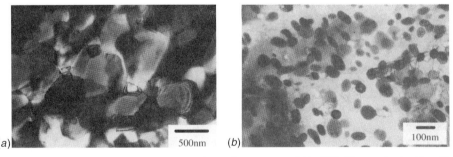

Fig. 19.9 Transmission electron micrographs of the microstructure of Zn–22% Al superplastic alloy. The average grain size in (a) is 1.3 μm; the microstructure is a mixture of Al- and Zn-rich phases. (b) Zn-rich precipitates with an average size below 100 nm are widely distributed within the Al-rich phase. (From Tanaka et al., 2002; with permission.)

19.3 CADMIUM

> Because of their tendency to blacken in the presence of sulphides, the chrome yellows were abandoned by most of the Impressionists toward the end of the 1870s. Monet replaced them with the more stable cadmium yellows.
>
> —Nicolas Pioch, commentary on Claude Monet's "Bathing at La Grenouillère,"
> *http://www.ibiblio.org/wm/paint/auth/monet/early/bathing/*, December 2003

19.3.1 History of Cd

Compared to its sister element Zn, Cd is a relative newcomer in commerce. Cd metal was first produced in 1817, but during the following century Cd was used for little more than paint pigments. Not until the mid-twentieth century did Cd begin to be widely exploited in batteries, electroplating, pigments for polymers, alloying additions to other metals, and polymer stabilizers. With this greater use came increased awareness of Cd toxicity hazards, which has motivated recent measures to curtail Cd use.

19.3.2 Physical Properties of Cd

Like Zn, Cd has low melting and boiling points, an HCP crystal structure with a high c/a ratio, and it is anodic to steel in a galvanic couple. Cd is much less abundant than Zn; world Cd production is about 500 times smaller than Zn production.

19.3.2.1 Oxidation and Corrosion. Cd is stable in room-temperature air and water, protected by a thin, adherent layer of CdO. When heated in air, the oxide layer thickens and turns a yellowish-brown color. When heated to a few hundred degrees, enough Cd vapor is released to pose a toxicity hazard (Sec. 19.3.4). If heated to its boiling point (765°C) in air, the Cd vapor oxidizes with a reddish-yellow flame, dispersing as aerosol of potentially lethal CdO particles. Cd resists attack by aqueous alkali solutions, but it dissolves slowly in HCl and H_2SO_4 acid and is rapidly attacked by HNO_3 acid. Cd has a lower seawater electrode potential (−0.77 V) than Fe (−0.68 V), so Cd coatings act as a sacrificial anode on steel just as Zn coatings do.

19.3.2.2 Crystallography and Mechanical Properties. Cd is a soft, ductile HCP metal with a c/a ratio of 1.886. At room temperature, Cd's primary slip system is (0001) $\langle 11\bar{2}0 \rangle$. This basal slip occurs at a low τ_{CRSS} (<0.1 MPa) and is accompanied by twinning. The "tin cry" of twin formation is clearly audible when a piece of pure Cd is bent at room temperature. Secondary slip (τ_{CRSS} = 7 MPa) has been observed on $\{10\bar{1}1\}\langle 11\bar{2}0 \rangle$ and $\{11\bar{2}2\}\langle 11\bar{2}3 \rangle$ systems (Sec. 19.3.6). Cd of 99% purity has σ_{UTS} = 95 MPa with 45% tensile elongation. Room temperature is half the absolute melting temperature of Cd, so plastically deformed Cd recovers and recrystallizes rapidly at 20°C. This contributes to the metal's malleability, but it also means that pure Cd creeps substantially under load at room temperature. The activation energy for creep in Cd at room temperature closely matches the activation energy for self-diffusion in Cd. This suggests that nonconservative motion of jogs on screw-character dislocations (Sidebar 4.1) may play a major role in Cd's creep mechanisms. Above room temperature, several different twinning reactions occur in Cd.

19.3.3 Applications of Cd

19.3.3.1 Batteries. Ni–Cd rechargeable batteries were introduced to the consumer market in 1961 and now account for 70% of all Cd use. These compact, relatively inexpensive batteries power a wide range of consumer products (e.g., portable hand tools, radios, razors) with the following reaction:

$$\text{charged} \rightarrow \text{discharged}$$

$$2\text{NiOOH} + \text{Cd} + 2\text{H}_2\text{O} \rightarrow 2\text{Ni(OH)}_2 + \text{Cd(OH)}_2$$

During charging, O_2 is produced at the positive electrode in the Ni–Cd cell by the side reaction $4\text{OH}^- \rightarrow 2\text{H}_2\text{O} + \text{O}_2 + 4e^-$, and this O_2 migrates to the negative electrode to be consumed by the reaction $\text{O}_2 + 2\text{H}_2\text{O} + 2\text{Cd} \rightarrow 2\text{Cd(OH)}_2$. No appreciable O_2 pressure builds up during charging, so the cell can be sealed, making the battery more robust and versatile. Ni metal hydride batteries (Sec. 16.3.3.8) and Li ion batteries are lighter and more compact than Ni–Cd batteries, but they are also more expensive and have captured only a portion of the rechargeable battery market.

19.3.3.2 Electroplating. About 8% of Cd is used to electroplate steel, particularly fasteners, for corrosion protection. In many ways, Cd electroplating is similar to Zn electroplating (Sec. 19.2.3.1); however, Cd-plated surfaces have a lower electrical contact resistance, a lower coefficient of friction, and provide better corrosion protection in marine environments ($CdCl_2$ is less water soluble than $ZnCl_2$). The low-friction quality of electroplated Cd allows easier engagement and disengagement of threaded assemblies; this makes thread "jamming" less likely, even after months or years in one position. Low contact resistance combined with good solderability make Cd electroplating useful in the electronic, aerospace, and automotive industries. The major disadvantages of Cd electroplating compared to Zn electroplating are Cd's higher cost and its toxicity.

19.3.3.3 Cd as an Alloying Addition. Small quantities of Cd are used as alloying additions to other metals. Pure Cu will soften from recovery and recrystallization if heated for soldering or used at service temperatures above 100°C; however, Cu containing about 1% Cd maintains its work hardening during soldering and during prolonged service up to 150°C. Cu–Cd alloys are often used to make radiators and overhead wire supplying power to electric trains. Addition of 1% Cd to Cu decreases electrical conductivity by only 10%. Ag containing 10 to 15% Cd or CdO is less susceptible to arcing and erosion in electrical switch contacts. Cd is also present in certain fusible alloys (Sidebar 19.3), bearing alloys, and solders. Cd is a strong absorber of thermal neutrons, and an 80% Ag–15% In–5% Cd alloy is widely used for neutron absorption

---

SIDEBAR 19.3: FUSIBLE ALLOYS SOLVE MACHINING AND FABRICATION PROBLEMS

The low melting points of many Cd-containing ternary, quaternary, and higher-order alloys (Table 22.2) allow them to solve special problems in machining and fabrication. Bending thin-walled tubing often crimps or collapses the tube walls, but this can be avoided by filling the tubing with molten fusible alloy, bending the tubing after the alloy solidifies, then heating the tubing to melt the fusible alloy so that it can be reused. These alloys are weak at room temperature, so they do not greatly increase the force required to bend the tubing. Its presence distributes the force applied to the tubing and produces a smoother, more uniform bend in the tube wall. Fusible alloys can also aid in clamping delicate or irregularly shaped parts for machining so they will not be scored or bent by clamping jaws. When the part is "cast" into the fusible alloy, it is held firmly without risk of damage. Fusible alloys that expand on freezing grip the embedded part tightly. When the machining is finished, the fusible alloy is easily melted away.

in nuclear reactor control rods. Addition of up to 5% Cd to Au–Ag alloys gives the metal a greenish hue known to jewelers as *Greek gold*.

19.3.3.4 Cd Compounds. Twenty percent of Cd is utilized as compounds rather than metal. CdO catalyzes various organic reduction–oxidation, polymerization, and hydrogenation–dehydrogenation reactions. CdS and other Cd compounds are used as pigments in artists' paints, producing brilliant and stable reds, oranges, and yellows. Small concentrations are used as pigments and stabilizers in polymers (particularly in polyvinyl chloride). Cd sulfide, tungstate, borate, and silicate are light-emitting phosphors activated by electron beams that can be used in cathode ray tubes, x-ray instruments, luminescent dials, and fluorescent lights. CdS, CdS_xTe_{1-x}, CdSe, and CdS–InP are used to make semiconducting photocells for photographic exposure meters, light sensors in photocopiers, and power-generating cells for solar-powered systems. Although Cd compounds are used in a wide array of products, their inappropriate disposal or incineration poses toxicity hazards (Sec. 19.3.4).

19.3.4 Toxicity of Cd

Cd is not a nutrient required for either plants or animals. Trace amounts of Cd in food (10 to 20 µg/day), water, and air are thought to pose no health threat. However, long-term doses of Cd only moderately higher than the normal "background dose" cause measurable physiologic damage, including Cd accumulation in kidney and liver tissue that can lead to hypertension, kidney stones, and other renal problems. Cd also causes Ca loss in bone that can lead to osteoporosis. Zn and Fe are essential elements in enzymes, blood formation, and other processes in the human body. In persons with the relatively common conditions of Zn or Fe deficiency (Sec. 19.2.4), the body attempts to substitute Cd for the needed Zn or Fe, which substantially worsens the degree of injury. Inhalation or ingestion of 10 to 15 mg of Cd can cause lung or gastrointestinal disturbances, and doses of several hundred milligrams can be fatal. Chronic effects of long-term inhalation or ingestion are irreversible and include bone degeneration, renal tubular dysfunction, and emphysema. There are statistical correlations between long-term Cd exposure and prostate cancer.

Cd can enter the body from drinking water contaminated by Cd-bearing mine wastes, incineration of Ni–Cd batteries, Cd impurities in galvanizing Zn on water pipes, cigarette smoke, industrial processing of Zn and Cd, and coal-burning power plants. Cd has an unusually low boiling temperature, and welding or soldering Cd alloys can release large amounts of Cd vapor; several deaths have resulted from such incidents. The most famous case of Cd poisoning was the Japanese Itai-Itai disease, caused by irrigating rice paddies with Cd-bearing mine wastewater. This caused thousands of injuries (Fig. 19.10) and numerous fatalities over a period of years. The severity of injury was apparently worsened by widespread Zn and Fe deficiencies in the victims.

Rising concern about Cd toxicity prompted more stringent regulation of workplace Cd exposure and Cd release to the environment during the past quarter century. The U.S. Environmental Protection Agency and the European Community have both discussed further restrictions or bans on some Cd uses. Liability concerns motivated some former Cd producers to end their Cd sales. The remaining Cd producers are resisting efforts to eliminate or reduce Cd use, arguing that present control and recycling procedures have already sharply curtailed Cd releases to the environment, making bans unnecessary. Replacing rechargeable Ni–Cd batteries with non-rechargeable batteries could greatly increase the quantity of other types of battery materials in landfills and incinerators. Moreover, since Cd is present in Zn, Cu, and Pb ores, a complete ban on Cd use would require some means to store or stockpile the Cd now sold as a by-product from production of these metals.

19.3.5 Sources of Cd and Refining Methods

Cd is produced as a by-product of Zn, Cu, and Pb refining. The sulfate solutions used for electrolytic Zn production (Sec. 19.2.5) typically contain small amounts of $CdSO_4$ that must be

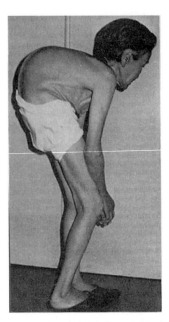

Fig. 19.10 Patient suffering from Cd poisoning (Itai-Itai disease). Her injuries resulted from Cd-contaminated mine wastewater used to irrigate rice crops in Japan in the 1950s. Cd poisoning can cause irreversible skeletal damage and chronic pain, resulting in spinal column collapse and bone fractures from slight stress, as in coughing. Cd also causes kidney and liver damage. (From Friberg et al., 1971, p. 134; with permission.)

removed from the solution to achieve proper electrolysis for Zn production. Zn dust is added to the Cd-containing solution to precipitate a sludge of reduced Cd (and Cu) metal on the Zn particles. This sludge is separated from the $ZnSO_4$ solution and further processed to remove Cu and other impurities prior to electrolytic reduction to Cd metal. Smaller amounts of Cd are also produced by distillation from pyrometallurgical Zn and by collection of flue dusts from roasting and smelting of certain Cu and Pb ores.

19.3.6 Structure–Property Relations in Cd

19.3.6.1 Slip in Cd and Zn. Cd and Zn were the first HCP metals observed to slip in a direction other than the densest-packed $\langle 11\bar{2}0 \rangle$ direction. Cd and Zn can be made to slip on the $\{11\bar{2}2\}\langle 11\bar{2}3 \rangle$ (Fig. 11.2) if the Schmid factor (Sec. 4.6) for basal slip is near zero. This finding created quite a buzz among mechanical metallurgists when it was reported in 1957, because the conventional wisdom had been that only $\langle 11\bar{2}0 \rangle$ slip could occur in HCP metals. Since that time, however, slip on the $\{11\bar{2}2\}\langle 11\bar{2}3 \rangle$ has been observed in several other HCP metals (Be, Mg, Co, Zr, Hf), usually with rather high τ_{CRSS} values (Sec. 11.3.2.2).

More recent study has shown that slip on the $\{11\bar{2}2\}\langle 11\bar{2}3 \rangle$ system in Cd, Mg, and Zn undergoes an anomalous increase in τ_{CRSS} with increasing temperature. When tensile stress is applied parallel to a Cd single crystal's c axis, the Schmid factor on the primary and secondary slip systems [i.e., the (0001) $\langle 11\bar{2}0 \rangle$ and $\{10\bar{1}1\}\langle 11\bar{2}0 \rangle$] is zero. In that orientation, only the $\{11\bar{2}2\}\langle 11\bar{2}3 \rangle$ slip system has a nonzero Schmid factor, and the specimens show a maximum τ_{CRSS} value at about 425 K (Fig. 19.11). This behavior contrasts sharply with the usual steady decline in τ_{CRSS} as temperature increases (shown by the dashed line in Fig. 19.11 and also in Figs. 10.2 and 11.3). The τ_{CRSS} maximum in Cd is believed to result from dissociation of edge dislocations (beginning at about 150 K) on the $\{11\bar{2}2\}\langle 11\bar{2}3 \rangle$ into a glissile (i.e., able to move) dislocation with its slip direction parallel to the basal plane and a sessile (i.e., unable

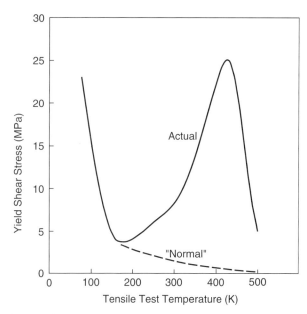

Fig. 19.11 Anomalous maximum in single-crystal Cd shear yield stress. Cd crystals were oriented to produce low Schmid factors on the primary slip plane (0001), so these data indicate ease of dislocation motion on secondary slip systems. The τ_{CRSS} of most metals decreases steadily with increasing temperature as shown by the dashed curve labeled "Normal." Secondary slip in Cd single crystals deviates sharply from the dashed curve, due to "locking" of dissociated dislocations operating in the slip system. (Data from Ucar, 1999, and Tonda and Ando, 2002.)

to move) dislocation with its slip direction perpendicular to the basal plane. This situation "locks" the edge portions of the $\{11\bar{2}2\}\langle11\bar{2}3\rangle$ dislocation loops and leaves the screw portions of the $\{11\bar{2}2\}\langle11\bar{2}3\rangle$ dislocation loops able to move only by a thermally activated double-cross-slip process, which requires temperatures above 400 K.

19.4 MERCURY

"The Dormouse is asleep again," said the Hatter, as he poured a little hot tea on to its nose.

—Lewis Carroll, "A Mad Tea-Party," *Alice's Adventures in Wonderland,* Holt, Rinehart and Winston, New York, 1961

19.4.1 History of Hg

Humanity's contact with mercury has been long and fraught with problems. Although the Mad Hatter's deranged behavior in Lewis Carroll's story was fictional, Hg has caused very real injuries to miners, hatters, and many other users since the first Hg metal was refined over 2500 years ago. Hg is easily separated from its ore, and its liquid state at room temperature is both fascinating and useful, but all told, the damage wrought by its toxicity (Sec. 19.4.4) probably outweighs its value for applications such as batteries, dental amalgams, precious metals recovery, and lighting. Hg is a metal in serious decline. World Hg production peaked at 10,400 tons in 1971, but concerns about its toxicity have driven production down to only 1490 tons in 2001.

19.4.2 Physical Properties of Hg

Hg's most salient characteristics are its low melting and boiling temperatures. Its ionization energy is the highest of all metals, and its electrical and thermal conductivities are about 50

354 Zn, Cd, AND Hg

times lower than those of Cu. Hg is a relatively noble metal that oxidizes only slightly at room temperature. Many other metals dissolve readily in liquid Hg, forming amalgams with a variety of engineering uses. Even at room temperature, Hg's vapor pressure is appreciable, and this volatility, combined with its toxicity, make Hg a potentially dangerous element to handle.

19.4.2.1 Ionization Energy and Crystal Structure. Hg's filled $5d$ and $6s$ subshells make the atom electronically stable. Hg's first ionization energy is high; 1007 kJ/mol is needed to remove one of the $6s$ electrons from each atom, leaving little net energy left to form bonds with neighboring atoms. Metallic Hg has a net bonding energy of only 68 kJ/mol [much lower than those of Al (324 kJ/mol) and W (849 kJ/mol)]. This bonding energy is too low to stabilize any crystal structure at room temperature, and the rhombohedral (hR1) crystal that finally forms below $-39°C$ has mixed metallic–covalent bonding.

19.4.2.2 Vapor Pressure of Liquid Hg. Hg's boiling point is well above room temperature, but significant amounts evaporate under ambient conditions. The saturation vapor pressure of Hg rises rapidly with increasing temperature (Table 19.2). Hg vapor is toxic, and sufficient vapor enters the air over liquid Hg at 20°C to injure persons breathing that air for prolonged periods (Sec. 19.4.4); the level of risk is greatly magnified if the Hg is warmed.

19.4.2.3 Oxidation and Chemical Reactivity. Hg shows little oxide formation when left exposed to air at room temperature. When heated, HgO forms more rapidly, but large amounts of oxide do not accumulate until the metal is near its 357°C boiling point. Above 400 to 500°C, HgO decomposes again to Hg vapor and O_2, making Hg ore easy to reduce. At 20°C, Hg reacts slowly with dilute HCl and H_2SO_4 acids, but it dissolves in HNO_3 acid and in warm, concentrated HCl and H_2SO_4 acids. Hg is slightly soluble in pure H_2O (60 µg/L). The relative inertness of Hg to many chemicals allows it to be used as a liquid electrode in some industrial and laboratory electrolytic separations (Sec. 19.4.3).

19.4.2.4 Alloying Behavior with Other Metals. Liquid Hg dissolves many other metals, and its high electronegativity promotes formation of numerous intermetallic compounds. Liquid solutions of Hg with other metals are called *amalgams* (Sibebar 19.4). Heavy metals tend to form amalgams more readily than lighter metals. *True amalgams* are essentially no different from the thousands of other liquid solutions of metallic elements, although Hg's ability to dissolve other metals at room temperature is certainly unusual. *Plastic amalgams* are suspensions of solid particles of other elements or intermetallic compounds in liquid Hg or a saturated Hg liquid solution. Hg intermetallic compounds have a variety of unusual crystal structures (Sec. 19.4.6).

19.4.3 Applications of Hg

A metal that is liquid at ambient temperatures is useful for thermometers, electrical switches, dental amalgams, and electrodes that can absorb the materials evolving on their surfaces. In the

TABLE 19.2 Saturated Hg Vapor Pressure in Air at 1 Atm Pressure at Various Temperatures

Temperature of the Hg and Air (°C)	Equilibrium Hg Partial Pressure (Pa[a])	Hg Content in the Air (g/m^3)
0	0.026	0.00238
10	0.070	0.00604
20	0.170	0.01406
30	0.391	0.03144
100	36.841	2.40400

Source: *Ullmann's* (2003b); with permission.

[a] One atmosphere of pressure is 101,330 Pa.

SIDEBAR 19.4: AMALGAMS USED FOR DENTAL FILLINGS

Hg and Ag form the amalgam (Hg alloy) used in dental fillings (Fig. 19.12). Immediately after fine Ag powder is mixed with liquid Hg, the mix is a puttylike plastic amalgam of Ag particles suspended in a liquid Hg solution. Over a period of minutes, the Ag reacts at room temperature to form the ε and γ intermetallics, which are solid at 20°C. If sufficient Ag is present, the mixture becomes mostly solid, although some liquid inclusions may remain trapped within the solid amalgam. Alloys for dental "fillings" contain Hg and Ag, plus additional alloying elements [e.g., Sn, Cu, and/or Zn (Sec. 18.3.3.2)] to improve strength and corrosion resistance and to reduce shrinkage as the amalgam solidifies. The plastic nature of freshly mixed amalgam allows it to be molded to fill the tooth cavity tightly, and the reaction to form the intermetallic compound(s) rapidly hardens the filling into a wear-resistant, corrosion-resistant alloy.

Abrasive wear of the amalgam and galvanic couples with other metals in the mouth release small amounts of metal into the saliva, which raises concern about possible Hg toxicity. However, it is unclear whether significant amounts of Hg are released and whether much of that metal is absorbed by the body. Some persons have gone so far as to have all their Hg-based dental fillings removed and replaced by Hg-free alternative filling materials such as Ag–Sn alloys. Most dental experts argue that amalgam fillings release so little Hg that no adverse effects are detectable and that replacing amalgam fillings is unnecessary and risks damaging healthy teeth.

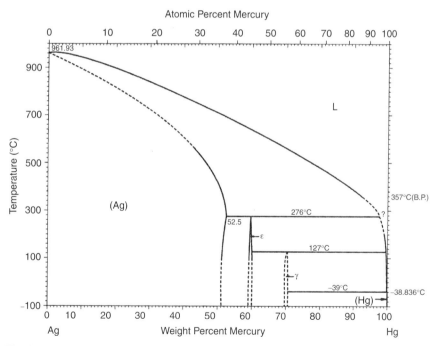

Fig. 19.12 Ag–Hg equilibrium phase diagram. (From Massalski, 1986, p. 31; with permission.)

chlor-alkali electrolysis of NaCl solutions, Cl_2 gas evolves on Ti anodes, and Na metal forms on a flowing Hg cathode, forming an Hg–Na amalgam. This amalgam is removed from the cell, reacted with water to produce NaOH and H_2 gas, and the purified Hg is returned to the cell. Hg has been used for thousands of years to extract microscopic flecks of Au and Ag from crushed ore containing these precious metals. The precious metal amalgams are then heated, vaporizing the Hg and leaving behind a nonvolatile precious metal residue (Sec. 18.4.4). Careless use of Hg in Au and Ag mining has been notorious for contaminating air and water, and this use has declined sharply. The high ionization energy of Hg causes excited Hg of emit high-energy (ultraviolet) photons. This ultraviolet light can be absorbed by phosphors and reemitted as white light in fluorescent lights and Hg vapor lamps. Hg oxide was once widely used in batteries and paint pigments, but less toxic alternative materials have replaced them. Hg compounds are used as fungicides, catalysts, disinfectants, and laboratory reagents.

19.4.4 Toxicity of Hg

Hg is toxic. Poisoning is most often caused by inhalation of metal vapor or ingestion of mercury compounds, the most troublesome of which are monomethylmercury compounds (MeHg). Acute exposure to high doses of inhaled Hg causes pneumonitis. If metallic Hg is swallowed, less than 0.01% is absorbed in the gastrointestinal tract, but soluble Hg salts (e.g., $HgCl_2$) cause severe gastrointestinal damage, acute renal failure, and cardiovascular collapse. The lethal dose for ingested Hg salts is approximately 30 to 50 mg of Hg per kilogram of body weight. Although numerous instances of acute Hg poisoning have occurred, the greater problem is chronic toxicity resulting from sustained low dosages over extended periods. The body requires about 10 weeks to reduce Hg concentrations to half their initial levels. The greatest damage occurs in the central nervous system and kidneys. Moderate levels of Hg accumulation cause headaches, mood changes, and irritability; higher doses and/or longer-term exposure cause tremor and sometimes violent muscle spasms. More serious intoxication causes major kidney damage, bladder inflammation, severe personality changes, depression, memory loss, and insomnia.

Two separate Hg poisoning incidents in Iraq in the 1950s and 1960s killed 459 persons and injured thousands more when seed grain coated with organomercury fungicides was diverted to bakeries to make bread. Hg poisoning killed 111 residents of Minamata, Japan in 1952 and seriously injured hundreds more. The Minamata victims consumed fish contaminated with Hg effluent from a nearby chemical plant using Hg salts as catalysts. Other severe Hg poisonings have occurred, but the broader concern is the effect of low-level doses acquired by millions of people from traces of MeHg in seafood. Some of the Hg released into rivers, lakes, and oceans is methylated by bacteria in sea and lake bottom sediments to form MeHg. This MeHg enters plants and small organisms at the lowest levels of the food chain and accumulates in progressively higher concentrations in small fish and then in larger predatory fish. Some predatory fish display MeHg body burdens 10^5 times higher than the MeHg concentration in the water they occupy. Humans who eat these MeHg-contaminated fish are susceptible to chronic low-level Hg poisoning. MeHg is, unfortunately, transferred efficiently from the gastrointestinal tract to the bloodstream. Toxicity concerns have resulted in large reductions of Hg use in batteries, thermometers, paints, and pesticides. Some governments have gone so far as to ban all Hg use within their jurisdictions. It was once thought that Hg was so essential to some applications (e.g., dental fillings and alkali chloride electrolysis) that finding suitable substitutes would be exceedingly difficult; however, some alkali chloride electrolysis and dental fillings restorations are now being performed by Hg-free processes. Although continuing use of small amounts of Hg is expected for the foreseeable future, the goal of some environmentalists to ban all Hg use may soon be technologically possible. Even if such bans were implemented, environmental Hg contamination would continue from combustion of coal containing trace amounts of Hg.

19.4.5 Sources of Hg and Refining Methods

The first reduction of HgS to metal may have been performed by heating the ore in brushwood fires. The heat separated the Hg and S, and the Hg vapor condensed on the unburned twigs, allowing it to be collected later from the ashes. Modern methods of Hg production are more efficient (and safer for production workers and the environment!), but the basic principle remains

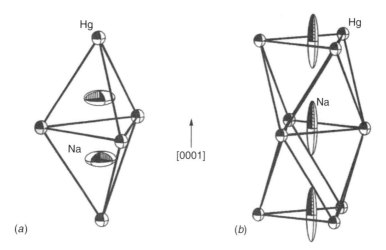

Fig. 19.13 Atom positions in the hexagonal α-Na₃Hg intermetallic compound: Hg atoms are arranged in the relative positions of a closest-packed structure, but with unusually large spacing between the Hg atoms. Na atoms occupy every interstitial site in the Hg lattice. (*a*) Na atoms in tetrahedral sites; (*b*) Na atoms in octahedral sites. The ellipsoidal markers designating Na atom positions indicate their range of thermal vibration displacements. (From Deiseroth and Toelstede, 1992; with permission.)

the same. HgS is separated from gangue by flotation and heated in a current of air to produce Hg vapor and SO_2. The Hg vapor is condensed. Scrap Fe or CaO is sometimes added to the process to react with the S, producing FeS or $CaSO_4$ rather than SO_2. Hot air can be blown through the crude Hg to oxidize impurities such as Fe, Cu, Zn, and Pb, allowing them to be skimmed from the surface of the metal; Hg is relatively noble, so little Hg oxidizes in this step. Final purification is achieved by distillation. Hg is transported in steel flasks that hold 34.5 kg of Hg. About half of Hg demand is met by recycled metal.

19.4.6 Structure–Property Relations in Hg

19.4.6.1 Na–Hg Amalgam. The Na–Hg system contains seven different intermetallic compounds, including one with a curious pseudo-close-packed crystal structure. Na and Hg react exothermically to form Na_3Hg, which melts peritectically at 34°C. The α-Na_3Hg intermetallic has Hg atoms in the approximate positions of an hexagonal close-packed structure but with large separations (>0.5 nm) between nearest-neighbor Hg atoms; the normal 12-coordinated atomic radius for pure Hg is only 0.151 nm. Na atoms occupy each tetrahedral and octahedral void in this widely spaced Hg lattice (Fig. 19.13).

ADDITIONAL INFORMATION

References, Appendixes, Problem Sets, and Metal Production Figures are available at
ftp://ftp.wiley.com/public/sci_tech_med/nonferrous

20 Al, Ga, In, and Tl

20.1 OVERVIEW

The group 13 elements aluminum, gallium, indium, and thallium have a ground-state s^2p^1 outer electron structure. With only one electron in an unfilled subshell, their properties might be expected to resemble those of the group 1 metals, which also have one unpaired outer electron. However, hybridization to a s^1p^2 structure provides three bonding electrons for the metallic state, giving these metals higher melting points and elastic moduli than alkali metals (Table 20.1).

Al is the most abundant of all metals, comprising 8.3% of Earth's crust. The other group 13 metals are far scarcer: Ga (19 ppm), In (0.24 ppm), and Tl (0.7 ppm). Worldwide Al use exceeds 25,000,000 tons per year, second only to iron and steel. Primary Al (produced from ore) meets about 75% of this demand with the balance supplied by secondary metal (recycled). Ga, In, and Tl are consumed in much smaller quantities; the combined production of all three metals is less than 1000 tons/yr. Al possesses ductility, high thermal and electrical conductivity, an adherent, protective oxide layer, and the ability to achieve high strength by alloying. These attributes, combined with relatively low cost, have made it invaluable for aerospace, terrestrial, and marine vehicles; electric power transmission; cans and foils; architectural trim; reflective coatings; and myriad other applications. The other three group 13 metals are used primarily in electronic devices, solders, and fusible alloys.

20.2 ALUMINUM

> For much of the 19th century, aluminum was considered a precious metal, expensive and difficult to produce. Emperor Napoleon III used sterling silver or gold tableware for routine entertaining, reserving his aluminum tableware for honored guests on special occasions.
>
> —Sheryll Luxton, Scott Community College, private communication, 2003

20.2.1 History of Al

To mid-nineteenth century metallurgists, today's annual consumption of Al (10 million cubic meters) would seem startlingly large. At that time, Al was produced by reducing $AlCl_3$ with K or Na metal; it was expensive and available only in small quantities. Al's status as a "boutique" metal changed abruptly in 1886–1888 with the nearly simultaneous development of the Hall–Héroult process for electrolytic Al reduction, the Bayer process for inexpensive production of Al_2O_3 from bauxite ore, and the dynamo for large-scale electric power generation. These breakthroughs allowed Al's exceptional physical properties (Fig. 20.1) to be applied to a host of engineering problems and made Al the leading nonferrous metal.

20.2.2 Physical Properties of Al

20.2.2.1 Mechanical Properties. Persons handling ultrapure Al (99.9999% purity) for the first time often question whether the metal is really Al because it is so much softer ($\sigma_{yield} = 12$ MPa)

Structure–Property Relations in Nonferrous Metals, by Alan M. Russell and Kok Loong Lee
Copyright © 2005 John Wiley & Sons, Inc.

TABLE 20.1 Selected Room-Temperature Properties of Ga, In, and Tl with a Comparison to Cu

	Ga	In	Tl	Cu
Valence	+3	+3	+1, +3	+1, +2
Crystal structure at 20°C	Orthorhombic (oC8)	Tetragonal (tI2)	HCP[a]	FCC
Density (g/cm^3)	5.91	7.31	11.85	8.92
Melting temperature (°C)	30	156	304	1085
Thermal conductivity (W/m·K)	33.5	81.6	46.1	399
Elastic (Young's) modulus (GPa)	9.8	10.6	7.9	130
Coefficient of thermal expansion (10^{-6} m/m·°C)	18.0	24.8	28.0	16.5
Electrical resistivity (μΩ·cm)	17.4	8.37	18.0	1.7
Cost ($/kg), large quantities	400[b,d]	150[c,d]	1250[b,d]	2

[a] Tl transforms to BCC above 230°C.
[b] Electronic grade: 99.9999+% purity.
[c] 99.97% purity (cost of ultrahigh-purity In is approximately $300 per kilogram).
[d] Ga, In, and Tl prices are subject to large fluctuations.

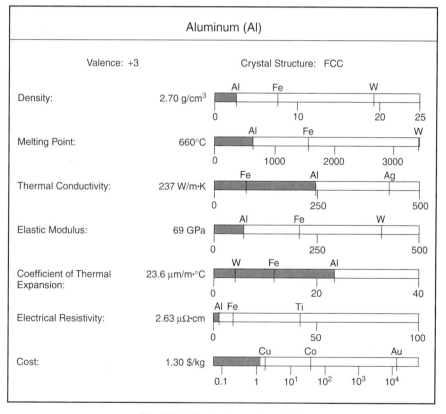

Fig. 20.1 Physical properties of Al.

than familiar Al alloys. At a more "normal" purity (99.0%), yield strength rises to 40 MPa, and some precipitation-hardening Al alloys can achieve σ_{yield} as high as 625 MPa, a remarkable 50-fold increase over ultrapure Al. The FCC crystal structure of Al has a low critical resolved shear stress (1.0 MPa) for {111}⟨110⟩ slip; thus, single crystals of high-purity Al have yield strengths of only 2 MPa. Al's partially directional bonding gives it a high stacking-fault energy

(150 mJ/m^2). This makes stacking-fault ribbons between partial dislocations (Sec. 4.5) only one or two atom spacings wide. Such narrow stacking faults can collapse by thermal fluctuation, which makes cross-slip easier in Al than in many other FCC metals. Unlike some other FCC metals (e.g., Cu, Sec. 18.2.6.1), impurity solute atoms usually do not lower stacking-fault energy in Al (see, however, Sec. 20.2.4.1).

Pure Al has excellent ductility; 35 to 50% tensile elongation is typical at room temperature. As with most FCC metals, there is no ductile–brittle transition; Al remains ductile at cryogenic temperatures with tensile elongation actually increasing somewhat below $-200°C$. Work hardening raises Al's strength quite substantially. Commercial purity Al (99.60% pure) has a yield strength of 27 MPa when fully annealed, but if cold worked by swaging or rolling to 75% reduction in area, σ_y increases to 125 MPa. In the Ludwik equation for uniform plastic deformation, $\sigma = \sigma_0 + K\varepsilon^n$ with stress (σ) increasing with true strain (ε); σ_0 (yield stress) and K are constants, and n is the strain-hardening exponent. Typical n values for Al and its alloys lie between 0.18 and 0.24, with K decreasing as n increases. Ultrahigh-purity Al does not maintain work hardening at room temperature; in the absence of impurity atoms the metal gradually recovers and recrystallizes. Although Al alloys can achieve high strength at room temperature, tensile and creep strength decline sharply above 200°C. Above 260°C, Al also slips on {100} and {211} planes.

20.2.2.2 Corrosion Behavior. When exposed to O_2, Al forms an amorphous oxide layer about 5 nm thick; if heated in air, the oxide layer may thicken to about 20 nm. The oxide is strongly adherent and protective in many corrosive environments. High-purity Al (99.99+% purity) has considerably better corrosion resistance than commercial-purity Al (99.5%), which in turn has better corrosion resistance than most Al alloys. A less protective mixed oxide scale forms over second-phase particles in impure Al, and these particles form microscale galvanic cells with the pure Al around them, accelerating corrosive attack (Fig. 20.2). Pure Al resists corrosion in water, seawater, ammonia, most food and beverage solutions, many organic acids, and even concentrated nitric acid. However, the Al_2O_3 surface layer is attacked by most mineral acids (e.g., HCl, HF, HBr, dilute HNO_3, and H_2SO_4 acids), some aqueous salts (e.g., Cu and Ni salts, $ZnCl_2$), and caustic solutions (e.g., NaOH, KOH). Al products (e.g., sheet on airplane wings and fuselage) are sometimes clad with a thin outer layer of high-purity Al to protect the underlying Al alloy from corrosion; such clad structures combine the high strength of the underlying alloy with the superior corrosion resistance of high-purity Al on the surface. Al_2O_3 films are often used to protect other metals from corrosion by aluminizing their surfaces; Al coatings and intermetallic compounds perform well in air up to about 1100°C (Sec. 25.4.3).

20.2.2.3 Reflectance. Al has high reflectance for visible and infrared light (Fig. 20.3). Al can be deposited on glass by physical vapor deposition (Sec. 8.7.3) to produce smooth surfaces; a 100-nm-thick deposit is sufficient to achieve maximum reflectance. Al coatings are used for nearly all vanity and optical mirrors (Sidebar 20.1). Deep luster automobile finishes contain high-purity Al powder to reflect light.

20.2.2.4 Electrical and Thermal Conductivity. Al's electrical and thermal conductivities are among the highest of all materials. Although group 11 metals (Ag, Cu, Au) have conductivities higher than Al's, they are heavier and more expensive than Al, making them less desirable for many applications. The electrical resistivity and thermal conductivity of pure Al over a range of temperatures are plotted in Fig. 20.5a and b. Increased electron–phonon interactions at higher temperatures raise electrical resistivity as temperature rises. Thermal and electrical conductivities are highest for high-purity Al and decrease as impurity content rises. In commercial purity Al, the most common impurities are Fe and Si, which raise electrical resistivity (Fig. 20.6).

20.2.2.5 Volume Change with Temperature. Al undergoes a large volume change at 660°C when the liquid (2.37 g/cm^3) freezes to the solid (2.55 g/cm^3). Careful attention to sprue and riser

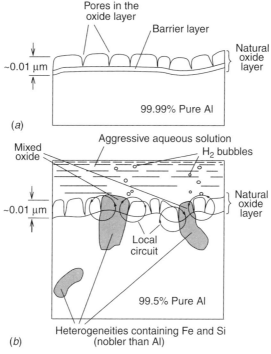

Fig. 20.2 (*a*) Naturally occurring oxide layer on high-purity Al. The thicker outer layer contains some cracks and porosity, but a thin, impervious barrier layer beneath protects the underlying metal from direct contact with air or aqueous solutions. (*b*) Corrosion of 99.50% purity Al containing Fe and Si impurities. The impurity sites form a mixed oxide scale that is less protective than pure Al_2O_3, allowing air and aqueous solutions to react with the metal. The impurity phases are more noble than the Al matrix, forming a local galvanic circuit at the site of each impurity precipitate. (Redrawn from Totten and MacKenzie, 2003, Vol. 1, p. 60.)

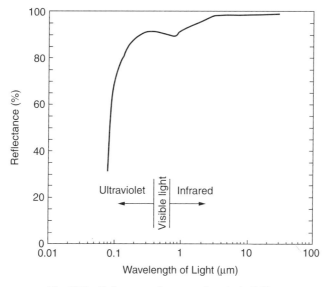

Fig. 20.3 Reflectance of vacuum-deposited Al film.

SIDEBAR 20.1: ALUMINIZED MIRRORS OF THE KECK TELESCOPES

Al's high reflectance and oxidation resistance are exploited in PVD Al coatings (Sec. 8.7.3) on the mirrors of reflecting telescopes. Two of the largest of these is a pair of identical 10-m-diameter mirrors in the Keck Observatory telescopes located at the 4100-m elevation of Mauna Kea volcano in Hawaii. It is difficult to make mirrors this large from a single piece of glass, so they are comprised of 36 smaller mirror segments arranged to act as one optical piece (Fig. 20.4). Al's superior reflectance in the infrared (Fig. 20.3) makes these telescopes even more efficient at infrared wavelengths than they are with visible light. In ordinary vanity mirrors, the Al layer is on the back surface of the glass, and the viewer looks through the glass to see the reflected image. In telescope mirrors, however, the Al is on the mirror's front surface to avoid the light loss and distortion that occur when light passes (twice) through the glass. This improves optical resolution, but makes the mirror vulnerable to scratches and dust accumulation. Telescope mirrors are usually recoated periodically to "refresh" the reflective surface.

Fig. 20.4 The technician perched in the central hole lends scale to the main mirror of the 10-m-diameter Keck astronomical telescope. The reflective coating is a thin film of Al metal vapor deposited onto the low-expansion glass of the mirror. The mirror is comprised of 36 hexagon-shaped segments precisely aligned to form a single optical surface. (Courtesy of L. Kraft, W.M. Keck Observatory; with permission.)

design is vital in Al castings to avoid shrinkage cavities. The coefficient of thermal expansion (COTE) for Al (23.6 μm/m·°C) is substantially higher than in more refractory commercial metals such as steel (11 to 12 μm/m·°C) or Ti (8.4 μm/m·°C). Al's large COTE causes considerably more shrinkage as the metal cools to its room-temperature density of 2.70 g/cm^3.

20.2.2.6 H Solubility in Al. Although most gases have little or no solubility in Al, molten Al can dissolve substantial quantities of H (Fig. 20.7). H solubility drops sharply when the metal solidifies. Al reacts with H_2O to liberate H:

$$3H_2O(vapor) + 2Al \rightarrow 6H(\text{dissolved in Al}) + Al_2O_3$$

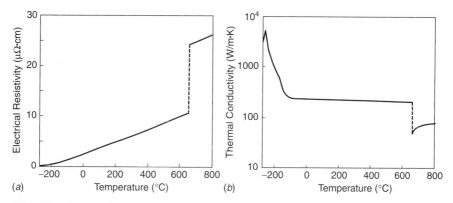

Fig. 20.5 Electrical resistivity and thermal conductivity of high-purity Al at various temperatures. A sharp discontinuity occurs in each curve at 660°C (the melting point). Al is a superconductor below −272°C.

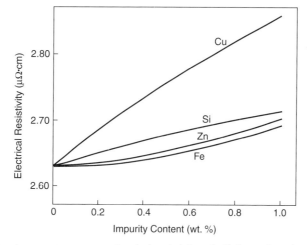

Fig. 20.6 Variation in room-temperature electrical resistivity of Al for various impurity species and concentrations.

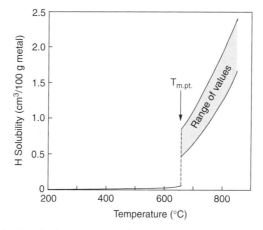

Fig. 20.7 H solubility in Al under 1 atm pressure of pure H. Various studies give a range of values for the solubility of H in liquid Al, depicted by the shaded area. H solubility decreases sharply upon solidification at 660°C.

Consequently, the potential exists for molten Al to dissolve H from atmospheric humidity or from H_2O adsorbed on the surfaces of the solid Al or crucible. Dissolved H forms gas bubbles when the metal solidifies, which lower strength and can compromise leakproof performance for beverage cans, cast vessels, and housings. This problem can be mitigated by bubbling inert gas through the molten metal. Although N_2 and O_2 do not dissolve appreciably in Al, they form Al_2O_3 and AlN, which can be detrimental (Sec. 20.2.8.3). Solid Al is susceptible to H damage if exposed to humid air during heat treatments. Sulfur compounds in the furnace atmosphere can breach the Al_2O_3 surface layer, producing H surface blisters by the foregoing reaction.

20.2.2.7 Nuclear Properties. Al has a low thermal neutron absorption cross section (0.21 barn), making it useful in fission reactors. It was used for most fuel cladding and internal core structures in early reactors; however, Al has largely been replaced by Zr alloys (Sec. 12.3.3.2) in modern commercial power reactors. Some research and test reactors still use Al core structures.

20.2.3 Properties and Applications of Pure Al (1xxx Series Wrought Alloys, 1xx.x Casting Alloys)

Although commercial purity Al is rather weak, it has excellent conductivity, formability, weldability, and corrosion resistance. Commercially pure Al, referred to as 1000 series wrought metal or 100.0 series cast metal (Sidebar 20.2), is used to produce electrical wire, heat-exchanger tubing, capacitor foil, cladding for Al alloys, chemical equipment, railroad tank cars, reflectors, and kitchenware. Ultrapure Al (up to 99.9999+% purity) is used for electronic circuit vias, contact pads, semiconductor dopant, and group 13–group 15 semiconducting intermetallic compounds. Most commercial-purity Al contains about 99.3 to 99.7% Al, with Fe and Si present as the major impurities (Sec. 20.2.7) and smaller amounts of Cu (0.05 to 0.10%). Fe has only 0.05% solid solubility in Al at 655°C, and this value drops to just 0.006% at 500°C; consequently, commercial-purity Al always contains scattered Al_3Fe precipitates. Si has greater solid solubility

---

SIDEBAR 20.2: Al ALLOY DESIGNATIONS

About 80% of all Al use is for wrought products (plate, sheet, extrusions, wire, etc.). Wrought Al alloys are identified by a four-digit code number in the Aluminum Association classification system (Table 20.2). The first digit designates the group. The second digit designates modifications to the original alloy, with zero indicating the original alloy and other numbers later versions. The last two digits are arbitrary numbers identifying the alloy (or in unalloyed Al, the purity). For example, in wrought alloy 6061, the first digit indicates that the alloy contains Mg and Si as the major alloying additions; the second digit indicates that this is the formulated composition when the alloy was introduced; and the last two digits indicate the nominal content of 0.8 to 1.2% Mg, 0.4 to 0.8% Si, and other alloying additions (e.g., Cu, Fe, Mn, Cr, Zn, and Ti). The last two digits are arbitrary; they do not express the weight percentages of alloying additions. In unalloyed Al, a designation such as 10<u>50</u> means that no less than 99.<u>50</u> wt% of the alloy is Al; maximum permissible amounts of specific impurities (e.g., Fe, Si, Cu, Mn, etc.) are also stipulated.

About 20% of Al is used for castings. The Aluminum Association designation for casting alloys is a three-digit/decimal point/one-digit number (Table 20.3). The first digit indicates the major alloying elements. In the 1yy.x series, the yy digits indicate Al purity above 99.00% (e.g., 130 indicates 99.3% pure Al). In all other casting alloys, the yy digits are arbitrary designators for composition that bear no numerical relation to alloying addition percentages. The digit after the decimal point indicates whether the metal is a casting (0) or an ingot (1). Prefix letters indicate modifications of existing alloys; other designations (Tables 20.4 to 20.6) are the same for wrought and cast products.

TABLE 20.2 Alloy Groups of Wrought Al Alloys

Group	Major Alloying Addition(s)	General Characteristics; Typical Uses
1xxx	Unalloyed Al of 99.00% purity or higher	Good corrosion resistance and formability, high conductivity, fairly low strength; architectural trim, heat exchangers, chemical equipment, electric power transmission lines, reflectors, lithographic sheet, kitchen foil
2xxx	Cu	High strength, good machinability, precipitation hardenable, lower corrosion resistance; aircraft and highway vehicle structures
3xxx	Mn	Moderate strength, good formability; furniture, storage tanks, cooking utensils, window frames, highway signs, roofing
4xxx	Si	Lower melting point and lower coefficient of thermal expansion, some are precipitation hardenable; filler material for brazing and welding, anodized components
5xxx	Mg	Good strength and weldability, excellent corrosion resistance in marine environments; ornamental trim, ships, cryogenic vessels, street lights
6xxx	Mg, Si	Good formability and weldability, precipitation hardenable, high corrosion resistance; highway vehicles, bridges, welded construction, extrusions, tooling plate
7xxx	Zn	Precipitation hardenable, high strength-to-weight ratio; aircraft structures, mobile equipment
8xxx	Other element(s)	(e.g., Al–Li, Al–Sn, Al–Ni–Fe, and others)

TABLE 20.3 Alloy Groups of Cast Al Alloys

Group	Major Alloying Addition(s)	General Characteristics; Typical Uses
1yy.x	Unalloyed Al of 99.00% purity or higher; yy digits designate purity level	Highest conductivity and ductility, low strength; conductor bars for electric motors
2xx.x	Cu (extra low Fe)	Heat treatable, high strength, mediocre corrosion resistance; pistons, cylinder heads, valve bodies, gears
3xx.x	Si, with added Cu and/or Mg	The most widely used casting alloys, good castability, heat treatable, higher strength than 4xx.x; machine tool parts, aircraft wheels, pistons, transmission casings
4xx.x	Si	General-purpose casting alloys, best castability, non-heat-treatable, good corrosion resistance; intricate castings with thin sections, housings, frames, engine parts
5xx.x	Mg	Medium strength, non-heat-treatable, good corrosion resistance; marine components, food-processing vessels, architectural trim
6xx.x	(unused series)	
7xx.x	Zn	Natural aging alloys, capable of producing good surface finish and good corrosion resistance, more difficult to cast
8xx.x	Sn	Specialty alloys; bearings and bushings
9xx.x	Other element(s)	

in Al, and commercial-purity Al typically contains elemental Si precipitates only when Si content exceeds 0.25%. Al_3Fe precipitates (and sometimes Si precipitates) are readily visible with optical metallography (Fig. 20.8). The precipitates, combined with the hardening effect of the small amounts of impurities in solid solution, produce modest strengthening. Since precipitates are much less harmful to electrical conductivity than solid-solution atoms, metal used for electrical

TABLE 20.4 Al Alloy Basic Temper Designations

Designation	Meaning
F	As fabricated; no special control over thermal or work-hardening conditions
O	Annealed; wrought alloys heated to achieve recrystallization or cast alloys annealed for improved ductility and dimensional stability
H	Strain hardened; wrought alloys strengthened by cold work; this designation is always followed by two or more digits to indicate the degree of strain hardening (see Table 20.5)
W	Solutionized and unstable at room temperature, used to indicate that the alloy has been solutionized at high temperature and quenched to preserve the metastable solid solution condition, applied only to alloys that spontaneously age harden at room temperature
T	Thermally treated to achieve a specific, stable heat treatment other than F or O, always followed by one or more digits (see Table 20.6)

TABLE 20.5 Al Alloy Strain-Hardening Designations

Designation	Meaning
H1x	Strain hardened only; the degree of strain hardening is indicated by the second (x) digit, which ranges from quarter-hard (H12), half-hard (H14), three-quarters hard (H16), to full-hard (H18). Full-hard is achieved by about a 75% reduction in area at room temperature
H2x	Strain hardened followed by partial annealing; achieved by cold working beyond the desired strength level, followed by sufficient annealing to attain quarter-hard (H22), half-hard (H24), three-quarters hard (H26), or full-hard (H28) condition
H3x	Strain hardened followed by stabilization; used for age-softening Al–Mg alloys that are strain-hardened, then heated at low temperatures to increase ductility and stabilize mechanical properties (H32, H34, H36, H38)

TABLE 20.6 U.S. Al Alloy Thermal-Treatment Designations

Designation[a]	Meaning
T1	Cooled from the fabrication temperature and naturally aged (i.e., aged at room temperature)
T2	Cooled from the fabrication temperature, cold-worked, and naturally aged
T3	Solution-treated, cold-worked, and naturally aged
T4	Solution-treated and naturally aged
T5	Cooled from the fabrication temperature and artificially aged (i.e., aged at elevated temperature)
T6	Solution-treated and artificially aged
T7	Solution-treated and stabilized by deliberate overaging
T8	Solution-treated, cold-worked, and artificially aged
T9	Solution-treated, artificially aged, and cold-worked
T10	Cooled from the fabrication temperature, cold-worked, and artificially aged

[a] Second and third digits are sometimes added for more specific thermal treatment descriptions.

wire may have small B or Ti additions to enhance precipitation of impurities. Commercial-purity Al responds well to strain-hardening treatments. Al 1100-O (fully annealed, 99.00% pure) has $\sigma_y = 35$ MPa, $\sigma_{UTS} = 90$ MPa, and 35% tensile elongation. The same metal in the H18 condition (fully strain-hardened) has $\sigma_y = 150$ MPa, $\sigma_{UTS} = 165$ MPa, and 3% tensile elongation. Work hardening is lost to recovery and recrystallization at elevated temperatures (Fig. 20.9).

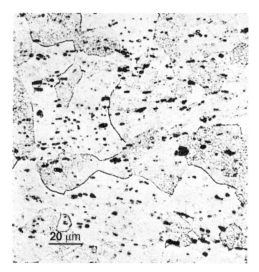

Fig. 20.8 Optical micrograph of 1100–O Al showing black FeAl$_3$ precipitates. The metal was cold-worked to break up the large, scriptlike precipitates that formed in the cast metal, then annealed to recrystallize it. (From ASM, 1972, p. 242; with permission.)

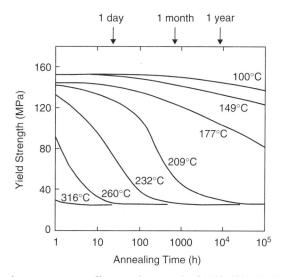

Fig. 20.9 Annealing time–temperature effects on the strength of 1100–H18 Al. Cold-worked specimens were annealed at the temperatures marked, and post-anneal room-temperature yield strengths are plotted here.

20.2.4 Properties and Applications of Wrought Al Alloys

20.2.4.1 Al–Mg Alloys (5xxx Series Wrought Alloys). Mg is one of the few elements with high solid solubility in Al (others include Ag, Li, and Zn). Mg's equilibrium solid solubility is nearly 15% at 450°C, although solubility approaches zero at 20°C. The Al–Mg equilibrium phase diagram (Fig. 11.9) appears to offer good potential for precipitation hardening, but the strengthening effect of such precipitates is poor. Consequently, the primary strengthening effect resulting from Mg additions to Al is (metastable) solid-solution hardening.

Most 5xxx series alloys contain 1 to 5% Mg in solid solution. Cr and/or Mn additions of 0.12 to 0.25% are often present as grain refining additions. Cr has a low diffusion rate in Al and forms fine dispersed phases in wrought products that retard nucleation and growth of grains. These are simple alloys not designed for heat treatment or precipitation hardening. The presence of Mg improves seawater corrosion resistance and lowers density slightly (to 2.65 g/cm^3 with 5% Mg). Mg lowers stacking-fault energy (Sec. 20.2.2.1) somewhat in Al, which raises the temperature of dynamic recovery and allows the alloy to retain work hardening to higher temperatures before recovery and recrystallization occur. This makes 5xxx alloys suitable for hot oil or asphalt applications and for storage tanks for heated products. Alloys with Mg content below about 4% usually do not form Mg_2Al_3, although Mg_2Al_3 can precipitate on grain boundaries if the metal is held between 120 and 200°C. This can cause intergranular corrosion. The 5xxx series alloys are seldom used in the H1 condition because they gradually form Mg_2Al_3 precipitates along slip bands at room temperature after cold work. The alloys are often put into the H3 condition (cold work followed by annealing at about 250°C) to force a fine dispersion of Mg_2Al_3 precipitates throughout the grain, a microstructure less susceptible to stress corrosion cracking. In a fully annealed condition (O) these alloys have σ_y values ranging from 40 to 160 MPa, with higher Mg content giving higher strength; some 5xxx series alloys have σ_y values as high as 340 MPa in the fully hard condition (H38) (Fig. 20.10). The 5xxx alloys generally provide better welding performance than precipitation-hardening Al alloys (Sidebar 20.3).

20.2.4.2 Al–Mn Alloys (3xxx Series Wrought Alloys). Additions of 0.5 to 1.2% Mn to Al form fine $(Mn,Fe)Al_6$ dispersoids that strengthen the metal and inhibit grain nucleation and growth. Some Mn remains in solid solution, further strengthening the metal. The 3xxx alloys are stronger than 1xxx alloys while maintaining good formability and excellent weldability and corrosion resistance. These alloys are often used for beverage containers, cooking utensils, architectural trim, extruded window frames and gutters, and welded pressure vessels. Many 3xxx alloys contain 0.5 to 1.0% Mg to complement the action of the Mn. These alloys can be solutionized at 600°C, followed by cold work and a 340°C anneal to form dispersoids (Sec. 20.2.8.2) on dislocations, pinning them and retarding recovery and recrystallization. Strengths of 3xxx alloys are intermediate between 1xxx and 5xxx alloys.

20.2.4.3 Al–Si Alloys (4xxx Series Wrought Alloys). Al and Si form a simple eutectic phase diagram (Fig. 20.11) with 1.65% solubility of Si in Al and 0.5% solubility of Al in Si at the

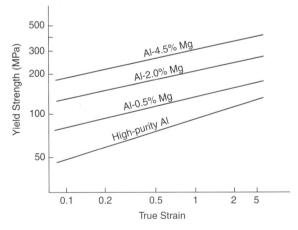

Fig. 20.10 Strain-hardening response from cold-rolling high-purity Al and Al containing varying amounts of Mg (wt%). True strain = 1.15 ln (initial thickness/final thickness). (Adapted from Hatch, 1984, pp. 105–133.)

SIDEBAR 20.3: WELDABILITY OF Al ALLOYS

Al alloys can be divided into two broad categories: the heat-treatable, precipitation-hardening alloys and the non-heat-treatable alloys. The latter group are strengthened only by cold work and solid-solution strengthening. Although the absence of precipitate-forming elements in the 1xxx, 3xxx, 4xxx, and 5xxx non-heat-treatable alloys means that they cannot achieve the high strengths possible from precipitation hardening, it also means that they have generally better welding performance. Alloy additions used for precipitation hardening (e.g., Cu, Zn, Mg + Si, etc.) can cause liquation or hot cracking during welding. In addition, a precipitation-hardened alloy will lose much of its strength in the heat-affected zone (HAZ) because the heat coarsens or dissolves the precipitates. Thus welded parts must be solutionized, quenched, and aged after welding, or special thicker weldments are needed to compensate for the loss of strength in the HAZ. In non-heat-treatable alloys, HAZ damage is limited to recovery, recrystallization, and grain growth, so their strength drop is smaller.

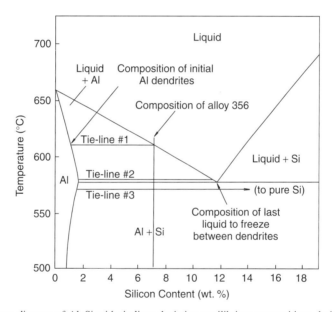

Fig. 20.11 Phase diagram of Al–Si with tie-lines depicting equilibrium compositions during the freezing of alloy 356 (Al–7% Si).

577°C eutectic temperature. These alloys have good fluidity and produce eutectic microstructures (Fig. 20.12) that are well suited to casting (Sec. 20.2.5.1) but less useful for wrought alloys, due to their limited ductility. Consequently, 4xxx wrought alloys are used primarily for brazing materials and welding filler rods, where the low eutectic temperature and good fluidity of Al–Si alloys provide liquid metal available for back-filling as the joined metals cool. This confers excellent hot cracking resistance. The corrosion resistance of Al–Si alloys is equal to that of 1xxx alloys.

20.2.4.4 Al–Cu Alloys (2xxx Series Wrought Alloys). Cu provides both solid solution and precipitation strengthening in Al. At elevated temperatures more than 5 wt% Cu can dissolve in solid Al, but solid solubility diminishes to near zero at room temperature (Fig. 20.13). This allows

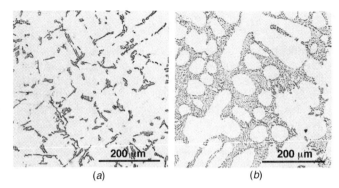

Fig. 20.12 (a) Microstructure of slow-cooled sand mold casting of alloy 356-F, an Al–7% Si–0.3% Mg alloy. Note the large Al dendrites and the coarse, acicular Si, Fe_3SiAl_{12}, and $Fe_2Si_2Al_9$ precipitates in the interdendritic regions. (b) Microstructure of slow-cooled sand mold casting of alloy 356-F with Na addition (0.025%) to refine the Si precipitates, enhancing strength and toughness. (From ASM, 1972, p. 258; with permission.)

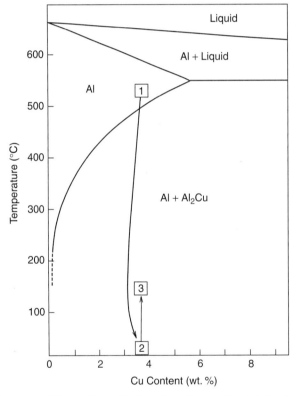

Fig. 20.13 Al-rich portion of the Al–Cu equilibrium phase diagram. Precipitation hardening can be performed on Al alloys containing a few percent Cu by (1) holding the alloy in the Al solid-solution phase field until a homogeneous, single-phase solid solution is achieved; (2) quenching the alloy to preserve this structure as a metastable phase at room temperature; and (3) aging the alloy between 130 and 200°C to precipitate a second phase.

alloys containing a few percent Cu to be solutionized, quenched, and aged to form nanoscale precipitates that greatly increase strength (Sec. 3.5). The typical precipitation-hardening heat treatment is depicted by arrows on Fig. 20.13 for a binary Al–3.5 wt% Cu alloy:

- The metal is first solution heat-treated in the Al solid solution phase field at about 520°C. At this temperature Cu is completely soluble in Al, and the Al_2Cu intermetallic particles present in the cast metal dissolve, distributing Cu atoms into random substitutional lattice sites in the FCC Al. Homogenization of the as-cast alloy into a single-phase solid solution typically requires 30 to 75 minutes in an air furnace. At this temperature, roughly one lattice site in 10^4 is vacant (Sec. 2.2).
- After a homogeneous solid solution is attained at 520°C, the metal is quenched to produce a metastable, supersaturated homogeneous solid solution at 20°C. The phase diagram indicates that the alloy should segregate into nearly pure $Al + Al_2Cu$ at room temperature, but diffusion is so slow at 20°C that it takes several days to achieve useful levels of strengthening and many years to form the equilibrium Al_2Cu phase.
- After quenching, the metal is heated to 130 to 190°C. This increases diffusion rates so that precipitates can form much more rapidly in the supersaturated solid solution (several hours to a few days). Diffusion is accelerated by the high quenched-in vacancy concentration. The precipitates that form initially are not Al_2Cu (often called θ *phase*); they are nonequilibrium structures known as GP1 zones, GP2 zones (or θ″), and θ′. The best mechanical properties are produced by aging the metal long enough to form these intermediate precipitates, but halting aging before the equilibrium Al_2Cu phase forms.

A thoughtful observer of this process might ask why the metal is not just solutionized at 520°C and then quenched directly to the aging temperature, eliminating the room-temperature step. This can be done, but quenching to 130 to 190°C requires that oil, fused salts, or fusible metals be used instead of water. These liquids are more costly and require temperature-regulating systems to hold them at the proper temperature. In addition, the higher temperature and lower heat capacities of these quench media slow the quench rate, risking partial formation of θ phase during cooling.

The aging of Al–Cu alloys is more complex than the phase diagram would indicate. Several different precipitate structures have been observed in supersaturated Al–Cu solid solutions during aging. Not all of these structures occur at all aging temperatures, but the general nature of the precipitation process can be summarized as

supersaturated solid solution → GP1 zones → GP2 zones (also called θ″) → θ′ → θ

At low aging temperatures (e.g., below about 100°C), GP1 zones form as Cu atoms congregate into disks one atom thick and 3 to 5 nm in diameter. (The letters "GP" stand for Guinier and Preston, scientists who performed pioneering studies on these structures.) GP1 zones do not have the composition or the crystal structure of the equilibrium Al_2Cu phase; they form on {100} planes in Al, and they are coherent with the Al matrix (Sec. 3.5). Since Cu atoms are smaller than Al atoms, this coherency strains the surrounding Al lattice, allowing GP1 zones to be observed by diffraction contrast in the transmission electron microscope. The lattice strain also impedes dislocation motion, so GP1 zone formation hardens the metal (Fig. 20.14). The composition of these zones is difficult to measure and has not been clearly determined; some studies report Cu contents below 25 at%, while others indicate they approach 100%. Between 100 and 130°C, GP1 zones can grow to become a few atoms thick (0.4 to 0.6 nm) and up to 10 nm in diameter (Fig. 20.15a). Above 190°C, GP1 zones are unstable and redissolve into the Al matrix.

GP2 zones (also called θ″) form at somewhat higher temperatures (about 130°C or above). They are 1 to 4 nm thick and 10 to 100 nm in diameter. As the particles grow, their Cu content

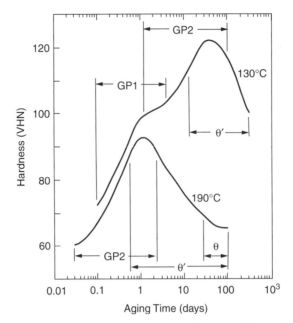

Fig. 20.14 Hardness versus aging time in a binary Al–4 wt% Cu alloy solutionized, quenched, and aged for varying times at 130 and 190°C. The precipitate structures observed are marked by witness lines to each curve. Maximum hardness occurs on both curves when a mixed GP2 + θ′ microstructure is present.

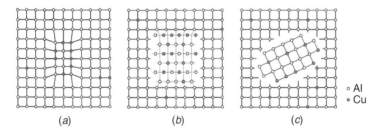

Fig. 20.15 Schematic depiction of (a) coherent, (b) semicoherent, and (c) incoherent precipitates in an Al–Cu alloy. Actual precipitate particle sizes are larger than shown here. (From Callister, 2003, p. 374; with permission.)

increases. They have a tetragonal crystal structure that remains coherent with the matrix, and they strain the lattice more severely than GP1 zones, providing greater hardening. If the metal is aged above 190°C or held at lower temperatures for long times, a different phase called θ′ appears. The θ′ does not grow from preexisting GP zones; it nucleates heterogeneously and has a tetragonal crystal structure with a lower c/a ratio than that of GP2 zones. The θ′ particles are semicoherent with the matrix (Fig. 20.15b) and larger than GP zones. They have thicknesses of 10 to 15 nm and diameters of 10 to 600 nm. An Al alloy whose precipitates are mostly or entirely θ′ particles is said to be *overaged*, because its strength is lower than in an alloy with GP2 zones. The θ′ precipitates' semicoherency makes them less effective than GP zones as strengtheners. When θ′ precipitates' replace GP zones, dislocation motion is no longer inhibited by lattice strain near the precipitates. Optimal strength is achieved at the transition point where dislocations stop shearing through precipitates and begin forming Orowan loops around precipitates (Fig. 20.16); this transition correlates with the GP2 → θ′ transition in precipitate structures.

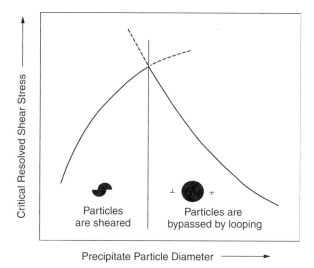

Fig. 20.16 Critical resolved shear stress (τ_{CRSS}) in an Al alloy versus precipitate size. Maximum τ_{CRSS} occurs at the transition point where dislocations no longer shear through the precipitates but begin to form Orowan loops around the particles. Sheared particles are shown in Figure 20.24a.

A microstructure containing mostly GP zones permits dislocations to shear through the zones, and once this shearing begins, deformation tends to become localized on a few shear bands, causing large dislocation pile-ups at grain boundaries. This in turn can cause intergranular fracture. The θ' precipitates resist dislocation shearing and are more stable at elevated temperatures, but they are larger and more widely spaced. Although a θ' precipitate is a more formidable obstacle to dislocation motion than a GP zone, there are far fewer θ' particles than GP zones, so as θ' replaces GP2 zones, the metal's critical resolved shear stress drops. The ideal microstructure for maximum strength would be precipitates large enough to resist shearing by dislocations and spaced so closely (about 10 nm) that Orowan looping around them requires high energies. Such an ideal precipitate structure has yet to be achieved in Al, and in real-world alloys, maximum strength occurs with mixed GP2 + θ' precipitates (Fig. 20.14).

If aged above 190°C or held at lower temperature for extended times, the alloy forms θ precipitates, the equilibrium Al_2Cu intermetallic compound. This phase can grow from θ' precipitates already present in the matrix, or it can nucleate directly in the matrix. When θ precipitates appear, the metal is severely overaged and much weaker than optimally aged material. The θ particles can become large enough to be visible by optical microscopy; they are incoherent (Fig. 20.15c) and poor strengtheners. The θ precipitate forms in slow-cooled Al–Cu castings.

Modern 2xxx alloys are actually quaternary or higher-order alloys containing Mg and Mn, Si, or Ni as well as Cu, but the basic precipitation hardening processes are similar to those of binary Al–Cu alloys. Mg accelerates precipitation hardening in Al–Cu alloys and raises strength levels in optimally aged metal. In alloys containing Mg, the final precipitates are precursors of the Al_2CuMg intermetallic compound called S' and S. These precipitates nucleate heterogeneously on dislocations, so cold work after quenching (T8 treatment) enhances precipitate formation. These precipitates grow in $\langle 001 \rangle$ directions as laths on Al $\{210\}$ planes. The faster aging possible in these alloys vis-á-vis binary Al–Cu alloys is a great practical advantage for heat treatment. It allows precipitation hardening to be performed at room temperature (Sidebar 20.4) over a period of weeks, or artificially in several hours (which is much faster than the several days needed for Al–Cu binary alloys). Faster aging greatly improves furnace throughput in heat-treating operations. The mechanical properties of 2014 alloy are shown in Fig. 20.17 and Table 20.7 for varying heat treatments.

SIDEBAR 20.4: NATURAL AGING IN Al AIRCRAFT RIVETS

Cu–Al alloys such as 2014 and 2017 are sometimes used without artificial aging. When solutionized and quenched, these alloys form GP zones at room temperature over a period of several days. The yield strengths of such naturally aged metal are lower than in artificially aged metal (Table 20.7), but naturally aged metal has high fracture toughness and fatigue resistance. Such alloys can be used in rivets to fasten Al-alloy aircraft skin to wings and fuselage. The rivets are solutionized, quenched, and immediately placed in refrigerated storage at −70°C to prevent the natural aging that would occur at room temperature. The rivets are removed from cold storage shortly before they are used. The metal is soft and ductile in the as-quenched condition, so the rivets drive easily without cracking. As the assembled aircraft sits at room temperature, the rivets age naturally, raising their strength over a period of weeks, during the aircraft's final assembly.

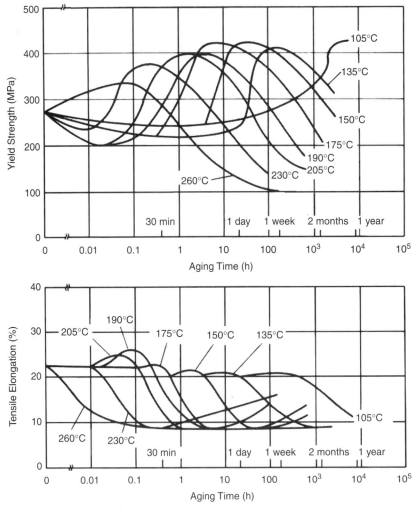

Fig. 20.17 Aging effects on yield strength and tensile elongation in Al 2014 (Al–4.5% Cu–0.85% Si–0.8% Mn–0.7% Fe–0.5% Mg) sheet. (From Davis, 1993, p. 311; with permission.)

TABLE 20.7 Comparison of Mechanical Properties of Al Alloy 2014 Determined by Tensile Testing

	Yield Strength (MPa)	Ultimate Strength (MPa)	Elongation (%)
Annealed (2014-O)	100	185	18
Naturally aged (2014-T4)	290	425	20
Artificially aged (2014-T6)	415	485	13

20.2.4.5 Al–Mg–Si Alloys (6xxx Series Wrought Alloys). Al–Mg–Si alloys (the 6xxx series) contain 0.6 to 1.2 wt% Mg and 0.4 to 1.3% Si and precipitate metastable forms of Mg_2Si. The alloys usually contain 0.2 to 0.8 wt% Mn, Cr, and/or Cu for grain size control and solid-solution hardening. Alloys such as 6061 and 6063 are particularly useful for hot extrusion because they can be quenched as they exit the extrusion die and moved immediately into an aging furnace, avoiding the cost of a postextrusion solutionizing/quenching process. Mg_2Si precipitation occurs more slowly than Al_2Cu precipitation, so air quenching is often sufficient to cool thin extrudates; this reduces quenching distortion.

Examples of precipitate formation in 6xxx alloys are shown in Figs. 20.18 and 20.19. The GP zones in these alloys do not produce significant coherency strains; the alloys are hardened by the energy required for dislocations to shear through the needle-shaped precipitates (called β' and β''). The composition and mechanical properties of 6061, the most widely used 6xxx alloy, are presented in Table 20.8. Although age-hardened 6061 is not quite as strong as 2xxx alloys, its superior corrosion resistance, formability (Sidebar 20.5), and weldability make 6061 one of the most versatile and heavily used Al alloys.

20.2.4.6 Al–Zn–Mg Alloys (7xxx Series Wrought Alloys). The Al–Zn–Mg ternary system provides the greatest precipitation-hardening effect of all Al alloys. This property combined

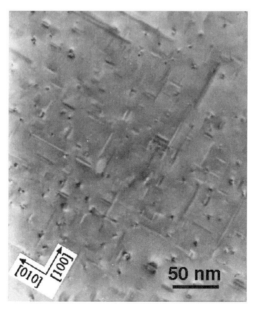

Fig. 20.18 Transmission electron micrograph of mixed GP zone and β' Mg_2Si precipitates in Al–0.62 wt% Mg–0.37 wt% Si alloy solutionized, quenched, and aged at 150°C for 20 minutes. The precipitates are needle-shaped, with their long dimensions parallel to the $\langle 100 \rangle$ directions in the Al matrix. (From Ikeno et al., 2001; with permission.)

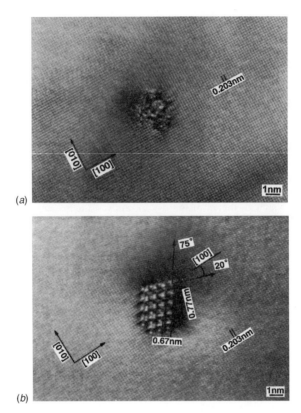

Fig. 20.19 High-resolution transmission electron micrographs of needle-shaped Mg_2Si precipitates in Al–0.62 wt% Mg–0.73 wt% Si alloy solutionized, quenched, and aged at 150°C. The precipitates' long dimensions were oriented perpendicular to the image plane, so they are seen here in cross section looking down the needle's axis. (a) A GP zone that has not fully crystallized; (b) the more fully developed monoclinic Mg_2Si structure (β'') in intermediate precipitates. (From Ikeno et al., 2001; with permission.)

TABLE 20.8 Composition and Mechanical Properties of 6061 Al Alloy

Composition: Al–0.8 to 1.2% Mg–0.4 to 0.8% Si–0.7% Fe–0.15 to 0.4% Cu–0.15% Mn–0.04 to 0.35% Cr–0.25% Zn–0.15% Ti (other elements 0.05% maximum each, 0.15% maximum total).

	Yield Strength (MPa)	Ultimate Strength (MPa)	Elongation (%)
Annealed (6061-O)	55	125	25
Solution-treated and naturally aged (6061-T4)	145	240	22
Solution-treated and artificially aged (6061-T6)	275	310	12
Solution-treated, cold-worked, and artificially aged (6061-T81)	360	380	15
Solution-treated, artificially aged, and cold-worked (6061-T913)	455	460	10

with densities slightly lower than those of 2xxx series alloys makes these alloys particularly useful in weight-critical structures in aircraft. The precipitation sequence in 7xxx alloys is

$$\text{supersaturated solid solution} \rightarrow \text{GP zones} \rightarrow \eta' \rightarrow \eta \ (MgZn_2)$$

SIDEBAR 20.5: EXTRUSION OF HOLLOW SHAPES

Extrusion of tubes and other hollow shapes is a more difficult processing challenge than extrusion of simple I-beam or rod shapes. Hollow shapes can be extruded by positioning a mandrel inside, but not touching, the extrusion die outlet (Fig. 20.20). This forces the billet (which must have a thick-walled hollow shape) to flow around all sides of the mandrel, producing a tube. An alternative method for extruding tubing is the porthole die (Fig. 20.21), in which the mandrel is connected directly to the die itself on several supports (five in the example shown in Fig. 20.21) that force the metal to divide into multiple streams, which then rapidly diffusion bond to their neighboring streams as it flows past the mandrel. This bonding process occurs inside the die in the absence of air, so the metal bonds easily to its neighboring streams, producing nearly perfect joints without oxide formation. Tubes tens of meters long with various cross-section shapes can be extruded in this fashion, and the product is so perfectly seamless that it is difficult to find the joining line where neighboring streams of flowing metal bonded to one another.

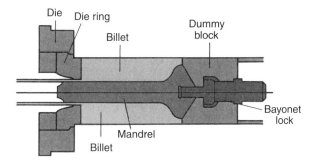

Fig. 20.20 Extrusion of a hollow-shaped billet around a mandrel. The metal is forced to flow between the die and the mandrel as the ram pushes the dummy block and billet from right to left through the die. The extruded product is tubular. (From Aluminum Association, 1995, p. 4–4; with permission.)

Fig. 20.21 Porthole die used to form tubular extrusions. The die and mandrel are all one piece. Metal from a solid cylindrical billet divides to flow through five openings "upstream" of the mandrel and then bonds to its neighboring streams as it is forced to flow around the outside of the mandrel. Since no air is present inside the die, the surfaces of the five separate streams form no oxide layer, and they bond instantly to the atomically clean neighboring streams as the metal exits the die around the mandrel. (From Aluminum Association, 1995, p. 4–5; with permission.)

The GP zones in Al–Zn–Mg alloys are spherical and coherent. Since the GP zone–matrix interfacial energy is quite low, GP zones as small as 3 nm in diameter can form in alloys aged at 20 to 120°C. The GP zones can serve as nuclei for the semicoherent η' phase, or η' can nucleate directly on dislocations and subgrain boundaries. In over-aged alloys, incoherent η phase grows from η' particles or nucleates directly on grain boundaries or other incoherent precipitate–matrix boundaries. Both GP zones and η' are sheared by dislocations; the energy to shear an η' precipitate is about twice the energy to shear a GP zone. Antiphase boundary (APB) energy (Sec. 25.6.1 and Fig. 25.4) is fairly high (360 mJ/m^2) in sheared η' precipitates.

Duplex heat treatments are often used in 7xxx alloys. Aging may begin at room temperature for three to five days, followed by an additional two days at a higher temperature (e.g., 120°C). The initial natural aging forms a high density of small GP zones with nearly uniform sizes. The elevated temperature aging then allows some of those GP zones to grow by Ostwald ripening (i.e., "cannibalizing" atoms by diffusion from neighboring GP zones), producing a range of GP zone sizes that more effectively impede dislocation motion. An alternative duplex heat treatment (16 hours at 80°C followed by 24 hours at 150°C) produces a coarser η' precipitate structure that is not quite as strong.

Early 7xxx series alloys tended to form precipitate-free zones (PFZs), regions near grain boundaries with low precipitate concentrations, and were vulnerable to stress corrosion cracking (Fig. 20.22). These problems were reduced by adding Cu to the alloys. Cu contributes some solid-solution strengthening, but its main function is to promote precipitation in all parts of the grain, reducing PFZ size (Fig. 20.23). When Cu content is less than 1%, the precipitation process remains the same as in Al–Zn–Mg alloys; however, when Cu content rises to 1.5 to 2%, as it does in some widely used 7xxx alloys (Table 20.9), the precipitation process is altered. Cu enters the precipitates, raising the temperature needed for GP zone formation and lowering the coherency of the precipitates at peak strength. Mg(Zn,Cu,Al)$_2$ forms in preference to MgZn$_2$. However, in these alloys Cu can cause solidification cracking, making the highest-strength Al–Zn–Mg–Cu alloys essentially unweldable.

Despite their high cost, Al–Sc–Zr 7xxx series alloys have experienced growing use (Sec. 23.6.2). Their recrystallization resistance during solutionizing allows them to maintain fine grain sizes and improve weldability in (Cu-free) Al–Zn–Mg alloys that are otherwise difficult to weld.

20.2.4.7 Al–Li Alloys. Li is one of the few elements with extensive solid solubility in Al. At 600°C, 16 at% (4.7 wt%) Li can dissolve in solid Al. Adding Li to Al has two highly

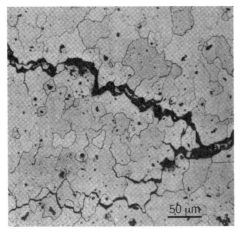

Fig. 20.22 Intergranular stress corrosion cracking in a 7075–T6 extrusion. Note how the fracture follows the grain boundaries through the metal. (From ASM, 1972, p. 255; with permission.)

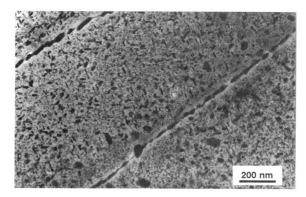

Fig. 20.23 Transmission electron micrograph of 7150–T651 Al alloy (Al–6.2% Zn–2.3% Cu–2.3% Mg–0.12% Zr). Most of the small, dark, platelike particles are η' precipitate; the balance are η phase. Note that even though the grain boundaries running diagonally through the image are decorated with precipitates; there are no significant PFZs. (From Callister, 1997, p. 321; with permission.)

TABLE 20.9 Composition and Mechanical Properties of 7075 and 7001 Al Alloys

7075 Composition: Al–5.1 to 6.1% Zn–2.1 to 2.9% Mg–0.5% Fe–1.2 to 2.0% Cu–0.3% Mn–0.18 to 0.28% Cr–0.4% Si–0.2% Ti (other elements 0.05% maximum each, 0.15% maximum total).

	Yield Strength (MPa)	Ultimate Strength (MPa)	Elongation (%)
Annealed (7075-O)	100	225	17
Solution-treated and artificially aged (7075-T6)	500	575	11
Solution-treated and overaged (7075-T73)	435	500	13

7001 Composition: Al–6.8 to 8.0% Zn–2.6 to 3.4% Mg–0.4% Fe–1.6 to 2.6% Cu–0.2% Mn–0.18 to 0.35% Cr–0.35% Si–0.2% Ti (other elements 0.05% maximum each, 0.15% maximum total).

	Yield Strength (MPa)	Ultimate Strength (MPa)	Elongation (%)
Annealed (7001-O)	150	255	14
Solution-treated and artificially aged (7001-T6)	625	675	9
Solution-treated, cold-worked, and overaged (7001-T75)	500	580	12

desirable effects; every 1 wt% Li added raises elastic modulus by 6% and lowers density by 3%. However, Al–Li alloys have seen only limited use because they cost three to five times as much as ordinary Al alloys, and some challenges remain in providing consistently high ductility and fracture toughness. Al–Li alloys' two-phase microstructures contain AlLi precipitates (δ) in an Al matrix. The AlLi precipitates have a large misfit with the Al matrix and tend to nucleate heterogeneously on grain boundaries in slow-cooled and overaged alloys. However, a metastable precipitate Al_3Li (δ') also forms readily in these alloys, and early precipitates are predominantly δ'. The δ' precipitates have a low interfacial energy with the matrix (about 14 mJ/m^2), so they can easily nucleate homogeneously in the matrix, forming spherical precipitates (Fig. 20.24a).

The δ' precipitates are coherent and have a low lattice mismatch (only 0.08%) with the matrix; consequently, they do not introduce significant elastic strain in the surrounding matrix as GP zones do in Al–Cu alloys, but harden the metal by providing shearable barriers to dislocation motion. Since δ' is an ordered intermetallic, a dislocation moving through a δ' particle forms an

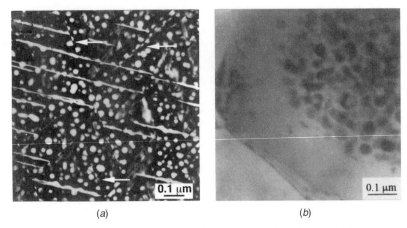

Fig. 20.24 (*a*) Dark-field transmission electron micrograph of spherical δ' (Al$_3$Li) precipitates (white spheres) and θ' (long, white streaks) in Al–2.5% Cu–1.84% Li–0.14% Zr alloy. Some δ' precipitate shearing is evident (arrows). (*b*) Bright-field transmission electron micrograph of spherical δ' (Al$_3$Li) precipitates (gray spheres) in Al–3.4% Li–1.0% Mg–0.3% Cu–0.25% Zr alloy. A precipitate-free zone has formed near the grain boundary in the left and lower left portions of the image. [(*a*) Used with permission of Woo and Kim, 2002; with permission. (*b*) From Zhen et al., 2002; with permission.]

APB that can be restored to its lowest-energy structure only by passage of a second dislocation on the same slip plane (Sec. 25.6.1 and Fig. 25.4). Each dislocation pair produces an offset (Figs. 20.16 and 20.24*a*) in the previously spherical δ' particles, which reduces the precipitate area through which subsequent dislocations must shear. This behavior means that every pair of dislocations traveling on a slip plane makes it easier for following dislocation pairs to travel on that same plane. This tends to encourage large numbers of dislocations to travel on the same slip plane, localizing strain in narrow slip bands within each grain. This is not ideal for good tensile ductility, but it does confer excellent resistance to tensile crack growth under cyclic loading (i.e., fatigue). Advancing cracks follow these slip bands through the grains, and cracks must make abrupt direction changes at each grain boundary. Such abrupt changes often cause crack tips to meander into less favorable orientations for continued expansion as they enter new grains, slowing crack advance.

Slow cooling or overaging in Al–Li alloys tends to form the equilibrium δ precipitates on grain boundaries, and the high elastic strain that these δ particles create in the nearby Al matrix generates mismatch dislocations (Sec. 9.5.2.1). These mismatch dislocations tend to dissolve metastable δ' precipitates near grain boundaries, making their Li atoms available for diffusion to the growing δ precipitates on the grain boundaries. This produces PFZs near grain boundaries (Fig. 20.24*b*) that can concentrate slip in PFZs, causing dislocation pile-ups at the brittle AlLi δ precipitates on grain boundaries. This, in turn, leads to intergranular fracture, poor ductility, and low fracture toughness. Such problems plagued early trials of Al–Li alloys. Cu and Mg alloying additions to these alloys reduce δ formation on grain boundaries and to maintain fine grain sizes to mitigate intergranular fracture tendencies. Cu and Mg lower Li solubility in Al and create Al$_2$CuLi (called T1) and Al$_2$CuMg (S$'$) precipitates. These additions solve some problems, but create new ones, making Al–Li alloy development an ongoing challenge. Despite these difficulties, Al–Li alloys are used in some aircraft structures where specific strength and stiffness are crucially important. Two widely used Al–Li alloys are 2090 and 8090 (Table 20.10).

20.2.5 Properties and Applications of Cast Al Alloys

In some ways, Al is well suited to casting. It has a fairly low melting point, low solubility for all gases except H$_2$, and good fluidity. Its low density simplifies handling both the molten

TABLE 20.10 Composition and Mechanical Properties Measured Parallel to the Direction of Rolling of 2090 and 8090 Al–Li Alloys

2090 Nominal Composition: Al–2.7% Cu–2.2% Li–0.12% Zr.
8090 Nominal Composition: Al–2.45% Li–1.3% Cu–0.95% Mg–0.12% Zr.

	Yield Strength (MPa)	Ultimate Strength (MPa)	Elongation (%)	K_{IC} (MPa$\sqrt{m}$)
Solution-treated, cold-worked, and artificially aged (2090-T83)	515	550	6	44
Solution-treated, cold-worked, and artificially aged to maximum hardness (8090-T84)	400	480	4	75
Solution-treated, cold-worked, and artificially aged (underaged) (8090-T81)	325	415	11	95–165

metal and final product. However, Al also poses special problems in casting caused by its high solidification shrinkage (3.5 to 8.5%, depending on the alloy) and tendency to form H_2 bubble porosity. In addition, cast Al alloys are susceptible to distortion during stress-relief anneals, and their mechanical properties are generally inferior to those of wrought alloys.

The most common casting methods for Al alloys are sand mold casting (small production numbers), permanent mold casting (production numbers of a few thousand), and pressure die casting (production numbers $>\sim 10^4$). Although several dozen Al casting alloys exist, relatively few are used in large quantities. Alloys 208 (4% Cu–3% Si), 213 (7% Cu–2% Si), 356 (7% Si–0.3% Mg), and 413 (12% Si–2% Fe) dominate Al sand casting. Alloys 319 (6.3% Si–3.5% Cu) and A332 (12% Si–1% Cu–1% Mg–2.5% Ni) are the most commonly used permanent mold casting alloys, and 380 (8.5% Si–2% Fe–3.5% Cu) and 413 are most widely used for pressure diecasting. Al casting alloys have varying castability; 3xx alloys are generally easiest to cast followed by 4xx, 5xx, 2xx, and 7xx, listed in order of decreasing castability.

20.2.5.1 Al–Si (4xx.x) and Al–Si–Mg–Cu (3xx.x) Casting Alloys. Al–Si casting alloys have good fluidity, weldability, corrosion resistance, and wear resistance. In addition, their COTEs are lower than most other Al alloys. These alloys also have disadvantages; the presence of hard Si particles makes them difficult to machine, and they require either rapid cooling or chemical modification (described below) to achieve good ductility.

The 4xx.x casting alloys typically contain from 5 to 12% Si and small amounts of Fe. The 3xx.x alloys are Al–Si alloys with Mg, Cu, and/or other alloying additions to increase strength by solid solution and precipitation hardening. These alloys have high fluidity and excellent corrosion resistance. Si has an exceptionally low coefficient of thermal expansion (COTE) (Sec. 21.2.2), so its presence lowers the COTE of eutectic alloys to about 19.6 μm/m·°C. In hypoeutectic Al–Si alloys (Si content < 12%), nearly pure Al dendrites are the first metal to freeze, and these dendrites have the composition shown at the left end of lever law tie-line 1 in Fig. 20.11 for a 7% Si alloy. As cooling progresses, Si content increases in the liquid metal remaining in the interdendritic spaces until it reaches the eutectic composition shown at the right end of tie-line 2 in Fig. 20.11. The last liquid at the eutectic must segregate into two phases as it freezes, nearly pure Al and nearly pure Si, as shown by tie-line 3 in Fig. 20.11. If the casting cools slowly, large, fragile Si and Al–Fe–Si intermetallic precipitates form between the Al dendrites (Fig. 20.12*a*). If the casting cools rapidly or is given Na additions, the eutectic metal forms small, closely spaced Al and Si regions like those shown in Fig. 20.12*b*. A finer eutectic structure improves strength, ductility, and fracture toughness, so considerable effort has been made to find methods that favor that structure, even in castings where rapid cooling is not possible. Na modification is often used to refine the eutectic structure in hypoeutectic alloys with Si contents above 7%. Adding Na metal (0.025% or less) or Na salts to the molten Al–Si alloy just prior to casting suppresses Si crystallization and effectively shifts the Al–Si eutectic point to lower temperatures and higher Si contents. This produces a greater number of nucleation sites in the undercooled metal, promoting a finer dispersion of Al and Si in the last metal to freeze, even when cooling rates are fairly slow.

Na modification mitigates the problems associated with coarse microstructures in Al–Si alloys, but it introduces new problems. Adding Na salts to molten Al also adds other, unwanted elements to the metal. Na is a volatile, reactive metal (Sec. 10.2.2) that floats on Al and its fluxes; it can rapidly be lost to evaporation and oxidation before the metal is poured. Introducing Na can cause turbulence in the metal that raises H content. Excess Na (overmodification) causes the Si particles to coarsen and denudes the outer regions of each eutectic grain of Si. For these reasons, Sr, Ca, and Sb have all been tried as alternative modifiers in Al casting alloys. They vaporize more slowly and are more "forgiving" if too much is added. Al–Sr master alloys are now used widely to modify Al–Si casting alloys. If Na, Sr, and Ca modifying additions are used, P content is usually held to 5 ppm or lower in hypoeutectic and eutectic alloys. P reacts preferentially with the modifier to form phosphides, which removes the modifying metal from solution and nullifies the modification effect. Curiously, P additions (15 to 300 ppm) are sometimes deliberately added to the less commonly used hypereutectic Al–Si alloys to produce Si particles that are more rounded and less angular.

The strength of Al–Si casting alloys can be enhanced by Mg additions (about 0.35%) or Cu additions (1.8 to 4.5%). These make the alloy heat-treatable (Sec. 20.2.4.4) by precipitation hardening and improve machinability, but their addition degrades castability and ductility. Corrosion resistance is relatively good in Al–Si–Mg alloys, but Al–Si–Cu alloys are more vulnerable to corrosion. Bi or Pb are sometimes added to form insoluble inclusions to improve machinability.

20.2.5.2 Al–Cu (2xx.x) Casting Alloys. The first Al casting alloys developed were Al–Cu alloys. Although they provide good precipitation hardening response (Sec. 20.2.4.4), they are vulnerable to hot tearing as the casting cools, and they require generous feeding of liquid metal to avoid shrinkage cavities. The alloys are also prone to stress corrosion cracking. Modern Al–Cu casting alloys contain other metals (Mg, Ni, Ag) to ameliorate some of these tendencies, allowing these alloys to be used for such diverse applications as engine pistons and cylinder heads and soleplates on clothing irons. When heat treated, these are the strongest casting alloys; some premium 2xx.x alloys can be precipitation hardened to achieve $\sigma_y = 480$ MPa, $\sigma_{UTS} = 550$ MPa, with 10% tensile elongation.

20.2.5.3 Al–Mg (5xx.x) Casting Alloys. Al–Mg alloys find applications where the cast part needs high corrosion resistance, good machinability, low density, and a surface amenable to anodizing. As with wrought Al–Mg alloys (Sec. 20.2.4.1), 5xx.x alloys can be strengthened only slightly by heat treatment, so solid-solution hardening is their primary strengthening mechanism. These alloys require more careful attention to cooling rates to avoid forming undesirable amounts of Mg_2Al_3 precipitates. High Mg contents (up to 10%) can lower density to 2.53 g/cm^3, but high Mg content increases the potential for stress corrosion cracking from intergranular Mg_2Al_3.

20.2.5.4 Al–Zn–Mg (7xx.x) Casting Alloys. Al–Zn–Mg alloys can be naturally aged over a period of weeks following casting, producing a moderately strong product that has good corrosion resistance and machinability without a postcasting heat treatment. However, strengths are substantially less than those seen in wrought Al–Zn–Mg alloys. The alloys are susceptible to overaging if heated in service, and they often suffer from hot cracking problems in permanent molds.

20.2.6 Other Uses of Al

20.2.6.1 Al in Steel. Aluminum ferrosilicon is used to deoxidize liquid steel. The Al and Si react with dissolved O (and to a lesser extent N) in the molten steel to prevent gas bubble formation as the steel solidifies. The Al_2O_3 and AlN precipitates that form in the steel also retard recrystallization. Deoxidized steel is often called "killed" steel, since deoxidizing additions quiet the bubbling turbulence that would otherwise occur from the reaction $FeO + C \rightarrow Fe + CO(g)$.

20.2.6.2 Al Compounds. The spectrum of engineering applications for Al chemicals is truly vast. Several million tons of alumina (Al_2O_3) are used annually as a major constituent in cement, abrasives, refractory brick, and ceramic armor. Alumina serves as a substrate for microelectronics packaging and for such diverse products as spark plug insulators, phosphors in fluorescent lights, flame retardants, catalyst supports, toothpaste, and dental implants. Alumina is present in many commercial glasses along with Al borate. Al_2O_3 fibers and particles are used as filter media and as reinforcing particles in composite materials. Hollow-fiber Al_2O_3 membrane units are used in water desalination. When processed to achieve near-zero porosity, alumina is transparent and can be used for windows in high-temperature applications where glasses would sag or devitrify.

Alums are used for plaster, tanning, and dyeing. Alkoxides are added to varnishes. Anhydrous $AlCl_3$ is present in petroleum cracking catalysts, rubber, and antiperspirants. $Al(OH)_3$ is a desiccant, a soft abrasive, and an inexpensive filler in plastics, rubber, and paper. Al hexahydrate is a wood preservative, an antiseptic, and a cosmetic additive. Al nitrates are used as corrosion inhibitors and leather tanning agents. Al sulfate is a particularly versatile compound; applications include use as a flocculant in water and sewage treatment, in fireproofing and waterproofing fabrics, in soaps, as a soil conditioner to lower pH, and as an intermediate in production of other chemicals.

20.2.7 Sources of Al and Refining Methods

Bauxite ore is used for nearly all Al production. The Al-bearing constituent of bauxite is nominally $Al(OH)_3$; however, AlOOH is often present in many deposits. World reserves of bauxite are enormous, estimated at 20 to 60 billion tons. Bauxite typically contains SiO_2 (1 to 15%), Fe_2O_3 (7 to 30%), TiO_2 (3 to 4%), plus small amounts of various fluorides, phosphates, and oxides (Sec. 20.3.5). In the Bayer process, bauxite is calcined to dehydrate it, then reacted with aqueous NaOH solution at 120 to 250°C under high pressure. This forms soluble $NaAlO_2$ that is separated from the various insoluble non-Al compounds (collectively called "red mud"). Al is then precipitated from the $NaAlO_2$ solution in cycles as $Al(OH)_3$, which is calcined at 1200°C to yield about 99% pure Al_2O_3.

About 15% of the Al_2O_3 produced by the Bayer process is used to make refractories, abrasives, and Al chemicals (Sec. 20.2.6.2), and the other 85% is reduced to Al metal. Direct electrolysis of Al_2O_3 is impractical since its melting point is 2045°C. In the Hall–Héroult process, Al_2O_3 is dissolved at a concentration of 2 to 6% in molten cryolite (Na_3AlF_6) to lower the melting temperature so that electrolysis can be performed at 950°C (Fig. 20.25). Cell amperages up to 300,000 A are used. Al metal production is energy intensive, so Al reduction operations are located where electric power costs are low. The electrolysis proceeds by a series of intermediate reactions that are equivalent to an overall reaction:

$$2Al_2O_3 \rightarrow 4Al + 3O_2$$

The O_2 is deposited on the C anode and reacts with it to produce CO_2 gas; thus, anodes are consumed gradually as the cell operates. For best efficiency, it is desirable to maintain the smallest possible gap between anode and cathode. However, if the gap is too narrow, turbulence from gas evolution and magnetic convection can induce undesirable liquid Al metal "bridges" that short circuit the electrodes. Steady improvements in cell instrumentation and design have lowered the energy required to produce Al from 20 kWh/kg in 1940 to 12 kWh today.

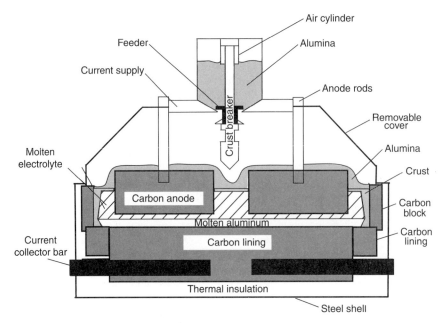

Fig. 20.25 Electrolytic reduction cell for converting Al_2O_3 to Al metal. Cells are about 10 m long, 4 m wide, and 1.2 m high. Electrodes are graphite, and the 3- to 25-cm thick liquid Al metal layer sinks to the bottom of the cell because it is denser than cryolite. A solid cryolite crust forms over the liquid alumina–cryolite electrolyte and along the walls of the chamber. This reduces air infiltration, minimizes heat loss, and prevents rapid erosion of the chamber walls. New alumina is fed from above with the aid of the crust-breaker tool. (Redrawn from Totten and MacKenzie, 2003, Vol. 2, p. 28.)

The Al accumulated is removed once every day or two from the cells and transported to a holding furnace where recycled Al scrap is often added. At this stage the metal contains nonmetallic inclusions, dissolved H, Na, Ca, and other impurities. These are removed by filtration, fluxing, and bubbling an Ar–Cl_2 gas mixture through the metal. These steps lower Na and Ca levels to 1 to 2 ppm and reduce H to about 1 mL (gas) per kilogram of metal. At this point, the metal is 99.50 to 99.80% pure; the principal impurities are Fe and Si. This purity level is sufficient for most uses; however, purity levels of 99.98 to 99.99% can be attained by subsequent refining using either electrolysis on the molten Al or unidirectional solidification or zone melting to segregate impurities into the liquid phase. Ultrapurity Al (99.9999% pure) for electronic use is produced by multiple cycles of the electrolytic and/or segregation methods.

20.2.8 Structure–Property Relations in Al

20.2.8.1 Crystal Structures in Liquid Al. It was once widely assumed that liquid metals were an amorphous, chaotic jumble of atoms that rapidly shift positions with their nearest neighbors. However, it now appears that this model is simplistic. Liquid metals contain enormous numbers of small crystallites with distinct crystal structures. Although the bonds between these crystallites are transient and rapidly shifting, the crystal structure within them is surprisingly like that of a solid. The lack of stable bonding between crystallites means that liquid metals cannot support a load, but they do show broad peaks in x-ray, electron, and neutron diffraction studies. Diffraction studies show that crystallites in pure liquid Al undergo three distinct "phase changes": FCC → BCC at 800 to 900°C, BCC → FCC at 1150 to 1250°C, and FCC → HCP at 1390 to 1450°C.

These structures affect liquid Al's viscosity, oxidation kinetics, density, and electrical conductivity. Impurities alter these liquid metal crystallite structures and their transitions, which in turn affects the metal's ability to distribute alloying additions uniformly throughout the liquid. Most Al binary phase diagrams depict a completely homogeneous liquid state, but it is likely that many systems actually contain various liquid phases (Sec. 22.4.2.1). For example, Al–Si alloys are now known to contain a surprising array of "new" phases that are stable in the liquid state but not in the solid state. These may help to explain some aspects of modified Al–Si alloy casting behavior (Sec. 20.2.5.1) that seem inconsistent with a simple eutectic phase diagram. In Al containing 10 to 12% Si, a cubic χ phase with the possible formula Al_6Si appears to exist in the liquid, and it can be retained as a metastable solid phase. Several other phases (Fig. 20.26) exist in hypereutectic alloys. It has long been known that Al–Si cast microstructures change as the metal's pouring temperature changes (i.e., the amount of superheat). Such behavior is hard to explain with the simple equilibrium phase diagram shown in Fig. 20.11; however, it is consistent with the liquid-phase eutectics and peritectics shown in Fig. 20.26.

Surfing through a CD of binary phase diagrams gives the impression that the important binary systems have all "been done." However, it may be that the blank fields labeled "liquid" in many of these diagrams are not perfectly homogeneous metal solutions but are, instead, equivalent to the blank spots on ancient maps labeled "terra incognita" and "here be dragons." The white space atop most phase diagrams may contain a complex and potentially useful menagerie of liquid phases.

20.2.8.2 Dispersoids in Al. Small amounts of Cr, Mn, or Zr are often added to Al alloys to form sub-micrometer-sized dispersoids in the metal. These elements diffuse slowly in Al, so they have difficulty forming large particles; this has the desirable effect of forming large numbers of small particles [e.g., $Al_{20}Cu_2Mn_3$, $Al_{11}Cr_2$, $Al_6(Fe,Mn)$, $Al_{12}Mg_2Cr$, and Al_3Zr]. Small dispersoids (less than about 400 nm) retard recrystallization by pinning subgrain boundaries and slowing subgrain coalescence, and they have low solubility in the Al matrix at higher temperatures, so

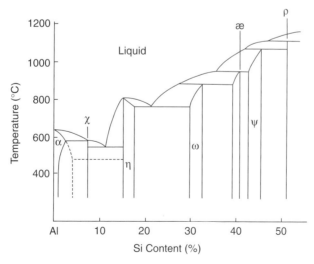

Fig. 20.26 Nonequilibrium phase diagram of metastable phases in the Al–Si system for cooling rates typical of casting conditions. Metastable phases are χ (cubic, Al_6Si), η (rhombohedral, Al_4Si), ω (hexagonal, Al_7Si_3), æ (tetragonal, $Al_{62}Si_{38}$), ψ (tetragonal, Al_3Si_2), and ρ (tetragonal, AlSi). Note the differences between this diagram and the equilibrium diagram in Figure 20.11. Much of the hypereutectic structure shown on this diagram lies above the liquidus line on the equilibrium Al–Si phase diagram. (Redrawn from Totten and MacKenzie, 2003, Vol. 2, p. 102.)

they resist dissolution during heat treatments and hot work. When a dispersoid-containing Al alloy is rolled or drawn, dispersoids become "strung out" along the boundaries of the pancake- or rod-shaped grains. This grain shape favors transgranular fracture in worked Al, resulting in high fracture toughness. However, if too many dispersoids are present, they actually lower fracture toughness by nucleating microvoids at the dispersoid–matrix interface that can connect with one another to make fracture easier. For this reason, dispersoid-formers are limited to 0.1 to 0.4%.

Coherent dispersoids such as Al_3Zr are preferred boundary pinning agents because they either (1) force the interface to shift from coherent to semicoherent or incoherent if the boundary breaks away from the dispersoid, or (2) require the dispersoid to redissolve into the matrix before the boundary can break away. Both of these changes require considerable energy, making breakaway from coherent particles more difficult. Zr has the additional advantage of forming unusually small dispersoids (often less than 50 nm in diameter). In some alloys Al_3Zr dispersoids promote precipitation of hardening phases on dislocation loops they induce in the adjoining Al matrix (Sec. 9.5.2.1 and Fig. 9.9). In Al–Li alloys, heterogeneous precipitation of δ' occurs on dislocation loops around Al_3Zr dispersoids, and the same effect occurs around S' and θ' phases in Al–Li–Cu–Mg–Zr alloys. Al_3Sc dispersoids appear to be the "ultimate dispersoid" (Sec. 23.6.2), but Sc's extraordinarily high cost sharply limits its use in Al alloys.

20.2.8.3 *Oxide Inclusions in Al.* Al_2O_3 originating in crucibles containing liquid Al can enter the metal. These so-called "old oxides" are undesirable because they reduce fatigue life in cast and wrought products and cause pinhole defects in sheet and foil. Old oxides can sometimes be filtered from liquid metal before it is cast, but even if they are present in the solidified alloy, they are large enough to be detectable by microscopy and nondestructive evaluation (NDE) techniques.

A second type of oxide defect is more insidious. These are the so-called "young oxides" that form on the surface of liquid Al (usually, as it is being poured) and become folded into the metal by turbulence when the flow velocity exceeds 0.5 m/s. These oxides (Fig. 20.27) are extraordinarily thin (tens of nanometers thick), and in regions where they become folded atop one another, there is little or no bonding at the Al_2O_3–Al_2O_3 interface. There is often a thin wedge of trapped air between the two oxide surfaces. Such defects act as cracks in the solid metal. They tend to reside at grain boundaries, since growing dendrites cannot penetrate the oxide–air layer in the solidifying metal. Small sheets of young oxides are often below the detection limit of NDE methods, and their extreme thinness makes them difficult to see metallographically.

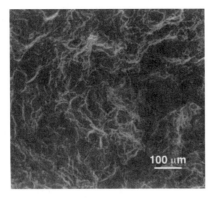 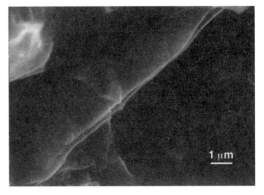

Fig. 20.27 Young Al_2O_3 film at the fracture surface of an Al casting seen at low (left) and high magnification (right). Such young oxides films can lower the bending strength of Al castings by as much as 90%. (From Tiryakioglu et al. 1996; with permission.)

Bending strengths of Al castings can be reduced by as much as 90% by young oxides, and young oxides can reduce fatigue life by orders of magnitude since parts essentially contain large cracks before their service life begins. Even when young oxides are present in smaller amounts, they cause much greater dispersion in all mechanical properties, degrading reliability and forcing wasteful "overdesign" of parts. For these reasons, pouring velocities of liquid Al must be kept below 0.5 m/s to avoid turbulence in sprues and mold cavities that forms young oxide inclusions.

20.3 GALLIUM

> Gallium was predicted as eka-aluminum by Mendeleev in 1870, and was discovered by de Boisbaudran in 1875. ... [T]he striking similarity of its physical and chemical properties to those predicted by Mendeleev did much to establish the general acceptance of the Periodic Law; indeed, when de Boisbaudran first stated that the density of Ga was 4.7 g·cm^{-3} rather than the predicted 5.9 g·cm^{-3}, Mendeleev wrote to him suggesting that he redetermine the figure (the correct value is 5.904 g·cm^{-3}).
>
> — Greenwood and Earnshaw (1997)

20.3.1 History of Ga

Mendeleev predicted the existence and properties of Ga in 1870, and it was first isolated five years later. Its first engineering application was as an alloying additive to Pu in nuclear weapons (Sec. 24.5.3.3). Ga had no significant commercial use until 1970, when microelectronic device fabricators began using semiconducting GaAs and GaP compounds. These semiconductors have performance advantages over Si devices (Sec. 22.2.4), but they are costly to produce. The only substantial use for Ga is fabrication of light-emitting diodes, lasers, microprocessor chips, and related semiconductor devices. These are high-value products, but their fabrication consumes only small amounts of Ga.

20.3.2 Physical Properties of Ga

Ga's unusual crystal structure (oC8, Fig. 1.5) provides each Ga atom with one nearest neighbor (0.244 nm away) and three pairs of other neighboring atoms 0.270, 0.273, and 0.279 nm away. Ga has a mixed metallic–covalent bonding that could be described as "Ga$_2$ molecules" using their single p electrons to bond covalently, embedded within a metal lattice. Ga's thermal expansion coefficient and electrical resistivity are highly anisotropic due to the low symmetry of its orthorhombic cell. Solid Ga has low atomic packing efficiency; Ga actually contracts 3.4% upon melting. Ga melts only a few degrees above room temperature, but its boiling point is 2420°C. Ga's low melting point poses challenges in its handling and transport (Sidebar 20.6). Ga's partially covalent bonding makes it brittle, but it possesses metal-like conductivity.

20.3.3 Applications of Ga

Approximately 98% of Ga production is used to make light-emitting diodes (Sidebar 20.7), diode lasers, electroluminescent devices, and specialty microprocessor chips. Since all of these require ultrapure material to function properly, nearly all Ga is processed to exceptionally high purity. The intermetallic compounds are produced by reacting them at elevated temperature with the appropriate ultrapure elements. The remainder of their fabrication is quite similar to that of Si chips (Sec. 21.2.4): growing single crystals of the compound, cutting that crystal into chips, masking and doping it with suitable impurities to achieve n- and p-type semiconducting regions, and assembling it with interconnecting wires and vias to produce the final components. The efficiency of LEDs exceeds 30 lu/W, and their operating life is considerably longer than incandescent lights. For these reasons LEDs are finding growing use in such diverse products as

SIDEBAR 20.6: TRANSPORT AND STORAGE OF Ga

Most Ga is used in semiconducting electronic applications, which require $6N$ to $8N$ purity (i.e., 99.9999 to 99.999999% pure). Since Ga melts at only 30°C, its handling and transport pose unusual challenges. If ultrapure Ga melts during transport, the liquid metal picks up contaminants from its container. Thus, ultrapure Ga shipments are planned carefully to keep the metal continuously cool in transit. In hot weather the shipping container is actively cooled. Air transport is preferred to shorten transit time and minimize melting risk. Surprisingly, Ga shipments pose a potential threat to aircraft. Liquid Ga forms a low-melting eutectic with Al that drops Al's strength to near zero. Since nearly all structural members in aircraft are Al alloys, a liquid-Ga spill in an airplane's cargo hold could conceivably destroy the aircraft in flight. For this reason, the International Air Transport Association mandates that Ga air cargo be packaged in a minimum of seven layers of containment, and limits are imposed on the quantity of Ga permitted on an aircraft.

SIDEBAR 20.7: LIGHT-EMITTING DIODES

Light-emitting diodes (LEDs) convert electrical power to light. The LED contains a junction between n- and p-type doped regions in a GaAs (or similar group 13–group 15 intermetallic compound) single crystal (Fig. 20.28). When current flows through the LED, electrons and holes (Sec. 21.2.4) meet near the junction of the doped regions, where they annihilate one another. The energy released from this mutual annihilation of charge carriers is emitted as light. The wavelength of the light can be varied by altering the LED's composition: pure GaAs (infrared), GaAsP (red), GaP (orange–red), GaAlAsP (red to yellow, high brightness), InGaN (green–blue), GaN (blue). Blue sources can be covered with a phosphorescent panel to emit white light, or white can be achieved by combining three or four LEDs of different colors. LEDs are reliable, efficient devices that require only low voltage power and generate little heat. These lights can also be used with batteries and pedal-powered generators to provide home lighting to Third World residents, who are often restricted to kerosene lanterns or disposable batteries for illumination.

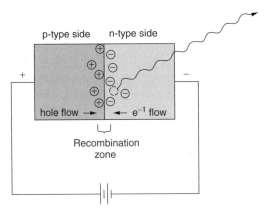

Fig. 20.28 Schematic depiction of hole–electron recombination to produce light in a GaAs light-emitting diode (LED). The voltage gradient across the LED generates holes at the left side of the GaAs crystal and free electrons at the right side of the crystal. These charge carriers move to the recombination zone, where they annihilate each other and emit photons of light.

flashlights, display lighting, and traffic signals. In addition to its use in semiconducting devices, Ga has a few minor uses in specialty dental alloys, Pu alloys, and as a substitute for Hg in high-temperature thermometers.

20.3.4 Toxicity of Ga

No conclusive evidence exists for Ga toxicity in humans. The principal Ga product, GaAs, is toxic due to its As content, but no injury has ever been reported from handling Ga metal. Ga uptake from the gastrointestinal tract is quite low, and it appears to serve no role in human nutrition. Laboratory animals given large intravenous doses of Ga compounds suffered kidney and nerve damage, but these results are probably not meaningful predictors of the effects of much smaller dosages obtained from routine handling of Ga. Until more data are available, persons working with Ga are advised to handle it as if it were toxic.

20.3.5 Sources of Ga and Production Methods

Trace amounts of Ga are present in a wide array of minerals, but nearly all Ga is produced as a by-product of bauxite processing in alumina production from the circulating liquors used in the Bayer process reaction: $Al_2O_3 + 2NaOH \rightarrow 2NaAlO_2 + H_2O$. Ga is extracted from this aqueous solution by fractional precipitation, electrolysis, or reaction with chelating agents. Most bauxite contains only about 0.005% Ga, but such huge quantities of alumina are produced that potential Ga recovery could be as high as several thousand tons/yr. Demand for Ga is too low to support such output, so Ga separation is performed only on bauxite containing the highest Ga content, and even in those ores, only part of the Ga is recovered. Ga is present in the fly ash from burned coal, but Ga content (0.01 to 0.1%) is too low for economic extraction. About 200 tons of Ga is produced annually, primarily in China, Germany, Japan, and Russia. Ultrapurity (99.9999+%) is achieved by fractional crystallization and distillation, zone refining, and single-crystal growth.

20.4 INDIUM

Es zeigte keine Thalliumlinie, dagegen eine indigoblaue bisher unbekannte Linie.

—Ferdinand Reich and Hieronymus Richter

20.4.1 History of In

Reich, a colorblind spectroscopist, employed Richter to assist him with color determinations; together they examined the emission from a mineral expected to show the green Tl line, but instead, they saw an unexpected, intense blue spectral line. This led to their discovery in 1863 of a new element, indium, named for the indigo blue of its characteristic spectral line. Indium is a soft, ductile, low-melting metal. Its early engineering applications were primarily in Au dental alloys and engine bearings. After World War II, indium use expanded into fusible alloys, glass–metal seals, and nuclear control rods, but in recent decades electrical and electronic devices have become the dominant use. Indium is produced primarily as a by-product of Zn, Pb, and Sn production.

20.4.2 Physical Properties of In

Indium is tetragonal (tI2), a slight distortion from the BCC structure ($a_0 = 0.458$ nm, $c_0 = 0.494$ nm). It is highly ductile, even at cryogenic temperatures, and tolerates almost indefinite rolling reduction since it does not work harden at 20°C. Indium is quite soft ($\sigma_{UTS} = 3$ MPa with 85% reduction in area). During deformation, indium emits a characteristic "tin cry" from twin formation. Indium is a reasonably good electrical and thermal conductor. Its low melting point and its ability to wet glass and bond well to metal make it an efficient glass–metal sealing agent. Indium does not oxidize in room-temperature air, and it reacts only slowly with most dilute mineral acids. The high thermal neutron cross section of indium (190 barns) makes it a useful control rod material in nuclear reactors. Indium is a superconductor below 3.37 K.

20.4.3 Alloys of In

Indium alloys have a wide range of applications. Indium alloyed with Pb, Sn, Cd, and/or Ag is used in some fusible alloys (Sec. 22.4.3) and low-melting solder pastes. Indium coatings on Pb–Sn bearings improve fatigue resistance and oil films retention (Sec. 18.3.3.4). Indium-coated bearings were a key factor in producing reliable high-performance engines for World War II military aircraft. Indium additions to Au dental and jewelry alloys improve machinability and resistance to discoloration. An alloy of 80% Ag–15% In–5% Cd is used in neutron-absorbing control rods for nuclear reactors. Indium is used as a corrosion inhibitor in alkaline dry-cell batteries, allowing their Hg content to be reduced.

20.4.4 Compounds of In

Transparent In–Sn oxide coatings sputtered onto glass surfaces make them electrically conductive for use in liquid-crystal displays, television and computer screens, low-pressure Na lamps, and windows and windshields that can be defogged by ohmic heating. Indium orthoborate serves as a phosphor in CRT displays. In–Ga–As–P laser diodes emit light at an infrared wavelength well matched to the wavelength of maximum transparency in glass communication fibers; and In–Ga–As photodetectors are often used to read laser signals at amplifying and receiving stations. In–Sb charge-coupled devices provide good infrared response for video cameras.

20.4.5 Sources of In and Production Methods

Fumes, dusts, slags, and residues from Zn mining provide most of the world's In. Leaching, precipitation, and cementation processes concentrate the In, and if ultrapurity is needed (e.g., for semiconductor applications), electrorefining is used, followed by vacuum distillation and zone refining. As with most by-product metals, large price fluctuations are common. Indium prices varied by a factor of 6 during the last two decades of the twentieth century.

20.5 THALLIUM

> The partial substitution of thallium to this high-T_c mercury-based oxide increased the (superconducting transition temperature) to 138 K for a nominal composition of $Hg_{0.8}Tl_{0.2}Ba_2Ca_2Cu_3O_{8+\delta}$.
>
> — Lee (2004)

20.5.1 Properties and Uses of Tl Metal

Tl is a soft, reactive HCP metal that oxidizes slowly in room temperature air. Pure Tl has an elastic modulus similar to that of softwood, and its ultimate tensile strength is only about 10 MPa. Tl deforms by slip on $\{10\bar{1}0\}$ and (0001) planes as well as by twinning. Tensile elongations in excess of 100% are typical at room temperature. Tl has an elevated-temperature BCC phase that contracts 3.2% upon melting. Although pure Tl is easily formed, it is so soft and reactive that it has no structural uses. The Tl–Hg eutectic temperature (−59°C) is 20°C lower than the freezing point of pure Hg, and Tl was once used as an alloying addition to fusible alloys. Tl additions can harden Pb and Cd for use in bearings, and Tl additions to Cu, Ag, and Au switch contacts reduce sticking tendencies. However, in recent years, the soaring price of Tl, combined with concerns over its toxicity (Sec. 20.5.3), has essentially halted its use in ordinary metallurgical applications.

20.5.2 Properties and Uses of Tl Compounds

The first large commercial use for Tl was $Tl_2(SO_4)$, the tasteless, odorless active ingredient in rat and insect poison. However, this compound caused so many accidental, homicidal, and suicidal injuries and deaths to humans (Sec. 20.5.3) that it has been banned in many countries. Tl's primary use today is in electronic and optical devices. Tl added to Se semiconducting rectifiers increases their resistance to reverse-bias current flow. Tl bromide–iodide is used for lenses in infrared detectors and "night vision" goggles. The electrical resistance of Tl sulfide decreases when it absorbs infrared light, making it useful in photoconductive cells. Tl combined with S or Se and As produces glasses with melting temperatures of only 125 to 150°C and high refractive indices. Tl additions to Ba–Cu–oxide-type superconductors raise the superconducting transition temperature (T_C), including one of the highest T_C values (138 K) ever measured, at 1 atm. Radioactive ^{201}Tl ($t_{1/2} = 3.04$ days) binds strongly to heart muscle and decays by x-rays and γ rays. These emissions are used to produce images of the heart. Less ^{201}Tl reaches heart muscle with partially obstructed blood flow, so these regions appear dark on the images.

20.5.3 Toxicity of Tl

Tl is a potent neurotoxin. Only 600 mg of Tl sulfate in an adult causes death by respiratory paralysis. Tl compounds cause intoxication much more frequently than Tl metal. Tl reacts with a variety of proteins and enzymes in the body, impairing the ATP reaction and damaging mitochondria in axons. Tl poisoning can result from ingestion, inhalation of dusts, or skin contact. Sublethal doses can cause blindness; birth defects; memory loss and emotional changes; and severe nerve, skin, and cardiovascular damage. The excretion rate is slow, Tl persists in the body for days or weeks, and damage is cumulative. Stringent safety precautions are essential for all operations involving Tl production, processing, and disposal.

20.5.4 Sources of Tl and Refining Methods

Nearly all Tl production occurs as a by-product of Zn production with small quantities also derived as Cu and Pb by-products. Tl content is so low in these ores that its recovery is a minor activity, usually performed by leaching Tl from flue dusts collected from smelters. In many mines, recovery of As, Cd, Bi, Se, and other elements is also performed in parallel with Tl recovery, and specific steps for recovery vary with each source. Tl is generally precipitated from the leachate and separated from other trace constituents by volatilizing As and other compounds. Final reduction to Tl metal is performed by electrolysis of aqueous solutions. Since the electronics industry is a major user of Tl, purities as high as 99.9999% are needed, and these ultrapurities are achieved by multistage electrorefining, fractional precipitation, and zone melting. Total world Tl production is estimated at only 15 tons/yr. Germany, Belgium, Peru, Russia, and France are the largest producers. During the last two decades of the twentieth century, the market price of Tl climbed from $16.50 to $1250 per kilogram. This steep price increase resulted from expanding use of Tl in high-temperature superconductors, magnetic applications, and optical and electronic devices. Since Tl is produced exclusively as a by-product metal, supply does not increase greatly when demand rises.

ADDITIONAL INFORMATION

References, Appendixes, Problem Sets, and Metal Production Figures are available at
ftp://ftp.wiley.com/public/sci_tech_med/nonferrous

21 Si, Ge, Sn, and Pb

21.1 OVERVIEW

The group 14 elements range from nonmetallic C through the metalloids Si and Ge to largely metallic Sn and fully metallic Pb. Si and Ge are brittle semiconductors with predominantly covalent bonding; Sn and Pb are conductive, ductile metals at room temperature. The group 14 elements' outer s^2p^2 electron structure hybridizes to provide four bonding electrons, allowing fourfold coordination of the diamond cubic structure of Si, Ge, and Sn's low-temperature allotrope. The abundances of the group 14 elements vary greatly. Si comprises 27.2 wt% of Earth's crust, second only to O; Ge (1.5 ppm) and Sn (2.1 ppm) are much scarcer. Pb (13 ppm) is the endpoint of several heavy-element radioactive decay chains, making it the most abundant of the heavy elements ($Z > 71$).

It is an understatement to call Si a key industrial element. Si is a major constituent of concrete, glass, ceramics, clays, electronic components, and silicone polymers. The heavier elements of group 14 pale in comparison to the immense importance of Si, but they also serve a number of important engineering functions in batteries, corrosion-resistant coatings, bearings, solders, and electronic devices.

21.2 SILICON

> As carbon is the element of life, silicon is the element of the earth.
> —J. B. Calvert, in "Silicon," *http://www.du.edu/~jcalvert/phys/silicon.htm*, 2003

21.2.1 History of Si

Silicates are the major constituent of most rocks, clays, and soils. They were the first inorganic materials used by humankind, and they still dominate much of human activity. Use of pure Si began in 1823 when the first elemental Si was produced by reduction of K_2SiF_6 with K metal. Si metal is used as an alloying additive in Fe and Al, and ultrapure Si is the primary material for microelectronic device fabrication. Despite the great economic and social value of Si-based communication and computation equipment, silicate materials remain Si's most important use. Anyone who lived during the 1940s knows from personal experience that civilization can function without Si-based electronic systems. However, the Si-based materials in concrete, brick, stone, glass, abrasives, and refractories are essential to sustain Earth's enormous human population.

21.2.2 Physical Properties of Si

Si is a classic metalloid. Its physical properties are dominated by its covalent bonding, particularly at lower temperatures (Fig. 21.1). Although Si looks like a normal metal in visible light, it is transparent over a wide range of infrared wavelengths. At room temperature, Si is brittle with

Structure–Property Relations in Nonferrous Metals, by Alan M. Russell and Kok Loong Lee
Copyright © 2005 John Wiley & Sons, Inc.

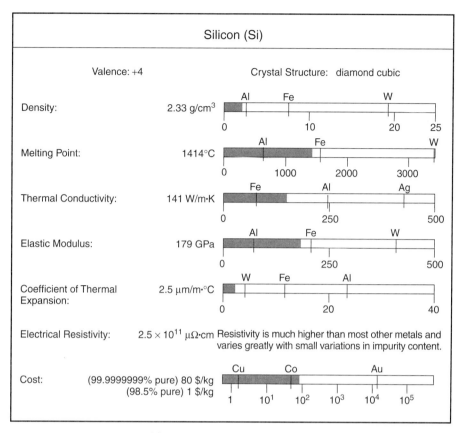

Fig. 21.1 Physical properties of Si.

a fracture strength of 80 to 130 MPa. Small single crystals of Si cleave along crystallographic planes, but fracture is conchoidal in larger specimens. Above 700°C, Si becomes ductile and a better electrical conductor (Fig. 21.2). Si's covalently bonded diamond cubic (cF8) crystal structure (Fig. 1.6) gives it good thermal conductivity and a low coefficient of thermal expansion. The cF8 structure has low packing efficiency, and Si actually becomes denser (2.51 g/cm^3) when it melts. When exposed to air, Si forms an amorphous SiO_2 layer 2 to 3 nm thick. This oxide coating protects Si from further oxidation at temperatures below 900°C. Measurable oxidation begins above 900°C, forming a vitreous coating of silica on the metal surface that gradually thickens and crystallizes as the temperature rises. Above 1400°C, Si also reacts with the N_2 in air to form SiN and Si_3N_4. The oxide layer resists attack by all the common mineral acids except HF acid, which dissolves SiO_2. Mixed HF and HNO_3 acids attack Si aggressively. Hot aqueous alkalis and halogens react with Si. Liquid Si reacts with most commonly used crucible materials, but ZrO_2 and certain transition-metal borides have good compatibility.

Si is a semiconductor with a 1.11-eV bandgap between the valence and conduction bands. Thus, either impurity atoms or elevated temperatures are needed to give Si appreciable electrical conductivity (Fig. 21.2). Si can assume either n-type (where free e^- charge carriers predominate) or p-type conductivity (where electron-deficient bonds called *holes* are the majority charge carriers). Si undergoes a series of crystal structure transformations at high pressure (Appendix A, *ftp://ftp.wiley.com/public/sci_tech_med/nonferrous*) that give it progressively higher coordination numbers (4 → 6 → 8 → 12) and more metallic character as pressure increases. The hardness of pure Si (12 GPa) is limited by its transformation to the softer β-Sn tetragonal crystal structure at

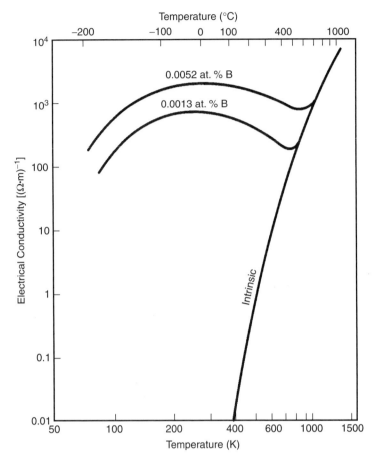

Fig. 21.2 Variation of electrical conductivity with temperature in high-purity Si (intrinsic) and in Si doped with 13 ppm B and 52 ppm B (extrinsic). The intrinsic conductivity of Si increases by about 10^6 between 100 and 1100°C as thermal excitation promotes an exponentially increasing number of electrons from the valence band to the conduction band. B dopant provides more charge carriers and increases conductivity. (From Callister W, 1997, p. 610; with permission; originally from Pearson and Bardeen, 1949.)

12 GPa (Fig. 21.3). Vapor deposition of Si below about 500°C produces amorphous Si, which is metastable at room temperature.

21.2.3 Si as an Alloying Addition to Other Metals

Si is a useful deoxidizer for steel, and steel is often "killed" by adding a ferrosilicon alloy (available in a wide range of Si–Fe ratios, most often 75% Si–25% Fe) to the steel shortly before it is poured. This minimizes gas bubble formation during solidification. Fully killed plain carbon steel typically contains 0.1 to 0.3% Si. Si–Mn alloys (silicomanganese) are often used to deoxidize steel; these two metals produce a manganese silicate precipitate in the steel and are more effective at deoxidizing the metal when used together than is either element alone. High-Si steels (0.5 to 5% Si) acquire a strong texture during rolling that aligns the easy magnetization direction favorably for higher efficiency in electric motors and transformer laminates. Stainless steel often contains about 1% Si, and gray cast iron contains 1 to 3.5% Si. About 3.5 million tons of ferrosilicon (Si content basis) is used each year in iron and steel production. SiC and CaSi alloys are sometimes added to steel in addition to or in place of ferrosilicon.

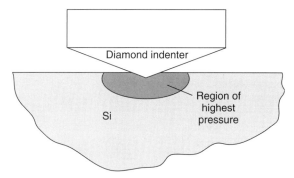

Fig. 21.3 Microhardness and nanohardness indentations in Si indicate a hardness of 12 GPa, because the stress of the indenter tip transforms Si's diamond cubic crystal structure to the softer β-Sn tetragonal high-pressure phase at 12 GPa pressure.

Many Al alloys used for casting, brazing, and weld-filler material contain 2 to 12% Si (Sec. 20.2.5). Si increases liquid Al's fluidity, improves weldability and wear resistance, and delivers corrosion resistance comparable to that of commercial-purity Al. There are, however, some drawbacks to Si additions. The hard Si particles in Al–Si alloys make them more difficult to machine, and the alloys require either rapid cooling or chemical modification to achieve good ductility. The 6000 series of wrought Al alloys contain less Si (0.4 to 1.3%) to form metastable precursors to Mg_2Si for precipitation hardening (Sec. 20.2.4.5). Various other metals contain Si alloying additions, usually to improve wear resistance, strength, or creep strength [e.g., the AS series of Mg alloys (Sec. 11.4.3.1) and silicon bronzes (Sec. 18.2.3.6)].

21.2.4 Si-Based Electronic Devices

Perhaps the greatest technological innovation of the past half century is the computation–communication revolution brought about by microelectronic devices based on semiconducting integrated circuits. The dominant material for fabricating these devices is ultrapure, single-crystal Si doped with ppm levels of group 13 and 15 elements. In an undoped state, pure Si has only *intrinsic* electrical conductivity, achieved by thermally activated promotion of a small number of valence band electrons to the conduction band. Si's intrinsic conductivity is low but rises sharply with increasing temperature (Fig. 21.2). However, when dopants of elements with +5 valence such as P are added as substitutional atoms in the Si lattice, four of the dopant atom's valence electrons bond covalently with the four neighboring Si atoms, but the fifth dopant atom electron needs only a small amount of thermal energy (available well below room temperature) to break free from the dopant atom and move through the lattice in response to a voltage gradient (Fig. 21.4a). This provides many more free electrons than the small number intrinsically available, raising Si's conductivity *extrinsically* by several orders of magnitude. A similar effect occurs when dopants of elements with +3 valence (e.g., B) are added to Si, but in that case the extrinsic conductivity arises from the missing electron needed as the three valence electrons of the dopant atom attempt to form covalent bonds with the four nearest-neighbor Si atoms (Fig. 21.4b). Since B has only three electrons to bond with its four neighboring atoms, a hole results. Holes move through the metal under a voltage gradient, although their mobility is somewhat lower than that of free electrons. This raises the conductivity of the Si, as shown in Fig. 21.2. The complex circuitry of a modern microprocessor is fabricated by a multistep process of masking, etching, and doping selected areas and patterns on the surface of an ultrapure Si single crystal. Conductive pathways on these microprocessors connect diodes and transistors about 100 nm wide, producing microprocessor chips (Fig. 21.5) that perform 4 billion computation steps per second (4 gigaflops). Such computational speed is vastly superior to that of the

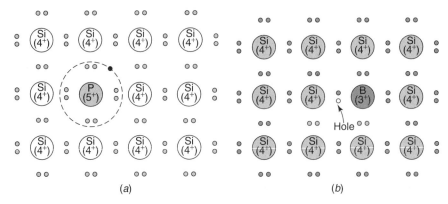

Fig. 21.4 Extrinsic conductivity can be produced in pure Si by doping it with small concentrations of substitutional atoms. (*a*) If P is added to the lattice, only four of its five valence electrons are needed to form covalent bonds with neighboring Si atoms. The fifth electron requires very little thermal excitation energy to break free from the P atom and act as a charge carrier. (*b*) If B is added to the lattice, its three valence electrons are insufficient to bond with all four Si nearest neighbor atoms, and this defective bond (called a *hole*) moves through the lattice as a positive charge carrier. (From Callister, 1997, pp. 606, 608; with permission.)

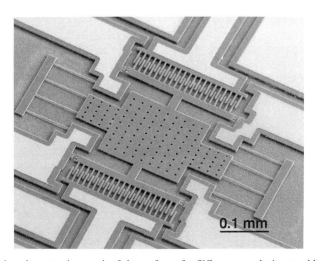

Fig. 21.5 Scanning electron micrograph of the surface of a SiC resonant device capable of operating up to 950°C. The various layers of masking, etching, and doping develop vertical relief on the initially flat surface of the chip. The substrate is Si; the white areas have been coated with Ni metal. (From Roy et al., 2002; with permission.)

first electronic digital computer (the Atanasoff–Berry computer) built in the late 1930s, which performed 3.75 computations per second.

21.2.5 Applications of Si Compounds

Although metallic Si is a key material in both the electronic and metallurgical industries, the most important applications of Si are those of its various compounds, particularly the silicates. Silicates are an indispensable part of roadways, buildings, windows, insulation, and refractories. Other Si compounds are used in abrasives, polymers, and high-temperature materials.

21.2.5.1 Silicates in Cement and Glass. Cement is a mixture of finely ground, dried powders of CaO, SiO$_2$, Al$_2$O$_3$, Fe$_2$O$_3$, MgO, and CaSO$_4$. In the presence of water, these compounds undergo a series of hydration reactions that harden the powder into a cohesive mass. When sand and gravel are mixed with cement and water, the blended mass is initially quite plastic and easily formed. Over time, the material hardens to produce the familiar concrete that is such an integral part of the built environment. Concrete is inexpensive and has high compressive strength. World cement production is almost 2 billion tons per annum.

Pure SiO$_2$ readily forms glass, a supercooled liquid that is transparent, formable when hot, and relatively inexpensive. At equilibrium the tetrahedral SiO$_4^{4-}$ building block of SiO$_2$ forms crystalline solids, but silica crystallizes slowly, allowing silica glass to be shaped at high temperature, then cooled too rapidly for the material to crystallize. Since the softening temperature of vitreous (glassy) silica is high (above 1600°C), SiO$_2$ is often blended with CaO, Na$_2$O, and/or other oxides to lower the forming temperature to 600 to 1000°C, making these glasses much simpler and more economical to melt and fabricate. Over 120 million tons of glass is produced annually for windows, bottles, insulation, communication fibers, lenses, catalyst supports, and many other applications.

21.2.5.2 Other Silicate Uses. Most natural stone has a large silicate content and is used in massive blocks in construction, as gravel for roadways, and in aggregate and sand for concrete and mortar. Silica and silica-rich clays are used in water filters and ceramic objects such as bricks, pottery, refractory brick, crucibles, and whiteware (sinks, toilets, bathtubs). High-purity silica crystals (quartz) can be grown from aqueous solutions at high temperatures and pressures to produce quartz oscillators for timepieces, transducers, and electronic filters. Silica gels are a microporous, amorphous SiO$_2$ with high surface area. Gels are used as desiccants, food additives, insulators, filters, and catalyst supports.

21.2.5.3 SiC. SiC can be produced by reducing SiO$_2$ with an excess of coke (C) in an electric arc furnace. At the 2000 to 2500°C temperatures achieved by the arc, the reactions are SiO$_2$ + 2C → Si + 2CO and Si + C → SiC. SiC is a hard (22 GPa) material that forms a protective SiO$_2$ layer that resists oxidation to about 1000°C. SiC also resists attack by all mineral acids except H$_3$PO$_4$. SiC is stable at high temperatures, finally decomposing by Si loss at about 2700°C. SiC has numerous crystal structures, all of which are stacking variations of the hexagonal wurtzite structure (ZnS-type) or diamond-cubic-like cF8 structure. The hardness and stability of SiC (often called Carborundum, a trade name) make it an excellent abrasive material. Although it is not as hard as some other abrasives, it is inexpensive, and fine particles of SiC have the ability to fracture as they are used, exposing sharp new edges to sustain the cutting action. SiC is also used extensively as a deoxidizing addition to iron and steel (Sec. 21.2.3). Global SiC production is about 1,000,000 tons per annum, with half of the product used as abrasive and half used to deoxidize steel.

21.2.5.4 Metal Silicides. Metal silicides are used as protective coatings for elevated-temperature oxidation resistance, thin-film coatings for electronic devices, and electric-resistance-furnace heating elements. Silicides have ordered intermetallic structures that maintain high strength and high elastic moduli at elevated temperature, and their Si content forms a protective SiO$_2$ coating that minimizes oxidation at high temperature. MoSi$_2$ is a particularly useful silicide because its high melting point (2030°C), good oxidation resistance, and high electrical conductivity make it a robust furnace heating element and a useful protective coating. Other silicides are used in smaller, niche applications, such as Nb silicide coatings to minimize oxidation of Nb rocket and jet engine components.

21.2.5.5 Silicone Polymers. Methyl silicone rubber, fluid, and resin compounds are synthesized into high-molecular-weight thermoplastic and thermosetting polymers. Silicone polymers have several advantages over C-based polymers, including high-temperature stability, low-temperature flexibility, low surface energy, and high gas permeability. Silicones also have excellent biocompatibility and electrical qualities, but they are more expensive than C-based polymers, which limits their use to especially challenging applications that are difficult to solve with C-based polymers. World silicone production (about 500,000 metric tons per year) is less than 0.5% of world C-based polymer production.

21.2.6 Sources of Si and Production Methods

Si is produced in electric arc furnaces by reduction of quartzite or sand with coke, converting SiO_2 to Si and CO. Since SiO_2 is an ubiquitous mineral, Si can be produced almost anywhere in the world. Most Si production is located at sites with a favorable combination of electric power cost, labor and tax costs, and proximity to customers. Four-fifths of all Si metal production is ferrosilicon (usually 55 to 80% Si) for steelmaking. Scrap Fe is added to the electric arc furnace for ferrosilicon production. About one-fifth of Si production is 98.5% purity Si, which is produced in the same manner (but without the scrap iron additions) for use in Al–Si alloys and the chemical industry.

Less than 1% (about 40,000 tons/yr) of world Si production is ultrapure Si (99.9999999 + % purity) for electronic applications. To achieve such high purity, volatile Si compounds (e.g., $SiCl_4$ and $SiHCl_3$) are purified by multiple distillations, deposited as Si by chemical vapor deposition or reduced with ultrapure Zn or Mg. The resulting Si metal is melted and grown into single-crystal boules (often as large as 380 mm in diameter) by either Czochralski crystal pulling or float-zone melting (Sec. 8.2.11). Producing such large single crystals is made even more challenging by the additional requirement that the crystal be dislocation-free. Although this is now achieved routinely by careful control of growth rates and geometry, it is a remarkable material-processing accomplishment. These large single crystals are sliced into thin sheets, cut into small squares or rectangles, and fabricated into electronic devices. For special applications it is possible to produce Si containing only one impurity atom per 10^{12} Si atoms. Although ultrapure Si comprises only about 1% of total Si production, the high cost of the ultrapure material makes it roughly equal in total dollar value to the other 99% of metallic Si production used in the steel, Al, and chemical industries.

21.2.7 Structure–Property Relations in Si

21.2.7.1 Strained Si for Faster Microprocessor Speeds. The operating speed of a microprocessor is limited by how rapidly transistors can switch on and off. Operating speed can be improved either by reducing the distance the current travels or by increasing the rate at which current flows. In the early years of microelectronics development, operating speed was improved by making the feature size on integrated circuits smaller and smaller, thereby shortening current travel distances. However, it is also possible to increase the rate of current flow in doped Si by inducing a residual tensile or compressive strain in the lattice. Electrons travel more rapidly parallel to a tensile strain, and holes travel more rapidly parallel to a compressive strain. Thus, performance would be enhanced if it were possible to stretch the n-doped regions while compressing the p-doped regions.

Ge atoms (atomic radius = 0.122 nm) are larger than Si atoms (0.118 nm). Si and Ge form a continuous series of solid solutions, so an 85% Si–15% Ge solid solution has an average interatomic spacing slightly larger than that of pure Si. If a thin layer of n-type Si is deposited atop an 85% Si–15% Ge substrate, the pure Si remains coherent with the substrate to minimize interface energy. (The coherent interface has a lower energy than an incoherent interface.) This strains the Si in the plane of the interface (Fig. 21.6), and the strained Si lattice has a 70% faster signal speed parallel to the interface. A 70% faster signal speed can increase computer chip processing speed by as much as 35%, and this improvement is achieved without reducing the feature size of the integrated circuit. Other methods exist to induce tensile strain, such as depositing Si_3N_4 over the Si to induce residual strain from differences in the coefficients of

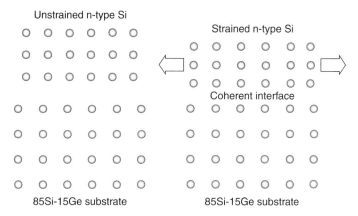

Fig. 21.6 Depiction of one method to induce a residual tensile strain within a Si layer in a microelectronic device. Ge atoms are larger than Si atoms, so when n-type Si is deposited atop a solid solution of 85% Si–15% Ge, the Si is placed in biaxial tensile strain (parallel to the interface) to maintain a coherent interface with the substrate. The proportions and the degree of mismatch of the interatomic spacings are exaggerated in this drawing; actual strain is less than 1%. The signal speed in the strained Si is 70% higher than in unstrained Si, allowing processor speed to increase by up to 35% without requiring reduced feature size (i.e., the size of the transistors and vias). (Redrawn from *http://www.research.ibm/resources/press/strainedsilicon/*, 2004.)

thermal expansion of Si and Si_3N_4. Compressive stresses can be induced by cutting trenches on both sides of a p-type Si region, then filling the trenches with Si–Ge to "squeeze" the p-type region for improved hole mobility.

21.3 GERMANIUM

> Brattain put a ribbon of gold foil around a plastic triangle, and sliced it through at one of the points. By putting the point of the triangle gently down on the germanium, they saw a fantastic effect—signal came in through one gold contact and increased as it raced out the other. The first point-contact transistor had been made.
>
> —ScienCentral, Inc., and the American Institute of Physics,
> *http://www.pbs.org/transistor/background1/events/miraclemo.html*, 2003

21.3.1 History of Ge

In contrast to Si, Ge is relatively scarce, and geologic processes have not worked to concentrate it effectively in minerals. Consequently, Ge was not isolated in metallic form until 1886. Prior to 1945, world production of Ge metal and compounds totaled less than 200 kg/yr. Ge vaulted to sudden fame by Brattain's discovery of the transistor in 1947, and transistors and diodes were often fabricated from Ge in the 1950s and 1960s. Si largely replaced Ge in electronic device fabrication during the 1970s, but Ge finds uses in infrared optics, polymerization catalysts, solar cells, and specialty microprocessors.

21.3.2 Physical Properties of Ge

Ge's physical properties are similar to Si's (Table 21.1). Ge's covalently bonded cF8 crystal structure (Fig. 1.6) permits no appreciable dislocation motion at room temperature, so Ge is brittle, with a fracture strength of about 100 MPa. Ge has a small bandgap (0.67 eV), which gives it semiconducting behavior and optical transparency over a wide range (7 to 14 μm wavelengths) of infrared wavelengths. The low packing efficiency of the cF8 structure gives Ge a 6.6% volumetric contraction upon melting. Ge is stable in air, and temperatures above

TABLE 21.1 Selected Room-Temperature Properties of Ge with a Comparison to Si and Cu

Property	Ge	Si	Cu
Valence	+2, +4	+4	+1, +2
Crystal structure at 20°C	Diamond cubic (cF8)	Diamond cubic (cF8)	FCC
Density (g/cm^3)	5.32	2.33	8.92
Melting temperature (°C)	945	1414	1085
Thermal conductivity (W/m·K)	60.2	141	399
Elastic (Young's) modulus (GPa)	79.9	179	130
Coefficient of thermal expansion (10^{-6} m/m·°C)	5.75	2.5	16.5
Electrical resistivity (μΩ·cm)	4.6×10^{7a}	2.5×10^{11a}	1.7
Cost ($/kg), large quantities	700^b	80^c	2

^{a} The electrical resistivities of both Ge and Si vary by several orders of magnitude with small variations in impurity content due to the doping effect.
^{b} Price of Ge metal of 99.9999% purity. Since most Ge is used in optical and electronic devices, high-purity oxide and metal are the most commonly traded commodity. Prices for lower-purity Ge metal are moderately lower.
^{c} Price of Si for 99.99999999% purity for use in microelectronic devices. Prices for lower-purity Si are much lower (about $1 per kilogram).

400°C are required to form a perceptible oxide layer. The oxide is protective against further oxidation in dry air but is destroyed by humid air. Two oxides exist, GeO_2 and GeO; GeO_2 is more stable than GeO, which sublimes at 710°C. Ge resists attack by HCl and HF acids, but it dissolves slowly in HNO_3 and H_2SO_4 acids. Ge can be dissolved rapidly in alkaline solutions if an oxidizing agent (e.g., H_2O_2) is present.

The intrinsic conductivity of high-purity Ge is low, but it can be doped with group 13 or 15 elements to produce extrinsic p- or n-type conductivity, making it possible to fabricate transistors, diodes, and entire microprocessor circuits from Ge. At high pressure, Ge undergoes the same crystal structure transformation as Si from the diamond cubic (cF8) structure to the more metallic tetragonal (tI4, β-Sn) structure.

21.3.3 Applications of Ge

21.3.3.1 Ge in Optical Materials. Ge metal is transparent to infrared light in the range of wavelengths emitted by objects at room temperature (8- to 12-μm wavelengths). Metallic Ge has a high refractive index (4.00 at 10.6-μm wavelength), and Ge lenses and windows are used in military and security surveillance "heat-sensing" cameras. Warm objects (e.g., people, animals, engines) appear bright to such cameras, and they require no visible light to operate. Ge can also form nonoxide glasses with As, Sb, and Se for use in infrared optics. In the telecommunications industry, $GeCl_4$ is reacted with the surface of glass fibers to raise the refractive index of the outer portion of the fiber. These graded-index fibers act to confine the signal-carrying laser beam to the central portion of the glass fiber, which reduces loss and signal distortion.

$Mg_{28}Ge_{10}O_{48}$ and $Mg_{56}Ge_{15}O_{66}F_{20}$ are used as phosphors in fluorescent lights. Although these materials are more expensive than As-based phosphors, they eliminate the toxicity concerns (Sec. 22.2.6) associated with As use in accidental breakage and routine disposal. $Bi_4Ge_3O_{12}$ emits visible light when excited by γ rays, allowing its use in high-energy γ-ray scintillator detectors.

21.3.3.2 GeO Catalysts and Ge Alloying Additions. GeO_2 and $NaH_3Ge_2O_6$ polymerization catalysts are used to produce polyesters such as PET (polyethylene terephthalate) that are used widely for beverage containers and textiles. Ge catalysts cost more than other catalysts (e.g., Ti-based compounds) but are preferred for premium-grade products because they produce

> **SIDEBAR 21.1: Ge ADDITIONS TO SUPPRESS FIRESCALE IN STERLING SILVER**
>
> For centuries, sterling silver (92.5% Ag–7.5% Cu) was considered the premier metal for tableware (Sec. 18.3.3.3). Its magnificent luster cannot be matched by other metals. However, the Cu in sterling silver oxidizes when it is heated for brazing and soldering, forming CuO or Cu_2O at and near the surface. Hot, dilute H_2SO_4 acid is sometimes used to remove the Cu oxides from the surface, but this treatment leaves unalloyed Ag over a layer of Ag + Cu oxide mixture. If metal in this condition is heavily polished, the soft, unalloyed Ag can be abraded away, exposing the dark stain of firescale lying underneath. Firescale not only mars the appearance, but solder joints attempted on firescaled metal may fail, and firescale thicker than about 25 μm often cracks because it is less ductile than the underlying metal.
>
> Until recently, the only way to avoid firescale in sterling silver was to cover the item with flux before heating or to anneal and solder in O-free environments. Such methods are time consuming and expensive. However, a new alloy called *bright sterling* (92.5% Ag–6.3% Cu–1.2% Ge) solves the firescale problem. Ge acts as a powerful deoxidizer in the metal, forming harmless Ge oxides with the O that diffuses into the metal so that no Cu oxides form. Bright sterling can be freely brazed and soldered in air. In addition, the metal tarnishes much more slowly in service, and the consumer can usually remove the tarnish that does form simply by wiping the surface with a clean cloth. This eliminates the frequent cleaning with polishing agents required to keep traditional sterling silver bright. The primary drawback to bright sterling is Ge's high cost, which makes the alloy 7 to 10% more expensive than ordinary sterling.

colorless, transparent fibers and plastics. Ge alloying additions to sterling silver improve its hot oxidation resistance (Sidebar 21.1).

21.3.3.3 Ge in Microelectronic Devices. Ge has electron mobility three times higher than that of Si, which gives it an inherently faster switching speed for transistors. In fact, transistors and diodes made during the 1950s and 1960s commonly used Ge rather than Si. However, Si-based semiconductors have three major advantages over Ge-based materials. Si is less expensive, Si's larger bandgap makes it more damage resistant at elevated temperatures, and it is easier to form an oxide layer on Si during microelectronic fabrication processing. These advantages led to the nearly total replacement of Ge by Si in microelectronic devices during the 1970s. Ge use in electronic devices is now quite small, although devices containing Si–Ge solid-solution semiconductors are growing in use. The high charge carrier mobility of Ge allows these Si–Ge semiconductors to process higher frequencies than those of conventional Si semiconductors, which is particularly useful in wireless communication devices. GaAs can also process high frequencies (Sec. 22.2.4), but GaAs is more costly than Si–Ge chips. Since Si–Ge devices are built with small, thin Si–Ge deposits on Si substrates, the mass of Ge used is quite small. Ge also finds use as a substrate for multiple-layer thin-film photovoltaic devices.

21.3.4 Sources of Ge and Refining Methods

"Rich" Ge deposits contain only about 300 to 600 ppm Ge, so ore cannot be processed economically solely for its Ge content. Most Ge is produced as a by-product of Zn refining, although smaller amounts are recovered from Pb, Sn, and Cu–Co ores. Ge is precipitated from solutions being prepared for electrolytic production of Zn or from flue dusts in pyrometallurgical ore treatments. Its separation and purification processes are somewhat complex and are based on fractional distillation of $GeCl_4$, which is then hydrolyzed with H_2O to produce GeO_2 or reduced by H to produce metal. High-purity Ge metal is produced by multiple zone-refining passes, and single crystals are made by the Bridgman or Czochralski method, similar to the preparation of electronic-grade Si. Production of Ge oxide and metal totaled a mere 68 tons in 2002. Over half of this production was in the form of GeO_2 for catalysis and infrared optical fibers and lenses. Most of the remainder was metallic Ge, much of it 99.9999+% purity for electronic use.

21.4 TIN

> Tin was vital to the ancients [for] making bronze. No civilization could thrive very long without it. Julius Caesar knew of the importance of British tin when he invaded the island in 55 B.C. After the conquest of Britain, the Romans were in control of most of the world's supply of the metal. Tin was one commodity that made it essential for the Romans to conquer and keep this remote part of their vast empire.
>
> —"The Legendary Tin Mines of Cornwall,"
> http://myron.sjsu.edu/romeweb/ENGINEER/art12.htm, 2004

21.4.1 History of Sn

Tin has served humankind for thousands of years. It is easily separated from its ores and readily alloyed with Cu to make bronze. Bronze was produced in Egypt in about 3000 B.C. by co-reduction of Cu and Sn ores. Chinese and Japanese metallurgists produced the first pure Sn in 1800 B.C. Uncounted thousands of soldiers died in the ancient world battling over strategic Sn mines for production of bronze, the strategic alloy of antiquity. Although the Iron Age supplanted the Bronze Age long ago, Sn remains a valuable metal today, useful in corrosion-resistant coatings, alloys, solders, and bearings.

21.4.2 Physical Properties of Sn

Sn is a soft, lustrous, low-melting metal (Fig. 21.7) with good corrosion resistance. Liquid Sn wets the surfaces of other metals well, allowing its use as a corrosion-resistant coating and a constituent of many soldering alloys. Sn's ability to retain oil films on its surface make it well suited for use as a bearing material. Although pure Sn is too weak for structural use, Sn is used as an alloying addition to harden other metals, such as Cu, Pb, Ti, and Zr.

21.4.2.1 Crystal Structures and Properties of Liquid Sn. Descending through group 14 in the periodic table, one passes through the cF8 metalloids Si and Ge, which are semiconductors with bandgaps of 1.11 and 0.67 eV, respectively. Both Si and Ge can be transformed to a tetragonal (tI4) structure at high pressure (Appendix A, *ftp://ftp.wiley.com/public/sci_tech_med/nonferrous*). The next element in the group is Sn. Below 13°C, Sn also has the cF8 structure (α-Sn, gray tin) and a small bandgap (0.08 eV). α-Sn transforms to the tI4 structure (β-Sn, "white tin") above 13°C. α-Sn's small bandgap makes it a semiconductor, but at 0°C thermal excitation promotes large numbers of electrons to the conduction band, and α-Sn's electrical resistivity (300 $\mu\Omega \cdot$ cm) is many orders of magnitude lower than that of Si or Ge. β-Sn is a normal metal with electrical resistivity of 12.6 $\mu\Omega \cdot$ cm. Sn's α → β transformation is sluggish; extended time at temperatures well below 13°C is needed to form significant amounts of α-Sn. The presence of small "seed" regions of α-Sn greatly accelerates β-Sn's transformation, so a specimen cycled repeatedly through the transformation forms β-Sn much more rapidly from the nucleating action of α-Sn vestiges. α-Sn (density = 5.76 g/cm^3) occupies more volume than β-Sn (7.29 g/cm^3), which causes β-Sn objects to suffer a gradual disintegration and crumbling action as they transform to the more voluminous, brittle α-Sn (Sidebar 21.2).

Sn melts at 232°C and boils at 2623°C, forming diatomic molecules of Sn_2 vapor. Sn's high boiling point keeps vapor pressure over liquid Sn low (Sidebar 21.3), even when the metal is several hundred degrees above the melting point (e.g., 1.3×10^{-7} atm at 1000°C). Liquid Sn's viscosity is unusually low (1.85×10^{-3} Pa·s at 232°C), two to three times lower than the

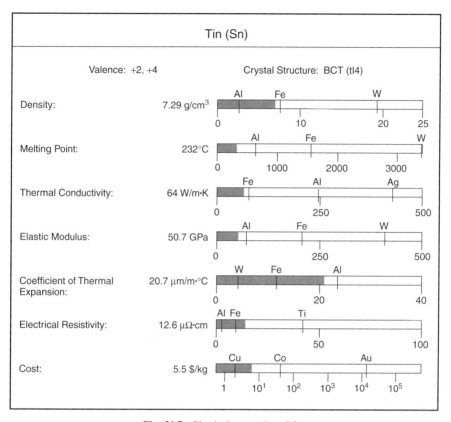

Fig. 21.7 Physical properties of Sn.

SIDEBAR 21.2: ROLE OF Sn'S β–α PHASE TRANSFORMATION IN NAPOLEON'S RUSSIAN CAMPAIGN

From 1805 to 1812, Napoleon's army conquered most of Europe. In June 1812, Napoleon led an army of 600,000 across the Russian border. The overmatched Russian army of 200,000 conducted a series of battles and strategic withdrawals that delayed but could not stop Napoleon's march toward Moscow. In September, Napoleon's army entered Moscow, only to find the city largely deserted and partially burned by the retreating Russians. As winter set in, Napoleon was unable to supply his troops so far from his eastern most secure outposts, and he began a long, disastrous westward retreat. In the cold and deep snow, 500,000 of Napoleon's troops perished from exposure, starvation, and harrying attacks from Russian Cossacks.

A metallurgical problem compounded the miseries of the retreating French. The buttons on their uniforms and coats were Sn castings, and in the severe cold, the Sn buttons began to transform from the ductile β-Sn (tI4) to the brittle α-Sn (cF8). With buttons disintegrating from tin pest, the soldiers faced the Russian winter winds with their coats agape and their trousers sagging. The soldiers scrounged whatever attire they could find in the farms and villages they passed. One Russian observer described Napoleon's passing army as "a mob of ghosts draped in women's cloaks, odd pieces of carpet, or coats burned full of holes." Modern Sn castings contain small amounts of Sb or Bi to prevent transformation to α-Sn, but two centuries ago the tin pest mechanism was not understood, and regrettably for the soldiers, methods for its remediation were unknown.

SIDEBAR 21.3: FLOAT GLASS ON MOLTEN Sn

Sn remains liquid over a wide range of temperatures (232 to 2623°C), which makes it useful in producing sheet glass. At 1000°C soda-lime glass has a viscosity of about 10^3 Pa·s, which allows it to flow relatively freely. When the glass cools to 600°C, its viscosity increases to about 10^8 Pa·s, making it nearly rigid. This viscosity change allows molten glass to be spread atop a pool of molten Sn metal at 1000°C, flowing to form a thin sheet with smooth, flat surfaces. The molten glass has a density of 2.4 g/cm^3, so it floats on the dense (6.5 g/cm^3) liquid Sn (Fig. 21.8). When cooled to 600°C, the glass becomes sufficiently rigid to support its own weight without deforming, and it is rolled off the Sn surface to annealing ovens to cool gradually to room temperature. Sn has almost no chemical interaction with glass at these temperatures, and its low vapor pressure minimizes losses. Sheet glass produced in this manner has smooth, nearly defect-free surfaces that require no subsequent polishing. Float glass production is the primary method used by the glass industry to manufacture window glass.

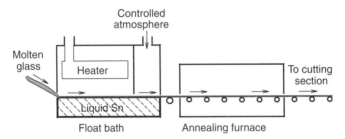

Fig. 21.8 Float glass production. Glass is spread atop a pool of molten Sn in an inert atmosphere at 1000°C and flows under the force of gravity to form a layer a few millimeters thick. As the glass moves left to right across the Sn, its temperature is reduced to 600°C, at which point it is sufficiently rigid to be rolled off the Sn onto rollers for annealing and cooling to room temperature. Both surfaces of the glass are so smooth and flat that it can be cut and used as sheet glass without polishing. (Redrawn from Barry and Thwaites, 1983, p. 50.)

viscosities of Al, Ag, and Zn at their melting points. This allows liquid Sn to flow freely into molds, aiding its castability. The surface tension of Sn is also low (0.58 N/m at 232°C).

21.4.2.2 Corrosion. Sn forms an oxide layer a few nanometers thick in dry air. Humidity, elevated temperature, and the presence of common impurities (Bi, Tl, Fe, Sb) accelerate the SnO_2 growth rate. At 200°C pure Sn shows noticeable discoloration after an hour's exposure to normal air. The oxide layer grows with parabolic kinetics and is protective even after the Sn has melted. Liquid Sn does not dissolve appreciable amounts of O or N below 1000°C, thus porosity from bubble formation in Sn castings is not a problem. However, exposing Sn to air at temperatures above 1000°C causes rapid oxidation. Sn shows almost no corrosion in pure water, and its corrosion resistance in seawater and mild acids is generally good. Sn corrodes more rapidly in strong mineral acids, particularly in HNO_3 acid. The reaction rates in HCl and H_2SO_4 acids are usually two to three orders of magnitude higher if the acid contains dissolved O. Since Sn is frequently used as a protective coating on other metals, its corrosion must often be considered in the context of galvanic couples formed with the other metal(s). Sn is more noble than Fe but is anodic to Cu.

21.4.2.3 Mechanical Properties. The α-Sn allotrope is brittle and difficult to fabricate into test specimens for mechanical properties testing. Pure β-Sn is an exceptionally weak metal. A 1-cm-diameter β-Sn rod can be bent by applying only moderate finger pressure, producing high-frequency clicking sounds (tin cry) from twin formation. At room temperature, pure β-Sn has a yield strength of 2.5 MPa and an ultimate tensile strength of 14 MPa. Sn's Young's modulus (50 GPa) is surprisingly high for such a weak metal. Room temperature is 60% of the homologous melting temperature of Sn, and the metal creeps rapidly under load at room temperature. Thus, β-Sn is anelastic at 20°C and cannot sustain loads in the elastic regime for an extended period. β-Sn recrystallizes immediately after cold work, and cold work can lead to accelerated grain growth that causes "work softening" in pure Sn and Sn-rich alloys (e.g., pewter and Pb–Sn solder). Below −120°C, β-Sn (which is metastable at this temperature) becomes much less ductile, a change associated with a reduction in the number of available slip systems.

21.4.3 Applications of Sn

21.4.3.1 Sn Coatings. Sn can be electroplated onto low-carbon steel to form a strongly adherent coating that resists corrosion. The largest use for tin plate is in food packaging; Sn prevents steel cans from rusting, even when somewhat acidic foods are stored in them for several years. Sn is electrodeposited from metallic Sn electrodes onto pickled, lightly cold-worked (1 to 4% reduction by rolling) steel sheet as the steel moves rapidly through $SnSO_4$, $SnCl_2$, or SnF_2 aqueous solutions. The Sn coating is typically 0.5 to 2.0 μm thick; the coating is usually made thicker on one side of the steel sheet than on the other. As electrodeposited, Sn has a dull appearance; this is usually "flow-brightened" by inductively heating the Sn-plated steel to melt the Sn momentarily and impart a highly reflective surface. The metal is then plunged into cool water. This melting also produces a thin $FeSn_2$ intermetallic layer at the Sn–Fe interface that improves surface hardness and corrosion performance. An electrochemical or chemical passivation treatment is usually given in aqueous dichromate solution to stabilize the surface; however, concern over Cr hazards has prompted research efforts to replace the Cr solutions with $Zr(SO_4)_2$ or other compounds. The Sn-plated steel is then formed into cans by soldering and/or welding the end caps in place or by deep drawing. For use with certain foods, the cans may be coated with a protective lacquer to minimize corrosion or dissolution of Sn into the food. A variety of other products also use tin plate, including batteries, automotive parts, and signs.

Sn alloys are also applied to steel and to other metals. Pb–Sn and Pb–Sn–Sb (terneplate) corrosion-resistant coatings containing 8 to 25% Sn are applied to mild steel by hot dipping. The Sn and Sb are present to promote wetting, since pure Pb will not wet steel. Terneplate layers are typically 0.3 to 2 mm thick and are sufficiently ductile and adherent to tolerate postcoating deep-drawing operations. Terneplate is used for roofing, gutters, architectural trim, conduit, electronic chassis, and junction boxes. Sn–Zn and Sn–Cd coatings are applied to steel electrolytically for corrosion protection in wet, marine environments. The 65Sn–35Ni intermetallic compound can be electroplated onto ferrous and nonferrous metals to produce an especially hard surface (VHN = 700) that is both wear and corrosion resistant. It provides a surface that solders well and is often used as an undercoat for Au plating. In addition to hot dipping and electroplating, Sn alloys are also applied by thermal spray and by tumbling in a slurry of Sn alloy powder, using mechanical alloying media such as glass beads and aqueous "promoter" solutions to improve adherence.

21.4.3.2 Sn-Rich Alloys. Although pure Sn is too soft to be a structural metal, alloying additions can impart useful strength to the metal. Alloys known as *pewter* have typical compositions of about 92 wt% Sn, 6 to 7 wt% Sb, and 1 to 2 wt% Cu. Pewter has been used for more than 2000 years; centuries ago, pewter often contained significant amounts of Pb, but modern pewters hold Pb content below 0.5% to avoid toxicity problems (Sec. 21.5.4). Pewter is readily joined by soldering, and it is exceptionally malleable, allowing forming by casting, hammering, spinning, or rolling. Pewterware's attractive surface finish, weight, and easy formability make it popular for tableware, vessels, candlesticks, jewelry, figurines, and medallions. Pewter alloys' strengths are similar to pure Ag and pure Au (hardness = 20 to 30 VHN). Thus, pewter's hardness is much improved over that of pure Sn, but it is still a relatively soft metal, and it cannot serve for components needing substantial strength.

A wide array of Sn-rich alloys having greater alloying additions than pewter (e.g., 70 wt% Sn–30 wt% Sb) are used to make die-cast objects. Sb is the most frequently used alloying addition, strengthening the castings by solid-solution hardening and by forming SbSn intermetallic compound. Sn-rich alloys have excellent fluidity, allowing the liquid metal to fill intricately shaped molds completely with channels as narrow as 0.5 mm. The low melting temperatures of Sn die-casting alloys make for long die life, low energy costs, and minimal shrinkage; however, the strengths of the castings are substantially lower than those of Zn, Mg, and Al die castings.

21.4.3.3 Bronze. Cu will dissolve a maximum of about 15 wt% Sn at 500°C (Fig. 18.8), and a family of wrought and cast Cu–Sn alloys called *bronze* has been used continuously for 5000 years. Early bronzes played a major role in advancing civilization, and the alloys are still widely used today for bearings, springs, bushings, piston rings, gears, and bells. Bronze is harder than brass and has high fatigue strength (Sec. 18.2.3.6). Tensile strengths as high as 1000 MPa are possible in cold-worked phosphor bronze.

21.4.3.4 Solders. The "workhorse" materials for many soldering applications in the twentieth century were eutectic or near-eutectic Pb–Sn alloys (Fig. 21.9). Pb–Sn alloys have several attractive attributes: low cost, low eutectic temperature (183°C), good wettability on many common workpiece metals, and a choice between eutectic freezing or a wide freezing range which produces "pasty" solid–liquid mixtures. Although Pb–Sn solders are still widely used for many

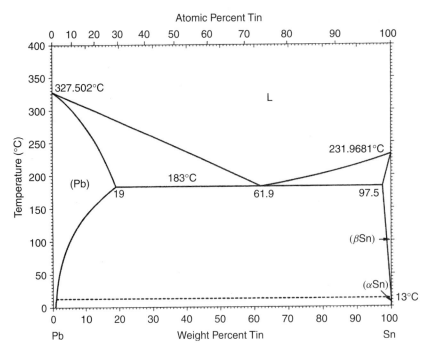

Fig. 21.9 Pb–Sn binary equilibrium phase diagram. Many solders are used with compositions near the eutectic point, although off-eutectic compositions provide a wide freezing range, which is useful for some applications. (From Massalski, 1986, p. 1848; with permission.)

applications (e.g., body panel soldering on automobiles), their use for joining tin plate and Cu pipe and for electrical circuit connections (Sidebar 21.4) is falling due to concerns about Pb toxicity (Sec. 21.5.4).

21.4.3.5 Bearing Alloys. Bearings provide a low-friction, compliant surface to hold rotating or sliding parts in alignment (Sec. 18.3.3.4). Sn alloys have several desirable properties for bearings. They are corrosion resistant; hold oil films well; are soft enough to conform to minor imperfections in shaft geometry; and trap (embed) tiny dirt and debris particles, preventing them

---

SIDEBAR 21.4: Pb-FREE SOLDERS

There is growing concern about contaminating groundwater with Pb leached from discarded electronic devices. For this reason, Pb-free solder development programs have been promoted by Japanese, European, and North American electronics trade organizations. Many manufacturers have replaced Pb–Sn solders with the widely available near-eutectic 95 to 98% Sn–5 to 2% Ag alloys. But the binary Sn–Ag solders are not only more expensive than Pb–Sn solders, they also have a higher eutectic temperature (221°C for Sn–Ag versus 183°C for Pb–Sn) and mediocre shear strength and thermomechanical fatigue resistance. Many of the electronic components being assembled are semiconductor devices that can be damaged by high temperatures, so the higher eutectic temperature poses quality assurance problems. In addition, the soldered devices may be used in environments such as automobile engine compartments, where wide temperature changes and inertially imposed stresses are common; this makes strength and fatigue resistance important performance criteria. These problems can be ameliorated by small additions of Cu, Sb, Bi, In, Zn, Fe, or Co to Sn–Ag solders to lower the ternary or quaternary eutectic temperatures by several degrees and to improve mechanical properties by refining the as-solidified microstructure (Fig. 21.10).

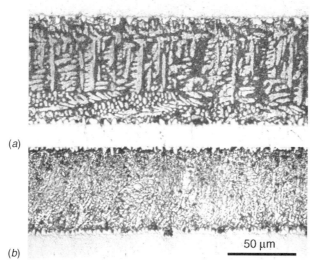

Fig. 21.10 Microstructure of Pb-free solders. (*a*) The ternary eutectic (Sn–3.7 wt% Ag–0.9 wt% Cu) forms relatively coarse dendrites. (*b*) Substitution of Co for some of the Cu (Sn–3.7 wt% Ag–0.6 wt% Cu–0.3 wt% Co) refines the microstructure, improving mechanical properties. (Courtesy of I. Anderson, Ames Laboratory of the U.S. Department of Energy; with permission.)

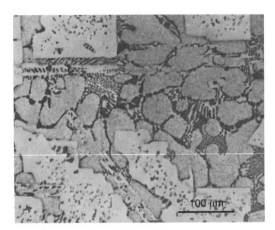

Fig. 21.11 Microstructure of Babbit bearing alloy 65% Sn–18% Pb–15% Sb–2% Cu. The lighter phases are SbSn and Cu_6Sn_5 intermetallics; the darker regions are Sn-rich solid solution with some interdendritic eutectic. (From ASM, 1972, p. 319; with permission.)

from acting as abrasive particles between the moving parts. White metal bearings (often called *Babbit bearings*) are approximately 70 to 80% Sn, with 3 to 5% Cu and varying percentages of Pb and Sb. These alloys contain a soft, Sn-rich solid-solution matrix phase plus hard Cu_6Sn_5 and SbSn intermetallic phases (Fig. 21.11). The softer phase wears rapidly at first as the bearing is "broken in," leaving microscopic plateaus of the harder phase. This allows oil to reside in the depressions, feeding extremely thin oil films across the surfaces of the high points. These alloys are relatively soft and must be supported by a stronger backing metal. Al–Sn alloy bearings are also widely used. Al and Sn have complete liquid solubility but no solid solubility, so pure Sn is the last metal to freeze, forming between the Al grains. Low-Sn compositions (e.g., Al–6% Sn) are strong enough to operate without backing metal, but they have low conformability. Al–Sn alloys containing 20 to 40% Sn provide better conformability but are weaker and require backing metals.

21.4.3.6 Other Sn Alloys. Sn is also used in a variety of other metallurgical applications. Fusible alloys, formulated to have exceptionally low melting points, often contain significant amounts of Sn (Sec. 22.4.3 and Table 22.2). Nb_3Sn is a superconducting intermetallic compound used in strong electromagnets in medical imaging devices and particle accelerators (Sec. 13.4.2.5). Addition of 0.1 to 0.5% Sn to gray or nodular cast iron promotes pearlite formation. About 4% Sn powder can be blended with bronze powder to accelerate sintering during powder metallurgy (P/M) fabrication. Steel P/M fabricators often use a 40% Sn–60% Cu sintering aid. Ag–Sn amalgams are used in dental reconstructions (Sidebar 19.4).

21.4.3.7 Sn Compounds. Sn forms a greater variety of organometallic compounds (*organotin compounds*) than does any other metal. One particularly useful family of organotin compounds has the general structure R_2SnX_2, where R is an alkyl group (e.g., methyl, butyl, or octyl) and X is a laurate, maleate, or acetate. These compounds stabilize polymers such as polyvinyl chloride (PVC) against discoloration and embrittlement when exposed to oxygen, sunlight, or elevated temperature. More than 40,000 tons/yr of organotin compounds is used as catalysts, insecticides, fungicides, wood preservatives, and accelerants for silicone polymer curing. Although Sn is considered a nontoxic metal, some organotin compounds are highly toxic. In 2003, use of antifouling organotin additives in the paint on ships' hulls was banned worldwide, and the environmental hazards associated with other organotin compounds are coming under increasing scrutiny. Several inorganic Sn compounds also see substantial use, including $SnCl_2$ (a reducing agent in organic synthesis), SnF_2 (an anticaries additive in toothpaste), and SnO_2 (a ceramic pigment and glaze opacifier and a conductive coating on glass to melt frost from automobile windshields).

21.4.4 Sources of Sn and Refining Methods

The only commercially important Sn ore is SnO_2. Several sites around the world have deposits with the necessary 0.1 to 0.3% SnO_2 concentration in the ore for economically feasible recovery, and these are mined by both underground and surface methods. SnO_2's high density allows gravity separation from the gangue by processing crushed ore slurries with vibrating jigs and shaker tables. This low-grade Sn concentrate is then roasted to drive off volatile elements (e.g., S, As), followed by acid leaching to dissolve impurities. The concentrated SnO_2 is usually reacted with FeS or $CaSO_4$ to form volatile SnS, which separates the Sn from other minerals whose sulfides are not volatile. The SnS vapor is then reacted with air to produce SnO_2, which is in turn reduced with C in an electric arc furnace by the reaction $SnO_2 + C \rightarrow Sn + CO_2$. Unfortunately, concentrated ore usually contains silicate minerals, and SnO_2 reacts readily with silica to form Sn silicates. Thus, a large fraction of the Sn is usually lost as silicates in the slag, and this material must be retreated to recover its Sn content.

Crude Sn metal obtained from the electric arc furnace is typically 92 to 99% Sn, and this metal is further refined by passing steam or air through the melt to coagulate much of the insoluble Fe and other transition-metal impurities plus As and Sb intermetallics, so they can be filtered from the liquid Sn. The metal is then cast into ingots that are slowly heated over the range 230 to 300°C to selectively melt out higher-purity Sn, leaving behind most of the remaining Fe. The Sn metal is then purged of Cu, As, Pb, and Bi impurities by adding S, Al, Cl_2, and Ca or Mg to the molten metal. This metal is usually sufficiently pure for most applications, but electrorefining can be used to achieve still higher purity. Sn recycling rates are high; nearly 60% of Sn is recovered and resmelted.

21.4.5 Structure–Property Relations in Sn

21.4.5.1 Sn Whisker Formation in Electronic Circuits. Sn whisker growth from Sn-plated surfaces in electronic assemblies is a problematic and incompletely understood metallurgical phenomenon. These "extrusions" of pure, defect-free, single-crystal Sn filaments (Fig. 21.12) have diameters ranging from 6 nm to 6 μm. They grow from a previously smooth surface, and in some systems grow more than 1 mm long, causing short-circuits by contacting nearby components. This, of course, poses a major threat to electronic system reliability, and the consequences are particularly severe in military and spacecraft electronic systems. Sn whiskers are usually absent during initial quality assurance checks; they develop in service, sometimes after a period of years. They have been observed to form with one of five different crystallographic directions along their length.

One theory to explain their origin postulates that they are fed by long-range diffusion caused by local microstresses in the Sn layer. These stresses are relieved by ejection of Sn atoms at sites where screw dislocations intersect the free surface. Studies have shown that whisker growth is absent in material maintained above 18°C and is often seen in material cycled repeatedly through Sn's $\alpha \rightarrow \beta$ transformation temperature. Stresses arise in the Sn from (1) volume expansion accompanying transformation to the α phase, and (2) thermal stresses in the Sn film resulting from its attachment to a substrate metal (often Cu) with a different coefficient of thermal expansion. These stresses do not easily alter the overall shape of the Sn grains in the film, because each grain is covered with (and constrained by) a SnO_2 layer that is stronger than Sn metal. Whiskers may arise from the metal adjacent to grain boundaries (which are known to behave superplastically in β-Sn near room temperature) to relieve these internal stresses.

Fig. 21.12 SEM micrographs of Sn whiskers protruding from a Sn-plated surface in an electronic device. (Courtesy of F. Laabs and T. Ellis, Ames Laboratory, U.S. Department of Energy.)

21.5 LEAD

I really must protest the relentless attempts by second-rate authors and poets to defame the wonderful metal lead. Athletes succumbing to exhaustion are said to have "leaden limbs"; the grief-stricken have "leaden hearts"; and, God help us, how many thousands of times have overcast skies been described to us as "leaden"? At least the rock group Led Zeppelin had the decency to shield their affront in a double entendre.

—Dillon Harris, Ames High School, 2004

21.5.1 History of Pb

Pb has solved engineering problems for over 5000 years. Pb is reasonably abundant, inexpensive, corrosion resistant, and ductile (Fig. 21.13). Pb alloys are strong enough to bear moderate loads. Early civilizations used Pb sheet and castings for ornaments and structures. Engineers of the Roman Empire made extensive use of Pb pipes for water handling more than 2000 years ago. New uses for Pb were developed over the centuries to include solder, roofing, stained-glass window cames, bearings, and guttering. A century ago, more tons of Pb were produced each year than any other nonferrous metal; it has since been surpassed by several metals (e.g., Al, Cu, Zn, Mn).

By the mid-twentieth century, Pb had a wide range of applications in batteries, paints, ceramics, antiknock compounds for gasoline, solders, sheet, and ammunition. More recently, however, concerns over Pb toxicity have driven steadily tightening regulatory restrictions that sharply curtailed many uses for Pb. However, rising demand for Pb–acid batteries partially offset reduced Pb use in other areas. These changes made Pb a much more "one-dimensional" element, with more than 80% of all Pb used for batteries. After peaking at 3.5 million tons/yr in 1980, primary Pb production diminished to current levels somewhat under 3 million tons/yr. Nearly 3 million additional tons of Pb are recycled each year. Most battery Pb is now recycled (the United States has a 97% recycling rate), which minimizes environmental dispersal of Pb and has so far allowed Pb–acid batteries to avoid major regulatory proscriptions.

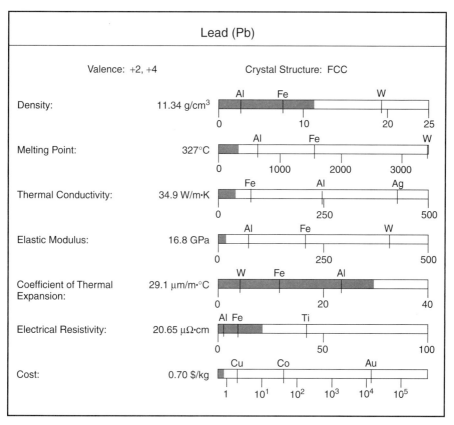

Fig. 21.13 Physical properties of Pb.

21.5.2 Physical Properties of Pb

21.5.2.1 Corrosion. Pb does not tarnish in dry air, but air of normal humidity causes slow oxidation, forming a gray patina of mixed oxide, carbonate, and sulfate that protects the underlying metal from further oxidation. Most Pb salts have exceptionally low water solubility, which makes massive Pb metal quite persistent. Vestiges of ancient Roman Pb plumbing remain today, and the Pb sheet roofs on some European cathedrals have lasted for hundreds of years. Pb is used widely for handling hot, concentrated H_2SO_4 acid and also resists attack by H_3PO_4 and CrH_2O_4 acids. $PbCl_2$ is only slightly water soluble, and HCl acid attacks Pb slowly. HNO_3 acid, however, forms highly soluble $Pb(NO_3)_2$, and dilute nitric acid attacks Pb aggressively. Pb is also dissolved by various organic acids, such as acetic acid, which can cause dangerous Pb contamination of fruit juices or wine stored in Pb containers. Pb generally shows good resistance to corrosion in moist soils, allowing it to be used as protective sheathing on buried cables and for underground pipes.

21.5.2.2 Mechanical Properties. Pure Pb is a soft, ductile FCC metal that slips on the $1/2\{111\}\langle 110\rangle$ slip system. Pb plastically deforms easily at room temperature; the metal recovers and recrystallizes rapidly, allowing extensive deformation without risk of fracture. Room temperature is 50% of Pb's homologous melting temperature, so published values for elastic modulus, yield strength, and ultimate tensile strength do not accurately predict long-term stress–strain behavior in Pb. In 99.99% purity Pb, a sustained tensile stress of only 1 MPa elongates the metal 0.5% during a 500-day period at 25°C; a 9-MPa tensile stress can be borne briefly, but this stress causes failure by creep in 1 hour at 25°C.

21.5.2.3 Pb Alloys. Alloying additions are often used to raise the yield strength, fatigue strength, and creep strength of Pb. Pb atoms are larger (atomic radius = 0.175 nm) and more electronegative than most metal atoms, so only a few elements (e.g., Bi, Hg, In, Tl) have extensive solid solubility in Pb at 20°C. These high-solubility elements are seldom used to make commercial Pb alloys, due to their high cost, toxicity, and/or poor strengthening effect. All other commonly available alloying elements have less than 2% solid solubility in Pb at room temperature. Some elements (e.g., Zn, Cu) even display liquid-phase miscibility gaps with Pb. The most widely used Pb alloying addition is Sb, which forms a simple eutectic with Pb (Fig. 22.3). Cu, Ca, Sr, Li, and Te have low solid solubilities in Pb (much less than 1 wt% at 20°C) but still provide useful solid solution or precipitation hardening in small concentrations (0.03 to 0.20 wt%).

21.5.3 Applications of Pb

21.5.3.1 Pb–Acid Batteries. The dominant use of Pb is manufacture of Pb–acid batteries, which produce electricity at about 2 V per cell by cycling between PbO_2 and $PbSO_4$:

$$PbO_2 + H_2SO_4 + 2H^+ + 2e^- \rightarrow PbSO_4 + 2H_2O \quad \text{positive electrode}$$

$$Pb + SO_4^{2-} \rightarrow PbSO_4 + 2e^- \quad \text{negative electrode}$$

The type of Pb–acid battery used in cars and trucks to start the engine has the Pb alloy electrodes covered with a sponge or paste of Pb (− electrode) or PbO_2 (+ electrode) to increase surface area. This design maximizes high power delivery for short time periods, even if the temperature is well above or below 20°C. These batteries are "sprinters," designed to deliver a large surge of power for a few seconds to start an engine; they are not intended to be totally discharged. Once a vehicle's engine starts, the alternator powers the vehicle's electronics and accessories, and the battery is no longer needed. A second type of Pb–acid battery (the deep-cycle battery) powers electric vehicles such as forklifts and golf carts and provides emergency backup power to critical facilities such as airports, nuclear power stations, computer operations, and "load-leveling" power storage by electric utilities. Batteries for these applications are "long-distance runners"; they deliver more total power over a period of hours but cannot match the intense power burst of batteries designed to start internal combustion engines. Deep-cycle batteries contain thicker plates with less surface area and can be discharged and recharged completely hundreds of times without damage.

Pure Pb is too weak to be used in battery grids; Pb–Sb or Pb–Ca alloys are used. Pb–Sb battery grids are easily die-cast, display good corrosion resistance, have strong adhesion between the PbO_2 paste and the grid, and are sufficiently strong to withstand the stresses produced by corrosion, creep, and the expansion and contraction of the active mass during charging and discharging. Pb–Sb alloys in batteries tolerate deep discharge well, but require more maintenance, particularly addition of water as the battery ages. Several decades ago, Pb–Sb alloys contained 5 to 8 wt% Sb, but Sb content has been declining steadily in modern alloys to about 0.6 to 3% Sb. Lower Sb contents reduce the tendency of the battery to discharge during storage, but low-Sb alloys require other alloying additions (Cu, As, S, Sn, or Se) to maintain good castability and strength. High-Sb alloys (Sb >3.5 wt%) are strengthened primarily by the eutectic microstructure (Fig. 22.4), whereas the principal strengthening effect in low-Sb alloys is precipitation hardening (Fig. 21.14).

Use of Pb–Ca battery alloys has risen sharply during the past several years. When stored, batteries containing Pb–Ca alloys lose their charge much more slowly than batteries with Pb–Sb alloys. This allows batteries to be charged at the factory and stored for several months before sale, eliminating the hazard and inconvenience of filling the battery with H_2SO_4 acid when it is installed in the vehicle. Pb–Ca battery grids can be designed so that no water additions are needed throughout the battery's service life. Ca has low solubility in Pb, but small amounts (0.03 to 0.15 wt%) produce substantial hardening (Table 21.2) by forming Pb_3Ca precipitates. Ca is a

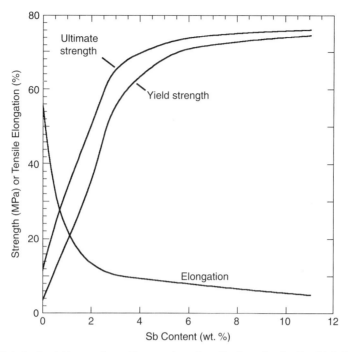

Fig. 21.14 Variation in yield strength, tensile strength, and tensile elongation for Pb with 0.15% As, 0.05% Cu, and varying Sb contents rapidly cooled to room temperature and aged at room temperature for 30 days prior to tensile testing.

TABLE 21.2 Comparison of Mechanical Properties of Pure Pb and Some Common Pb Alloys (wt%)

Property	High-Purity Pb	Pb–0.06% Cu	Pb–1% Sb	Pb–0.04% Ca	Pb–0.08% Ca–1.0% Sn
Tensile strength (MPa)	12.2	17.4	20.9	27.9	59.7
Elongation (%)	55	55	35	34	15

Source: Ullmann's (2003b).

potent strengthener in Pb, but it is somewhat difficult to use because it reacts readily with O, Sb, As, and S, precluding Pb_3Ca precipitate formation. CaO, CaS, and Ca intermetallics also reduce Pb's fluidity, making it more difficult to cast. For this reason, many Pb–Ca alloys also contain 0.02 to 0.03 wt% Al to form a thin Al_2O_3 skin on the surface of the molten Pb to reduce Ca oxidation in the liquid metal. Pb–Ca–Sn alloys have better castability than binary Pb–Ca alloys, although ductility is lower. Pb–Ca alloys can also be fabricated as cast slab, rolled into sheet, and pierced to expand it into mesh; this is less expensive than die-casting Pb–Sb alloys.

21.5.3.2 Pb Cable Sheathing, Sheet, and Pipe. Pb's easy fabricability and good corrosion resistance make it an excellent sheathing material to protect buried and overhead cables from corrosion. Pure Pb or Pb alloys containing small amounts (<1 wt%) of Cu, Sn, Sb, or Cd are specified for sheathing applications, with the exact composition chosen to optimize fatigue strength, vibration damping ability, and corrosion performance. Large alloy additions are avoided because they lack the high ductility needed to fabricate the sheath. Pb–Cu alloy sheath is often extruded continuously around electric power cables. Higher fatigue strength can be achieved in Pb–Cu sheathing by small additions of Sn and As. Pb sheathing use has declined in recent decades, due to competition from polymers, but Pb outperforms polymers in some applications, and a complete switch to polymer sheathing is not anticipated. Cu has low solid solubility in Pb, but small amounts (up to 0.08 wt%) are often added to Pb as a hardener. Cu refines grain size in Pb and raises yield strength, fatigue strength, creep resistance, and corrosion resistance (Sidebar 21.5). Pb–Cu alloys are often specified for acid-handling tanks and pipes and for Pb roofing. Pb sheet also finds use as a sound-deadening layer in walls and as a radiation barrier.

21.5.3.3 Solders, Coatings, and Bearing Alloys. Pb–Sn alloys make excellent solders for electrical and mechanical joining (Sec. 21.4.3.4), but their use is declining as users switch to Pb-free solders (Sidebar 21.4). Pb–Sn alloys make useful corrosion-resistant coatings for steel (terneplate, Sec. 21.4.3.1). Sn–Pb alloys (white metal) were once the dominant alloy for bearings (Sec. 21.4.3.5), but they are losing market share to cast or P/M Cu–Pb or bronze–Pb bearings with 20 to 40% Pb content. Modern automobile and truck engines require fatigue strengths and elevated-temperature strengths higher than white metal can deliver; consequently, Cu–Pb and bronze–Pb compositions (usually electroplated with Pb–Sn or Pb + In overlayers) are finding increasing use.

SIDEBAR 21.5: Cu ADDITIONS TO IMPROVE Pb CORROSION RESISTANCE

High-purity single-phase metals are usually most resistant to aqueous corrosion. Second-phase particles (e.g., Al_2Cu precipitates in Al–Cu alloys) typically establish local galvanic couples at the metal surface that accelerate corrosion. In Pb–Cu alloys, however, the presence of second-phase precipitates of essentially pure Cu actually improves corrosion resistance in H_2SO_4 acid and sulfate solutions (Fig. 21.15). The Cu is cathodic to the Pb matrix, forming a galvanic couple that forms a $PbSO_4$ corrosion product on the surface. Since Pb's corrosion resistance relies on formation of insoluble reaction products at the surface, the Cu actually improves corrosion resistance by thickening the $PbSO_4$ barrier. If this coating is damaged, it re-forms quickly. For this reason, Pb–0.06% Cu alloys (often called *chemical lead*) are used in preference to pure Pb for handling H_2SO_4 and sulfate solutions.

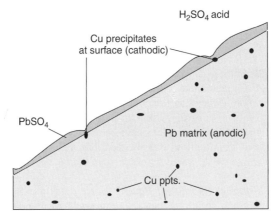

Fig. 21.15 Cu precipitates (typically 1 to 5 μm in diameter) in chemical lead (Pb–0.06 wt% Cu) form local galvanic couples on the metal's surface in H_2SO_4 acid. The Pb is anodic in the Cu–Pb galvanic couple, which thickens the protective layer of $PbSO_4$ on the metal's surface. $PbSO_4$ has very low solubility in the acid and forms a protective barrier; the galvanic action of the Cu phase thickens the barrier and improves corrosion resistance.

21.5.3.4 Ammunition. The easy formability, low cost, and high density of Pb make it the preferred metal for most ammunition. Pb ammunition is a blend of primary and recycled Pb, so compositions vary somewhat. Sb content usually ranges between 0.5 and 8%, and As is commonly added in lesser amounts for further hardening. Pb shot is produced by dropping molten Pb alloy through the air into water (formerly in 40-m-tall towers, but now often with much shorter drops). For this process at least 1% Sb is helpful to produce rounder shot. Use of Pb shot for shotgun hunting is prohibited in many hunting areas because waterfowl swallow pebbles to aid digestion and often die from inadvertently ingesting Pb shot residing in the mud of marshlands that are hunted frequently. Steel, Bi, and W-polymer composites are the most common replacement materials; however, they are not entirely satisfactory substitutes. Steel and Bi have a lower density than Pb and decelerate faster after leaving the gun. Bi and W are more costly than Pb. Steel shot causes increased barrel wear in older guns not designed for its use, and steel shot ricochets off hard surfaces with more energy than that of Pb.

21.5.3.5 Other Metallurgical Pb Uses. Pb (0.5 to 7 wt%) is sometimes added to steel, brass, and bronze to improve machinability. Pb is insoluble in these metals and forms small particles of pure Pb that serve as chipbreakers and as an in-situ lubricant during cutting operations. This lowers fabrication costs but adds a Pb toxicity hazard to melting, fabrication, and disposal; consequently, use of these leaded, free-machining alloys is diminishing. Pb is also present in many low-melting alloys (Sec. 22.4.3 and Table 22.2) for fire sprinkler systems, machining fixtures, and pipe-bending operations. Pure Pb, Pb–Sb, Pb–Ca, Pb–Ag, and Pb–As alloys are used as electrodes in electrorefining and electroplating operations. Pb containing 1 to 6% Ag is used for soldering above 300°C to achieve stronger joints with excellent corrosion resistance. Pb–0.06 wt% Te alloys are useful for shielding from neutron radiation; the alloy has good strength and shows minimal transmutation radioactivation even after exposure to heavy neutron fluences. The high density, easy formability, and low cost of Pb make it a useful weighting material in counterweights and sailboat keels (Fig. 21.16 and Sidebar 21.6).

21.5.3.6 Pb Compounds and Chemicals. About half a million tons of Pb is used each year in chemicals and compounds. A few decades ago, the most widely used Pb compounds were tetraethyllead, $Pb(C_2H_5)_4$, and tetramethyllead, $Pb(CH_3)_4$, antiknock additives that raise gasoline's

Fig. 21.16 The high density, low cost, corrosion resistance, and easy formability of Pb alloys make them the material of choice for many weighting applications. In sailing craft, a heavy keel is essential to counteract the tipping tendency of wind forces on the sails. Large craft use Pb alloy keel weights with a mass of several tons. (Courtesy of A. Skarin, Spray Norlin, *http://hem.spray.se/ake.skarin/*, 2003; with permission.)

> **SIDEBAR 21.6: Pb KEEL WEIGHTS IN SAILBOATS**
>
> The forces on the sails of sailing craft in a strong wind total many thousands of newtons. Since this force is applied high above the water level, it produces a large moment acting to tip the boat. The boat's hull requires a heavy mass well below the waterline to counteract this tipping force. Pb alloys are often used to provide this mass since they are dense, inexpensive, corrosion resistant, and easy to cast.

octane rating. Production of these compounds alone consumed about 450,000 tons of Pb in 1970, but concerns about air pollution and poisoning of catalytic converters gradually led to total prohibition of Pb in motor vehicle gasoline in Japan, the United States, and most of Europe (aviation gasoline is exempt from the ban in several nations). Although these chemicals are still used in less-developed nations, complete cessation of production is eventually foreseen. In the mid-twentieth century, basic Pb carbonate, $2PbCO_3 \cdot Pb(OH)_2$, was widely used as a white pigment in paints. This compound has largely been replaced by TiO_2, due to greater awareness of the toxicity hazards posed by Pb-bearing paint chips and airborne paint dusts (Sec. 21.5.4).

Although uses of many Pb compounds have been curtailed, some other Pb chemicals are still used widely. Lead acetate trihydrate $Pb(C_2H_3O_2)_2 \cdot 3H_2O$ is used as a dye mordant, a water repellant, and a processing agent in the cosmetics industry. $PbCO_3$ continues to be used as a production catalyst and stabilizer for polymers, a ceramic glaze, and a high-pressure lubricant. Various Pb halides are used as catalysts for polymer production, additives to infrared transmitting optical fibers, and fluxes for brazing and soldering Al, cast Fe, and brass. Pb oxides are used in large quantities as the active paste in Pb–acid batteries (Sec. 21.5.3.1), as pigments in anti-corrosion paints for steel, as additives in optical glass, as vulcanizing and stabilizing agents in rubbers and polymers, as high-temperature lubricants, and in capacitors and mixed oxide transducers (lead zirconate and lead titanate). Pb silicates serve as glazes and frits for ceramic and glass production. Like Sn, Pb forms a wide array of organometallic compounds, and although tetraethyl and tetramethyl lead are headed toward extinction, several other organolead compounds are used as antifungal agents, cotton and wood preservatives, rodent repellants, and lubricants.

21.5.4 Toxicity of Pb

The toxic effects of Pb were known to the ancients; Hippocrates wrote of "lead colic" in 370 B.C. However, the full extent of Pb poisoning hazards was not understood until the twentieth century. Pb poisoning can result from drinking contaminated water; inhalation of airborne dust; skin contact with dusts, contaminated soil, or Pb compounds; and ingestion of contaminated food or paint chips/dust. Children are sometimes attracted to the sweet taste of $PbCO_3$ compounds in peeling paint. Pb is metabolized as if it were Ca or Fe, so Pb uptake increases in persons with Ca or Fe dietary deficiencies. Pb causes neurobehavioral deficits in growing children, damages the kidneys (possibly leading to hypertension), causes sterility in men and miscarriage in women, inhibits hemoglobin and globulin formation (inducing anemia), and may be carcinogenic. Pb can cross the placental barrier and injure the fetus. Fetuses and younger children are most vulnerable to injury and permanent impairment from Pb poisoning. Harmful levels of Pb in the body usually cause no immediate symptoms unless the dose is quite high. Persons with Pb levels high enough to cause injury are often initially unaware of any problem. At high exposure levels, Pb severely damages the brain and kidneys, causing permanent neurological dysfunction, weakness or tremor in the extremities, personality disorders, paralysis, coma, and death.

Pb's unusually wide dispersal in the environment elevates concerns about its effects. Many toxic metals pose a significant threat only to workers directly exposed in the workplace or in relatively rare instances of release and concentration in the environment (Secs. 19.3.4 and 22.2.6). However, Pb has been so widely dispersed in exhaust gases from gasoline-burning engines, paints, water pipes, solders, and other products that most persons in the developed world have elevated Pb in their blood and bones. It is unclear whether the moderately elevated Pb levels seen in the general population cause significant harm, but medical studies suggest that the risk to children in particular is sufficient to justify prevention of Pb contamination wherever possible, even if such actions are costly. These findings motivated numerous regulatory restrictions and bans on Pb use during the past quarter century.

21.5.5 Sources of Pb and Refining Methods

Pb is the most abundant heavy element in Earth's crust (Sec. 21.1), and it has been concentrated fairly efficiently by geologic processes. The most important Pb mineral is PbS, although some Pb is recovered from $PbCO_3$ and $PbSO_4$. Pb-bearing minerals are dense, and coarse particles can be sorted from gangue by gravity concentrators such as jigs and shaking tables. Finer particulates can be separated by flotation. PbS concentrate (which also contains some Fe, Cu, and Zn sulfides) is roasted to convert it to PbO, Pb silicate, and Fe, Cu, and Zn silicates. This material is then reduced with coke in a blast furnace. The blast furnace produces crude Pb containing Cu, As, Sb, Sn, Bi, and Ag impurities. Much of the Zn is recovered as vapor from the blast furnace exhaust gases for later use in purifying the Pb.

As the impure, molten Pb tapped from the blast furnace cools, many impurities segregate and float to the surface because their solubility in Pb decreases with decreasing temperature. Fe and Zn compounds, arsenides, antimonides, and Cu and Pb sulfides are simply skimmed away. S is added to the metal when it cools to about 340°C to further lower its Cu content by forming insoluble Cu_2S. Sb, As, and Sn impurities can be removed either by agitated fluxing with NaOH and $NaNO_3$ or by blowing air or steam through the metal to oxidize these impurities preferentially. Some PbO is also formed and removed with the Sb, As, and Sn oxides, which are recovered as by-products. Ca or Mg is next added to the molten Pb to remove Bi by forming solid Bi intermetallic compounds. The Pb's precious metal impurities (Ag, sometimes with traces of Au) can be captured by addition of Zn metal powder, forming a skimmable Zn–Ag(–Au) crust. The residual Zn in the metal is either oxidized with Cl_2 or vaporized by pulling a vacuum over the molten Pb. For high-purity Pb, remaining trace impurities (As, Sb, Sn, and Zn) can be removed with stirred NaOH fluxing. Since refining crude Pb requires many steps, efforts have been made to simplify refining by using electrolysis to remove the impurities in a single step. About a third of world Pb production is refined electrolytically, but electrolytic refining poses new problems with disposal of effluent and separation of the impurities in the electrode slime. For these reasons, electrolytic refining of Pb has not become the dominant process as it has in other refining operations (e.g., Zn).

21.5.6 Structure–Property Relations in Pb

21.5.6.1 Effect of Atmosphere on Fatigue Life of Pb. The fatigue strength of metals is often lower in corrosive environments. In the case of pure Pb, the presence of O_2 in the test atmosphere accelerates fatigue crack growth and lowers Pb's fatigue strength. Figure 21.17 shows the relation between strain amplitude and the number of loading cycles required to cause fatigue failure in high-purity polycrystalline Pb samples tested in pure O_2, in air, and in high vacuum. Pb's fatigue life in vacuo is 60 times longer than its fatigue life when O_2 is present. Such behavior is common in many metals. As the test begins, the surface of the Pb oxidizes preferentially beside grain boundaries, which lowers the grain boundary strength and leads to crack formation. Vacancies caused by dislocation motion during fatigue strain diffuse to the grain boundaries, further promoting intergranular crack growth. As the crack opens during the tensile part of each cycle, O_2 flows into the opening to form a new monolayer of PbO on crack surfaces. This stabilizes the crack by preventing crack "healing" when the crack closes during the compression part of the cycle.

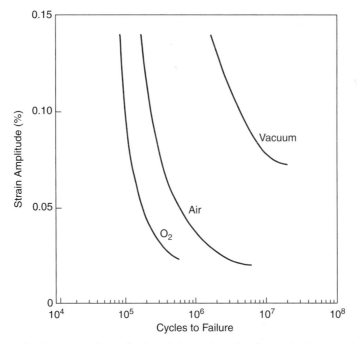

Fig. 21.17 Relation between strain amplitude and the number of cycles required to cause fatigue failure in high-purity Pb at 22°C. Fatigue strength decreases in the presence of O_2 during testing. The air and O_2 pressures were 1 atm, and the vacuum level was 2×10^{-6} torr. The loading frequency was 8.3 Hz. (Adapted from Snowden, 1964.)

During rapid fatigue loading/unloading (above about 200 Hz), the presence of O_2 is much less harmful to fatigue life. This suggests that a certain minimum time is necessary for O_2 to move into the crack while it is open (during the tensile portion of the cycle) in order to degrade fatigue life seriously. This conclusion is also supported by the results from "interrupted O_2 exposure" during vacuum fatigue testing. In that test the specimen is fatigue tested under vacuum, but the test is paused from time to time, and the specimen is surrounded by O_2 during the pauses. After the O_2 is removed by the vacuum pump, the fatigue testing resumes. Samples tested in this manner show results similar to those of simple continuous vacuum tests, indicating that to have a deleterious effect, the O_2 must be present during the crack-opening phase of fatigue strain.

ADDITIONAL INFORMATION

References, Appendixes, Problem Sets, and Metal Production Figures are available at
ftp://ftp.wiley.com/public/sci_tech_med/nonferrous

22 As, Sb, Bi, and Po

22.1 OVERVIEW

The group 15 elements arsenic, antimony, and bismuth (As, Sb, Bi) and group 16 element polonium (Po) are generally considered to be metals, although they occupy "frontier territory" on the periodic table, adjacent to the nonmetals (Fig. 1.1). Their bonding is partially covalent, and their crystal structures have low symmetry, low packing efficiencies, and directional bonding. The group 15 metals are brittle, highly anisotropic, and poor conductors (Table 22.1). Group 16's only metal, Po, has radioactive isotopes that are all short-lived. Only subgram quantities of Po have ever been available for study, and little is known about its properties.

As, Sb, and Bi all possess an hR2 crystal structure that gives each atom three nearest neighbors within a sort of "puckered sheet" stacking arrangement (Fig. 22.1); each atom has three additional second-nearest neighbors in the adjacent sheet. This structure is consistent with the trivalent s^2p^3 outer electronic structure of the elements. The group 15 elements also have several nonmetallic allotropes, some of which are equilibrium structures at high pressures (Appendix A, *ftp://ftp.wiley.com/public/sci_tech_med/nonferrous*). Po is the only element with a simple cubic crystal structure under ambient conditions.

These elements are rather scarce. As comprises 1.8 ppm of Earth's crust, Sb has a lower abundance (0.2 ppm), and Bi (0.008 ppm) is scarcer than Pt. However, geological processes have concentrated the group 15 elements efficiently in certain minerals, and they are often available as by-products from smelting of other metals. Consequently, they are far less costly than other rare metals. Extraordinarily small traces of Po exist in Earth's crust (3×10^{-10} ppm) from radioactive decay of U. $^{210}_{84}$Po was first extracted from pitchblende but is now more easily made by neutron bombardment of $^{209}_{83}$Bi.

These metals' high electronegativities (Fig. 1.14) allow their compounds to be easily reduced to metals. In fact, As, Sb, and Bi are sufficiently electronegative to be found occasionally in native form. The group 15 metals have several engineering uses, mostly as alloying additions to other metals such as Pb and Sn. Arsenic is toxic; Sb and Bi have lesser, but still significant toxicity hazards.

22.2 ARSENIC

A fair-skinned, 35-year-old male's symptoms include skin lesions and numbness and tingling in his toes and fingertips, progressing slowly in the ensuing weeks to involve the feet and hands in a symmetric "stocking-glove" fashion. In the past 2 to 3 weeks, the tingling has taken on a painful, burning quality, and he has noted weakness when gripping tools.... He has been a carpenter since completing high school 17 years ago.... The patient's urine revealed 6,000 μg per liter as total arsenic; normal is <50 μg/l.

—Case Studies in Environmental Medicine: Arsenic Toxicity, U.S. Department of Health and Human Services, Course S3060, *http://www.atsdr.cdc.gov/HEC/CSEM/arsenic/arsenic.pdf*, October 2000

Structure–Property Relations in Nonferrous Metals, by Alan M. Russell and Kok Loong Lee
Copyright © 2005 John Wiley & Sons, Inc.

TABLE 22.1 Selected Room-Temperature Properties of As, Sb, Bi, and Po with a Comparison to Cu

Property	As	Sb	Bi	Po	Cu
Valence	+5, +3, +2, −3	+5, +3, −3	+5, +3, −3	+6, +4, +2, −2	+1, +2
Crystal structure at 20°C	Rhombohedral	Rhombohedral	Rhombohedral	Simple cubic[a]	FCC
Density (g/cm^3)	5.78	6.68	9.81	9.14	8.92
Melting temperature (°C)	816[b] (38.6 atm)	631	271	250± 4	1085
Thermal conductivity (W/m·K)	50	24.4	7.9	NA[c]	399
Elastic (Youngs) modulus (GPa)	∼8[d]	∼54.7[d]	∼34.0[d]	∼102[e]	130
Coefficient of thermal expansion (10^{-6} m/m·°C)	5.6	9.0	13.4	NA[c]	16.5
Electrical resistivity (μΩ·cm)	33.3	41.7	117	42	1.7
Cost ($/kg), large quantities	1	1.90[f]	8	NA[g]	2

[a] Simple cubic Po transforms at ∼ 75°C to a slight rhombohedral distortion (hR1) of the simple cubic structure.
[b] As sublimes at 1 atm pressure at 613°C; melting requires imposition of high pressure.
[c] Physical property data are difficult to obtain for Po since only milligram quantities have been produced. The metal's intense radiation causes self-heating and rapid accumulation of radiation damage.
[d] The large elastic anisotropy of As (and to a lesser extent, Sb and Bi) causes the Young's modulus value to vary, depending on texture (see Sec. 22.2.2).
[e] The bulk modulus (56 GPa), C_{11} (113 GPa), and C_{12} (28 GPa) have been calculated for α-Po by *ab initio* methods (D. Roundy, personal communication). These allow estimation of Young's modulus using the expressions $E = c_{44}(3c_{12} + 2c_{44})/(c_{12} + c_{44})$, and $c_{44} = 0.5(c_{11} - c_{12})$. There are no reports of experimental determination of Young's modulus in Po.
[f] Sb prices often undergo large and rapid fluctuations; the cost shown here is approximately in the middle of the recent historical price range.
[g] Po is rarely sold commercially; it is available by special order from Oak Ridge National Laboratory.

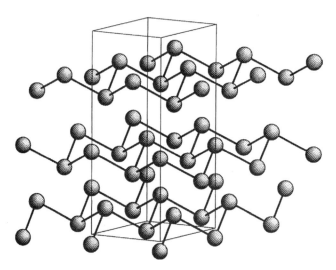

Fig. 22.1 Rhombohedral crystal structure of α-As at room temperature. Sb and Bi have similar structures, differing only in bond lengths and minor shifts in bond angles (≤1.2°). This structure provides three nearest-neighbor atoms within the puckered horizontal sheets and three second nearest-neighbor atoms in the adjacent sheet. This structure causes large anisotropies in the elastic moduli of the group 15 metals. (From Cahn and Greer, 1996, p. 37; with permission).

22.2.1 History of As

Arsenic was known to the ancients. Writings from the fifth century B.C. describe As compounds as useful medicines and effective poisons. It is difficult to determine when As metal was first produced. The work of early chemists was poorly documented and often written in code or deliberately falsified to withhold secrets from rivals. However, it appears that the first As metal was produced in A.D. 1250 by reacting As_2S_3 with soap. As_2O_3 was a frequently used poison throughout the Middle Ages because it is tasteless, odorless, and fatal in small quantities (6 to 20 mg) (Sec. 22.2.6). In an era when early death was commonplace and autopsies unknown, murder by poisoning was difficult to detect. Today, As is used in wood preservatives, herbicides, insecticides, decolorizing additives to glass, metal pickling solutions, electrolyte additives for Zn purification, alloys of Pb and Cu, and semiconductor electronic components. The toxicity of As poses major hazards to its users, and As use is declining in response to more stringent governmental controls on As in the workplace and the environment.

22.2.2 Physical Properties of As

Although As is metallic, its partly covalent bonding plays a major role in determining its physical properties. When vaporized in vacuum or inert gas, As forms tetrahedral As_4 molecules rather than the monoatomic vapor produced by most metals. In addition to the equilibrium metallic phase, As also forms metastable, nonmetallic crystal structures. The equilibrium crystal structure (Fig. 22.1) allows little dislocation motion, making pure As and As-rich alloys brittle. The bonds between the sheets are so weak that they have been compared to van der Waals bonds. As has the greatest elastic anisotropy of any metal; the Young's modulus perpendicular to the sheets (the vertical direction in Fig. 22.1) is only 7.26 GPa, but it is 81.8 GPa parallel to the sheet plane. As's Poisson's ratio varies from less than -1 to more than $+0.5$, depending on crystallographic orientation. All group 15 metals have large elastic anisotropy, but As's is the most extreme. The electronic band structure of As contains a small band overlap that provides fewer free electrons than "good" metals. This makes As (and also Sb and Bi) a rather poor electrical and thermal conductor, particularly at low temperatures. In most metals heat is conducted primarily by free electrons with a smaller contribution from lattice vibrations (phonons). In the group 15 metals, however, the limited number of free electrons lowers the electronic component of thermal conductivity to a level roughly equal to the phonon contribution.

When exposed to dry air at room temperature, As forms a thin, protective oxide layer. However, humid air causes a thicker oxide to form, gradually blackening the metal. At elevated temperature, As poses a toxicity hazard because the metal sublimes and forms toxic As_2O_3 vapor. This reaction is phosphorescent above 300°C. Exposure to higher heat poses the additional risk of fire. As burns brilliantly while emitting toxic oxide particles and vapor to the surrounding air.

22.2.3 Use of As as an Alloy Addition

Although pure As metal has no structural applications, it is a useful alloying addition in both Pb and Cu. Pb plates in Pb–acid batteries sometimes contain 0.5 to 0.75% As to strengthen them and improve their stability during long-term use. Addition of 0.5 to 2% As to Pb shot hardens the shot and improves the sphericity of the pellets. Arsenic has extensive solid solubility in Cu, and As additions of about 0.5% (arsenical copper) improve corrosion resistance and raise the recrystallization temperature, which increases elevated temperature strength.

22.2.4 GaAs Semiconductors

The dominant semiconductor materials in electronic systems are doped Si and SiGe, but for some special applications, the III–V intermetallic compound GaAs is the preferred material. Signal speed is almost five times faster in GaAs than in Si, allowing use of higher frequencies without overheating the chip. GaAs delivers a better signal-to-noise ratio than Si and tolerates elevated temperature and radiation damage better than Si. Despite these impressive advantages, GaAs is much less frequently used than Si or SiGe because it is more difficult and expensive to process. Use of GaAs for microprocessor chips, photocells, and transducers is a high-value, low-volume use of As. Ultrapure, single-crystal GaAs is used in these devices, and although only tens of tons of As are consumed by this application each year, the economic value of the product makes it a significant part of the As industry.

22.2.5 Applications of As Compounds

Over 97% of all As is used in nonmetallic form. Throughout much of the twentieth century, As was used extensively as a wood preservative to impart long-term decay resistance in exposed or buried structures. Arsenic compounds (chromated copper arsenate or ammonium copper arsenate) can be forced deep into dimension lumber in high-pressure, high-temperature autoclaves, producing wood that poisons any microbes or fungi attempting to consume it. However, As-treated lumber presents a toxicity hazard to carpenters and homeowners, and its use was prohibited in the European Community several years ago. In 2003 the U.S. wood preservative industry terminated sales of As-treated wood for residential use. These actions limit As-treated wood in much of the world to industrial and military applications, substantially reducing global As consumption.

Use of As_2O_3 and calcium arsenate was widespread in the early and mid-twentieth century. These chemicals are effective against insect pests in coffee plantations and rice paddies and helped to control boll weevil damage to cotton crops in Mexico and the United States. However, unintended environmental damage and numerous instances of As poisoning in farmworkers prompted drastic reduction of such use. As compounds still find applications in the glass industry as a decolorizing and fining agent, in electrolytes for electrolytic recovery of Zn, and in metal pickling baths. Small amounts of As compounds are also used in catalysis and pharmaceuticals (As^{5+} is much less toxic than As^{3+}).

22.2.6 Toxicity of As

As^{3+} ions bind to enzymes in the body, disabling the enzymes and interfering with cell respiration. Arsenic causes lung and skin cancer and is suspected of causing other cancers. Arsenic can be ingested by drinking water from wells in As-bearing rock strata, by inhaling arsine gas (AsH_3) or powders of As metal or compounds, by skin contact with arsenated wood or other As compounds, or by gastrointestinal ingestion. Although As metal is nontoxic, it is readily converted to toxic As compounds once inside the body. A partial list of As effects includes stomach and intestinal lesions, bleeding, and perforations; vascular tissue damage; heart arrythmia; gangrene of the extremities ("blackfoot syndrome"); dark patches on the skin; respiratory tract inflammation; liver cirrhosis and necrosis; renal insufficiency and failure; birth defects; lethargy, headaches, delirium, coma, and convulsions; and death. As is gradually removed from the body by the kidneys, and patients usually recover from low-level As poisoning, although permanent tissue damage and elevated risk of later cancer are common. Chronic exposure to low levels of As can cause cancer years or even decades later. This litany of medical horrors motivated governments and industry groups to limit As use and to monitor drinking water for As content. As use is generally declining worldwide, but continued use within tighter controls is anticipated rather than a complete ban.

22.2.7 Sources of As and Stabilization for Safe Disposal

China operates mines for the sole purpose of As recovery, but most As is a by-product from Cu, Pb, and Zn ores, where As sulfides are often present. Since As has a low market value and presents a toxicity hazard and environmental costs, it is often viewed as an unwanted tramp element in these ores. Arsenic sulfides and oxides are more volatile than most minerals, so they enter the flue gases from roasting and smelting operations, where they can be (indeed, must be!) condensed as As_2O_3, "white arsenic."

As metal is relatively easy to produce simply by heating As_2O_3 with C under an alkali carbonate protective flux. However, most As is used as As_2O_3 or in other compounds. With declining demand for As products, some mines regard As as toxic waste, converting As_2O_3 flue gas condensate to a stable iron arsenate intermetallic compound that can be buried in tailing heaps without risk of leaching into groundwater. Arsenic disposal, rather than sale, is becoming more common as demand for As declines.

22.3 ANTIMONY

> The use of makeup is also said to stem from witchcraft where the painting of one's face was believed to ward off evil. Mascara was particularly a charm inasmuch as it is made of antimony, an old witch metal.
> —Brother Michael John Nisbett, *http://www.nisbett.com/symbols/makeupbody_paint.htm*, 2003

22.3.1 History of Sb

Sb has been used by humankind since antiquity. Stibnite (As_2S_3) was prized as eye makeup by ancient Mediterranean cultures and was probably used by such notables as Jezebel and Cleopatra. A Chaldean vase from 4000 B.C. was cast from Sb, and Chinese Sb artifacts have been dated to 3000 B.C. Ancient Egyptian and European writers often referred to Sb as a variety of Pb, and it was not recognized as a separate element until the sixteenth century. Today, Sb is used widely as an alloying addition to Pb and Sn for batteries, ammunition, bearings, solders, and pewter. Sb compounds serve as flame retardants, as decolorizing agents in glass, in semiconductor devices, as opacifiers in ceramics and polymers, and in pigments.

22.3.2 Physical Properties of Sb

Sb's hR2 crystal structure (Fig. 22.1) makes it brittle and anisotropic, much like As (Fig. 22.2 and Sec. 22.2.2). In the vertical direction in Fig. 22.1, Sb's Young's modulus is 33.9 GPa, but it is 125 GPa in a direction inclined 62° from the vertical axis. Pure Sb metal has no structural uses, but it is used frequently to strengthen Pb and Sn alloys. Various nonmetallic, metastable forms of Sb exist. These are somewhat similar to the P allotropes, and revert gradually to metallic Sb at room temperature; if heated, the transformation is rapid. The stability of these metastable phases can be precarious; one metastable Sb allotrope is reported to be explosive if impurities are present. Sb remains bright and untarnished in dry air, humid air, and in water. The metal resists attack by dilute HNO_3 and HCl acids and by concentrated HF and H_2SO_4 acids. Sb metal will burn in air if heated to 500 to 600°C.

22.3.3 Use of Sb as an Alloying Addition

Sb is too brittle to be useful in the pure state, but it is a valuable alloying additive for several Pb and Sn alloys. The most familiar example of these is the Pb–Sb alloy used for grids in Pb–acid storage batteries for cars and trucks. The ultimate tensile strength of pure Pb is only 17 MPa, but addition of 6 wt% Sb raises this to 30 MPa. Pb–Sb forms a simple eutectic system (Fig. 22.3). The eutectic microstructure of antimonial Pb (Pb–11.1 wt% Sb) (Fig. 22.4) raises both ultimate and creep strength. Creep is a common problem for Pb in automotive batteries because underhood temperatures often exceed 100°C and stress results from vibration and electrode volume changes during charge–discharge cycles. Use of Pb–Sb alloys in Pb–acid storage batteries (typically, 2 to 5 wt% Sb) is gradually diminishing in favor of Pb–Ca alloys (Sec. 21.5.3.1). Antimonial

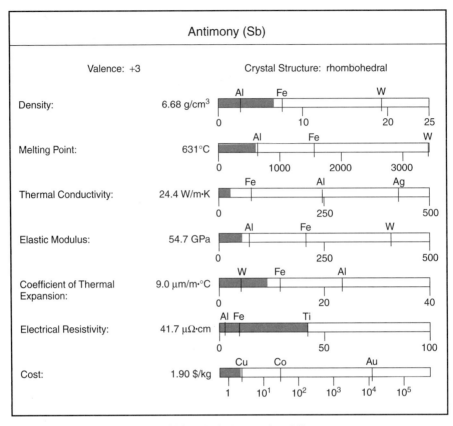

Fig. 22.2 Physical properties of Sb.

Pb alloys continue to be used for ammunition, bearings, solder, and acid-resistant pipes. High-Sb-content Pb alloys (up to 25 wt% Sb) are used in the printing industry for type metal; Sb lowers the melting point, raises wear resistance, and reduces shrinkage during freezing (pure Sb contracts only 0.8% during freezing).

The alloys known as *pewter* are comprised of Sn containing 7 to 20% Sb and 2% Cu (Sec. 21.4.3.2). Pewter is used to produce cast objects such as lamps, vases, and tableware; it retains Sn's luster while improving strength. Sb is also used to strengthen Pb–Sn solders for auto body panel assembly, seams on "tin" cans, and plumbing. Sb is often added to Pb-free solders (Sidebar 21.4).

22.3.4 Applications of Sb Compounds

High-purity Sb is used in manufacturing III–V semiconducting compounds such as AlSb, GaSb, InSb, and GaInAsSb for light-emitting diodes, photocells, and lasers. Sb can also be used as a dopant in n-type Si semiconductors, where it diffuses more slowly than P, thereby improving elevated temperature stability. Bi–Sb alloys have useful thermoelectric properties (Sidebar 22.1). Although comparatively small Sb tonnages are used in the electronics industry, the high value of these products makes this an economically important Sb use.

Tens of thousands of tons of Sb_2O_3 are used annually in combination with halogen-containing polymers to make flame-retardant plastics. The heat of a fire forms oxide halides and Sb trihalides (e.g., $SbCl_3$) that char rather than release volatile gases. The charred surface acts as a heat shield to delay combustion of the underlying polymer. These flame-retardant polymers are used

ANTIMONY

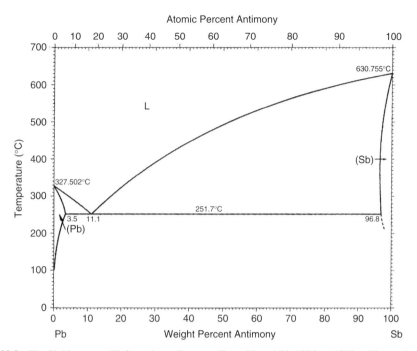

Fig. 22.3 Pb–Sb binary equilibrium phase diagram. (From Massalski, 1986, p. 1843; with permission).

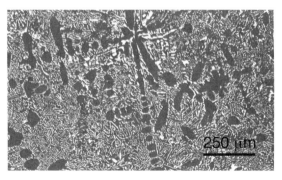

Fig. 22.4 Microstructure of eutectic antimonial Pb alloy (Pb–11.1 wt% Sb). The dark phase is Pb-rich; the light phase is Sb-rich. This microstructure raises both the yield and creep strengths of the alloy vis-á-vis pure Pb. Some alloy segregation is evident in this sample, forming large Pb phases along with the eutectic structure. (From ASM, 1972; with permission).

in furniture fabrics, draperies, automotive seats and moldings, electrical equipment housings, wire sheathing, building panels, and even lifeboat hulls. Sb_2O_3 is also used in pigments and opacifiers in ceramics, as an additive to remove color and bubbles from glass or television tubes and computer monitors, and as a catalyst for producing polyesters.

22.3.5 Toxicity of Sb

Sb has toxic effects quite similar to those of As (Sec. 22.2.6). However, the risk of Sb poisoning is fairly low because Sb metal is more resistant than As metal to oxide formation in air and aqueous solutions. Sb_2O_3 has low water solubility. The use of pewter for tableware appears

> **SIDEBAR 22.1: Sb AND Bi IN THERMOELECTRIC GENERATORS**
>
> When a temperature gradient exists across a material, the Seebeck effect generates a voltage between the hot and cold regions. The Seebeck effect results from an imbalance in charge carrier drift velocities between the hot and cold regions. Electrons in the hot region have high drift velocities and tend to "expand" from the high-temperature region to the low-temperature region. "Cold" electrons in the low-temperature region move more slowly toward the hot region. The result is a net charge imbalance (i.e., a voltage gradient) between the hot and cold regions. The same effect produces a reverse bias in p-type material where the charge carriers are positive. The Seebeck effect can be used to produce electric power from a heat source. Alternatively, it can also be used "in reverse" to provide refrigeration when external power is supplied. Thermoelectric devices can be designed to exploit this effect, using a structure (Fig. 22.5) that connects the hot and cold regions. A high voltage is desirable between the hot and cold regions in such devices, and this can be expressed by a material's Seebeck coefficient, measured in $\mu V/°C$.
>
> The figure of merit for the performance of a thermoelectric device (Z) can be calculated from
>
> $$Z = \frac{(\alpha_p - \alpha_n)^2}{[(\lambda_p \rho_p)^{0.5} + (\lambda_n \rho_n)^{0.5}]^2}$$
>
> where α_p and α_n are the Seebeck coefficients of the p- and n-type legs, λ_p and λ_n are the thermal conductivities, and ρ_p and ρ_n are the electrical resistivities. Z has units of $°C^{-1}$; and the higher Z is, the greater the efficiency of the device. This expression for Z shows that the best device performance occurs for materials with a large difference between the n- and p-type Seebeck coefficients, low thermal conductivities, and low electrical resistivities.
>
> It is difficult to produce materials with low electrical resistivity *and* low thermal conductivity. The best combination of these two properties occurs in semiconductors with heavy atoms and complex crystal structures. Doped n-type $Bi_2(Te_{0.9}Se_{0.1})_3$ and p-type $(Bi_{0.25}Sb_{0.75})_2Te_3$ semiconductors deliver some of the highest Z values known for devices used near room temperature. The heavy atoms and complex crystal structures in these materials cause phonon scattering, which lowers λ. Thermoelectric materials are used to cool small objects such as electronic circuit components and hold them at a steady temperature within $\pm 0.1°C$; they produce electric power in remote sites; and they actively cool portable coolers for consumer use. They are simple, reliable devices with no moving parts.
>
> Bi–Sb–Te–Se materials are unsuitable for high-temperature use, and other materials (unfortunately, with lower Z values) have been developed for those applications. Thermoelectric generators based on doped Si–Ge semiconductors use the heat from radioactive decay in $^{238}PuO_2$ (Sec. 24.5.2) to supply the power for deep-space missions such as *Voyager*, *Galileo*, and *Cassini*. Over 250 million thermoelectric operating hours have been logged on these space missions without a single device failure.

to pose no toxicity hazard because foods and beverages do not leach Sb from the pewter. SbH_3 (stibine gas) is a potent toxin requiring stringent precautions; however, governmental restrictions on handling and release to the environment of other Sb compounds and metal are far less stringent than those placed on As. Nevertheless, Sb metal and compounds have caused numerous cases of illness and death, and skin contact, ingestion, and inhalation of powders should be minimized. Sb carcinogenicity in humans is suspected but not demonstrated.

22.3.6 Sources of Sb

World production of Sb totaled 141,000 tons of contained Sb content in 2002, which is much greater than that of its congener elements As (35,000 tons/yr) and Bi (3900 tons/yr). Its use is

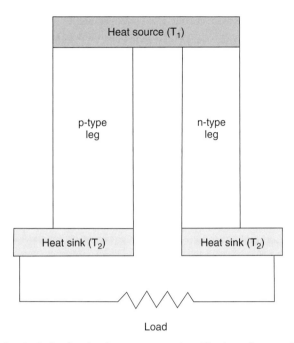

Fig. 22.5 Thermoelectric device for electric power generation. The thermal energy from the heat source produces a temperature gradient in both the n- and p-type legs of the device. The temperature gradient produces a voltage that delivers power to the load. Ideal materials for the two legs would have low electrical resistivity, low thermal conductivity, and a large voltage between the hot and cold temperatures. $(Bi_x,Sb_{1-x})_2Te_3$ is one of the most efficient thermoelectric materials available for this purpose.

approximately 40% metallic Sb and 60% Sb compounds. China is the dominant world supplier. Sb_2S_3 (stibnite) is the most important Sb-bearing mineral. Sb metal can be obtained by reducing stibnite with scrap Fe. Sb_2O_3 is sometimes recovered from the flue dust generated by smelting other metal ores (especially Pb). Sb_2O_3 can be reduced by C to produce Sb metal. In contrast to the declining As market, there is no movement to diminish or eliminate Sb use, and global Sb production has increased gradually in recent years.

22.4 BISMUTH

... which brings us finally to bismuth, the last outpost of nuclear stability.

—Erling Jensen, Iowa State University, from a 1969 physics lecture

22.4.1 History of Bi

Bi metal is metallurgically and chemically similar to its group 15 congeners, but it has a number of intriguing and useful differences. $^{209}_{83}Bi$ has the distinction of being the heaviest nonradioactive nuclide; elements 84 and above have no stable nuclei. [^{209}Bi is actually very slightly radioactive, but its decay half-life (1.9×10^{19} years) is a billion times longer than the age of the universe, so for all practical purposes, it is stable.] Bi is one of the few metals that expand upon freezing. It has exceptionally low thermal and electrical conductivity for a metal, and the electrical conductivity of the solid is actually lower than that of the liquid.

22.4.2 Physical Properties of Bi

22.4.2.1 Freezing Expansion. Pure Bi expands 3.3% during solidification. Most metals contract as they freeze because the regular periodic atom arrangement of a solid lattice has a higher density than that of the more chaotic atom arrangement in liquid metal. Bi's odd freezing expansion might be thought to result from its rhombohedral crystal structure (Fig. 22.1), which provides a less efficient packing structure than the more common BCC, FCC, and HCP structures. However, As and Sb have nearly identical hR2 crystal structures, and they contract during freezing. Although the issue needs further study, it appears that Bi's freezing expansion is not caused by its solid structure but by its liquid structure. It has long been known that there is something peculiar about liquid Bi. The electrical resistivity of liquid Bi is high; the liquid metal is a semiconductor. This is believed to result from the presence in the liquid of Bi atom clusters 2 to 10 nm in size. Such clusters have been observed in Bi thin films on graphite substrates, and ab initio calculations indicate that these clusters would probably possess a densely packed structure, such as HCP or FCC. Thus, current models of liquid Bi describe a sort of "cluster soup" consisting of densely packed Bi crystallites a few nanometers in size, and although there would be interactions between the clusters, the stronger bonding would occur within the clusters. This model is consistent with the high electrical resistivity of liquid Bi; it would also explain the volume expansion during freezing as densely packed liquid clusters transform into less efficiently packed rhombohedral solid crystals.

22.4.2.2 Bi Diamagnetism and Hall Effect. Bi has the highest diamagnetism of any metal. The specific susceptibility is -16×10^{-9} m^3/kg at 20°C. In a diamagnetic material, no atom dipoles exist unless an external magnetic field is imposed on the metal; when that is done, dipoles are induced in the individual atoms aligning antiparallel to the external field vector. If placed near a strong magnetic field, Bi is attracted to regions where the field is weak; however, the effect is small. In the presence of an external magnetic field, the electrical resistivity of Bi increases more than that of any other metal (i.e., Bi has the largest Hall effect).

22.4.3 Alloys of Bi

Although pure Bi is too brittle for any structural use, several alloys exploit Bi's low melting temperature and freezing expansion. The exceptionally low melting points of fusible alloys (Table 22.2) make them useful for automatic fire sprinkler systems, safety-release plugs for storage tanks, and electric fuses. The ductile Bi–Pb–Sn–Cd alloys melt below 100°C and expand during freezing; these alloys are used to fill the interior volume of thin-walled tubes temporarily so that they can be bent without buckling or wrinkling (Sidebar 19.3); after the tubes are bent, the alloy is melted out in boiling water. A noneutectic alloy of 48% Bi–28% Pb–15% Sn–9% Sb expands substantially as it solidifies, which permits it temporarily to be cast around irregularly shaped or fragile objects to grip them for machining. This alloy is also diamagnetic, making it useful to mount magnet arrays where the individual magnets need to be isolated from one another.

The noneutectic 40% Bi–60% Sn alloy has only 0.01% contraction during freezing, making it ideal for accurate impressions of casting molds that reproduce the original or model in fine detail without dimensional distortion. Eutectic 57% Bi–43% Sn alloy also has low expansion

TABLE 22.2 Fusible Alloys[a]

System	Eutectic Composition (wt%)	Eutectic Temperature (°C)
Bi–Cd	60Bi–40Cd	144
Sn–Pb–Bi	49Sn–41Pb–10Bi	142
Bi–Sn	57Bi–43Sn	139
In–Sn	52In–48Sn	117
Bi–Pb–Sn–Cd	50Bi–26.7Pb–13.3Sn–10Cd	70
Bi–Pb–In–Sn–Cd	44.7Bi–22.6Pb–19.1In–8.3Sn–5.3Cd	47

[a] These alloys are used for their low melting points.

behavior and is used for dental crown and implant molds. Some Bi–Pb–Sn alloys wet glass and enameled surfaces and are used to form vacuum-tight seals between metal and glass.

Bi is sometimes a minor ingredient in solders. Bi and Pb are insoluble in steel and Al alloys, where they serve as "chipbreakers" to facilitate machining. Bi's unusually low thermal neutron absorption cross section (0.034 barn) has led to its use in experimental liquid-metal-cooled nuclear reactors; however, no commercial reactors use Bi coolant.

22.4.4 Toxicity of Bi

Although Bi is a potentially toxic element, the danger from handling the metal is low because it does not usually react with water or body fluids to form soluble compounds. Most Bi poisoning cases result from problems with administration of Bi-containing pharmaceuticals. There have been cases where water-insoluble Bi compounds were converted by bacteria in patients' intestines to water-soluble compounds, causing acute poisoning and sometimes death. Water-soluble Bi compounds cause many of the same symptoms seen in Pb and Hg poisoning: damage to bone, skin, kidneys, and the central nervous system. Chelating compounds have proven effective in binding to Bi ions and accelerating their excretion. There are no indications that Bi causes cancer or birth defects.

22.4.5 Sources of Bi and Production Methods

Bi is produced primarily as a by-product of Cu and W mining in China, the largest producing nation. In other producing countries it sometimes occurs with Pb, Mo, and Sn, but it is seldom recovered alone. Since Bi is a by-product resource, Bi prices vary widely. Production often continues unabated, even when Bi is in surplus, and shortages do not necessarily trigger expanded production. During the latter part of the twentieth century, Bi metal prices ranged from a low of $2.86 per kilogram to a high of $26.40 per kilogram. Even the high end of this price range is a remarkably modest cost for an element that is among the scarcest on Earth (Appendix C, *ftp://ftp.wiley.com/public/sci_tech_med/nonferrous*). The specific steps used to recover Bi from its ore vary with the primary metal in the ore. Bi is often leached from complex ore concentrates with HCl acid, yielding $BiCl_3$ that can be reduced with scrap Fe to yield Bi metal. Bi can be purified quite efficiently by vacuum distillation. About half of Bi use is for nonmetallic Bi compounds for products in the chemical, pharmaceutical, and cosmetics industries.

22.4.6 Structure–Property Relations in Bi

When a solid metal is immersed in a liquid metal, the liquid will sometimes penetrate rapidly along the solid grain boundaries and embrittle the solid. This has been observed in several systems, such as liquid Sn and Zn in solid Fe–Si alloys, liquid Ga and In in solid Al, and liquid Bi in solid Ni and Cu. It is an important engineering phenomenon since numerous systems and processes involve solid–liquid metal contact (e.g., Na-cooled nuclear reactor structures). The liquid Bi–solid Cu case has been particularly well studied.

When tensile-tested at 300°C, polycrystalline Cu normally elongates about 35%, then fails by ductile, transgranular fracture (Sec. 5.2.4). However, when Cu tensile tests are performed with the specimens immersed in liquid Bi at 300°C, they elongate only 1% before intergranular fracture occurs. This change results from Bi diffusing along grain boundaries (Fig. 22.6), forming a Bi-rich, metastable, amorphous layer that forms a wedge between the Cu grains. This amorphous layer lowers the interfacial energy along the Cu grain boundaries. When the Cu is under no load, the Bi advances along the grain boundaries at about 1.1 μm/min at 300°C; however, tensile stress across the grain boundary accelerates Bi penetration to 1 mm/min. Scanning electron microscope examination of Cu specimen fracture surfaces embrittled by liquid Bi (Fig. 22.7) shows that small crystals of Bi form on the Cu grain boundaries. In addition, x-ray photoelectron spectroscopy shows that large portions of the Cu grain boundary surface contain

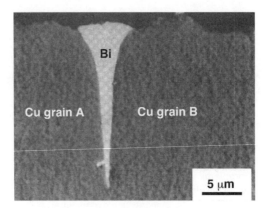

Fig. 22.6 Micrograph of Bi penetration at a grain boundary in polycrystalline Cu immersed in liquid Bi at 600°C for 4 hours. No tensile load was applied to the Cu during its immersion. (From Joseph et al., 1998; with permission.)

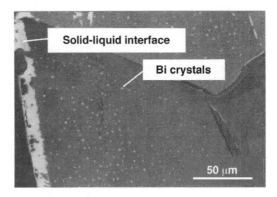

Fig. 22.7 Back-scattered electron scanning electron micrographs of Bi crystals on a Cu grain boundary exposed by intergranular fracture. The Cu specimen was immersed in liquid Bi for 16 hours at 600°C prior to fracture. The white spots are Bi-rich; the gray region contains 50 to 60% Bi (balance Cu) to a depth of several atoms. (From Joseph et al., 1998; with permission.)

50 to 60% Bi to a depth of several atom layers. This Bi-rich material is present even in regions far from the nearest Bi crystal.

Bi weakens the Cu by replacing the strongly bonded, ductile Cu–Cu interface at the grain boundaries with liquid (at elevated temperatures) or brittle solid (at room temperature). This behavior is surprising because the maximum solid solubility of Bi in Cu is only 30 atomic ppm, and the maximum solid solubility of Cu in Bi is even lower. Bi penetrates Cu grain boundaries because the surface energies at the grain boundaries alter the behavior that would be expected in the bulk material. Figure 22.6 shows that Bi does not penetrate the bulk of the Cu grains exposed at the Cu–Bi interface; attack is restricted to the grain boundaries.

22.5 POLONIUM

22.5.1 Discovery and Production of Po

The initial discovery of Po in 1898 is a famous example of scientific persistence. Marie Curie processed large quantities of U ore in an immensely labor-intensive effort, using the newly discovered property of radioactivity to monitor the effectiveness of the various reactions in concentrating element 84. Her eventual success earned her a Nobel Prize for Chemistry and the right to name the new element for her native Poland. The three longest-lived Po isotopes are $^{208}_{84}$Po ($t_{1/2} = 2.9$ years), $^{209}_{84}$Po ($t_{1/2} = 102$ years), and $^{210}_{84}$Po ($t_{1/2} = 0.38$ year). Unfortunately, the fast-decaying ^{210}Po isotope is the one present in U ore and the easiest isotope to produce artificially. The longer-lived ^{208}Po and ^{209}Po isotopes can be produced only by proton or deuteron bombardment reactions, while $^{210}_{84}$Po can be produced by neutron capture in $^{209}_{83}$Bi. High neutron fluences are much more easily produced than high proton or deuteron fluences, so ^{210}Po is the only isotope available in milligram quantities. Thus, most study of Po has been performed on a "hot" isotope, making many properties impossible to measure.

22.5.2 Crystal Structure and Applications of Po

Po has the simple cubic structure (cP1) (Fig. 22.8) at ambient temperature and pressure. [High-pressure structures of Ca and As are also cP1 (Appendix A, *ftp://ftp.wiley.com/public/sci_tech_med/nonferrous*).] The stability of the simple cubic structure in Po is attributed to the $6s^2 6p^4$ outer electron structure. Ab initio calculations show that the filled s band lies 10 to 15 eV below the Fermi energy and is not involved in bonding; however, the three $6p$ bands are only two-thirds filled, which allows some partial directional bonding. The computed electron density distribution shown in Fig. 22.9 shows higher electron densities along [100] directions associated with this covalent bonding.

With such a short half-life, ^{210}Po generates tremendous radioactive decay energy; 1 g of freshly prepared ^{210}Po metal produces 141 W of power from its alpha decay. Specific power for radioisotopes can be calculated from

$$P = \frac{2119.3 Q m}{A t_{1/2}}$$

where P is power (W), Q the energy per disintegration (MeV), m the mass of the sample (g), A the atomic weight (mass number), and $t_{1/2}$ the half-life (years). Po's potent self-heating effect makes study of its properties challenging because the radiation damage and heat break

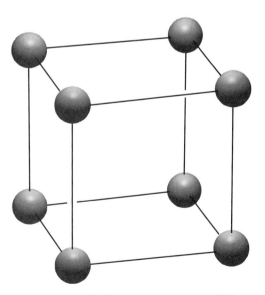

Fig. 22.8 Simple cubic crystal structure of α-Po at room temperature. This structure provides six nearest-neighbor atoms. Po is the only element with this structure at ambient temperature and pressure.

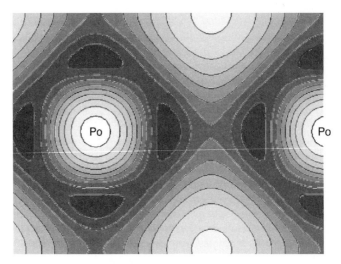

Fig. 22.9 Density distribution map of $6p$ electrons calculated for the {100} plane of α-Po. The designation "Po" locates atom nuclei. Dark areas correspond to regions with high electron density; lighter areas are regions of low electron density. The high electron density between the two nuclei correlates with the partially directional bonding in α-Po. (Courtesy of D. Roundy, University of California–Berkeley; with permission).

bonds and alter crystal structures. Despite these difficulties, an allotropic structure change has been observed near 75°C from cP1 to the rhombohedral hR1 structure. A fair amount has been learned about the chemistry of dilute Po aqueous and metallic solutions, although the mechanical properties of Po remain unexplored.

^{210}Po is an almost pure alpha emitter, which makes it a convenient power supply for spacecraft because heavy shielding is not needed to protect ground crews prior to launch. ^{210}Po was used as the power source in early thermoelectric power supplies for short-duration space missions. The 1970 Soviet *Luna 17* and *21* lunar rover missions used ^{210}Po decay heat to keep equipment warm during the long lunar nights. However, ^{238}Pu (Sec. 24.5.2) has replaced ^{210}Po for most space power applications. ^{210}Po has also been used as a convenient neutron source for research and for initiators in atomic bombs (Sec. 11.3.3.2), although the latter use has been superseded by other methods. The short half-life of ^{210}Po makes it an extremely potent radiologic toxin; the maximum permissible body burden has been set at 6.8×10^{-12} g.

ADDITIONAL INFORMATION

References, Appendixes, Problem Sets, and Metal Production Figures are available at
ftp://ftp.wiley.com/public/sci_tech_med/nonferrous

23 Sc, Y, La, Ce, Pr, Nd, Pm, Sm, Eu, Gd, Tb, Dy, Ho, Er, Tm, Yb, and Lu

23.1 OVERVIEW

Scandium ($_{21}$Sc), yttrium ($_{39}$Y), lanthanum ($_{57}$La), cerium ($_{58}$Ce), praseodymium ($_{59}$Pr), neodymium ($_{60}$Nd), promethium ($_{61}$Pm), samarium ($_{62}$Sm), europium ($_{63}$Eu), gadolinium ($_{64}$Gd), terbium ($_{65}$Tb), dysprosium ($_{66}$Dy), holmium ($_{67}$Ho), erbium ($_{68}$Er), thulium ($_{69}$Tm), ytterbium ($_{70}$Yb), and lutetium ($_{71}$Lu) are known as *rare earth metals*. (The elements La through Lu are called *lanthanides*.) The name *rare earth* originated in the nineteenth century from the now antiquated term *earth*, meaning "oxide," and to distinguish them from "common earths," such as Al, Mg, and Ca oxides. In fact, rare earth elements are not rare. Even Tm, with the lowest abundance in Earth's crust of all naturally occurring rare earths, is considerably more abundant than familiar elements such as Bi, Cd, Hg, and I. The group 3 transition-metal members of the rare earths (Sc, Y, La) have an outer electron configuration of (rare gas core) $+ d^1 s^2$. Most of the remaining 14 metallic elements ($_{58}$Ce to $_{71}$Lu) maintain the $5d^1 6s^2$ bonding electron structure of La while filling the inner $4f$ electron subshell as the atomic number increases across the row. This gives most rare earths identical outer electron structures, causing their physical properties (Table 23.1) and chemical behaviors to be quite similar. Indeed, many of the early rare earth studies focused on the long and difficult quest to achieve good separation of the rare earth element mixtures found in their ores. This was finally achieved in the 1940s, and rare earth elements have found a remarkable number of engineering uses that exploit their unusual metallurgical, magnetic, optical, and catalytic properties.

23.2 HISTORY

> The positive charge in the nucleus of the atom and the number of electrons surrounding it rises by one unit for every step upwards in the element series. [For most elements] this additional electron usually forms part of the outermost shell of the atom, and since the chemical characteristics depend on the structure of the atom in just this part, the successive members in the series of elements can... be clearly distinguished from one another in respect to their chemical properties. But within the group of the rare earths, it is not the outermost electronic shell that is developed, nor the shell beneath it, but the one that underlies that.
>
> —Arne Westgren, Chairman of the Nobel Committee for Chemistry of the Royal Swedish Academy of Sciences, from his presentation of the 1951 Nobel Prize in Chemistry to Glenn Seaborg

All rare earth (RE) metals are highly electropositive, and the challenge of reducing them to metallic form was not overcome until the mid-nineteenth century. The oxides and other compounds were known earlier, but their similar chemical behavior made separation of individual

---

Structure–Property Relations in Nonferrous Metals, by Alan M. Russell and Kok Loong Lee
Copyright © 2005 John Wiley & Sons, Inc.

TABLE 23.1 Room-Temperature Properties of the Rare Earth Metals with a Comparison to Cu

	21Sc	39Y	57La	58Ce	59Pr	60Nd	61Pm	62Sm	63Eu	64Gd	65Tb	66Dy	67Ho	68Er	69Tm	70Yb	71Lu	29Cu
Valence	+3	+3	+3	+3, +4	+3	+3	+3	+3	+2, +3	+3	+3	+3	+3	+3	+3	+3	+3	+1, +2
Crystal structure[a] at 20°C	HCP	HCP	α-La	α-La[b]	α-La	α-La	α-La	α-Sm (hR3)	BCC	HCP	HCP	HCP	HCP	HCP	HCP	FCC	HCP	FCC
Density (g/cm³)	2.99	4.47	6.15	6.69	6.77	7.01	7.26	7.52	5.24	7.90	8.23	8.55	8.79	9.07	9.32	6.97	9.84	8.92
Melting temperature (°C)	1541	1522	918	798	931	1021	1042	1074	822	1313	1356	1412	1474	1529	1545	819	1663	1085
Thermal conductivity (W/m·K)	15.8	17.2	13.4	11.3(γ)	12.5	16.5	15[c]	13.3	13.9[c]	10.5	11.1	10.7	16.2	14.5	16.9	38.5	16.4	399
Elastic (Young's) modulus (GPa)	74.4	63.5	36.6	33.6(γ)	37.3	41.4	46[c]	49.7	18.2	54.8	55.7	61.4	64.8	69.9	74.0	23.9	68.6	130
Coefficient of thermal expansion (10⁻⁶ m/m·°C)	10.2	10.6	12.1	6.3 (γ)	6.7	9.6	11[c]	12.7	35.0	9.4[d]	10.3	9.9	11.2	12.2	13.3	26.3	9.9	16.5
Electrical resistivity (μΩ·cm)	56.2	59.6	61.5	82.8	70.0	64.3	75[c]	94.0	90	131	115	92.6	81.4	86.0	67.6	25.0	58.2	1.7
Cost for large orders ($/kg)	1750–12,000[d]	39–56[e]	4–17[e]	7–40[e]	8–19[e]	8–20[e]	NA[f]	18–32[e]	6000[g]	400[g]	270–320[h]	40–99[e]	1200[g]	160–295[g]	6500[g]	78–100[e]	7500[g]	2

[a] Many rare earths are polymorphic; see Appendix A (ftp://ftp.wiley.com/public/sci_tech_med/nonferrous) for a list of crystal structures at nonambient temperatures and pressures.
[b] See the text for a discussion of Ce room-temperature crystal structures.
[c] Estimated.
[d] Purity of 99.9 to 99.999%.
[e] Purity of 99 to 99.9%.
[f] NA, not commercially available, due to radioactivity and short half-life.
[g] Purity of 99.9%.
[h] Purity of 99 to 99.99%.

RE elements an extraordinarily difficult task. RE studies performed a century ago are replete with accounts of astonishingly laborious separation procedures. For example, one chemist used 15,000 recrystallization sequences to isolate pure Tm bromate. Separation of a few of the lighter REs (e.g., Ce and Eu) was more easily achieved, since these elements can be converted to valences other than +3, but no economical method of separating all REs was available until the mid-twentieth century.

The first major use of a RE compound came in the 1890s. Addition of 1% CeO_2 to ThO_2 produced an excellent catalyst for combustion of gas in lanterns that transformed the weak yellowish light of an ordinary gas flame into an intense white light (Sec. 24.3). The great commercial success of this material launched a global search for both Th and Ce, which are often found mixed in the same ores. Large deposits of Th + RE minerals were quickly discovered, but extraction of the Th left an excess of Ce and other RE compounds as a by-product. Experimentation with this by-product revealed that mixed RE ores could be reduced to form *mischmetal* (German for "mixed metal") that could be alloyed with Fe to make excellent pyrophoric flints for cigarette lighters. Fe–mischmetal flints are still used today. However, uses for mischmetal are somewhat limited, and REs were sparingly utilized prior to World War II.

During the 1930s and 1940s, it was discovered that RE elements are useful alloying agents for producing ductile cast iron and creep-resistant Mg alloys. These alloying uses increased demand for REs, but the greatest factor in expanding RE use was the U.S. Manhattan Project for atomic bomb production. Successful U fission requires removal of RE elements usually present in U and Th ores. In addition, RE elements comprise a large fraction of the fission product yield, and their separation during fuel reprocessing is an essential step in Pu and U recovery. For these reasons, methods for chemical separation of the REs suddenly became a matter of intensive study, and solvent extraction and ion-exchange techniques were developed to separate the individual RE elements in high-purity form. These processes not only met the needs of the Manhattan Project but also allowed the individual RE elements to be separated at reasonable cost and used individually. Today, RE production has risen to about 100,000 tons of oxides per annum. The greatest expenditures for RE elements are for phosphor production for television sets and computer display monitors, permanent magnets for electric motors and speakers, and automotive exhaust catalysts. The total global market for REs approaches $1 billion.

23.3 PHYSICAL PROPERTIES

RE metals have closely packed crystal structures (HCP, α-La, α-Sm, BCC, or FCC) at room temperature. Their electrical and thermal conductivities are unusually low. Sc, Y, and the heavier lanthanide metals ($_{62}$Sm to $_{71}$Lu) are reactive and gradually darken during exposure to air as oxide forms. The light lanthanides ($_{57}$La to $_{60}$Nd) form white oxide coatings that spall off and expose the underlying metal to continuing oxidation. The RE metals generally have low to moderate strength, low Young's moduli, and mediocre ductility, although La, Ce, Eu, and Yb are highly ductile.

Many properties of RE metals vary in a strikingly uniform fashion with increasing atomic number. Examples of this are the variations in atomic radii and melting temperatures depicted in Fig. 23.1. The regularity of change in the lanthanides is much more uniform than one sees in analogous series such as the actinide metals or the $3d$, $4d$, and $5d$ transition metals (Fig. 1.13). This behavior results from the increasing nuclear charge attracting all the electrons more strongly toward the nucleus, thereby reducing atomic radius. One might think that each additional electron in the $4f$ subshell would be balanced exactly by the added attraction of each additional proton, resulting in no net change in atomic radius. However, the $4f$ subshell is not spherically symmetric, which lessens its ability to screen the nuclear charge from the outermost electrons; consequently, atomic radii contract as atomic number increases (i.e., the lanthanide contraction, Sec. 1.6). Figure 23.1 shows that Eu and Yb are "outliers" on these plots, lying far from the

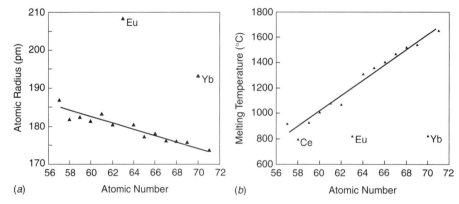

Fig. 23.1 (a) Variation in atomic radius with increasing atomic number across the lanthanide row. The divalency of Eu and Yb moves them far off the other lanthanides' trend line. (b) Variation in melting temperature of the lanthanides.

trend lines. This results from a shift of the $5d^1$ electron into the $4f$ subshell to achieve the lower-energy subshell structure of a half-filled (for Eu) or completely filled (for Yb) $4f$ orbital (Hund's rule):

$$_{57}\text{La} - (_{54}\text{Xe core}) + 4f^0 + 6s^2 5d^1$$
$$_{58}\text{Ce} - (_{54}\text{Xe core}) + 4f^1 + 6s^2 5d^1$$
$$_{59}\text{Pr} - (_{54}\text{Xe core}) + 4f^2 + 6s^2 5d^1$$
$$_{60}\text{Nd} - (_{54}\text{Xe core}) + 4f^3 + 6s^2 5d^1$$
$$_{61}\text{Pm} - (_{54}\text{Xe core}) + 4f^4 + 6s^2 5d^1$$
$$_{62}\text{Sm} - (_{54}\text{Xe core}) + 4f^5 + 6s^2 5d^1$$
$$\mathbf{_{63}Eu - (_{54}Xe\ core) + 4f^7 + 6s^2}$$
$$_{64}\text{Gd} - (_{54}\text{Xe core}) + 4f^7 + 6s^2 5d^1$$
$$_{65}\text{Tb} - (_{54}\text{Xe core}) + 4f^8 + 6s^2 5d^1$$
$$_{66}\text{Dy} - (_{54}\text{Xe core}) + 4f^9 + 6s^2 5d^1$$
$$_{67}\text{Ho} - (_{54}\text{Xe core}) + 4f^{10} + 6s^2 5d^1$$
$$_{68}\text{Er} - (_{54}\text{Xe core}) + 4f^{11} + 6s^2 5d^1$$
$$_{69}\text{Tm} - (_{54}\text{Xe core}) + 4f^{12} + 6s^2 5d^1$$
$$\mathbf{_{70}Yb - (_{54}Xe\ core) + 4f^{14} + 6s^2}$$
$$_{71}\text{Lu} - (_{54}\text{Xe core}) + 4f^{14} + 6s^2 5d^1$$

The presence of only two bonding electrons rather than three means that Eu and Yb are less strongly bonded than the trivalent REs; Eu and Yb more nearly resemble the divalent alkaline metals (e.g., Ba).

23.3.1 Strength and Plasticity

At room temperature all the REs possess one of the closely packed crystal structures (Fig. 1.2) (Table 23.1). The crystal structure of Sm is a particularly intriguing example of a complex closest-packing lattice; it has a (0001) stacking sequence of *ABCBCACAB*, a nine-layer repeating pattern. Like most metals with closely packed crystal structures, RE metals, possess reasonably good ductility. Their strengths and Young's moduli are considerably lower than those of most transition metals. Since REs are not used as structural materials, little effort has been made to develop high-strength RE alloys. Table 23.2 presents selected mechanical properties of pure RE metals. The purities, grain sizes, and prior treatment (e.g., annealed, as-cast, cold-worked) varied widely in these specimens, so direct comparisons of the various mechanical properties in Table 23.2 should be made with caution. Little mechanical testing has been done on Eu, which is quite reactive and soft (VHN = 17, a hardness similar to Pb). No mechanical properties data are available for radioactive Pm (Sec. 23.5). Systematic trends in strength and ductility are difficult to determine, due to variation in purities, grain sizes, and prior treatment of the specimens tested. Eu and Yb are clearly weaker than the other REs, but beyond that, the effect of purity differences and other factors confuses attempted comparisons. However, a definite increase in Young's modulus occurs with increasing atomic number. (Here, as with many RE properties, Sc and Y are best grouped with the heavy lanthanides.) The lanthanide contraction would be expected to increase bond strength as atomic number increases, and (Eu and Yb aside) that trend is clearly seen in the data in Table 23.2.

23.3.2 Corrosion

All RE metals oxidize in air, particularly in humid air, where a mixed oxide–hydroxide film forms. The RE metals' oxidation rates in air vary greatly. $_{63}$Eu corrodes particularly rapidly;

TABLE 23.2 Room-Temperature Mechanical Properties of Rare Earth Metals with a Comparison to Cu

Metal	Tensile Yield Strength, 0.2% Offset (MPa)	Ultimate Tensile Strength (MPa)	Tensile Ductility (% Elongation)	Young's Modulus of Elasticity (GPa)	Slip System(s)	Twinning System(s)
$_{21}$Sc	173	255	5	74.4	—	—
$_{39}$Y	129	176	34	63.5	$\{10\bar{1}0\}\langle 1\bar{2}10\rangle$ $(0002)\langle 1\bar{2}10\rangle$ $\{11\bar{2}2\}\langle \bar{1}\bar{1}23\rangle$	$\{11\bar{2}1\}\langle \bar{1}\bar{1}26\rangle$ $\{10\bar{1}2\}\langle \bar{1}011\rangle$ $\{11\bar{2}2\}$
$_{57}$La	125	130	8	36.6	—	—
γ-$_{58}$Ce	28	117	22	33.6	—	—
β-$_{58}$Ce	86	138	~20	—	—	—
$_{59}$Pr	73	147	15	37.3	—	—
$_{60}$Nd	71	164	25	41.4	—	—
$_{62}$Sm	68	156	17	49.7	—	—
$_{63}$Eu	—	—	—	18.2	—	—
$_{64}$Gd	15	118	37	54.8	$\{10\bar{1}0\}(0002)$	$\{10\bar{1}2\}\langle \bar{1}011\rangle$ $\{11\bar{2}1\}$
$_{65}$Tb	—	(696)^a	(16)^a	55.7	—	—
$_{66}$Dy	43	139	30	61.4	$\{10\bar{1}0\}\langle 1\bar{2}10\rangle$ $\{11\bar{2}2\}\langle \bar{1}\bar{1}23\rangle$	$\{11\bar{2}1\}\langle \bar{1}\bar{1}26\rangle$ $\{10\bar{1}2\}\langle 10\bar{1}1\rangle$ $\{11\bar{2}1\},\{10\bar{1}2\}$
$_{67}$Ho	222	259	5	64.8	$\{10\bar{1}0\}(0002)$	$\{11\bar{2}1\},\{10\bar{1}2\}$
$_{68}$Er	60	136	12	69.9	$\{10\bar{1}0\}(0002)$	$\{11\bar{2}1\},\{10\bar{1}2\}$
$_{69}$Tm	—	(539)^a	(26)^a	74.0	—	—
$_{70}$Yb	7	58	43	23.9	—	—
$_{71}$Lu	—	304	2	68.6	—	—
$_{29}$Cu	70	220	45	110	$\{111\}\langle 10\bar{1}\rangle$	$\{111\}\langle 11\bar{2}\rangle$

Source: Gschneidner and Eyring (1978).
^{a}No tensile data available; values in parentheses are compression test results.

the light lanthanides ($_{57}$La, $_{58}$Ce, $_{59}$Pr, and $_{60}$Nd) are also vulnerable to serious corrosive attack from prolonged exposure to air and will eventually oxidize completely, even in thick sections. These metals corrode rapidly in liquid water. RE metals heavier than Nd oxidize less rapidly than the light REs, and the REs heavier than $_{65}$Tb (including $_{39}$Y and $_{21}$Sc) can be stored in air indefinitely with formation of only a thin oxide patina. Exposure to air at elevated temperature promotes rapid oxidation of all REs and can even result in spontaneous ignition of finely divided metal. RE metals can be protected temporarily from corrosion by formation of a passivating surface layer produced by electropolishing in a chilled perchloric acid–methanol solution, but this layer is not sufficiently robust to provide long-term protection. Some protection can also be afforded by treating the metals with aqueous HF acid, which forms a (RE)F$_3$ coating.

23.3.3 Conductivity

RE metals have high electrical resistivities (Table 23.1). High resistivity is often seen in metals with low-symmetry crystal structures and a significant covalent component to their bonding (e.g., α-Mn or α-Pu), but the reasons for RE metals' high resistivities lie elsewhere and can be understood by recalling a basic expression for electrical resistivity:

$$\rho = \frac{1}{n|e|\mu_e} = \frac{E}{n|e|v_d}$$

where ρ is the electrical resistivity, n the density of charge carriers, $|e|$ the absolute value of the electrical charge on an electron, μ_e the electron mobility (equal to v_d/E, where v_d is the average drift velocity of the electrons and E is the electric field). The REs have closely packed crystal structures and predominantly metallic bonding, so the number of free electrons (n) is fairly high. But RE metals' conduction electrons interact with the incompletely filled inner electron orbitals of the metal atoms they pass while moving through the lattice, giving them a low μ_e value. This same effect also raises most transition-metal resistivities, but to a lesser degree. RE metals also experience a magnetic interaction between the spin of a moving conduction electron and the moment of the incompletely filled 4f subshell of the metal ion core. If all the ion-core magnetic moments were perfectly aligned in one direction, this interaction would have no effect on electron mobility; but above 0 K, the 4f moments are partially randomized by thermal vibrations, which scatters the moving electrons. These factors make μ_e (and hence v_d) quite low for RE metals, and their resistivities are roughly 50 times higher than those of good conductors such as Cu. Since thermal conductivity in metals is largely dependent on electron mobility, RE thermal conductivities are also low.

23.3.4 Magnetic Behavior

Most REs have a partially filled 4f subshell separated from the conduction band electrons. As the 4f subshell fills with increasing atomic number, the orbital angular momentum and spin momentum are not balanced. The 4f electrons, which do not overlap directly with those of neighboring atoms, experience an indirect exchange interaction with neighboring atoms that produces a variety of useful magnetic properties. Many RE metals and compounds are antiferromagnetic, ferrimagnetic, or ferromagnetic. All the strongest ferromagnetic materials known today are intermetallic compounds containing both transition metals (Fe, Co, or Ni) and RE metals. This combination of elements can produce a high magnetocrystalline anisotropy, which is the effect of the crystal structure resisting reversal of the magnetization vector in the crystal. High magnetocrystalline anisotropy correlates with high coercivity, and hence with a high-energy product (Sec. 16.2.2.4). Although many pure RE metals possess interesting magnetic behavior (e.g., Gd is ferromagnetic below 19°C), commercial RE magnetic materials are intermetallic compounds (Sec. 23.4.2.2).

23.3.5 Optical Behavior

The optical transitions of the REs are often confined to narrow ranges of wavelength, which makes them optically different from most transition elements. This behavior results from transitions in the $4f$ subshell, where electrons exist at discrete energy levels rather than in band structures. RE oxide additions impart beautifully pure colors to glass. For example, an Er_2O_3 addition of a few weight percent to otherwise clear glass alters the transmittance spectrum (Fig. 23.2), coloring it a delicate pale pink, a tint that cannot be produced by any other glass composition. The sharp energy steps of $4f$ excitations in RE atoms are exploited in color phosphors for television sets and computer monitors, lasers, and thermophotovoltaic devices (Sidebar 23.1).

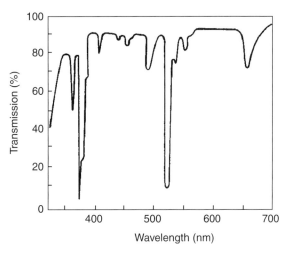

Fig. 23.2 Transmittance of Er_2O_3 in visible and near-ultraviolet wavelengths. Note the sharp absorption bands at 370- and 520-nm wavelengths. These result from electron excitations in the $4f$ subshell, which has discrete energy levels rather than a band structure. (From Gschneidner, 1981, p. 86; with permission.)

SIDEBAR 23.1: RE PHOSPHORS FOR TELEVISION AND COMPUTER DISPLAY MONITORS

In the 1950s, color television sets had strong, clear blues, reasonably good rendition of greens, but weak reds that resembled tomato soup with milk added. The only red phosphors then available [$Zn_{0.2}Cd_{0.8}S$:Ag and $Zn_3(PO_4)_2$:Mn] were not ideal for converting the high-energy electrons of cathode ray tubes (CRTs) into red light. In the 1960s several new red phosphors were discovered, all based on emission from the Eu^{3+} ion: most notably, Y_2O_3:Eu^{3+}, YVO_4:Eu^{3+}, and Y_2O_2S: Eu^{3+}. These phosphors produce sharp, almost monochromatic "spikes" of red color at just the right frequency to match the human eye's red color receptors, and they are brighter emitters than the older phosphors (Fig. 23.3). Eu is one of the least abundant and most expensive RE elements, but no other material can match Eu's spectral response, so Eu has dominated the red phosphor market for decades. The total amount of Eu used for this application is relatively small (about 40 tons/yr), but the commercial value of these phosphors is on the order of $300 million per year. Although flat-panel plasma and electroluminescent displays are gradually replacing CRTs, these newer displays also utilize Eu (and in some cases Er) phosphors (although often with different compositions), so it appears that RE phosphors will continue to be the materials of choice for the foreseeable future.

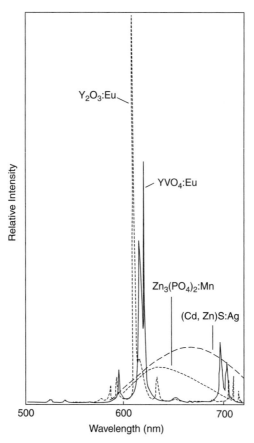

Fig. 23.3 Spectral energy distributions of several red-emitting CRT phosphors. Note the much higher relative energies of the Eu-phosphors and their narrow wavelength dispersion; these provide an intense, pure display of the wavelength that the human eye efficiently perceives as red. (From Gschneidner, 1981, p. 182; with permission.)

23.4 APPLICATIONS

REs are used in both mixed form (with RE elements mixed in the same proportion that they occur in the ore) and in the separated form, in which purities of about 95% or higher are achieved for one RE. On a tonnage basis, most RE materials are used in the mixed form; however, the commercial values of mixed versus separated REs are about equal. RE separation methods are now well developed, but they are sufficiently involved to make separated REs substantially more expensive (Table 23.1) than mischmetal ($5 to $6 per kilogram). As Table 23.1 shows, the prices of the light REs (La to Eu) are, on average, lower than the prices of the heavy REs, and the prices of the even-atomic-numbered elements are generally lower than those of the odd-atomic-numbered elements. These price differences result from variations in the abundance of the elements, although some metal prices reflect special difficulties in producing high-purity metal. Metallic REs and RE compounds are used in roughly equal amounts.

23.4.1 Mixed Rare Earths

Although exact compositions of mixed REs vary with each ore body, mischmetal is typically comprised of about 50% Ce plus lesser amounts of La, Nd, and the other light REs; the heavy

REs and Y comprise only a few percent of mischmetal. It is suitable for applications that exploit the pyrophoric nature of Ce or the strong reactivity of these elements with O and S. These mixed metals are used as minor alloying additions to cast iron to produce nodular iron (Sec. 11.4.3.4). They are also added to steel to avoid undesirable formation of long MnS inclusions, particularly in high-strength low-alloy steels. Mischmetal additions raise creep resistance in Mg alloys (Sec. 11.4.5.1), although Y is preferred for this use. An alloy of 70% mischmetal–30% Fe is used as a pyrophoric alloy for lighter flints. Compounds of MNi_5 (M = mischmetal) are used in Ni–metal hydride battery electrodes.

The mixed oxides are useful polishing compounds in the glass industry. CeO_2 reacts with the glass surface chemically as well as acting as an abrasive, providing combined chemical and mechanical polishing actions; the presence of the other RE oxides is harmless for this use, and it is less costly to leave them in the mix than to separate them. RE oxides also serve as both decolorizing and colorizing additives to glass; larger amounts can be added to increase the refractive index of optical glass. Mixed RE oxides are used in zeolite cracking catalysts for refining petroleum and can also be used in noncracking catalysts, including automotive exhaust catalytic converters.

23.4.2 Separated Rare Earths

The ability to achieve efficient separation of the individual RE elements has greatly expanded their engineering use. Many applications, particularly in optics and magnetism, require the presence of one (and only one) specific RE element. This is especially true for optical applications, where the presence of other lanthanide elements must be controlled to the ppm level. Solvent extraction and ion-exchange processes separate RE compounds, allowing efficient (though somewhat costly) separation of the REs.

23.4.2.1 Alloying Additions. RE metals' moderate to low strength, high reactivity, and relatively high cost make RE-rich alloys poorly suited for structural use. The first widely available RE metal was Y, which is a constituent in protective coating alloys with compositions Ni–Cr–Al–Y and Co–Cr–Al–Y. These coatings are often thermally sprayed onto substrate alloys to form a strongly adherent, diffusion-resistant oxide layer for use in oxidizing high-temperature environments. Y acts to prevent void formation at the oxide–substrate interface. Y makes a better addition than mischmetal for creep-resistant Mg alloys, and Y is added to Al power transmission lines along with Zr in small amounts (about 100 ppm) to improve conductivity. Y additions of about 200 ppm improve ductility in vacuum arc–melted Ti alloys. Ce additions (100 to 300 ppm) deoxidize and desulfurize Ni superalloys. La additions of 200 to 400 ppm in Ni and Co alloys improve cyclic oxidation resistance; La oxide enters the Ni and Co oxide layers and improves their adherence to the underlying metal. La is also used as $LaNi_5$ intermetallic in Ni–metal hydride batteries. Pure La gives better performance than the mischmetal generally used in these batteries by increasing the material's ability to absorb H. Sc additions to Al alloys increase yield strength for a wide range of the major commercial Al alloys. Sc acts to refine grain size and produce an Al_3Sc dispersoid with the highest known coherency mismatch (Sec. 23.6.2) of any Al alloy precipitate. Despite their high cost, Al–Sc alloys are used in a growing range of applications.

23.4.2.2 Permanent Magnets. The first commercially successful ferromagnetic RE material was introduced in the 1970s: $SmCo_5$, an intermetallic compound with a high theoretical maximum energy product (258 kJ/m^3) and a high Curie temperature (725°C) (Sec. 16.2.2.4). $SmCo_5$ outperforms the Fe-based and Al–Ni–Co magnets that preceded it, but it is costly, since both Sm and Co are expensive. More recently, two-phase lamellar microstructures of $SmCo_5$ and Sm_2Co_{17} have been developed with Fe, Cu, and Zr additions that produce a higher energy product than that of pure $SmCo_5$.

In the 1980s the ternary intermetallic compound $Nd_2Fe_{14}B$ was found to possess extraordinary ferromagnetic properties. Its theoretical maximum energy product (512 kJ/m^3) is considerably higher than that of $SmCo_5$, and $Nd_2Fe_{14}B$ is less expensive. The cost of Nd is about half that of Sm, and Nd comprises a lower weight percentage of $Nd_2Fe_{14}B$ than Sm's percentage of $SmCo_5$. Fe is far less expensive than Co, and although the price of B is high, it comprises only 1.0 wt% of the compound. The Curie temperature of $Nd_2Fe_{14}B$ is rather low (310°C), but this is not a limitation for room-temperature applications. $Nd_2Fe_{14}B$ permanent magnets have become the material of choice for small electric motors, speakers, and a variety of electronic devices; annual sales of $Nd_2Fe_{14}B$ now exceed $2 billion.

23.4.2.3 Magnetocaloric Materials. The magnetocaloric effect allows magnetic materials to transfer heat energy into or out of a system. This produces the same result as those of traditional fluid compression and expansion refrigeration technologies, but with a much simpler device. In a ferromagnetic material near its Curie temperature, imposing an external magnetic field causes the unpaired electrons in the ferromagnetic atom to align with the field. This change heats the ferromagnetic material by putting it into a lower-entropy state. If the external magnetic field is removed, the electrons return to a nonaligned, disordered state with higher entropy, and the ferromagnetic material cools. This behavior can be used, for example, to run an ordinary kitchen refrigerator by imposing an external magnetic field on a ferromagnetic material that has a Curie temperature near room temperature, then extracting the heat caused by the magnetic alignment and rejecting it to the room air. When the external magnetic field is turned off, the ferromagnetic material cools below room temperature, producing the refrigeration effect. By cycling the ferromagnetic material continuously, the system cools the refrigerator's interior. Magnetic refrigerators have important advantages over conventional refrigeration systems. They are quieter, their efficiencies can be up to 30% higher, and they use no environmentally harmful chlorofluorocarbons. However, the key to successful utilization of magnetic refrigeration lies in developing materials with Curie temperatures near the desired operating temperature having the largest possible magnetic entropy. As Fig. 23.4 shows, the RE metals have a substantial advantage over the 3d ferromagnetic metals (e.g., Fe) in this regard. Other factors, such as the heat capacity and cost of the material, must be considered in optimizing the design of a magnetic refrigerator, but it is clear that RE intermetallic compounds are the materials best-suited for this application (Sidebar 23.2).

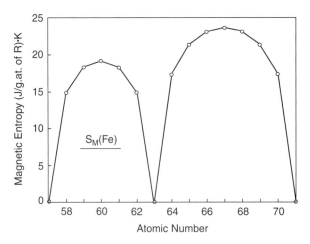

Fig. 23.4 Magnetic entropy of the lanthanide (R) metals versus atomic number. Metals with empty (La), half-filled (Eu), and completely filled (Lu) 4f orbitals have zero magnetic entropy. For comparison, the magnetic entropy of Fe (7.3 J/g.at.Fe·K) is marked. (From Bautista et al., 1997, p. 216; with permission.)

SIDEBAR 23.2: $Gd_5(Si_xGe_{4-x})$ FOR MAGNETOCALORIC REFRIGERATION

The intermetallic compound $Gd_5(Si_xGe_{4-x})$, where x is near 2, possesses a giant magnetocaloric effect that makes it an appealing material for magnetocaloric refrigeration (Sec. 23.4.2.3). This intermetallic compound can exist with either orthorhombic or monoclinic crystal structures at room temperature (Fig. 23.5), but the monoclinic structure is preferred for refrigerant use because its magnetic ordering entropy (Fig. 23.6) is more than twice as large as that of the orthorhombic phase. The monoclinic phase transforms to the orthorhombic phase upon cooling below 0°C by a combined magnetic and martensitic transformation. This transformation in $Gd_5(Si_xGe_{4-x})$ is unusual because it causes a volume contraction associated with the change in magnetic state. (Most martensitic transformations cause a volume expansion.) The orthorhombic phase can be converted to monoclinic phase again by warming the material above about 310 K.

Magnetic refrigerators employ a rotary design. A wheel containing $Gd_5(Si_{2.09}Ge_{1.91})$ powder passes through a gap in a strong permanent magnet. While passing through the magnetic field, the $Gd_5(Si_{2.09}Ge_{1.91})$ in the wheel heats as its ferromagnetic Gd atoms align with the magnetic field. Cooling water then extracts this heat from the $Gd_5(Si_{2.09}Ge_{1.91})$ powder. When the material leaves the magnetic field, it cools by the magnetocaloric effect, and a second stream of water is chilled by contact with the cold wheel. This water then cools the contents of the refrigerator. A magnetocaloric unit is virtually silent, and its energy efficiency is substantially higher than older compression–expansion cycle refrigerators.

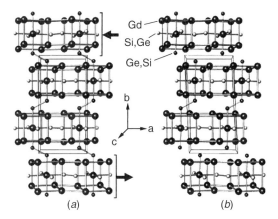

Fig. 23.5 Crystal structures of $Gd_5(Si_2Ge_2)$. Large black atoms are Gd; sites with small black atoms have a 60% probability of being occupied by Ge, 40% by Si; light gray atoms have a 60% probability of being Si and 40% Ge. (*a*) The orthorhombic crystal structure (left) forms in the as-cast material. Its magnetocaloric effect is less than half that of the monoclinic crystal structure shown on the right (*b*). Heat treatment of the orthorhombic structure (7 hours at 1300°C followed by fairly rapid cooling to room temperature) transforms most of the orthorhombic structure to the monoclinic structure. Unit cells are outlined on each drawing; both structures have 36 atoms/unit cell. The orthorhombic structure has all the bonds between layers intact, whereas only half of these bonds are intact in the monoclinic structure. (From Pecharsky et al., 2002; with permission from Elsevier.)

23.4.2.4 Magnetostrictive Materials.
When the magnetic energy of a material is partially dependent on its crystal structure, the material has the potential to serve as a transducer to convert electrical energy to mechanical motion, or vice versa. Elastic strains in such a crystal change the distances between the atoms in the crystal and thereby alter the magnetic energy of the crystal. Conversely, changing the external magnetic field applied to such a crystal will

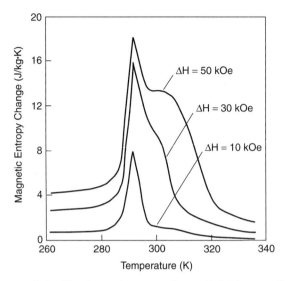

Fig. 23.6 Giant magnetocaloric effect (magnetic entropy change (ΔS_M) in monoclinic $Gd_5(Si_{2.09}Ge_{1.91})$ for three different external magnetic field strengths as a function of temperature. (Redrawn from Pecharsky et al., 2002; with permission from Elsevier.)

cause it to change its interatomic spacing (and thus the overall shape of the crystal), which allows electrical signals to be converted to mechanical motion. The magnetostrictive effect is useful in detecting small dimensional changes in an object, and it can also be used to convert electrical power to mechanical power. All magnetic materials have a magnetostrictive effect, but its magnitude varies widely. The ferromagnetic $3d$ transition metals (Fe, Co, Ni) achieve magnetostrictive strains of only about 5×10^{-5} at magnetic saturation. Such small strains make it difficult to design effective transducers. The RE metals Tb, Dy, and Sm have much larger magnetocrystalline anisotropies because their $4f$ electron distributions are highly anisotropic. At cryogenic temperatures RE metals and intermetallic compounds produce magnetostrictions as large as 10^{-2} (i.e., 1% strain). Although the effect diminishes at higher temperatures, RE materials can produce strains of 10^{-3} at 200°C. These materials (Sidebar 23.3) are useful for a wide range of products, including sonar systems for submarines; microactuators for producing

---

SIDEBAR 23.3: TERFENOL

Among the many RE magnetostrictive materials, one of the best performers is $Tb_{0.3}Dy_{0.7}Fe_{1.92}$, called *Terfenol-D* (named for terbium–ferrum–naval ordinance laboratory–dysprosium). Terfenol-D's electrical-to-mechanical magnetostriction generates strains 20 times greater than non-RE magnetostrictives and two to five times greater than traditional piezoceramics. A 60-mm-diameter rod of Terfenol-D is capable of generating over 220,000 N of dynamic force. Terfenol-driven sonar systems can generate acoustic pulses of 210 dB. This is a remarkably strong acoustic pulse. For comparison a jet engine produces about 120 dB; a gunshot at close proximity (a few centimeters from the muzzle) produces about 140 dB and causes immediate hearing damage. Every 10 dB corresponds to a 10-fold increase in acoustic power. Most Terfenol components are single crystals, grown along the [112] direction in twinned lamellae (Fig. 23.7). The principal drawbacks to Terfenol-D and related RE magnetostrictive materials are their brittleness and high cost. Tb and Dy are relatively expensive, and producing crystals with the appropriate crystallographic alignment adds to the processing cost.

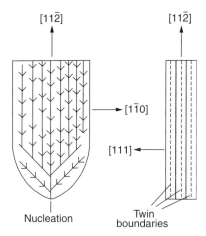

Fig. 23.7 Controlled crystal growth is used to improve the magnetostrictive performance of Terfenol-D. Single crystals are most easily grown in the [112] direction, forming twinned, dendritic sheets. Although the largest magnetostrictive effect occurs along the [111] direction in the cubic unit cell, the effect in the [112] direction is 83% as large, and production of single crystals with a [112] axis is much less costly. (From Bautista et al., 1997, p. 157; with permission.)

tiny, controlled displacements; vibration suppression systems; and "smart" aircraft surfaces that can change their shape in flight to provide the best aerodynamic performance as speed and altitude change.

23.4.2.5 Nuclear Applications. Certain isotopes of Eu, Sm, Gd, and Dy have exceptionally large capture cross sections for neutrons at thermal energies (about 0.025 eV). Thus, they are potentially useful materials to control power output from nuclear reactors. Many reactors use Gd as a safety material to shut down fission if reactivity is climbing out of control. When Gd and Dy absorb a neutron, they are converted to isotopes with much lower capture cross sections. They can therefore be used as *burnable poisons* that absorb substantial numbers of neutrons initially but become progressively less able to capture neutrons with time. They can be inserted along with fresh fuel that needs to have excess neutrons absorbed to control the reaction. As the reactor operates, the amount of fissionable ^{235}U in the fuel rods decreases and neutron-absorbing fission products accumulate, lowering reactivity. Simultaneously, the burnable poison is gradually transmuting to other isotopes, so fewer neutrons are absorbed. This allows the reactor core to maintain a more nearly constant reactivity. In contrast, the isotope Eu forms a succession of five isotopes by neutron absorption that all have high capture cross sections. This makes Eu an excellent material for adjustable control rods that maintain high neutron-absorbing power even after extended use in the reactor core. Certain RE elements have the highest thermal neutron absorption cross sections of all the elements, but they compete with other materials (e.g., B and Cd) for this application.

23.4.2.6 Corrosion Inhibition. RE metals have proven useful as corrosion inhibitors for other metals. In closed-loop saltwater circulation systems, addition of a few hundred ppm RE chlorides reduces corrosion of 2024 and 7075 Al alloys by a factor of 10. Posttest analysis of the Al specimens showed a layer of RE oxide on the surface that is believed to provide the protective effect. Similar benefits have been seen in mild steel, galvanized steel, and pure Zn immersed in ordinary tap water with 50 to 100 ppm $CeCl_3$ addition, where a yellowish film of R hydroxide or hydrated oxide forms on the metal surface. R metal additions (e.g., Y or Nd) to Mg alloys not only improve creep resistance but reduce corrosion rates significantly by formation of RE oxide and hydroxide surface films.

23.5 SOURCES

REs are present in a wide range of minerals, but the commercially significant minerals are (1) monazite, a mixture of Ce, Th, and light RE phosphates; (2) bastnasite, a mixed light RE fluorocarbonate; (3) China's ionic type I, which is over 60% Y; and (4) China's ionic type II, which is high in Nd and other light REs with about 10% Y. The ionic type ore reserves are considerably smaller than the monazite and bastnesite reserves. Monazite tends to settle as monazite sands in beaches and rivers, and deposits exist in India, South Africa, Brazil, Australia, and Malaysia. Monazite contains Ce, La, Nd, and Pr (in that order of abundance) plus 5 to 10% ThO_2 and 3% Y and other heavy REs. Bastnesite occurs in large deposits in California and China. Bastnesite is primarily light REs with little or no ThO_2 or heavy RE content. The absence of Th and its radioactive decay products makes bastnesite processing environmentally simpler, and China's bastnesite and ionic ores now supply about 90% of world RE production (100,000 tons/yr of RE oxides). The Chinese deposits also contain monazite, but only the bastnasite is mined commercially.

Roughly half of RE uses require metallic REs; reduction of RE oxides is difficult due to their stability and high melting temperatures. The preferred method for metal production is reduction of RE halides (usually chlorides) by fused salt electrolysis; however, that process works best at temperatures below about 1100°C, so it is most effective in production of mischmetal, La, Ce, Pr, and Nd. These metals constitute the bulk of RE metal production, so electrolysis is the dominant production method. Ca reduction of the fluoride is usually used to reduce the RE metals with higher melting temperatures.

One RE element, Pm, has no stable isotopes and is found in nature in only trace quantities from spontaneous fission of U in minerals. $^{145}_{61}Pm$ has the longest half-life ($t_{1/2} = 17$ years), but ^{147}Pm is more easily produced in gram-scale quantities (by neutron capture in $^{146}_{60}Nd$) and emits radiation with a low penetrating power, making it safer to handle. Pm is a product of U and Pu fission, and mixed isotopes can be recovered during reprocessing of spent reactor fuel. A few minor engineering uses exist for ^{147}Pm oxide as a low-energy β-radiation source. These include thickness gauges for paper and metal foil manufacture, excitation of phosphors for illumination, and "nuclear batteries" that use photocells to convert light from Pm-excited phosphors into electricity.

23.6 STRUCTURE–PROPERTY RELATIONS

23.6.1 Polymorphism in Pure Ce

The $4f$, $5d$, and $6s$ energy levels are all nearly equal in Ce, and thus small energy changes can cause numerous polymorphic transformations. The pressure–temperature phase diagram for pure Ce (Fig. 23.8) contains seven different crystal structures. Despite numerous studies over a 50-year period, some aspects of the diagram are still in dispute. Perhaps the most peculiar feature of Ce's phase diagram is the critical point near 600 K and 1.8 GPa. (Different studies place the critical point at somewhat different temperatures and pressures.) Ce has two FCC structures at normal pressure: a low-temperature FCC structure called α-Ce ($a_0 = 0.485$ nm at 77 K) and a higher (room)-temperature FCC structure called γ-Ce ($a_0 = 0.516$ nm at 298 K). In the temperature range 96 to 283 K, β-Ce with the α-La structure is the equilibrium phase. As pressure is increased, β-Ce gives way to one or the other of the two FCC phases, which are separated by a phase boundary from about 0.3 to 1.8 GPa. At the critical point, the pressure forces their two FCC lattice parameters to the same value, and the two phases become indistinguishable.

Although Fig. 23.8 gives the appearance of tidy, discrete phase fields, the Ce phase transformations are actually sluggish and hysteretic, making accurate determinations of the phase boundaries frustratingly difficult. One team stored a specimen that had initially been 100% β-Ce and reanalyzed it years later, finding that it had gradually transformed about a fourth of its volume to γ-Ce during its 20-year storage at room temperature. The expectation is that the specimen

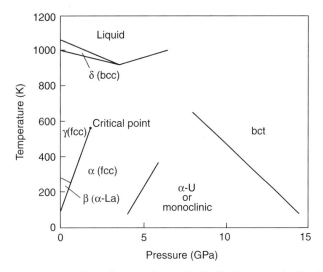

Fig. 23.8 Pressure–temperature phase diagram of pure Ce. In Ce the energy levels of the $4f$, $5d$, and $6s$ electrons are nearly equal, and a variety of phases exist as temperature and pressure change. At normal pressure, FCC α-Ce ($a_0 = 0.485$ nm at 77 K) is the equilibrium phase below 96 K. From 96 to 283 K, the α-La (double hexagonal close-packed) crystal structure is the lowest energy phase. From 283 to 999 K, another FCC phase γ-Ce ($a_0 = 0.516$ nm) is the equilibrium phase. In the narrow temperature range from 999 to melting at 1071 K, BCC δ-Ce is the equilibrium structure. The two FCC phases reach a critical point at 585 K and 1.8 GPa, where the α and γ lattice parameters become equal, making the two phases indistinguishable.

will eventually transform completely to γ-Ce but over a time period much longer than any scientist's life span! At higher pressures, α'-Ce with the α-U (oC4) structure (Fig. 24.3) and α''-Ce with a monoclinic structure (four atoms per unit cell) appear to coexist (i.e., phase boundaries are unclear, and different investigators report that one or the other of the phases is metastable) until pressure becomes high enough to stabilize a body-centered tetragonal (tI2, In-type) structure that persists to the highest pressures measured to date (about 60 GPa). High pressure localizes the $4f$ electron, producing partially covalent bonding and low-symmetry crystal structures.

The simplest explanation for Ce's multiple crystal structures and sluggish transformations is that α-Ce has the $4f$ electron localized within the atom, and γ-Ce promotes the $4f$ electron to the conduction band. In this model α-Ce would be trivalent, and γ-Ce would be tetravalent; however, analysis of magnetic susceptibility data, Hall coefficient measurements, and atomic radii suggests that a more accurate description of the Ce valence is α-Ce (+3.67 at 116 K and 1 atm pressure), β-Ce, and γ-Ce (+3.06 at 116 K and 1 atm). As pressure increases, the valence of α-Ce decreases toward a value of +3.26 at the critical point while the valence of γ-Ce increases to reach the same value, +3.26, at the critical point. The α'-Ce phase appears to be truly tetravalent and is a low-temperature superconductor. A clear model to explain all the Ce transformations has yet to be presented, but there is little doubt that Ce's many crystal structures result from a complex shifting of electrons between the $4f$, $5d$, $5p$, and $6s$ levels. The near equality of these subshells' energy levels makes Ce a metal that is often "on the verge" of transformation in many regions of its $P-T$ phase diagram. This behavior makes Ce difficult to produce in a phase-pure condition below room temperature. (Above room temperature, pure γ-Ce can be produced.) There are often two and sometimes three phases present in Ce, and special heat treatments are necessary to achieve phase purity at some temperatures.

Ce's electrical resistivity behaves oddly, declining only slightly with cooling from 400 to 50 K (most metals see a five- or tenfold reduction over the same temperature range), then dropping sharply below about 20 K. Mechanical properties of Ce show peculiar jumps and odd maxima and minima in the temperature dependence of strength, strain rate sensitivity, and ductility. These effects are at least partially attributable to stress-induced transformations that occur during tensile testing.

23.6.2 Al–Sc Alloys

Sc is expensive. Adding even small amounts of Sc to other alloys drives costs up dramatically. Adding 0.4% Sc + 0.1% Zr to commercial Al alloys, such as the 7000 series alloys, raises the total alloy cost to $15 per kilogram, 10 times higher than the price of commercial purity Al without Sc. But despite their high cost, Al–Sc–Zr alloys are used in several products, including jet fighters, handguns, golf clubs, and baseball bats. The Sc addition enhances Al alloy performance in several ways. Sc additions to Al alloys were first used in the former Soviet Union in the 1970s, but their widespread adoption began elsewhere in the 1990s.

Sc forms nanoscale Al_3Sc dispersoids that have the cP4 ($AuCu_3$-type) crystal structure, an ordered variant of the FCC structure (Sec. 25.6.2 and Fig. 25.6). The cP4 dispersoids and the FCC Al matrix form a coherent interface with a lattice parameter mismatch (1.5%) that induces heavy strain in the Al matrix. These dispersoids form in large numbers with small average diameters (Fig. 23.9); addition of 0.1% Zr inhibits high-temperature coarsening of the

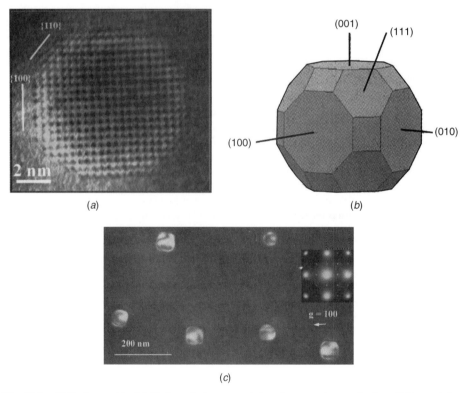

Fig. 23.9 Al_3Sc dispersoids. (a) High-resolution transmission electron micrograph of an Al_3Sc precipitate in Al–0.3 wt% Sc alloy aged 350 hours at 300°C. Note the faceting on the {100} and {110} planes, which accurately matches (b) the precipitate shape predicted by first-principles calculations of the interface energies between the Al matrix and the various crystallographic planes in Al_3Sc. (c) These dispersoids are remarkably resistant to coarsening. Even after prolonged aging at high temperature (120 hours at 400°C), the Al_3Sc dispersoids are still only tens of nanometers in diameter. In this alloy dispersoid diameter increases approximately as (aging time)$^{0.33}$. (From Marquis and Seidman, 2001; with permission.)

STRUCTURE–PROPERTY RELATIONS 449

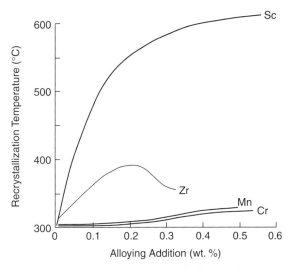

Fig. 23.10 Relation between recrystallization temperature and alloying additions in Al. Note that Sc has a much greater effect on raising recrystallization temperature than do the other alloying additions. Adding just 0.3 to 0.4% Sc makes the recrystallization temperature higher than the solutionizing temperature used for precipitation-hardening Al alloys. This allows a fine grain structure to survive the precipitation-hardening process. (Redrawn from *scandium.org/microbats.html*, 2003.)

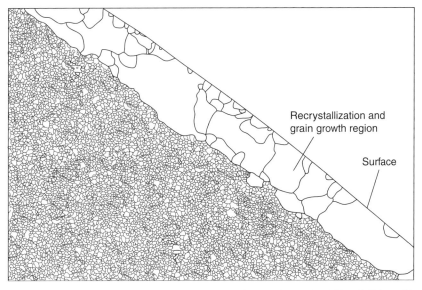

Fig. 23.11 Extruded Al 7000 series alloy in seamless tubing used to make Al baseball bats. Note the fine-grain size in the interior and the large grains that have recrystallized and grown at the surface. Large grains often exist at the surface of extruded or rolled metal because the degree of cold work is greatest at the surface, providing a greater driving force for recrystallization followed by grain growth. This sharply reduces yield strength in cold-worked metal. With Sc and Zr additions, the presence of Al_3Sc dispersoids suppresses recrystallization, maintaining fine-grain structure in both the interior and surface regions of the extruded tube. (Redrawn from *scandium.org/microbats.html*, 2003.)

$Al_3(Sc_{0.8}Zr_{0.2})$ dispersoids. The dispersoids and their associated strain fields are highly effective barriers to dislocation motion.

Sc also raises the recrystallization temperature of Al; its effect is much greater than that of any other alloying addition to Al (Fig. 23.10). Adding just 0.3 to 0.4% Sc makes the recrystallization temperature higher than the solutionizing temperature used for precipitation-hardening Al alloys. This allows metal with an initially fine grain structure to maintain its fine grain size during the solutionizing and precipitation-hardening processes. Even Al alloys that have been cold-worked 80% and held at 550°C do not recrystallize if they contain 0.3 to 0.4% Sc. In rolled and extruded products, where the heaviest cold work usually occurs near the surface, recrystallized (and hence weakened) metal is a common problem in fabricated Al alloys. Figure 23.11 shows a 50-mm-diameter seamless extruded Al 7000 series alloy tube used to manufacture baseball bats. Large, recrystallized Al grains are visible near the surface; these big recrystallized grains are an especially serious problem in thin-walled stock because they lower the yield point. The same alloy containing 0.4 wt% Sc suppresses recrystallization, maintaining a fine grain size in both surface and interior regions.

Sc promotes formation of unusually fine grain size in continuously cast Al, and the alloy maintains its fine grain size during subsequent welding, allowing crack-free welding of alloys that cannot normally be welded. Most of the highest-strength Al alloys are difficult or impossible to weld; Sc additions produce such fine grain size in the fusion and heat-affected zones of welds that these alloys become weldable. The fine grain size of Al alloys containing Sc allows the metal to be superplastically formed and also raises the yield point by the Hall–Petch effect (Sec. 3.2). Sc can be added to Al alloys containing Zn, Mg, and Li; but adding Sc to Al alloys containing Cu or Si is unproductive, because these elements react with Sc.

ADDITIONAL INFORMATION

References, Appendixes, Problem Sets, and Metal Production Figures are available at
ftp://ftp.wiley.com/public/sci_tech_med/nonferrous

24 Ac, Th, Pa, U, Np, Pu, Am, Cm, Bk, Cf, Es, Fm, Md, No, and Lr

24.1 OVERVIEW

The actinide metals progressively fill the $5f$ subshell with increasing atomic number, just as the lanthanides fill their $4f$ subshell (Sec. 23.1). However, $5f$ and $6d$ subshell energies are nearly equal, so the actinides lack the lanthanide's remarkable constancy in electronic structure, atom radii, and numbers of bonding electrons. As a result, physical and mechanical properties, crystal structures, and chemical behavior vary more among the actinides than among the lanthanides (Table 24.1).

The actinides' salient feature is their nuclear instability. With a few exceptions involving Th and U (Secs. 24.3.3 and 24.4.4), their engineering uses exploit nuclear properties. There are no stable actinide isotopes, although Th and U have such long decay half-lives ($>10^9$ years) that they have retained much of their primordial abundance (Appendix C, *ftp://ftp.wiley.com/public/sci_tech_med/nonferrous*). Thousands of tons of the lighter actinides ($_{90}$Th, $_{92}$U, $_{94}$Pu) have been mined or produced in fission reactors, but as atomic number increases, the quantities available decrease drastically: $_{93}$Np, $_{95}$Am, and $_{96}$Cm are available in kilogram quantities, $_{97}$Bk and $_{98}$Cf in hundreds of milligrams, $_{99}$Es in milligrams, and $_{100}$Fm in subnanogram quantities. The heaviest actinides ($_{101}$Md, $_{102}$No, $_{103}$Lr) have half-lives measured in hours and are produced in astonishingly small quantities, usually measured by counts of individual atoms. The heavier actinides' short half-lives make them intensely radioactive, and this imposes severe safety constraints on their handling.

The most important engineering attribute of the actinide elements is the ability of certain U and Pu nuclei to fission. U fission reactors generate 20% of the world's electricity, and the destructive power of U and Pu fission weapons (atom bombs) is a major factor in military and political planning. Fission reactions have even occurred naturally in ancient geological formations (Sidebar 24.1). Engineering use of actinide element compounds is extensive, but with the exception of nuclear weapons and the fuel rods in a limited number of fission reactors, no large-scale applications exist for actinide metals.

24.2 HISTORY AND PROPERTIES

Metallic $_{90}$Th and $_{92}$U were first produced in the mid-nineteenth century, but most knowledge of the actinides was acquired in the past century. $_{89}$Ac and $_{91}$Pa are U decay products, and they were known in the early twentieth century; however, until the discovery of the neutron in 1932, all elements heavier than $_{92}$U were unknown. Neutron capture in fission reactors can produce the elements from $_{93}$Np through $_{100}$Fm, but all neutron capture sequences inevitably lead to $^{258}_{100}$Fm, which has such a short half-life (a few seconds) that it prevents production of still heavier nuclides by further neutron capture. Because of this "fermium barrier," the three heaviest actinides have been produced only by nuclear fusion in particle accelerators.

The actinides are dense, reactive metals. Most actinides possess the close-packed crystal structures typical of metallic elements, but the most heavily used actinides, U and Pu, have

---

Structure–Property Relations in Nonferrous Metals, by Alan M. Russell and Kok Loong Lee
Copyright © 2005 John Wiley & Sons, Inc.

TABLE 24.1 Room-Temperature Properties of Selected Actinide Metals with a Comparison to Cu

Property	$_{89}$Ac	$_{90}$Th	$_{91}$Pa	$_{92}$U	$_{93}$Np	$_{94}$Pu	$_{95}$Am	$_{96}$Cm	$_{97}$Bk	$_{98}$Cf	$_{29}$Cu
Valence	+3	+2, 3, 4	+3, 4, 5	+3, 4, 5, 6	+3, 4, 5, 6, 7	+3, 4, 5, 6, 7	+2, 3, 4, 5, 6	+3, 4	+3, 4	+2, 3, 4	+1, +2
Crystal structure at 20°C[a]	FCC	FCC	BCT	Ortho	Ortho	Mono	α-La	α-La	α-La	α-La	FCC
Density (g/cm^3)	10.07	11.72	15.37	19.05	20.45	19.86	13.67	13.51	14.78	15.10	8.92
Melting temperature (°C)	817[b]	1755	1572	1135	644	640	1176	1345	1050	900	1085
Thermal conductivity (W/m·K)	12	77	47	27	6	6	10	10	10	10	399
Elastic (Young's) modulus (GPa)	—	72.4	—	208	—	94	—	—	—	—	130
Coefficient of thermal expansion (10^{-6} m/m·°C)	14.9	11.4	9.7	13.9	27.5	53.8	—	—	—	—	16.5
Electrical resistivity (μΩ·cm)	—	15.4	19.1	30.8	122	145	71	—	—	—	1.7

[a] Ortho, orthorhombic; mono, monoclinic; many of the actinides are polymorphic; see Appendix A for a list of high-temperature allotropes.
[b] Melting temperature reports vary; 1050°C is also reported.

> **SIDEBAR 24.1: GEOLOGICAL IN-SITU NUCLEAR REACTOR IN AFRICA**
>
> The west African nation of Gabon contains several U deposits that once fueled natural fission reactors operating deep underground. The minerals now present in these deposits show convincing evidence that extensive nuclear fission occurred ~1.8 billion years ago in concentrated UO_2 deposits. The isotopic mix in contemporary U is 99.3% ^{238}U and 0.7% ^{235}U. The ^{235}U isotope can readily be fissioned by neutrons slowed by collisions with atoms at ambient temperature, but ^{238}U cannot. U with only 0.7% ^{235}U cannot sustain a chain reaction with ordinary water serving as the moderator. However, ^{235}U decays faster than ^{238}U, so Earth's U isotopic ratio would have been about 3% ^{235}U when the Gabon deposits were laid down by geological processes 1.8 billion years ago. A 3% ^{235}U content *is* sufficient to sustain a fission chain reaction with a water moderator, and it appears that water seeped through fractures in the rocks to moderate a fission reaction. The fission process was apparently cyclic. When liquid water was present, it moderated the neutrons and sustained the fission reaction. However, the heat of the nuclear reaction would have boiled the water, driving it away from the U-rich regions. Without liquid water, the chain reaction halted, and the rock cooled. When the temperature dropped enough for liquid water to again penetrate the rock fissures, the reaction restarted, and the cycle would repeat. Over eons of time, the ^{235}U was gradually depleted to the point that the UO_2 and water could no longer sustain a fission reaction. Geologists have been able to reconstruct this scenario of fission activity because the Pt group, lanthanide, and inert-gas fission product elements are still locked in the rocks, and the isotopic ratios of these elements matches that of fission reactor waste. These ratios are quite different from the isotopic ratios seen for those elements elsewhere in Earth's crust. There is little doubt that the fission reaction occurred long ago, but it is a phenomenon that cannot recur since Earth's greater age has depleted the ^{235}U content of modern minerals.

low-symmetry crystal structures. Most actinide metals tarnish rapidly when exposed to air. Th forms a self-protective oxide layer, but other actinide oxide layers are less effective barriers to continuing oxidation. Those with short half-lives tend to self-heat from their radioactive decay energy. A kilogram of $^{239}_{94}Pu$ ($t_{1/2} = 24{,}100$ years) produces perceptible warmth from its own decay energy (2 W); a kilogram of $^{238}_{94}Pu$ ($t_{1/2} = 87$ years) has sufficient decay power (560 W) to melt itself.

The bonding participation of the $5f$ electrons is greater in the actinides than is the $4f$ participation in the lanthanides' bonding. $_{89}Ac$ and $_{90}Th$ could be considered $6d$ transition metals, because $5f$ electrons are absent in their ground states: (Rn core) $+\ 6d^17s^2$ for Ac and (Rn core) $+\ 6d^27s^2$ for Th. Their FCC crystal structures and melting temperatures make them more similar to transition metals than to the other actinides. $_{91}Pa$, $_{92}U$, $_{93}Np$, and $_{94}Pu$ have complex crystal structures because their $5f$ electrons participate in the metallic bond. The covalent, directional nature of bonds involving $5f$ electrons favors the low-symmetry tetragonal, orthorhombic, and monoclinic structures of Pa, U, Np, and Pu. A schematic binary phase diagram is shown in Fig. 24.1 depicting these crystal structures. In the higher actinides ($_{95}Am$ and above), the $5f$ electron energy levels drop below the Fermi energy (Sec. 1.5), the $5f$ subshell bands narrow, and the crystal structures become more typically metallic (α-La at low temperature and FCC at high temperature). Interestingly, the heavier actinides transform at high pressure to the partially covalent α-U structure (Appendix A, *ftp://ftp.wiley.com/public/sci_tech_med/nonferrous*) because the externally imposed reduction of interatomic spacings causes the $5f$ electrons to interact with neighboring atoms as they do at low pressure in U. By comparison, the lanthanides above Ce all have localized $4f$ electrons that do not participate in the metallic bond, and crystal structures are much more consistently densely packed, classic metallic structures across the lanthanide row. Under high pressure, however, the lanthanides also transform to more complex crystal structures due to $4f$ involvement in the metallic bond.

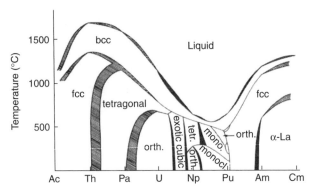

Fig. 24.1 Schematic depiction of the binary phase diagrams of the early actinide metals. The complex low-symmetry structures seen in Pa, U, Np, and Pu involve a directional, covalent bonding component from $5f$ electrons. The later actinides (Am and higher) revert to simpler, more densely packed crystal structures like those of the higher lanthanides because their $5f$ electron energies lie below the Fermi level and do not participate in bonding. (Adapted from Smith and Kmetko, 1983.)

24.3 THORIUM

Th compounds have been known for centuries, and Th metal was first produced in 1828. Th was little used until the invention of thoriated gas mantles in 1891 (Sec. 24.3.3), which were a huge commercial success at the time and still find limited use today. In the 1950s and 1960s, Th was intensively studied because of its potential to breed fissionable U for reactor fuel, but this process has never been adopted commercially. Use of Th compounds, primarily $Th(NO_3)_4$ and ThO_2, has always exceeded Th metal use. Demand for Th products peaked in the 1960s. Tighter environmental regulations governing use and disposal of Th products and waste have suppressed demand in recent decades.

24.3.1 Th Mechanical Properties and Oxidation Behavior

Th is FCC at room temperature, transforming to BCC at 1360°C, and melting at 1750°C. High-purity Th is exceptionally ductile, elongating over 80% in tensile tests and capable of 99% reduction in thickness by rolling at 20°C without stress-relief anneals. Th forms an oxide film that gradually darkens the metal's luster but is protective at room temperature; however, the metal oxidizes excessively at elevated temperature and must be protected during hot handling by fluxes or inert atmosphere. Pure Th's low yield strength (55 MPa), low elastic modulus (72 GPa), radioactivity (Sec. 24.3.4), and high cost ($35 per kilogram) preclude any structural uses. C is a potent interstitial strengthening impurity in Th; just 0.2 wt% C raises yield and ultimate tensile strength severalfold while reducing ductility somewhat. Th containing 0.1 to 0.2% C can be solutionized, quenched, cold worked, and aged to form ThC precipitates that raise yield strength to about 350 MPa with 10 to 20% tensile elongation. Several Th alloys have been studied for possible use in reactor fuel breeding (Sec. 24.3.2), but only one alloy (Mg–Th master alloy for Mg alloy production) is available commercially.

24.3.2 Nuclear Properties of Th

^{232}Th is the dominant isotope in natural Th. It is slightly radioactive, decaying by α emission with a 1.41×10^{10}-year half-life. Th isotopes are not fissionable by thermal neutrons, but ^{232}Th transmutes to ^{233}U by neutron capture:

$$^{232}_{90}\text{Th} + ^{1}_{0}\text{n} \rightarrow {}^{233}_{90}\text{Th}$$

$$^{233}_{90}\text{Th} \rightarrow {}^{233}_{91}\text{Pa} + \beta^{-} \quad (t_{1/2} = 22 \text{ min})$$

$$^{233}_{91}\text{Pa} \rightarrow {}^{233}_{90}\text{U} + \beta^{-} \quad (t_{1/2} = 27 \text{ days})$$

^{233}U has a moderately long half-life (159,000 years). It can be fissioned by both thermal neutrons and fast neutrons, so it can be used as reactor fuel and in nuclear weapons. ^{233}U has several advantages over ^{235}U as reactor fuel. It has a higher neutron yield from fission, which relaxes neutronic constraints on reactor design and operation; the fission products of ^{233}U contain a lower fraction of some of the nuclides that cause the greatest long-term reactor waste storage difficulties; and ^{233}U has several antiproliferation advantages over the ^{239}Pu produced by fast breeder reactors. The major drawback to use of Th to breed ^{233}U is that ^{233}U must be separated from used reactor fuel by a procedure that is more complex and costly than the competing U isotopic enrichment processes (Sec. 24.4.2.1). Rising projections of global U ore reserves and improved abilities to refine leaner U ores economically make Th breeding of reactor fuel uncompetitive. Although the technology to convert Th to ^{233}U is well developed, the process remains "a solution in search of a problem." Some nuclear weapons have been constructed with ^{233}U as the fissionable element, but ^{235}U fuel dominates commercial power reactor use.

24.3.3 Nonnuclear Applications of Th

Most commercial use of Th utilizes Th compounds rather than Th metal. The first major use for ThO_2 was the invention in 1891 of the incandescent gas light mantle. These mantles are produced from cotton woven into the shape of a pouch and soaked in $Th(NO_3)_4 + Ce(NO_3)_3$ solution. When dried and ignited in a stream of flowing natural gas or propane, the fabric rapidly burns away, leaving a delicate web of sintered nitrate residues that calcine to ThO_2 (99%) and CeO_2 (1%). ThO_2 has low thermal conductivity, and it supports and insulates small particles of CeO_2 that catalyze gas combustion, reaching higher temperatures than would be possible without ThO_2. The result is a combustion process that produces an intense white light rather than the weaker yellow light from uncatalyzed combustion. Thoria–ceria gas lamp mantles are still manufactured today for use in sites remote from electric power, and production of mantles accounts for a substantial fraction of total Th use. U.S. manufacturers have substituted Y for Th in mantle production to avoid the regulatory costs associated with Th products and wastes.

ThO_2 has the highest melting point of all binary oxides (3300°C), making it useful in specialty refractories. ThO_2 also serves as a dispersion-hardening additive to metal alloys to increase strength at elevated temperatures (e.g., thoria dispersion-hardened Ni alloys). Thoria particles pin grain boundaries and suppress grain growth, and they do not suffer from the dissolution and coarsening effects that weaken ordinary precipitation-hardened alloys. ThO_2 can be added to glass to increase the refractive index, and it serves as a catalyst in various chemical processes. Th nitrate improves arc stability in welding electrodes.

Total world demand for Th compounds is small (hundreds of tons per year). The use of Th metal is smaller still, and the end of commercial Th metal use may be near. Th metal is used to a small extent as an alloying additive for creep-resistant Mg alloys (Sec. 11.4.3.1) and as a coating for W filament wire in magnetron tubes for microwave emission, where its low thermionic work function increases electron emission rates.

24.3.4 Toxicity of Th

Freshly refined high-purity ^{232}Th is only weakly radioactive. The α-emission has low penetrating power and the half-life is very long, so the primary radiation hazard is inhalation of Th dust. However, trace impurities of other Th isotopes and other actinides can increase radioactivity. Th has 11 daughter products that all decay rapidly, emitting α, β, and γ radiation. Therefore, the activity of Th increases with time after it is refined. Decades-old Th is significantly more hazardous than recently refined Th. However, Th is a substantially less hazardous than other actinides, and its chemical toxicity appears to be low.

24.3.5 Sources of Th and Refining Methods

Th is not rare; it comprises 8.1 ppm of Earth's crust. It is a by-product from refining monazite sands for Ti, Zr, and lanthanide metal production. Demand for those metals greatly exceeds Th demand, so only a small fraction of available Th minerals are used. The high melting temperature and reactivity of Th metal make reduction of the oxide to metal a somewhat difficult task. For this reason the cost of Th metal is significantly higher than that of the nitrate. Increasingly stringent environmental regulations governing Th use and disposal have suppressed world demand for Th in recent decades. Normal supply-and-demand market forces play no part in determining the price of Th metal, since it is an "overly abundant" by-product with falling demand and potential long-term disposal liabilities for users. The price of Th metal reflects reduction and purification costs.

24.4 URANIUM

> Some recent work by E. Fermi and L. Szilard, which has been communicated to me in manuscript, leads me to expect that the element uranium may be turned into a new and important source of energy in the immediate future.
>
> —Albert Einstein, from a letter sent on August 2, 1939 to Franklin Roosevelt warning of the possibility of constructing "extremely powerful bombs of a new type" using U fission, National Atomic Museum, Albuquerque, NM, *http://www.atomicmuseum.com/*

24.4.1 History of U

Unlike most elements, U can be said to have been "discovered" twice, once in the chemical sense when Peligot reduced UCl_4 with K in 1841. The second discovery came in 1896, when Becquerel demonstrated the radioactive decay of U, evidence of U's nuclear instability. U's radioactivity was studied intensively during the early twentieth century, but little practical use was made of U's nuclear properties until 1939, when Hahn and Strassmann reported their U fission experiments. The first nuclear chain reaction and the first atomic bomb explosion followed within a few years. The tremendous energy release that occurs when the U atom is "split" dramatically altered military and geopolitical events for the remainder of the twentieth century.

24.4.2 Nuclear Properties of U

24.4.2.1 U Fission. U can fission when a neutron enters its nucleus. It is relatively easy for the neutron to penetrate the nucleus, since neutrons have no charge and are not repelled by the atom's negatively charged electron cloud or its positively charged nucleus. Fission produces several dozen different reaction products. An example of one such fission reaction is $^{235}_{92}U + _0n^1 \rightarrow ^{89}_{36}Kr + ^{144}_{56}Ba + 3_0n^1 + 215$ MeV. The 215 MeV of energy is divided between the emission of γ-rays and the kinetic energies of the fission products and neutrons. U metal or compounds that have had some of its atoms fission contain over 30 different fission product elements as substitutional or interstitial atoms or as gas bubbles. The fission products are intensely radioactive. The high-speed fission products escaping from the reaction produce high concentrations of lattice defects from their repeated collisions with U atoms (Sec. 7.5). Fission of ^{235}U produces an average of 2.4 neutrons per fission. Since more neutrons are emitted than are consumed in the reaction, it is possible to achieve a supercritical chain reaction in which the neutrons emitted by fission reactions cause the fission of ever larger numbers of other U atoms, leading to a violent explosion of the mass. Alternatively, the reaction can be controlled in a nuclear reactor such that 1.4 neutrons from each fission are lost to absorption by nonfission nuclear capture and to escape from the reaction volume, leaving just enough neutrons to sustain the fission reaction at a steady, controlled rate. Actually, such perfect balance would be nearly impossible to maintain were it not for the fortuitous presence among the fission products of delayed neutron emitters. These provide a "built-in delay" that gives reactor operators tens of seconds rather than microseconds to react to changing fission reaction rates in the reactor core.

Both slow and fast neutrons cause fission in ^{233}U and ^{235}U. ^{233}U can be produced by neutron capture in ^{232}Th (Sec. 24.3.2). ^{235}U occurs naturally as 0.72% of the U in Earth's crust. The most abundant U isotope, ^{238}U, comprises 99.27% of natural U but requires fast neutrons to induce fission, so it cannot be used to fuel conventional fission reactors. U with the natural isotopic blend is difficult to use in reactors because the concentration of ^{235}U atoms is low. It can be made to work if graphite or deuterium oxide is used to slow the neutron velocity (i.e., the neutron moderator), but ordinary water absorbs too many neutrons to sustain the chain reaction.

U is often "enriched" to increase the ^{235}U isotope content. Enrichment can be achieved by vaporizing UF_6 and pumping the vapor through a "cascade" of hundreds of barriers containing exceptionally small holes. The $^{235}UF_6$ molecule is slightly lighter than the $^{238}UF_6$ molecule, so by the effect of $E = \frac{1}{2}mv^2$, its velocity is, on average, slightly higher. Thus, each plate preferentially passes a slightly higher fraction of $^{235}UF_6$ vapor, and after hundreds of successive plates have been passed, the U is enriched to about 3 to 4% ^{235}U. This is high enough to fuel reactors cooled and moderated by ordinary water, which absorbs more neutrons than deuterium oxide. Continuing the enrichment process by passing the vapor through several thousand plates achieves enrichment ranging from 90 to 99% ^{235}U, which is optimal for use in atomic bombs. UF_6 is a corrosive gas that must be used at 70 to 80°C to remain a vapor. It reacts strongly with water and many organic compounds. Uniform holes smaller than 50 nm in diameter are needed to achieve good isotopic separation, and the barrier material must be Ni or Al to resist corrosive attack by the UF_6 vapor. The engineering challenges in constructing and operating a gaseous diffusion plant are prodigious, and the process is expensive. Magnetic separation can also be used to "sort" the two isotopes. During World War II, 10% of all the electricity generated in the United States was used to separate U isotopes with magnetic and UF_6 technologies to build the first atomic bomb. More recently, other methods have become available to enrich U, including preferential laser excitation and gas centrifugation. Each has achieved successful isotope separation, but only centrifugation has proven to be a practical replacement for the gaseous diffusion approach. Hundreds of tons of highly enriched ^{235}U exist in weapons and strategic materials stockpiles. In the 1990s thousands of weapons were retired from active military status; this made substantial amounts of highly enriched U available for use in reactor fuel and reduced the need to operate enrichment plants at full capacity.

24.4.2.2 Fission Weapons. Although difficult to achieve in practice, a fission explosion has only two requirements: (1) More than one neutron from each of the early fission events must induce fission in other atoms (if too many neutrons are absorbed or leak out of the mass, too few will remain to produce the exponentially rising number of fission events needed for an explosion), and (2) once initiated, fission reactions must increase rapidly in number, so most of the atoms fission before the mass flies apart. (If the fission proceeds too slowly, a weak explosion will scatter the fissionable material after only a small fraction of the mass undergoes fission.)

These requirements can be satisfied in a number of ways, but the most common approach is to construct a thick-walled hollow sphere (the *pit*) of ^{233}U, ^{235}U, or ^{239}Pu that is surrounded by conventional chemical high explosive charges (Fig. 24.2). When the chemical explosives fire, the pit is compressed from all sides and implodes. At the center of the pit is a neutron source that starts releasing neutrons at the right moment during the implosion (Sec. 11.3.3.2) to start the fission reactions. A certain minimum critical mass of fissionable material is needed to minimize loss of neutrons from the mass, because neutrons usually travel a few centimeters before causing another fission event. Too small a fissionable mass leaks so many unreacted neutrons that the chain reaction is not sustained. The force of the implosion raises the pit's density to more than 40 g/cm^3 (more than twice its normal density) within 1 to 4 μs. Since the critical mass of a fissionable material varies inversely as the square of its density, this sudden density increase sharply raises the efficiency of the fission multiplication effect, and a large fraction of the fissionable atoms are split before the mass scatters from the energy of its fission explosion. Efficiency is enhanced by (1) surrounding the core with a neutron reflector to return some of the neutrons escaping out the surface back to the pit, and (2) surrounding the reflector with an inertial tamper (mass) to resist dispersal of the U atoms before the maximum possible number of fission reactions has occurred. Part of the inertial tamper material can be natural or

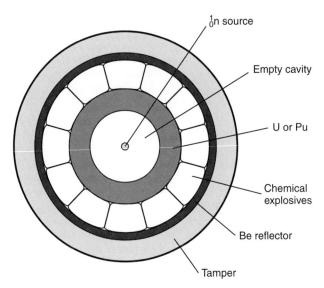

Fig. 24.2 Depiction of a commonly used design for fission weapons. A subcritical geometry of fissionable material (U or Pu) is present as a thick-walled hollow sphere surrounded by chemical explosives. Detonating the chemical explosives crushes the hollow sphere and briefly raises its density to over 40 g/cm^3, creating a supercritical mass. The neutron source at the center of the assembly initiates the fission chain reaction, which is enhanced by the neutron reflector and inertial tamping materials in the outer layers of the bomb.

depleted U, since the fast neutrons from the exploding mass will fission ^{238}U, adding additional energy to the explosion.

The critical mass varies with the isotope. For an uncompressed, unreflected sphere of ^{235}U, the critical mass is 52 kg. ^{233}U, which produces more neutrons per fission event, has a 16-kg critical mass. ^{239}Pu has a critical mass of only 10 kg. Although these masses sound large, U and Pu have densities of about 19,000 kg/m^3, so the volume of metal is fairly small. In actual weapons, reflectors are used and the implosion raises the density temporarily during the crucial microseconds that fission is occurring, so the actual mass of fissionable material needed to achieve criticality in a weapon is considerably smaller than the numbers cited above.

24.4.3 Physical Properties of U

24.4.3.1 U Crystal Structures and Mechanical Behavior. U is a hard, dense, ductile metal with three allotropes. The room temperature α-U crystal structure is orthorhombic (oC4), transforming at 668°C to the rather brittle tetragonal β-phase (tP30). The β phase in turn transforms to BCC γ-U (cI2) at 775°C, which melts at 1133°C.

The α-phase crystal structure (Fig. 24.3) consists of corrugated sheets of atoms somewhat similar to the structures of As, Sb, and Bi. Interatomic spacing within the sheets is shorter (0.28 nm) than interatomic spacing between the sheets (0.33 nm). The 5f electron participation in bonding favors the directional, partially covalent bonding seen in α-U. This odd crystal structure has low solid solubility for most solute metals and is highly anisotropic (Sidebar 24.2). It is surprising that α-U is ductile, deforming by a combination of slip and twinning. α-U can be deformed plastically at room temperature without fracture. The primary slip system is the (010)[100], which corresponds to the corrugated planes sliding past one another along the a_0 axis on Fig. 24.3. Acting alone, (010)[100] slip is insufficient to satisfy the von Mises criterion (Sec. 4.6); twinning is necessary to reorient the crystal to accommodate flow in neighboring grains. Twinning in α-U occurs on the (130) plane by shear in the [310] direction. The twinning

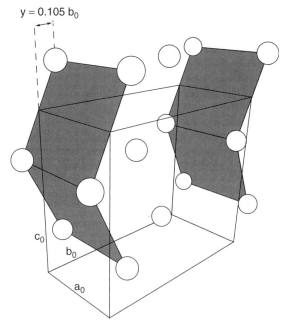

Fig. 24.3 Orthorhombic crystal structure of α-U. The atoms lie in corrugated sheets, with the interatomic spacing within the sheets considerably shorter than the spacing between sheets.

SIDEBAR 24.2: ANISOTROPY OF α-U

The low symmetry and partially covalent bonding of the α-U crystal structure make U's coefficient of thermal expansion highly anisotropic. In the direction parallel to the axis at 25 to 650°C:

- a axis [100]: 36.7×10^{-6} K^{-1}
- b axis [010]: -9.3×10^{-6} K^{-1}
- c axis [001]: 34.2×10^{-6} K^{-1}

When polycrystalline α-U is heated, each grain contracts along the [010] direction and expands along the [100] and [001] directions. When cooled, the [010] direction expands while the other two contract. The result is a highly nonuniform strain in each grain. This effect, combined with the anisotropy of α-U's elastic moduli, imposes large internal stresses on polycrystalline α-U when the temperature changes.

In an extraordinary experiment performed in the 1950s, a polycrystalline cylinder of U 50 mm tall was held in an inert atmosphere while the temperature cycled between 50 and 550°C a total of 1300 times. No phase change occurred, because α-U is the equilibrium phase up to 668°C; however, these large temperature changes produced high internal stresses between grains, and during the time at the high end of each temperature cycle, the metal partially relieved those stresses by creeping. The result is the extreme deformation (Fig. 24.4) of the original right cylinder into an irregular rod 120 mm tall with a heavily wrinkled surface and irregular dimensions.

motion in α-U is particularly complex, requiring small displacements in atom positions in addition to the simple shear of normal twinning. Thus, α-U achieves its high polycrystalline ductility in a manner similar to certain HCP metals (Sec. 12.3.5.1) that use twinning to compensate for inadequate numbers of independent slip systems.

The mechanical behavior of α-U is highly temperature sensitive. Cleavage fracture dominates deformation in commercial-purity U below about 0°C. At room temperature, the metal has a 0.2% offset yield strength of about 210 MPa and an ultimate tensile strength of 620 MPa, but these values decrease rapidly with rising temperature; at 600°C, $\sigma_{UTS} = 50$ MPa and σ_y is low. Tensile ductility is about 25% elongation at 25°C and increases to more than 60% at 600°C. After transformation to the tetragonal β-phase at 668°C, ductility drops to a few percent elongation, but strength increases to about 100 MPa. The γ phase is highly ductile but very weak. Pure U is sensitive to H impurity content; at room temperature, polycrystalline U with low H content displays 30% reduction in area ductility, but ductility drops to about 5% if the metal contains only 0.3 wt ppm H.

24.4.3.2 Alloys. The importance of U motivated exhaustive study of the mechanical properties of pure U. The severe anisotropy (Fig. 24.4) and moderate strength of pure α-U motivated U-alloy development. Solubility of nonactinide elements in α-U is low (<0.3%), and β-U will hold only slightly higher concentrations (about 1% at most) of solutes, so most U alloy work exploits high solid solubility in the BCC γ-U phase to strengthen the metal at room temperature. Commercial alloys such as U–2 wt% Mo, U–2.3 wt% Nb, U–6 wt% Nb, and U–0.75 wt% Ti all have single-phase BCC microstructures at high temperature and better strength and corrosion resistance than those of pure U.

U–0.75 wt% Ti provides an example of the types of heat treatments possible in U alloys possessing a high-temperature γ solid-solution phase field. As Fig. 24.5 shows, the U–Ti system has an eutectoid reaction at 0.75 wt% Ti. Slow cooling through this eutectoid point will produce a lamellar microstructure (analogous to Fe–C pearlite) of nearly pure α-U + U_2Ti (called δ *phase*). Quenching through the eutectoid point does not retain metastable γ phase but, rather, causes a martensitic reaction that produces α′, a phase similar to α-U. Unlike some martensites, α′-U is ductile, but its strength is only moderately higher than the strength of pure α. It is, however, metastable, and aging α′ at 200 to 400°C (Fig. 24.6) precipitates U_2Ti as GP zones in an α matrix to achieve $\sigma_y = 950$ MPa, $\sigma_{UTS} = 1550$ MPa, with 18% tensile elongation.

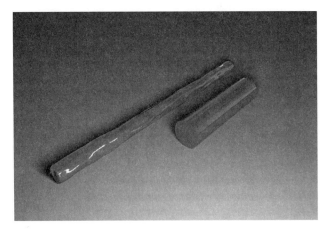

Fig. 24.4 Distortion induced in a polycrystalline α-U cylindrical rod subjected to 1300 cycles of temperature change between 50 and 550°C. The initial rod geometry was a right circular cylinder (right), and the shape of the rod after the thermal cycling is shown on the left. The large anisotropy of α-U coefficients of thermal expansion and elastic moduli induce thermal stresses in the material that cause creep deformation during the hot part of each cycle.

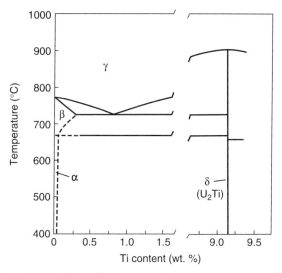

Fig. 24.5 U-rich portion of the U–Ti equilibrium phase diagram. Note the eutectoid at 0.75 wt% Ti and 720°C. (Redrawn from Eckelmeyer, 1982, p. 161.)

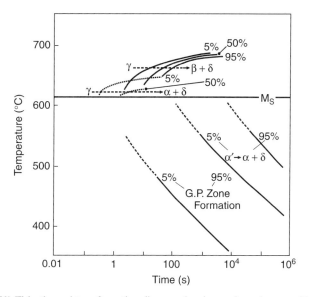

Fig. 24.6 U–0.75% Ti isothermal transformation diagram showing γ-phase decomposition (above 615°C) and aging of α′ martensite (below 615°C). The δ phase is the U_2Ti intermetallic compound. (Redrawn from Eckelmeyer, 1982, p. 167.)

For U used for fission weapons, additional constraints on alloy additions are imposed by the undesirable decrease in U atom density caused by alloy additions. Greater spacing between U atoms sharply reduces the efficiency of the fission chain reaction, so alloying additions that preserve the highest-possible U atom density are preferred. Although relatively little U metal is used as reactor fuel, the neutron absorption behavior of alloying elements is particularly important in U reactor fuel, since certain elements (e.g., B, Cd) "poison" the chain reaction by absorbing neutrons needed to sustain fission.

24.4.3.3 Corrosion. Pure α-U oxidizes readily in room-temperature air. An initially bright metal surface turns a golden color in a matter of hours, and after a few days, a porous black scale of mixed U oxide and nitride coats the surface. This coating is nonprotective, and corrosion continues indefinitely. U reacts strongly with liquid H_2O, evolving H_2 gas and forming UO_2. Finely divided U is pyrophoric; special precautions are needed to avoid igniting the metal during machining and welding.

Alloying considerably improves U corrosion resistance. Two particularly corrosion-resistant U alloys are U–6% Nb and "Mulberry" (7.5% Nb–2.5% Zr); these are sometimes referred to as "stainless U." U can be protected from corrosion by Ni and/or Zn electroplating. This is essential for uses that expose pure U to air, and it minimizes corrosion in U alloys. Even electroplated U alloys are still vulnerable to several vexing corrosion problems, such as stress corrosion cracking, H embrittlement, and pitting. For this reason, U should be considered a corrosion-sensitive metal, and its use in moist environments can be problematic.

24.4.4 Applications of U

24.4.4.1 Fission Weapons. Both ^{235}U and ^{233}U are used in fission weapons. Unalloyed U metal has the best neutronic properties for fission (i.e., smallest critical mass), because its density of U atoms is highest; however, alloys with low alloying additions (Sec. 24.4.3.2) are often chosen because they are stronger and more corrosion resistant. Certain weapons are subjected to severe stresses in service, and for these applications the superior strength of U alloys is an advantage.

24.4.4.2 Reactor Fuel. Early fission reactors commonly used U or U alloys as reactor fuel, and a small number of reactors still do. However, U metal suffers embrittlement and dimensional distortion after prolonged exposure to neutrons caused by radiation damage (Sec. 7.5) and accumulation of gaseous fission products and He. UO_2 reactor fuel is now used in nearly all commercial power reactors because it is more tolerant of radiation damage (particularly gas accumulations) and is less chemically reactive in the event of a cladding failure.

24.4.4.3 Ordnance. Isotopic enrichment processes generate "depleted" U as a by-product of the ^{235}U enrichment process. This U is much less useful for nuclear applications than other U, since it contains little ^{235}U. It does, however, still possess the high density and pyrophoric nature of all U, and these characteristics make it an effective armor-penetrating projectile (Sidebar 1.3). The high density of U delivers a greater shattering force to the impact site than lighter metals, and U burns fiercely, adding to the destructive effect. Smaller-caliber projectiles also use depleted U because its greater kinetic energy increases target damage (Sidebar 24.3). Only the U.S. military uses depleted U ordnance, and concerns about the health effects of U on nontargeted persons have generated a long-running debate about the advisability of depleted U use (Sec. 24.4.5).

24.4.4.4 Other Applications. The United States holds an inventory of 7×10^8 kg of depleted UF_6, accumulated as a by-product from U enrichment processing. There are a few uses for the U metal and oxide that can be produced from depleted UF_6 (e.g., coloring glassware, ammunition, and radiation shielding), but the great quantities available dwarf current demand. Efforts are ongoing to find other uses for the material in flywheels, aircraft landing gear counterweights, and high density concrete. Most uses exploit U's high density and low cost. These applications are relatively few in number, and concerns about U toxicity make potential users reluctant to select U over W, its main rival for such uses. The U.S. Department of Energy has initiated a $588 million project to convert all its depleted UF_6 inventory to U_3O_8 and HF acid by 2028.

24.4.5 Toxicity of U

24.4.5.1 Radiological Toxicity. All U isotopes are radioactive, emitting α-rays that have low penetrating power but deliver high radiation doses to tissue in direct contact with the U. Thus, U poses little threat from close proximity, but its inhalation can deliver a high radiation dose to lung

SIDEBAR 24.3: PHALANX DEPLETED-U ORDNANCE

Depleted U is the ammunition in the U.S. Navy's Phalanx close-in weapons system. Phalanx is a combination radar, computer fire control system, and rapid-fire machine gun used as the "last line of defense" against low-flying attack aircraft, cruise missiles, and related threats to surface naval craft. The device (Fig. 24.7) tracks potential targets with radar and infrared detectors, analyzes their threat with its computer system, and directs the fire of six, 20-mm-caliber machine guns operating in a rotating, Gatling-gun configuration. The weapon fires 4500 rounds per minute and is used during the final few seconds of an approaching weapon's flight as it nears the ship. Phalanx attempts to destroy the target with an overwhelming number of high-velocity projectiles. The high density and pyrophoric nature of depleted U delivers a greater destructive effect on the target than Pb ammunition. The effect of so many high-energy projectiles striking the target in such a short time has been likened to "shredding" the target.

Fig. 24.7 Phalanx close-in weapons system. This weapon can fire depleted U or W projectiles at a rate of 4500 rounds/min to destroy low-flying aircraft, cruise missiles, or mines that threaten naval vessels. The high density of the ordinance enhances the system's ability to destroy the target during the final few seconds before impact with the ship. (Courtesy of the Raytheon Corporation, *http://www.raytheon.com/newsroom/photogal/phalanx_h.htm.*)

tissue contiguous to U particles. Swallowing insoluble U compounds poses little risk, since U passes rapidly through the gastrointestinal tract. However, ingestion of soluble U compounds is hazardous since U can accumulate and injure certain organs with its radioactivity and its chemical toxicity. The 159,000-year half-life of ^{233}U makes it a much greater radiological hazard than the more widely used ^{235}U ($t_{1/2} = 704$ million years) or ^{238}U ($t_{1/2} = 4.47$ billion years). U miners face an additional radiological risk from U ore because one of the decay products of U is ^{222}Rn, a radioactive gas that accumulates in underground mines and causes lung cancer. U irradiated by neutrons in a reactor or in atomic bomb debris will contain dozens of intensely radioactive fission products that pose much more severe radiological hazards.

24.4.5.2 Chemical Toxicity. In addition to its radiological hazards, U is a chemical toxin. U damages the kidneys and has been observed to collect in the brain, testicles, and bones. The most harmful toxicological effect of U chemical poisoning at high doses is renal failure; lower doses may cause reduced fertility, birth defects, and skin or lung irritation. Accidental exposure of workers to UF_6 vapor in gaseous diffusion plants has caused serious lung and skin injury, but this is caused primarily by the reaction between UF_6 and H_2O vapor to form HF acid. The use of depleted-U ammunition in the Persian Gulf wars prompted debate about the medical consequences of dispersing U dust in the environment.

24.4.6 Sources of U and Refining Methods

Recoverable U reserves are widely distributed across the world. Pitchblende, the most common ore, contains a few percent mixed UO_2 and UO_3 that is concentrated and processed to produce U_3O_8. Mining is split between underground and open pit mines with the largest reserves in western Canada. Most U is processed further by enrichment (Sec. 24.4.2.1) to about 3 to 4% ^{235}U for use as fuel in reactors cooled and moderated by ordinary water. Nuclear weapons inventories began shrinking after the end of the Cold War in the early 1990s, and production of highly enriched U has been scaled back because the large inventories of highly enriched U now in storage from disassembled weapons can be "diluted" with unenriched U and used for reactor fuel. U metal is produced by reduction of the oxide with Mg or Ca metal. Unenriched U costs about $24 per kilogram, although lower prices are often cited for depleted U, the by-product of the enrichment process.

24.5 PLUTONIUM

"What on earth has happened?" said my mother, holding her baby tightly in her arms. "Is it the end of the world?" We knelt in the air-raid shelter, praying to God with all our hearts. Injured people came into the shelter one after another muttering, "Urakami is a sea of fire."

—Sachiko Yamaguchi, survivor of the Nagasaki atomic bomb, Nagasaki Atomic Bomb Museum, *http://www1.city.nagasaki.nagasaki.jp/nabomb/museum/m2-11e.html*, 2003

24.5.1 History of Pu

Inconsequential traces of Pu exist naturally from the occasional interaction of spontaneous fission neutrons with ^{238}U, but useful quantities of Pu can be obtained only by artificial synthesis. ^{238}Pu and ^{239}Pu were first produced in 1940–1941 in particle accelerators, and ^{239}Pu became the focus of immediate, intense interest when its fission behavior was measured. Tens of kilograms of ^{239}Pu were produced during the next four years as part of the U.S. Manhattan Project for atom bomb development by irradiating ^{238}U in fission reactors to breed ^{239}Pu:

$$^{238}_{92}U + {}_0n^1 \rightarrow {}^{239}_{92}U$$

$$^{239}_{92}U \rightarrow {}^{239}_{93}Np + \beta^- \quad (t_{1/2} = 23.5 \text{ min})$$

$$^{239}_{93}Np \rightarrow {}^{239}_{94}Pu + \beta^- \quad (t_{1/2} = 2.36 \text{ days})$$

The ^{239}Pu isotope has excellent fission characteristics, and its reasonably long half-life ($t_{1/2} = 24{,}100$ years) and nonpenetrating α-radiation make it a useful material for weapons and, to a lesser extent, for reactor fuel. The first Pu weapon was the atomic bomb detonated over Nagasaki, Japan, in 1945, and thousands of nuclear weapons based on Pu fission have been produced in the ensuing decades. Some experimental and prototype breeder reactors use ^{239}Pu fuel, although commercial power reactor fuel cycles are dominated by ^{235}U-based fuels.

24.5.2 Nuclear Properties of Pu

Some Pu isotopes have longer half-lives than ^{239}Pu, most notably ^{244}Pu ($t_{1/2} = 8.26 \times 10^7$ years), but these isotopes have less favorable fission behavior than ^{239}Pu, and ^{239}Pu is by far the most important isotope. ^{238}PuO$_2$ has been used as a heat source in thermoelectric power supplies for space missions (e.g., *Voyager, Galileo, Cassini*) because it has a short half-life ($t_{1/2} = 87.7$ years) and nonpenetrating α-radiation emission. A relatively small mass of ^{238}PuO$_2$ heats itself to over 1000°C and can be used without elaborate radiation shielding.

24.5.3 Physical Properties of Pu

Pluto was the mythological god of hell, and Pu metallurgists have often joked that the element's namesake was aptly chosen. As Fig. 24.1 shows, it is the heaviest actinide element with significant $5f$ involvement in its metallic bonding, and this gives it several complex crystal structures and unusual physical properties that make it one of the oddest metals in the periodic table.

24.5.3.1 Polymorphism and Thermal Expansion. Many metallic elements have two or more allotropes, but only Pu has six. As Table 24.2 shows, the low-temperature phases have complicated, low-symmetry crystal structures. The higher-temperature phases possess simple, classically metallic structures. At higher pressures, a hexagonal phase appears, ζ (hP8, $c/a = 1.657/2$). As Fig. 24.8 shows, some of the transformations between these phases cause large volume changes, and the metal actually contracts at the δ → δ′, δ′ → ε, and ε → liquid transformations. These volume changes cause problems in casting pure Pu, because they impose large

TABLE 24.2 Pu Crystal Structures

Phase	Stable Temperature Range (°C)	Space Lattice (Pearson Symbol)	Density (g/cm³)	Coefficient of Thermal Expansion (K^{-1})
α	Below 122	Simple monoclinic (mP16)	19.86	54
β	122 to 207	Body-centered monoclinic (mC34)	17.70	42
γ	207 to 315	Face-centered orthorhombic (oF8)	17.14	34.6
δ	315 to 457	Face-centered cubic (cF4)	15.92	−8.6
δ′	457 to 479	Body-centered tetragonal (tI2)	16.00	−65.6
ε	479 to 640	Body-centered cubic (cI2)	16.51	36.5

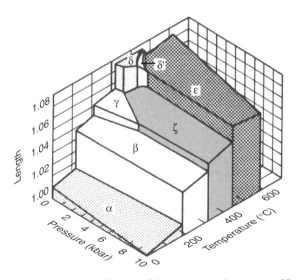

Fig. 24.8 Pressure–temperature–volume diagram of the seven crystal structures of Pu. (From Katz et al., 1986, Vol. 1, p. 603; with permission.)

stresses on the metal as it passes through the multiple phase transformations. Since α-Pu has almost no ductility, castings tend to crack rather than deform in response to these stresses. Consequently, pure Pu castings usually contain microcracks, and full density α-Pu castings are difficult to produce.

Both δ and δ′ phases have negative coefficients of thermal expansion due to thermally activated transfer of some outer electrons to the partially filled $5f$ subshell, reducing atomic radius. Since the density of the liquid is greater than that of the solid, ε-Pu floats on liquid Pu. The viscosity of liquid Pu at its melting point is the highest value (6.0 cP) for any pure liquid metal. Although the melting temperature of Pu is only 640°C, the boiling temperature is 3235°C, giving Pu a large liquid temperature range.

24.5.3.2 Conductivity. Pu is a poor conductor. The α phase has the lowest thermal conductivity of all metals, and α-Pu rivals α-Mn for lowest electrical conductivity. The electrical resistivity of pure α-Pu *decreases* as the metal warms from 100 to 400 K. In most metals higher temperatures increase the number of scattering events for electrons traveling through the lattice, raising resistivity at higher temperatures. Adding impurities to Pu usually lowers electrical resistivity, the opposite of the normal effect.

24.5.3.3 Mechanical Properties and Pu Alloys. α-Pu (99.88% purity) is a moderately strong metal with near-zero ductility. Yield strength (0.2% offset) is 270 MPa, ultimate tensile strength is 420 MPa, and Poisson's ratio is somewhat low at 0.18. The Young's modulus of α-Pu is 94 GPa; this drops to 40 GPa for β-Pu and 17 GPa for δ-Pu. Some ductility is seen in compression tests of α-Pu, length reductions of 9 to 12% occur before cracking begins, and test specimens show numerous twins. The β, γ, δ, and ε phases are all highly ductile. Many of the phases have a high strain rate sensitivity (Sec. 6.6). At 130°C, β-Pu's yield strength increases fivefold as the strain rate is changed from 3.3×10^{-5} s^{-1} to 1.7×10^{-2} s^{-1}.

Due to the military importance of Pu and the "difficult" properties of α-Pu, numerous Pu alloys have been developed to stabilize the elevated-temperature phases to room temperature. Few elements have sufficient solubility in β-Pu or γ-Pu to allow them to be stabilized to room temperature. One exception to this is Zr, which has been used to produce metastable β-Pu at room temperature; however, at room temperature, Pu–Zr alloys lack the high ductility seen in pure β-Pu in its equilibrium temperature range (122 to 207°C). Better ductility is achieved with the FCC δ phase, which can be stabilized at 20°C by adding a few atomic percent Ce, Al, Ga, Sc, In, or Am. These alloys have a single-phase FCC structure as the room-temperature equilibrium phase. (However, at low solute concentrations in some of these alloys, the δ phase can revert to α phase under pressure.) This produces ductile Pu and makes casting much more manageable. These alloys have lower Pu atom density than α-Pu, so their uncompressed, unreflected critical masses are higher. Bulk ε phase can be retained metastably at room temperature only by high alloy additions (10% or more) and ultrarapid quenching.

The most commonly used weapons alloy is a δ-phase alloy: Pu–3.0 to 3.5 at% Ga (0.9 to 1.0 wt%). This alloy is stable over a wide temperature range (−75 to 475°C), and its coefficient of thermal expansion is low. Alloy stability is vital for weapons use, since the Pu pit is machined to very close tolerances, and the distortion caused by a phase change seriously degrades explosive yield. Although it is more difficult to fabricate, pure α-Pu has been used in some weapons because of its high Pu atom density.

24.5.3.4 Corrosion. In low-humidity air, Pu darkens in a matter of hours and eventually forms a dark green, loose PuO_2 scale that is partially protective against further attack. Corrosion is much worse in humid air because water vapor forms a nonprotective mixed oxide and hydride. Pu metal reacts rapidly with liquid water. If left in air of normal humidity for a sufficiently long time, Pu metal will oxidize completely. Oxidation also poses a safety hazard, since the powdery oxide sloughs off easily and is highly toxic when inhaled (Sec. 24.5.4). The metal is pyrophoric in a finely divided state, and serious Pu fires have occurred in the process of machining Pu pits for nuclear weapon production. Alloys containing a few percent Zr, Al, or Ga reduce oxidation rates in air by one to two orders of magnitude. Electroplating Pu with Ni protects it from attack; however, it is difficult to achieve a 100% pore-free electroplated coating, so corrosive attack is still possible in "challenging" environments, such as immersion in water or condensation of liquid water on the surface.

24.5.4 Toxicity of Pu

Pu is one of the most toxic materials known. There are some indications that Pu is chemically toxic, but this effect is minor compared to its radiotoxicity. Pu is primarily an α emitter; its γ and neutron emissions are low. Since α particles have little penetrating power, Pu's primary health hazard is ingestion into the body. Alpha particles' penetrating power in tissue is only 30 to 40 μm; however, severe radiation damage occurs when Pu remains in prolonged contact with tissue because a small volume of tissue receives huge doses of α particles. Pu is poorly absorbed through the gastrointestinal tract. Even water-soluble salts pose a fairly low risk because they tend to bind with the contents of the stomach and intestines. Rapid death from gastrointestinal Pu poisoning requires gram-sized doses. Lower doses are injurious but do not do sufficient damage to be lethal during the relatively short time before the Pu is excreted.

Inhalation poses a much greater risk of injury, because inhaled Pu is likely to remain in the body for a long time, often permanently. Inhalation of 100 mg of fine Pu dust (1 to 3 μm) causes death from pulmonary edema in a few days. An inhaled dose of 20 mg will cause death by lung fibrosis in about a month. Lower inhaled doses are not immediately fatal but elevate lung cancer rates years after inhalation. Inhalation of 1 μg of Pu is thought to increase the lifetime risk of developing cancer by 1%. Inhaling 10 μg raises the lifetime risk of cancer from 20% (the normal cancer incidence rate for the general population) to 30%, although the cancer usually does not appear for 15 to 30 years. Inhaling 100 μg or more makes eventual development of lung cancer almost certain.

Pu rarely enters the bloodstream from accidental exposure; when it does, it concentrates in tissues that contain Fe, such as bone marrow, the liver, and the spleen. Once the Pu moves to these tissues, it tends to remain there for the life of the patient. Of these organs, bone marrow is the most sensitive to radiation damage. An intravenous dose of 1.4 μg of ^{239}Pu will impair immune system function and is likely to cause bone cancer within several years. Since 10 μg of PuO_2 is an amount too small to see with the unaided eye, Pu poisoning is an invidious threat, and Pu-handling facilities must take elaborate precautions to assure worker safety from both toxicity and criticality hazards (Sidebar 24.4).

SIDEBAR 24.4: THE CRITICALITY HAZARD IN Pu HANDLING

Although the critical mass of unreflected, uncompressed Pu metal is about 10 kg, the critical mass is much lower (about 500 g) in aqueous solutions due to the neutron moderating effect of the H_2O. One of the authors (A.R.) spent some years working in a Pu research facility that routinely handled Pu metal and solutions. In one of the operations at this facility, a technician was assigned a task involving aqueous solutions of Pu. Thinking he would accomplish the work more rapidly if he increased the solution concentration, he inadvertently produced a supercritical solution. When a beaker of the solution began to boil vigorously, the technician was momentarily baffled as to why the water had suddenly become so hot. He called his supervisor over to the glove box containing the Pu solution. The supervisor immediately recognized what was occurring, reached into the glove box, and deliberately knocked the beaker over, spilling its contents across the glove box floor. This dispersed the solution into a subcritical geometry and halted the fission reaction. The supervisor suffered acute radiation injury and required several months to recover. The technician died of a radiation overdose.

24.5.5 Sources of Pu and Refining Methods

Recovering Pu isotopes from spent reactor fuel is a daunting challenge. The fuel contains U, Np, Pu, and over 30 different fission product elements, most of which are strongly radioactive. The original fission product concentrations must be reduced by a factor of about 10^{10} in order to render the Pu safe enough to handle in a simple glove box environment and in applications where military personnel will be in close proximity to Pu weapons. Both aqueous and solvent extraction steps are used in a complex sequence of separation and precipitation reactions that are performed in heavily shielded, remotely manipulated processing bays. These procedures were developed successfully in a two-year time period during the Manhattan Project—truly a chemical engineering tour de force. Pu metal production is usually done by bomb reduction of Pu halides with Ca metal; maximum batch size is, of course, limited by criticality constraints.

There is no commercial Pu market. The risk of Pu diversion to clandestine nuclear weapons production makes Pu one of the most strictly controlled substances on Earth. World Pu inventories are estimated to total approximately 1500 tons. About 80% of this mass was produced as a by-product in power-generating fission reactors and has a less than optimal isotope content for weapons use. The balance was produced especially for use in weapons. These 1500 tons of Pu now exist in (1) weapons still in active military service; (2) mixed with U and fission products in unreprocessed, used reactor fuel; (3) in heavily guarded facilities where Pu from retired weapons and reprocessed fuel is stored; and (4) dispersed globally as fine particles scattered by aboveground testing of nuclear weapons. Pu produced for weapons is extracted from fuel that was given only a short "burn-up," so it contains ^{239}Pu concentration of 93% or higher and only a few percent ^{240}Pu and ^{241}Pu. The heavier Pu isotopes are undesirable in weapons due to their tendency to cause *preinitiation*, in which a spontaneous fission event in the heavier Pu isotopes causes premature detonation of the Pu before the optimal compression of the pit has occurred. This lowers explosive yield, but yield is still high enough to cause great damage. Thus, greater nuclear proliferation concerns exist for theft of weapons-grade Pu. Although power reactor-grade Pu can also be used to make atomic weapons, they would have lower explosive yield and are more difficult to handle safely, due to the presence of larger amounts of the heavier Pu isotopes and their decay products. Small inventories of ^{238}Pu are also held for use in thermoelectric generators for spacecraft.

24.5.6 Structure–Property Relations in Pu

24.5.6.1 Coring in Pu–Ga Alloys. The δ-stabilized Pu–Ga alloys (Sec. 24.5.3.3) have several advantages over pure α-Pu for weapons use. They can be cast without microcracking; they have no phase changes over a wide temperature range; they oxidize less rapidly; and although their strength is lower, they have better ductility and machinability. However, Pu–Ga alloys are susceptible to coring, the uneven distribution of Ga in the grains of the δ-stabilized alloy. As a Pu–3.5 at% alloy transforms from ε to δ during cooling (see inset in Fig. 24.9), the slope of the (ε + δ) phase field is such that the first δ phase to form will be higher in Ga than 3.5%, and the last δ phase to form will be lower in Ga. This tends to leave grain centers (the locations where the first transformation to δ occurred) Ga-rich and metal near grain boundaries (the last to transform from ε) Ga-lean. This is undesirable because the metal near the grain boundaries contains too little Ga to stabilize the δ phase fully. Thus, metal near grain boundaries is vulnerable to stress-induced transformation to α-Pu on a local scale during machining, which distorts the precise dimensions of the Pu pit and degrades explosive yield. Two options are available to minimize coring problems. The metal can be cooled through the (ε + δ) phase field so rapidly that diffusion has too little time to redistribute the Ga. This is difficult to achieve in massive sections, due to the low thermal conductivities of ε- and δ-Pu. Alternatively, a cored casting can be annealed in the upper part of the δ-phase field (about 450°C) to allow the uneven distribution of Ga to be homogenized by diffusion. Cast Pu pits are given an homogenization anneal prior to final machining.

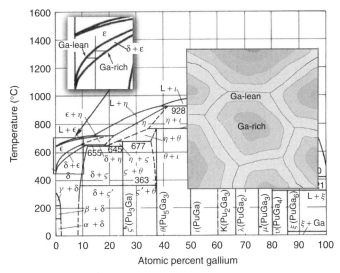

Fig. 24.9 Pu–Ga equilibrium phase diagram. The δ-phase field extends to below room temperature in this system, allowing formation of ductile FCC Pu alloys containing about 3 to 4 at% Ga. The insets show the cause of coring in cast Pu–Ga alloys. The first metal to transform during cooling from the ε region to the δ region tends to be Ga-rich, and the last δ to form (at a supercooled temperature below the δ + ε phase field) is the Ga-lean region near the grain boundaries. (Adapted from Ellinger et al., 1968.)

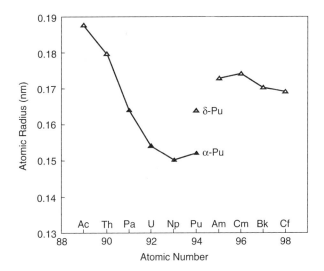

Fig. 24.10 Transition in the actinide series from delocalized $5f$ electrons that participate in metallic bonding to localized $5f$ electrons that do not participate occurs at Pu. The α-Pu atoms have delocalized $5f$ electrons with smaller radii, the δ-Pu atoms have partially delocalized $5f$ electrons, and the Am atoms have localized $5f$ electrons that lie entirely inside the atom, which enlarges the atomic radius.

24.5.6.2 Partial Delocalization of 5f Electrons in δ-Pu. As described in Sec. 24.2, the early actinides ($_{91}$Pa to $_{94}$Pu) display $5f$ participation in the metallic bond, while the heavier elements $_{95}$Am and higher have localized $5f$ electrons. The transition occurs because the increasing electrostatic attractions of the heavier nuclei pull the f electrons closer to the nucleus until finally a transition point is reached and the f electrons no longer overlap with those of neighboring atoms. Pu is located right at this transition point in the actinide series, and Pu shifts from delocalized f electrons to partially localized f electrons as the temperature rises. During the transition from α-Pu to δ-Pu, the metal increases 25% in volume as the $5f$ electron orbitals withdraw into the atom's outer radius. This places δ-Pu's atomic radius midway between that of the delocalized case of α-Pu and the fully localized case of Am (Fig. 24.10). This electronic transition causes many of the odd physical properties of Pu and gives the heavier actinides lower densities, lower valences, and simpler crystal structures than the lighter actinides (Table 24.1).

24.6 LESS COMMON ACTINIDE METALS

24.6.1 Actinium

Ac is a naturally occurring element, discovered in 1902 as a by-product of the separation of Ra from U ore. It is a decay product of $^{235}_{92}$U (Fig. 24.11). However, the 21.8-year half-life of its most stable isotope, $^{227}_{89}$Ac, limits its steady-state concentration in U ores to trace levels. The chemical behavior of Ac is nearly identical to that of its congener La, and since La and other rare earth metals are always present in U ores, separating Ac from U ore is difficult, akin to the separation of the rare earths (Sec. 23.2). It is easier to produce Ac synthetically by neutron bombardment of $^{226}_{88}$Ra in fission reactors, forming ^{227}Ra, which decays ($t_{1/2}$ = 42 min) to ^{227}Ac. Separating the ^{227}Ac from the remaining Ra and the other products of the neutron irradiation is a challenging task since ^{226}Ra is intensely radioactive, emitting toxic ^{222}Rn gas and rapidly damaging ion-exchange resins.

Despite these difficulties, tens of grams of Ac are produced annually, and small quantities of this have been reduced to the metallic state. Ac is a relatively soft, reactive FCC metal (Table 24.1). When exposed to air, a whitish layer of Ac$_2$O$_3$ forms rapidly on the metal that is partially protective against further oxidation. The mechanical properties of Ac metal have received little attention, and no engineering applications exist for Ac metal. Some prototype heat sources of Ac$_2$O$_3$ have been developed for possible use in thermoelectric power supplies for space missions to the outer solar system. Although $^{227}_{89}$Am is primarily a weak β emitter,

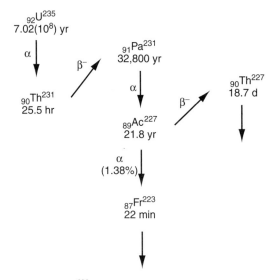

Fig. 24.11 The upper portion of the ^{235}U decay sequence produces the most stable isotopes of Pa, Ac, and Fr. Times indicated are decay half-lives.

its daughter products include five successive short-lived nuclides that are all α emitters. This gives Ac unusually high thermal power per gram without emission of penetrating γ or neutron radiation. Despite this apparent advantage, the $^{238}_{94}$Pu isotope has dominated thermoelectric applications because ^{238}Pu can be produced in kilogram quantities and at lower cost. Ac–Be neutron sources are sometimes used as laboratory-scale neutron generators; they provide a stronger neutron source than Ra–Be (Secs. 11.2.3.2 and 11.5.1.4) due to the large number of α-emissions from the Ac decay products.

24.6.2 Protactinium

Pa is a naturally occurring element with an abundance of about 1 ppt in Earth's crust. $^{231}_{91}$Pa is a decay product of $^{235}_{92}$U (Fig. 24.11). The difficulty of extracting Pa from mineral sources delayed its discovery until 1927. One early investigator processed 5.5 tons of U ore residues to recover 0.5 g of Pa! In the mid-1950s, the U.K. Atomic Energy Agency sponsored a large-scale recovery project that separated 127 g of pure ^{231}Pa that had become "concentrated" by the prior processing of 60 tons of U-bearing minerals. Since ^{231}Pa has a reasonably long half-life (32,760 years), that 127-g supply has been used, recycled, and reused repeatedly during the past several decades for scientific study.

^{231}Pa is an α emitter with a half-life and radiotoxicity similar to that of $^{239}_{94}$Pu. Consequently, Pa does not require elaborate remote handling procedures, but it still must be handled in glove boxes and with the same safety procedures used for Pu. No commercial use for Pa exists, and the element holds no engineering significance except for the role of ^{233}Pa as the intermediate nuclide in the production of $^{233}_{92}$U from $^{232}_{90}$Th (Sec. 24.3.2). The basic properties of metallic Pa have been determined (Table 24.1), and a few intermetallic compounds with Pt, Rh, Ir, and Be have been synthesized.

24.6.3 Neptunium

The light actinides with even atomic numbers (i.e., Th, U, and Pu) have greater nuclear stability and more utility than the odd-numbered elements (Ac, Pa, Np, and Am). Of the odd-atomic-number elements, Np is available in the largest quantities because it accumulates in reactor fuel from neutron capture reactions with ^{238}U and ^{235}U. These same reactions occur naturally in U ores, where occasional spontaneous fission events produce a ratio of about 1 Np atom per trillion U atoms. ^{237}Np has a half-life of 2.14×10^6 years, decaying by α emission to ^{233}Pa. Np metal has three allotropes: orthorhombic α-Np, stable below 280°C; tetragonal β-Np (280 to 577°C); and BCC γ-Np, stable to the 637°C melting point. Like Pu, Np has an exceptionally wide liquid temperature range; boiling at 4174°C. Np's thermal and electrical conductivity are low (Table 24.1) due to the $5f$ electron participation in bonding. Np is the densest (20.45 g/cm^3) of all actinide metals. Although substantial quantities of Np are available, its mechanical properties have received little study.

Small amounts of Np are used in some neutron detection devices, but its principal engineering use is as the precursor material for breeding ^{238}Pu radioisotope for thermoelectric power supplies: $^{237}_{93}$Np + $^{1}_{0}$n → $^{238}_{93}$Np; followed by the decay reaction, $^{238}_{93}$Np → $^{238}_{94}$Pu + β^{-} ($t_{1/2}$ = 2.1 days).

24.6.4 Americium

The first Am ($^{241}_{95}$Am) was produced in 1944 by Pu neutron capture in fission reactors. Several kilograms are produced annually as a nuclear fuel reprocessing by-product. Am is a ductile metal that tarnishes slowly in dry air. Am's $5f$ electrons lie below the Fermi level, so it has simpler, more typically metallic crystal structures (α-La $\xrightarrow{769°C}$ FCC $\xrightarrow{1077°C}$ BCC $\xrightarrow{1176°C}$ liquid). ^{241}Am is the most common Am isotope in spent reactor fuel. It is relatively "hot" with a 458-year half-life. Although a more stable ^{243}Am isotope exists ($t_{1/2}$ = 7370 years), the ready availability and stronger radiation of ^{241}Am make it more useful as a radiation source in commercial products. ^{241}Am decays by both α and γ radiation, which poses serious safety concerns for handling larger (gram+) quantities. ^{241}Am γ radiation provides a compact, inexpensive radiation source for portable gamma radiography of welds and as a thickness gauge in the manufacture of flat glass. The α radiation provides an ionization source for smoke detectors. All these applications use the more chemically stable AmO_2 rather than Am metal.

U.S. government fuel reprocessors sell AmO_2 to licensed commercial users at $1500 per gram. A typical household smoke detector contains 200 μg of $^{241}AmO_2$. The ^{241}Am α particles ionize the air in the detector, allowing a continuous current flow through the detection chamber. When smoke particles enter the space between the chamber's electrodes, they absorb much of the α radiation, reducing the ionization of the air and decreasing current flow. This triggers the alarm. Although inhalation of Am dusts and ingestion of soluble Am compounds are serious toxicity risks, the radiation dose to users of $^{241}AmO_2$ smoke detectors is near zero, since the detector case absorbs the α rays, and γ-ray intensity is quite low. Am particles lodged in lung tissue carry the same types of health risks as those of Pu (Sec. 24.5.4). Soluble Am is a bone-seeking element that concentrates in the skeleton.

24.6.5 Curium

Several isotopes of $_{96}$Cm are produced by multiple neutron capture events in Pu reactor fuel. However, the more stable isotopes, such as ^{248}Cm ($t_{1/2} = 340{,}000$ years) are somewhat difficult to produce. The most easily produced isotopes, ^{242}Cm and ^{244}Cm, have been produced in kilogram quantities, but they have short half-lives (^{242}Cm, $t_{1/2} = 163$ days; ^{244}Cm, $t_{1/2} = 18.1$ years). A single gram of ^{242}Cm produces 122 W of thermal power from its decay heat, and a fresh piece of ^{242}Cm oxide weighing only a few grams heats itself to incandescent temperatures. ^{242}Cm and ^{244}Cm oxides have been used as high-power heat sources for thermoelectric generators, but in recent years the easier availability of ^{238}Pu has largely replaced Cm use.

Cm metal has two allotropes; α-Cm (α-La structure) transforms to FCC β-Cm at 1277°C. Cm melts at 1345°C. The self-heating of the short-half-life Cm isotopes makes larger pieces oxidize rapidly in air. The mechanical properties of Cm metal are largely unknown, although its chemical behavior has been well studied because chemical properties are more easily determined from tiny quantities of material in remotely manipulated experiments.

24.6.6 Berkelium

Bk has only one isotope with a reasonably long half-life, $^{247}_{97}$Bk ($t_{1/2} = 1380$ years), but this isotope is difficult to produce in quantity. Consequently, most study of Bk has been performed on the more easily synthesized ^{249}Bk ($t_{1/2} = 320$ days). A total of 730 mg of ^{249}Bk has been isolated from reprocessed fuel rods in the U.S. High Flux Isotope Reactor over an 18-year period. Since this isotope's half-life is so short, only a few tens of milligrams of Bk have ever been available for study at any one time, and Bk has no engineering uses. The largest piece of Bk metal ever produced weighed 0.5 mg; very little is known about the metal's properties aside from its bulk modulus (30 ± 10 GPa) and crystal structures: α-La structure up to 977°C and FCC to the 1259°C melting point. Bk boils at 2625°C.

24.6.7 Californium

Isotopes with even atomic numbers tend to be more stable than odd-Z elements, and $_{98}$Cf is actually easier to produce in quantity than $_{97}$Bk. ^{249}Cf ($t_{1/2} = 351$ years) is produced at the rate of tens of milligrams per year at the U.S. High Flux Isotope Reactor. The world inventory of ^{249}Cf is now the better part of 1 g. ^{252}Cf can be produced in larger amounts (about 0.5 g/yr) by neutron capture, but it has a short α-decay half-life ($t_{1/2} = 2.64$ years). ^{252}Cf has a high spontaneous fission rate ($t_{1/2} = 66$ years), making it an intense source of neutrons that requires heavy shielding for even microgram quantities. Cf is the heaviest element that has engineering use. ^{252}Cf has been used in cancer treatment and in various experiments where a strong neutron source is required remote from reactors. These include neutron activation analysis of rocks and neutron radiography in the field to supplement γ-ray and x-ray radiography.

Production of Cf metal is chemically difficult. Specimens of metallic ^{252}Cf weighing about 10 mg have been made, often as thin films on substrates. These specimens have allowed determination of the crystal structure. Like the other heavy actinides, Cf has the α-La structure at lower temperatures and an FCC structure above 590°C. Cf metal melts at about 900°C and boils at 1470°C.

24.6.8 Einsteinium

Es was discovered by aircraft sampling the mushroom cloud of the first H-bomb test at Eniwetok Atoll in 1952. $^{253}_{99}$Es ($t_{1/2}$ = 20.5 days) was produced by rapid neutron capture of the ^{238}U in the bomb jacket. The ^{253}Es isotope is also the isotope most amenable to neutron capture synthesis in reactors, and ^{253}Es is produced at the rate of about 2 mg/yr in high-flux reactors. Isotopes of such short half-life impose severe problems in analysis. For example, their self-irradiation is so intense that they destroy their own crystal structures by radiation damage within an hour of solidification. All work must be done in hot cells by remote manipulation techniques. Diffraction analysis of very small Es metal samples indicates an α-La crystal structure. By heating Es with a transmission electron microscope beam and observing formation of "micropuddles," an 860°C melting temperature has been estimated.

24.6.9 Fermium

Fm is the heaviest element that can be synthesized by neutron capture in a reactor. The longest-lived isotope is $^{257}_{100}$Fm ($t_{1/2}$ = 100 days), which is produced in quantities of only 10^9 atoms/yr, but subnanogram quantities of the less stable ^{255}Fm ($t_{1/2}$ = 20.1 hours) have been produced as a β$^-$ decay product from ^{255}Es. No condensed, pure Fm metal has been produced, but a few experimental findings are available on the metal. Fm dissolved in molten La evaporates at a rate which suggests that the metal is divalent (determined by comparison with divalent and trivalent lanthanide evaporation rates).

24.6.10 Mendelevium, Nobelium, and Lawrencium

Because of the fermium barrier (Sec. 24.2), Md, No, and Lr can be produced only in particle accelerators, an energy-intensive process with an extremely low yield. Their short half-lives ($^{256}_{101}$Md $t_{1/2}$ = 78 min, $^{259}_{102}$No $t_{1/2}$ = 58 min, $^{262}_{103}$Lr $t_{1/2}$ = 3.6 h) make it impossible to accumulate large numbers of atoms by long-term operation of particle accelerators. Consequently, none of these elements has ever been seen, even in a microscope. All that is known of their physical properties are hints about their chemical behavior obtained from heroically difficult analyses involving chromatography, their action in ion-exchange columns, evaporative behavior from liquid metal solutes, and predictions based on theorists' calculations. Melting-point predictions have been made (Md, $T_{m.pt.}$ = 1100 K; No, $T_{m.pt.}$ = 1100 K; Lr $T_{m.pt.}$ = 1900 K). Even in *ab initio* calculations, the analysis is made more difficult because electron velocities in these elements are so high that relativistic effects play a large role in their electronic structures. Technological breakthroughs will be required before the bulk properties of these metals can be determined experimentally, and their prospects for engineering use appear remote.

ADDITIONAL INFORMATION

References, Appendixes, Problem Sets, and Metal Production Figures are available at
ftp://ftp.wiley.com/public/sci_tech_med/nonferrous

25 Intermetallic Compounds

> The major problem of strong intermetallics is their brittleness.
> —Gerhard Sauthoff, *Intermetallic Compounds*

25.1 OVERVIEW

Intermetallic compounds are comprised of two or more metallic elements with specific stoichiometries, such as Ni_3Al or $Be_{17}Nb_2$. Unlike most ordinary metals, intermetallics bond with substantial covalent and ionic components. Because their mixed bonding is more directional than that of ordinary metals, they are often lighter, stronger, stiffer, and more corrosion resistant than ordinary metals, particularly at high temperatures. Yet their uses are limited because they are usually brittle at room temperature, making them difficult to fabricate and vulnerable to fracture. These materials hold great promise for improving the performance of engines; aerospace, marine, and terrestrial vehicles; pumps; filters for dust-laden hot gas streams; boilers and heat exchangers; furnace components; tool and die parts; and materials-processing equipment in the chemical process industry.

Skeptics have sometimes joked that "intermetallic compounds are the materials of the future, and they always will be!" This cynical quip captures both the promise and the frustration of intermetallic compounds. Their excellent oxidation resistance, low densities, high strengths, and high stiffness, particularly at elevated temperatures, appeal to engineers challenged with improving performance. The magnitude of these advantages is substantial; for example, the elastic moduli of the titanium aluminide intermetallics are greater at 800°C than the elastic modulus of commercial Ti alloys at room temperature. However, their low ductility and poor fracture toughness have so far limited their use to "niche" applications such as trays for heat treatment furnaces, high-performance permanent magnets, and superconducting wire.

25.2 BONDING AND GENERAL PROPERTIES

Most engineering systems use metals comprised of one dominant metallic element with small amounts of other alloying elements added to improve strength, corrosion resistance, creep behavior, and other properties. Familiar examples include Al with 2 to 4 wt% Cu added to allow precipitation hardening or Fe with 11 wt% or more Cr added to improve corrosion resistance. In these and similar engineering alloys, the bonds between the atoms of the dominant element remain primarily metallic after alloying additions are made. If the alloying elements form an intermetallic compound as a second phase dispersed in the dominant metal matrix (e.g., GP zones within Al), the nondirectional bonding in the matrix preserves the ductility and toughness of the two-phase material. Unfortunately, these two-phase microstructures do not deliver the full measure of desirable properties that single-phase intermetallic compounds can offer. In intermetallic compounds, the nondirectional nature of the metallic bond is partially lost, and the bonds take on a mixed character, becoming partly metallic, partly covalent, and partly ionic.

Structure–Property Relations in Nonferrous Metals, by Alan M. Russell and Kok Loong Lee
Copyright © 2005 John Wiley & Sons, Inc.

TABLE 25.1 Generalized Properties of Metals, Intermetallic Compounds, and Ceramics[a]

Metals	Intermetallic Compounds	Ceramics
High densities	Intermediate densities	Low densities
Intermediate elastic moduli	Fairly high elastic moduli	High elastic moduli
Extensive ductility at RT	Little ductility at RT	No ductility at RT
Moderately high tensile and compressive strength at RT	Variable tensile strength, fairly high compressive strength at RT	Variable tensile strength, high compressive strength at RT
Fairly low hot strength	High hot strength	Very high hot strength
Mediocre/low hot oxidation resistance	Fairly high hot oxidation resistance	High hot oxidation resistance
High electrical conductivity	Moderately high electrical conductivity	Very low electrical conductivity
High-RT fracture toughness	Low-RT fracture toughness	Low-RT fracture toughness

[a] RT, room temperature.

This "mixed" bonding confers properties on intermetallics that are intermediate between those of metals and ceramics (Table 25.1).

25.3 MECHANICAL PROPERTIES

Among the valuable attributes of intermetallics for engineering use are their high specific strength (elastic modulus/density) and their high elastic modulus at elevated temperature. Elastic moduli of many intermetallics range from 150 to 400 GPa for materials with densities as low as 2 to 5 g/cm^3. High stiffness combined with low density is rare in normal metals (Be being the salient exception), and this combination of properties yields the specific stiffnesses listed in Table 25.2. High stiffness minimizes flexing and distortion of parts under load, making it easier for the design engineer to build assemblies of parts that function compatibly.

The ability to maintain a high elastic modulus at elevated temperature is particularly valuable, and here again, intermetallics have a major advantage over ordinary metals. Ordinary metal's elastic moduli decrease more steeply than intermetallics as the temperature rises (Fig. 25.1). This results from their partially nonmetallic bonding and their high melting temperatures.

25.4 OXIDATION RESISTANCE

Many intermetallics contain metals such as Al, Be, Ni, and Si that form strongly protective oxides. These can provide adherent oxide coatings to protect the intermetallic from oxidation at temperatures well above 1000°C. Some intermetallic compounds can operate for prolonged periods in oxidizing atmospheres at temperatures that would melt or ruinously oxidize all ordinary metals except Pt group metals.

TABLE 25.2 Comparison of Elastic Modulus, Density, and Specific Stiffness for Several Ordinary Metals and Intermetallic Compounds

Metal or Compound	Elastic Modulus, E, at 20°C (GPa)	Density, ρ (kg/m^3)	Specific Stiffness, E/ρ (MJ/kg)
W	411	19,250	21.4
Al	70	2710	25.8
Fe	208	7874	26.4
NiAl	184	5850	31.5
TiSi$_2$	272	4390	62.0
MoSi$_2$	440	6200	71.0
ZrBe$_{13}$	289	2720	106

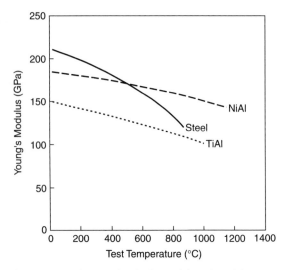

Fig. 25.1 The elevated temperature decrease in elastic modulus of steel is compared to NiAl and TiAl. These three materials all have similar melting temperatures, but the elastic moduli of the intermetallic compounds at high temperatures are greater than that of steel's.

The oxidation mechanisms in intermetallics are essentially the same as those in ordinary metals. Diffusion of metal atoms through the oxide layer to the surface and diffusion of O atoms through the oxide layer to the underlying metal cause progressive thickening of the oxide layer. Failure occurs either by cracking or spalling of the oxide layer or by operation at a temperature where diffusion becomes ruinously fast, converting an unacceptably large volume of metal to oxide. The principal oxidation advantage of intermetallics over ordinary metals is that the metals with the most adherent oxides can be combined to form intermetallics with higher melting temperatures and optimized oxide layer behavior. For example, Be in its pure form has an oxide layer that is protective to about 800°C, and Be melts at 1287°C. By comparison, the intermetallic compound $ZrBe_{13}$ has good strength and stiffness and useful oxidation resistance to 1250°C.

One unfortunate weakness in the otherwise excellent oxidation resistance of many intermetallics is the grain boundary hardening or *pest* effect at intermediate temperatures. The cause of this phenomenon is still debated and may vary from one material to another, but in many intermetallics it appears to occur by grain boundary diffusion of O, which causes selective oxidation of the material near grain boundaries (Fig. 25.2). This generates internal stresses from the volume change of transformation that catastrophically fragment the material after a few hours of exposure at 700 to 800°C. This damage occurs in materials that can withstand prolonged exposure to O_2 above 1000°C. This has been observed in silicides, beryllides, and aluminides, and there is some evidence that stress concentrations or Griffith flaws may participate in addition to (or in place of) grain boundary effects.

25.4.1 Oxidation Resistance of Silicides

The presence of Si in an intermetallic promotes formation of a silica layer that is usually protective. One of the best-performing intermetallic compounds at high temperature is $MoSi_2$, which is frequently used as a heating element in furnaces. $MoSi_2$ forms a crystalline SiO_2 oxide layer. Little Mo is present in this coating because Mo oxides are volatile and vaporize as the silica layer forms. $MoSi_2$ exposed to O_2 for 3000 hours at 1700°C develops only a 50- to 100-μm oxide layer thickness. Other silicides, such as SiC, also show minimal oxidation damage after prolonged exposure to air at high temperatures. SiC forms an amorphous silica

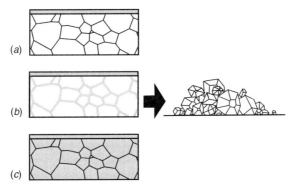

Fig. 25.2 Grain boundary hardening can fragment intermetallic compounds that oxidize at intermediate temperatures (*pest*). (*a*) At or near room temperature, a thin surface oxide layer forms. (*b*) At intermediate temperatures, fast diffusion along grain boundaries causes oxide to form along grain boundaries, but not in the centers of grains. This can induce internal stresses from volume changes associated with oxidation that fragment the material. (*c*) At high temperatures a thick, protective oxide layer forms, and the O that does diffuse through the surface layer is more uniformly distributed by bulk diffusion, avoiding the internal stresses present when grain boundary diffusion dominates, as in (*b*).

layer below about 1200°C that transforms to crystobalite at higher temperatures; both the glassy and crystalline silicas are protective.

25.4.2 Oxidation Resistance of Beryllides

Be intermetallics display generally good oxidation resistance between 1200 and 1500°C. Be often forms intermetallics with high-Be contents (M_2Be_{17}, MBe_{12}, MBe_{13}, where M is a transition metal such as Ti, Zr, Nb, Ta, or Mo). These intermetallics offer excellent oxidation resistance in intermetallics with densities ranging from 2.1 to 4.2 g/cm^3. The high coefficient of thermal expansion of BeO (9.3×10^{-6} °C^{-1}) matches the expansion coefficients of many metals and intermetallics better than do most other oxides. This allows BeO to resist spalling and cracking during thermal cycling. Despite these attributes, the beryllides have the drawbacks of high cost and toxicity (Sec. 11.3.4), as well as the low-temperature brittleness common to nearly all intermetallics.

25.4.3 Oxidation Resistance of Aluminides

Aluminides are particularly useful intermetallics for challenging oxidation environments. The low cost and low density of Al combine with its tenacious, diffusion-resistant Al_2O_3 film to provide protective coatings for metallic substrates (Sec. 16.3.3.6). A small army of scientists and engineers have striven for half a century to improve the room-temperature ductility of aluminides sufficiently to allow their use as bulk structural materials. Intermetallics such as $NbAl_3$ and $TaAl_3$ form tenacious alumina thin films that are strongly protective up to 1250°C. TiAl forms a mixed Ti and Al oxide film that is protective to about 1000°C, and NiAl forms a mixed NiO–Al_2O_3 layer that protects the underlying intermetallic to about 1200°C. FeAl has an oxide layer that is protective to about 1100°C. These intermetallics are of particular interest because they have some ductility at room temperature. Particularly good progress has been made in improving FeAl ductility (Sec. 25.8.3).

25.5 NONSTRUCTURAL USES OF INTERMETALLICS

In addition to their interesting structural properties, intermetallics have a wide range of applications that exploit their electronic, magnetic, and superconducting behaviors. Several semiconducting intermetallics (e.g., GaAs, AlP, GaSb, InSb, InAs) are used in microprocessors, photocells, and electroluminescent and laser devices. PbTe and Bi_2Te_3 are useful thermoelectric materials for solid-state refrigeration systems (Sidebar 22.1), and $Gd_5(Si_2Ge_2)$ is being studied for use in magnetic refrigerators (Sec. 23.4.2.3). Most ferromagnets with high-energy products are intermetallic compounds (Sec. 23.4.2.2); particularly notable among these is $Nd_2Fe_{14}B$, which has grown from its discovery in the mid-1980s to become a material with over $2 billion in annual sales. $SmCo_5$ has a somewhat lower-energy product than $Nd_2Fe_{14}B$ but a higher Curie temperature (725°C). Several superconductors with high critical field and critical current density performance are intermetallic compounds (Sec. 13.4.2.5), including Nb_3Sn, $AlNb_3$, and MgB_2.

All of these applications are tolerant of brittle materials, and hence intermetallic compounds perform well in these uses. However, their brittleness makes fabrication more difficult and costly. For example, hundreds of millions of $Nd_2Fe_{14}B$ magnets are used annually in automobiles, but the rejection rate of these magnets is high, due to fracture during fabrication and assembly. The large amounts of $Nd_2Fe_{14}B$ scrap produced by such fracture is difficult to recycle because it melts peritectically, preventing simple remelting for reuse.

25.6 STOICHIOMETRIC INTERMETALLICS

In their simplest form, intermetallics have a specific stoichiometry (e.g., Ti_3Al, NiAl) and ordered crystal structures. As Table 25.1 shows, many intermetallics possess highly desirable properties for engineering use, particularly at high temperature, and they would be much more widely used if they could be made more ductile at room temperature. Intermetallics' brittleness increases in low-symmetry unit cells with large numbers of atoms per unit cell. Consequently, most research has focused on intermetallics with simpler, high-symmetry crystal structures, since these come closest to possessing the desired ductility and fracture toughness.

25.6.1 Stoichiometric cI2 (CsCl-Type) Structure Intermetallics

One potentially useful intermetallic is NiAl, which has the cI2 (CsCl-type) crystal structure shown in Fig. 25.3. Although this structure resembles the BCC structure, the Al and Ni atoms are ordered such that each Al atom has eight Ni atoms as nearest neighbors, and each Ni atom has eight Al atoms as nearest neighbors. NiAl has bonding that is distinctly directional; the Ni–Al bonds are stronger than Ni–Ni and Al–Al bonds, as evidenced by the higher melting temperature and critical resolved shear stress of NiAl (Table 25.3).

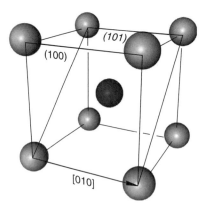

Fig. 25.3 cI2 (CsCl-type) crystal structure of NiAl intermetallic compound. The gray atom in the center of the unit cell is Ni, and the white atoms at the cell corners are Al. Atoms are shown disproportionately small for clarity. The two slip systems active at room temperature in this material, {100}⟨010⟩ and {101}⟨010⟩, are marked.

STOICHIOMETRIC INTERMETALLICS

TABLE 25.3 Comparison of High-Purity Al, Ni, and NiAl

Material	Crystal Structure	Melting Temperature (°C)	Active Slip System(s) at 20°C	Critical Resolved Shear Stress (MPa)
Al	FCC	660	$\langle 10\bar{1}\rangle\{111\}$	1
Ni	FCC	1455	$\langle 10\bar{1}\rangle\{111\}$	5
NiAl	cI2 (CsCl-type)	1638	$\langle 010\rangle\{100\}, \langle 010\rangle\{101\}$	30

Calculations and measurements of NiAl's electronic structure indicate Ni d and Al p hybridization along $\langle 111 \rangle$ directions between nearest-neighbor Ni and Al atoms. Instead of the approximately uniform distribution of bonding electrons seen in Al or Ni unit cells, NiAl has an electron deficiency near Ni and Al atoms and along $\langle 100 \rangle$ directions between like atoms. There is a corresponding increase in electron density between Ni and Al atoms along $\langle 111 \rangle$ directions, indicative of a strong covalent bond between nearest-neighbor Al and Ni atoms. These directional bonds along $\langle 111 \rangle$ are superimposed on a metallic bond present throughout the lattice and a weaker ionic repulsion along $\langle 100 \rangle$. These factors combine to produce highly anisotropic elastic constants, high along $\langle 111 \rangle$ and low along $\langle 100 \rangle$.

As Table 25.3 shows, the active slip systems in NiAl differ from those of FCC Al and Ni. The dominant FCC dislocation motion, $\frac{1}{2}\langle 10\bar{1}\rangle\{111\}$, is geometrically impossible in NiAl. The similarity of the NiAl structure to the BCC structure might suggest that one of the common BCC slip systems would be active in NiAl. The most common BCC Burgers vectors and slip planes are $\frac{1}{2}\langle \bar{1}11\rangle\{110\}$, $\frac{1}{2}\langle \bar{1}11\rangle\{211\}$, and $\frac{1}{2}\langle \bar{1}11\rangle\{321\}$. However, if NiAl slips in the $\frac{1}{2}\langle \bar{1}11\rangle$ direction, it causes Al atoms to be juxtaposed with other Al atoms and Ni atoms to be juxtaposed with other Ni atoms (Fig. 25.4), leaving the previously well-ordered crystal with a planar defect called an *antiphase boundary* (APB).

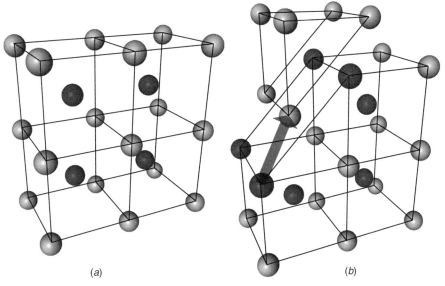

Fig. 25.4 (*a*) Atom positions in four unit cells of a perfect cI2 structure intermetallic compound. (*b*) Antiphase boundary (APB) present in the crystal after a dislocation with Burgers vector (arrow) has passed through the crystal. Note how like atoms become nearest neighbors along $\langle 111 \rangle$ directions across the APB. If a second dislocation were to pass through the lower crystal on the same slip plane and with the same Burgers vector as the first dislocation, the crystal would be restored to the condition shown in part (*a*).

To remove the APB and restore a perfect ordered structure, a second dislocation of the same type must pass through the crystal on the same plane to restore the original ordered structure. This coordinated movement of dislocation pairs on one slip plane is called a *superdislocation*. The repulsion between two dislocations of the same sign is minimized if the two dislocations are widely separated (Fig. 4.3a). However, the APB defect energy between the two dislocations is minimized by keeping them close together. These two opposing tendencies dictate that the two dislocations of a superdislocation will have the separation that minimizes the sum of the APB and strain energies. Note that this is similar to, but not the same as, the behavior of partial dislocations in HCP or FCC crystals separated by a stacking fault (Sec. 18.2.6.1).

APB energy is typically high in strongly ordered intermetallics. For example, in pure Cu, the energy of a twin is about 25 mJ/m^2, and the energy of a high-angle grain boundary is about 600 mJ/m^2. By comparison, the energy of a NiAl $\frac{1}{2}\langle 111\rangle\{110\}$ APB is about 1000 mJ/m^2. This high APB energy causes the two members of a superdislocation to travel close to one another in NiAl to minimize the APB area between them. It has been calculated that for energies larger than 400 mJ/m^2, the APB width between the two dislocations of a superdislocation will be less than 1 nm, which is difficult to resolve by transmission electron microscopy (TEM). This also makes it difficult for the dislocations to form and move, because the strain energy of the superdislocation is high when its two members are so close together. A TEM image of AuCu intermetallic containing APBs is shown in Fig. 25.5; the APBs are wide in this material, making them easy to observe by electron diffraction contrast.

When APB energy is high, it becomes difficult to form and move superdislocations; in such intermetallics, other slip systems with longer Burgers vectors are often the primary slip mechanism. The active slip systems in NiAl at room temperature are the $\langle 010\rangle\{100\}$ and $\langle 010\rangle\{101\}$ (Fig. 25.3). Each of these slip systems has a Burgers vector, **b**, equal to the unit cell edge length, which is longer than the $\frac{1}{2}\langle \bar{1}11\rangle$ Burgers vector. Since dislocation energy scales as **b**^{2}, dislocations with a long Burgers vector require greater energy to form and to move. At room temperature, the critical resolved shear stresses for slip in high-purity Al and Ni are 1 and 5 MPa, respectively, but in high-purity NiAl it is 30 MPa. Engineers view this as a mixed blessing; it makes NiAl intrinsically stronger than pure Al and Ni, but limiting slip to $\langle 100\rangle$ directions seriously impairs ductility. At room temperature, the ductility of polycrystalline NiAl is 1 to 2% elongation; above 400°C, polycrystalline NiAl becomes more ductile, and at 600°C, tensile elongation exceeds 40%.

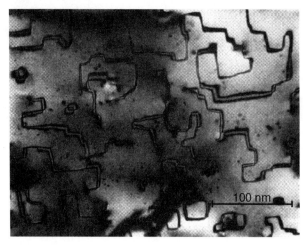

Fig. 25.5 Transmission electron micrograph of antiphase boundaries in AuCu intermetallic. The APBs are visible because the altered structure at the APB diffracts the electrons differently than the perfect crystal. (From Pashley and Presland, 1959; with permission.)

Large plastic strains are possible in single-crystal NiAl at 20°C when it is loaded with the tensile axis parallel to $\langle 111 \rangle$ or $\langle 110 \rangle$, because these orientations favor $\langle 100 \rangle$ slip (Sec. 4.6). However, plastic deformation in polycrystalline material is inherently more difficult to achieve. The shape change each grain must accomplish during deformation imposes strain requirements on its neighboring grains, which require slip in almost every conceivable direction in three-dimensional space to avoid forming cracks or voids at grain boundaries (Fig. 3.1). A grain that can slip only in the $\langle 100 \rangle$ direction cannot be elongated or compressed in that direction because the Schmid factors for such slip are zero. Consequently, the limitation of slipping only along $\langle 100 \rangle$ directions at room temperature makes polycrystalline NiAl fracture before substantial plastic flow occurs.

Five independent slip systems are required by the *von Mises criterion* (Sec. 4.6) in polycrystalline material to accommodate the complicated strain requirements imposed by grain boundaries. In this context, a slip system is defined to be independent if its operation changes the shape of the crystal in a way that cannot be duplicated by combinations of slip occurring on other slip systems. For the cI2 structure with active slip limited to the $\langle 010 \rangle\{100\}$ and $\langle 010 \rangle\{101\}$ slip systems, there are many particular slip plane–slip direction combinations to consider (e.g., [010] (100), [001] (100), [010] (101), [001] (110), etc.). When all these combinations are analyzed, only three of the requisite five independent slip systems are available, and polycrystalline specimens of this material cannot undergo extensive plastic deformation. True to von Mises's prediction, room-temperature tensile tests of polycrystalline cI2 intermetallics, such as NiAl and FeAl, show low ductility. If $\frac{1}{2}\langle \bar{1}11 \rangle\{110\}$ slip were also active in these materials, there would be five independent slip systems, but the high APB energies of strongly ordered cI2 intermetallics preclude extensive room temperature $\langle 111 \rangle$ slip. Some of the cI2 compounds that have lower ordering energies (e.g., CuZn, which has an APB energy of only about 100 mJ/m^2) have been observed to slip at room temperature on the $\frac{1}{2}\langle \bar{1}11 \rangle\{110\}$.

Low ductility and the low fracture toughness that accompanies it, make fabrication of parts to design dimensions difficult and pose major safety concerns in load-bearing structures subject to possible impact loading or thermal shock (Sidebar 11.1). NiAl, for example, would make an excellent material for combustion-zone turbine blades, which are currently fabricated from Ni superalloys (Sec. 16.3.3.6). The strength of NiAl actually increases as temperature rises over the range 20 to 800°C; the density of NiAl is 5.85 g/cm^3, considerably lower than Ni superalloy densities; NiAl's oxidation resistance is slightly better than that of Ni superalloys; and NiAl's thermal conductivity is three times greater than that of Ni superalloys. Yet NiAl turbine blades are not used in turbomachinery, primarily because of the material's poor fracture toughness at room temperature. Efforts have been made to modify NiAl, FeAl, and other cI2 intermetallics to achieve polycrystalline ductility, and these efforts have met with some success (Secs. 25.6 to 25.8).

25.6.2 Stoichiometric cP4 (AuCu$_3$-Type) Intermetallics

Another simple structure with good ductility potential is an ordered variation of the FCC structure, the cP4 AuCu$_3$ type of intermetallic (Fig. 25.6). This structure occurs in Ni$_3$Al, Ni$_3$Fe, Co$_3$Ti, Al$_3$Li, Al$_3$Sc, and Ni$_3$Si, all of which have potentially useful combinations of low density and/or high oxidation resistance. Dislocations in the cP4 structure have possible Burgers vectors in both the $\langle 100 \rangle$ and $\langle 110 \rangle$ directions (Fig. 25.6), but the dominant slip system at and moderately above room temperature is $\frac{1}{2}\langle \bar{1}10 \rangle\{111\}$, which produces an APB. Slip in the $\langle 100 \rangle$ direction would occur by perfect dislocations without APBs, but $\langle 100 \rangle$ slip is usually seen only at higher temperatures.

The von Mises criterion's requirement for five independent slip systems is satisfied by the $\frac{1}{2}\langle \bar{1}10 \rangle\{111\}$ slip system, and $\frac{1}{2}\langle \bar{1}10 \rangle\{111\}$ slip is the main deformation mechanism in FCC metals, which generally have excellent ductility at all temperatures. Single crystals of most cP4 intermetallics are highly ductile, as are polycrystalline cP4 intermetallics at high temperature. However, these materials are unsuitable for most engineering use because tensile loading at

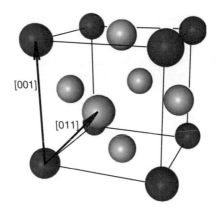

Fig. 25.6 Atom positions in the cubic cP4 structure, which is similar to the FCC structure of simple metals. Examples of ⟨001⟩ and ⟨011⟩ slip directions are marked.

room temperature causes almost immediate brittle intergranular fracture before much plastic flow occurs. Even though a sufficient number of independent slip systems is available to accommodate plastic flow, the cP4 intermetallics fracture along grain boundaries before extensive slip occurs. It appears that the intergranular fracture is caused by dislocation pile-ups at grain boundaries, generating stress concentrations that nucleate cracks. If the grain boundary is weak, the cracks propagate intergranularly in preference to plastic yielding within the grain. Several explanations for the low grain boundary strength of intermetallics have been proposed:

1. *Segregation of embrittling impurities at the grain boundaries.* In ordinary metals, highly electronegative elements (e.g., P or S) segregate at grain boundaries and draw bonding electrons to themselves. This weakens the metallic bond across the grain boundary, allowing brittle intergranular fracture. Auger electron spectroscopy of several high-purity intermetallic systems has shown that cP4 intermetallics with no detectable impurities still fracture intergranularly. This suggests that the boundaries are intrinsically weak; however, Auger does not detect H, so it is possible that H is weakening the boundaries.

2. *Hydrogen embrittlement.* Many intermetallics decompose H_2O vapor at the material's surface, and atomic H so produced diffuses rapidly along the grain boundaries, embrittling the metal and facilitating intergranular fracture. This phenomenon is particularly common in aluminides and silicides. Tensile tests of both cI2 and cP4 polycrystalline intermetallics often show much greater ductility in dry oxygen or vacuum environments than in humid air when the compound is "off-stoichiometry" (e.g., Ni_3Al with a composition of 76.6 at% Ni–23.4 at% Al, rather than 75 at% Ni–25 at% Al) (Sec. 25.9).

3. *Difficulty in nucleating dislocations at grain boundaries.* In ordinary metals, a dislocation pile-up on one side of a grain boundary causes dislocations to nucleate and move in the adjacent grain across the boundary (Fig. 3.3). This process produces the "geometrically necessary" dislocations needed to allow the many grains in a plastically deforming polycrystalline material to remain in contact with one another as deformation progresses (Fig. 3.1). Without these geometrically necessary dislocations, voids would form at grain boundaries, and when the voids grow large enough to become critical-size cracks, the metal fractures. TEM studies of intermetallics show that they have ordered structures within one atomic layer of a grain boundary. The high ordering energy of intermetallics may require such high energy for dislocation nucleation near the boundaries that fracture becomes the lower-energy failure mode.

4. *Electronic charge depletion at grain boundaries.* The metallic elements in intermetallics have unequal electronegativities, so their lattices have lower electron densities in interstitial volumes than one would see in ordinary metals. Such low electron densities may also occur at

intermetallic grain boundaries, reducing the number of bonding electrons present in the "gap" between the two misaligned lattices just as it does in the interstitial regions of a perfect crystal. Electron depletion at the boundary would reduce the boundary's cohesive strength.

25.6.3 Other Stoichiometric Intermetallics

In addition to the cI2 and cP4 structures, some other intermetallics possess relatively high symmetry structures, including the tP4 structure of TiAl and the cF16 structure of Fe_3Al. Although these structures display slight (1 or 2% elongation) polycrystalline ductility at room temperature, it is no greater than that of the cI2 and cP4 structures. The fracture modes of TiAl and Fe_3Al are often cleavage or transgranular rather than intergranular. With the exception of Fe_3Al, the prospects of their achieving useful amounts of room-temperature ductility are generally considered more remote than they are for cI2 and cP4 materials.

More challenging still is the task of achieving ductility in low-symmetry intermetallic crystal structures, some of which have remarkable combinations of high melting temperatures, high elastic moduli, and low densities. For example, $Be_{17}Nb_2$ (hR19) has a 1800°C melting point, a 320-GPa Young's modulus, and a density of only 3.23 g/cm^3. For Si_2V (hP9) the melting temperature is 1677°C, the Young's modulus is 331 GPa, and the density is 4.63 g/cm^3. However, producing such materials with good fracture toughness at room temperature will require major technological breakthroughs that are currently unforeseen.

25.7 NONSTOICHIOMETRIC INTERMETALLICS

With few exceptions (see, however, Sec. 25.9.2), polycrystalline stoichiometric intermetallic compounds have little or no ductility at room temperature. However, many intermetallics can maintain a single-phase microstructure over a range of compositions, as shown in the Co–Ti phase diagram in Fig. 25.7. It has been observed that "off-stoichiometric" compositions often possess much better ductility than perfectly stoichiometric material. For example, Co_3Ti (cP4) elongates 58% in tension when the composition is $Co_{0.80}Ti_{0.20}$, but ductility approaches zero as Ti content approaches 25 at%. Similar behavior is seen in several other intermetallics.

It might appear that these high-ductility off-stoichiometric intermetallics are the solution to intermetallics' room-temperature ductility problem. Unfortunately, the increase in ductility comes at a heavy price. Many of the most appealing characteristics of intermetallic compounds are severely degraded in off-stoichiometry material. In cI2 FeAl, for example, the tensile yield strength (σ_y) falls sharply as Al content decreases from 50 at% to 45 at% (Fig. 25.8), σ_y drops from about 1000 MPa to only 250 MPa. An additional drawback to off-stoichiometric intermetallics is that undesirable thermally activated processes proceed much more rapidly as composition deviates from stoichiometry. Movement of a vacancy in a fully ordered intermetallic (e.g., cI2) raises the crystal energy by placing like atoms in nearest-neighbor sites. A combination of six separate jumps for that vacancy are needed before the perfect ordering of the lattice can be restored, and the first three of those six jumps are endothermic. This makes diffusion much slower in ordered compounds than in ordinary metals. Stoichiometric intermetallics are inherently more resistant to diffusion-driven phenomena such as dislocation climb, grain boundary sliding, recovery, recrystallization, and grain growth. Thus, ordered intermetallics have superior high-temperature strength and creep resistance; deviations from stoichiometry compromise those attributes.

Not all intermetallics weaken as the stoichiometry deviates from the ideal value; in some intermetallics deviation from exact stoichiometry actually raises strength. In stoichiometric NiAl, $\sigma_y = 230$ MPa, but strength rises to over 1200 MPa as Al content increases to 53 at%. This hardening of Al-rich material results from extraordinarily large numbers of vacancies in the lattice. NiAl accommodates the shortage of Ni atoms in Al-rich material by leaving Ni sites vacant. Vacancy concentrations as high as 6% have been observed in Al-rich NiAl. (By comparison, equilibrium vacancy concentrations in ordinary metals are under 0.1%, even near the melting point.)

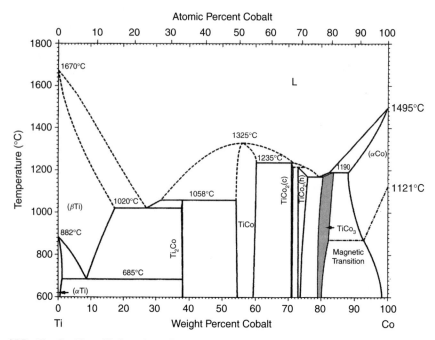

Fig. 25.7 The Co–Ti equilibrium phase diagram. Note the range of single-phase TiCo$_3$ compositions that are possible (shaded) for the TiCo$_3$ intermetallic compound. Ti$_{0.20}$Co$_{0.80}$ is not an equilibrium single-phase material at room temperature, but the transformation to two phases during cooling is sluggish, and the phase is easily retained metastably to 20°C. (From Massalski, 1986, p. 809; with permission.)

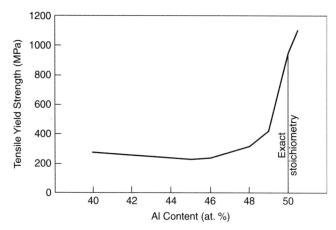

Fig. 25.8 Relation between Al content of cI2 FeAl intermetallic compound and the tensile yield strength. Specimens were annealed 5 days at 400°C before testing. Note the sharp decrease in yield strength as Al content decreases below the 50 at% value of the fully stoichiometric compound.

These high vacancy concentrations impede dislocation motion, providing a powerful "solid-solution hardening" effect. Ni-rich NiAl is also stronger than stoichiometric NiAl; σ_y increases to over 800 MPa as Al content drops to 42 at%. Unfortunately, deviations from stoichiometry in NiAl actually reduce ductility; the low ductility (1 to 2% elongation) of polycrystalline NiAl at 20°C becomes zero ductility for both Al- and Ni-rich compositions.

It appears that ductility improves in some off-stoichiometry intermetallics because deviation from the perfectly ordered structure of exactly stoichiometric materials relieves the problems of electronic charge depletion and high dislocation nucleation energies at grain boundaries. In stoichiometric Ni_3Al, the Ni content is 75% at the grain boundary and 75% far from any boundary. However, in Ni-rich Ni_3Al, the Ni content at the grain boundary is greater than the Ni content distant from a grain boundary. Thus, even small deviations from stoichiometry in the overall sample may cause large deviations in composition at the grain boundary. For this reason, small deviations in bulk composition can cause major changes in grain boundary cohesive strength. As the stoichiometry deviates from the ideal, the material behaves more like an ordinary metal; grain boundary strength rises while σ_y in the bulk of the grain decreases until the intergranular fracture seen in exactly stoichiometric material is replaced by ductile, transgranular failure in off-stoichiometry material.

25.8 INTERMETALLICS WITH THIRD-ELEMENT ADDITIONS

25.8.1 B Doping

Perhaps the most exciting finding of the past quarter century of intermetallics research was the discovery that small B additions to certain intermetallics greatly improve ductility, apparently by improving grain boundary cohesive strength. Although B improves ductility in Ni-rich Ni_3Al, the effect is highly sensitive to stoichiometry. As Fig. 25.9 shows, tensile elongations of 50% are seen at room temperature in polycrystalline 76.0 at% Ni–24.0 at% Al containing 0.05 wt% B. However, B provides no ductility improvement in 75.0 at% Ni–25.0 at% Al specimens. Fracture surfaces of $Ni_{76}Al_{24}$ + B tensile test specimens show ductile transgranular fracture, as opposed to the intergranular fracture seen in $Ni_{75}Al_{25}$.

The degradation of mechanical properties described previously for off-stoichiometry intermetallics still occurs when B is added to these materials, but the discovery of B's ductilizing effect in Ni_3Al moved intermetallics closer to widespread engineering utilization. It is important to understand why only Ni-rich Ni_3Al is ductilized by B additions. As stated previously, stoichiometric Ni_3Al has the same Ni/Al ratio in the bulk of the grains as at the grain boundaries. However, Ni-rich Ni_3Al has higher than overall Ni content at the grain boundaries and

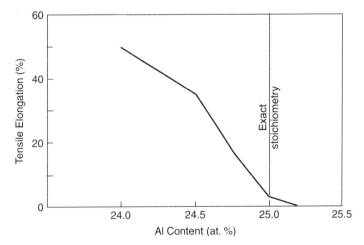

Fig. 25.9 Relation between Al content in Ni_3Al + B and tensile test ductility. All specimens contained 0.05 wt% B. For comparison, ductility in Ni_3Al at 76.0 at% Ni–24.0 at% Al with no B additions is only 2.5% elongation.

lower than overall Ni content distant from grain boundaries. It appears that B, which segregates preferentially to grain boundaries, slows H diffusion along grain boundaries and acts synergistically with the Ni enrichment at the grain boundary to provide additional electrons for atomic bonding across the boundary. This acts to lower the ordering tendencies of the material near the boundary that had already begun with the Ni enrichment. B may also lower the energy required to nucleate new dislocations at the grain boundary. Finally, there is some evidence that B slows the H_2O decomposition reaction at the surface of the metal that feeds H to the boundaries.

After the initial discovery of B's effect in off-stoichiometric Ni_3Al, B additions were tried in many other intermetallics. B was found to improve ductility greatly in the cP4 compounds Ni_3Ga, Ni_3Si, and Co_3Ti, but in Ni_3Ge and Pt_3Ga, there was no ductility improvement from B additions. In the cI2 compound FeAl, addition of 300 wt ppm B increases tensile elongation at room temperature in polycrystalline $Fe_{0.60}Al_{0.40}$ from 1.2% to 4.3% while shifting the fracture mode from intergranular to cleavage. In cI2 AlRu, B additions have little effect on strength or ductility. No cI2 compound has been observed to show the great improvement in ductility from dopants seen in some of the cP4 materials.

25.8.2 Other Dopants

Other third-element additions have been tried as potential ductilizing additives in Ni_3Al with mixed results. Be, Mn, Fe, Co, Cu, Zr, Pd, Ag, Hf, and Pt have all been shown to improve ductility in Ni_3Al somewhat, but not as much as B. Another dozen elements, including C, S, Sc, Ti, and V, provide no ductility improvements. Following a laborious, Edisonian approach, investigators have tried a broad range of "non-boron" third-element additions in Ni_3Al, but no unifying theory has come forward to explain why some element additions improve ductility, whereas others do not.

25.8.3 Use of Multiple Ductilizing Strategies

A sustained effort to develop commercially useful wrought structural materials from FeAl intermetallic compounds has been ongoing at the U.S. Oak Ridge National Laboratory since the mid-1980s. This work has shown that several factors must be addressed simultaneously to achieve satisfactory performance in FeAl-based intermetallics:

- Maintaining Al content near 38 at% to avoid intergranular fracture
- Producing a fibrous, elongated grain structure
- Producing and maintaining a fine grain size
- Deliberately oxidizing the surface to provide a protective coating
- Minimizing the concentration of quenched-in vacancies by slow cooling
- Adding several alloying elements

A family of materials has been developed with the general composition Fe (55 to 59 at%), Al (36 to 40 at%), Mo 0.2 at%, Zr 0.05 at%, C 0.13 at%, and B 0.24 at%. Mo improves high-temperature strength and creep resistance. C improves weldability, forms carbides, and reduces hot cracking tendencies. B improves room-temperature ductility by raising grain boundary cohesive strength. Zr forms borides to refine the grain size and retain a fibrous grain morphology. Processing protocols have been developed to achieve the desired grain size and shape and to avoid excessive numbers of quenched-in vacancies as the material cools from high temperatures. These materials' mechanical properties, oxidation and sulfidation behavior, creep performance, weldability, and castability are now well characterized. When preoxidized to form a protective oxide scale, these FeAl-based materials achieve 14.2% tensile ductility, $\sigma_y = 360$ MPa, and $\sigma_{UTS} = 940$ MPa. Yield strength changes relatively little between room temperature and 700°C. The best combinations of strength and ductility are seen in ultra-fine-grained material (about

3 to 4 μm average grain size), but these materials require more costly powder metallurgy and extrusion processing methods. A protective alumina-based scale forms during high-temperature exposure to air that allows hot working of the metal in air and provides useful high-temperature oxidation resistance to 1000°C. J_{IC} fracture toughness tests performed in air on Fe–35 at% Al produced estimated K_{IC} values of 33 MPa·m$^{-0.5}$, which is comparable to the K_{IC} of various Al alloys in widespread aviation use. K_{IC} values are rarely reported for most intermetallic compounds, because for binary stoichiometric materials they are typically quite low (5 MPa·m$^{-0.5}$ or lower).

25.8.4 Two- and Multiphase Structures

Yet another approach to providing better room-temperature ductility for intermetallic compounds is production of two- or three-phase microstructures that contain ordinary metal and ordered intermetallic phases. This strategy has been used to produce a system based on a mixed microstructure of Co_3AlC, Co solid solution, and cI2 CoAl. Co_3AlC has the E2$_1$ structure, which is the same as the cP4 structure with a C atom at the unit cell's center. It has the same $\{111\}\frac{1}{2}\langle\bar{1}10\rangle$ slip system as the cP4 structure, which makes it potentially capable of extensive plastic deformation. A microstructure with 52 vol% Co_3AlC, 42 vol% Co, and 6 vol% cI2 phase has a room temperature $\sigma_y = 460$ MPa with 8.6% tensile elongation. At 1000°C, this material has a σ_y value of 200 MPa. The presence of ordinary metals on one side of phase boundaries mitigates some of the problems associated with dislocation nucleation near grain boundaries, electronic charge depletion at boundaries, and inadequate numbers of independent slip systems. Of course, whether such materials can correctly be called *intermetallic materials* becomes a matter of semantics, since familiar alloys such as the Ni-base superalloys are also ductile solid solutions of ordinary metals strengthened by large volume fractions of intermetallic compounds.

25.9 ENVIRONMENTAL EMBRITTLEMENT

25.9.1 H Embrittlement

H plays a major role in embrittling many intermetallic compounds. Tensile tests and fracture toughness tests performed on aluminide and silicide intermetallics in H_2O-free atmospheres (e.g., pure, ultradry O_2 or high vacuum) show much higher ductility and toughness than do the same materials tested in normal air. In nonstoichiometric FeAl, for example, the K_{IC} fracture toughness of Fe–35% Al is 33 MPa·m$^{0.5}$ in air but 105 MPa·m$^{0.5}$ in ultradry O_2. Room-temperature tensile test elongation for Fe–35% Al is only 3.5% in air, but this increases to 13% in vacuum and 20% in dry O_2. H embrittlement occurs by the decomposition of H_2O by Al (or Si) at the surface by the reaction $2Al + 3H_2O \rightarrow 6H + Al_2O_3$.

Very little moisture is necessary to cause this H embrittlement effect. A "good" conventional vacuum of 1 mPa contains sufficient water vapor to embrittle Ni_3Al and FeAl. The atomic H formed by this reaction travels by grain boundary diffusion through the material. The resulting H concentration at grain boundaries draws bonding electrons away from the metal atoms at the boundary, reducing the boundary's cohesive strength. H has also been calculated to reduce the cleavage strength of the {100} planes of FeAl by 70%, favoring cleavage over ductile fracture. Since tests performed in ultradry O_2 show greater toughness and ductility than those performed in vacuum, it appears that O forms a protective oxide layer that inhibits H access to the crack tip. Test results from ultradry O_2 atmospheres are thought to give the most accurate indication of the inherent ductility of intermetallics.

A number of attempts have been made to counteract the H environmental embrittlement effect by alloying additions that will either interfere with the formation of H at the surface or impede its diffusion. Small amounts of B in the intermetallic can slow H diffusion, thereby reducing the environmental embrittlement effect. Addition of Cr to Ni_3Al forms a protective

Cr_2O_3 layer; preoxidation to build a protective coating inhibits H generation at the surface, and this approach has also been effective in FeAl.

25.9.2 Lack of Embrittlement in RM Intermetallics

An intriguing exception to the environmental embrittlement effect in intermetallics is the high ductility and high fracture toughness of cI2 compounds with the formula RM, where R is a rare earth element and M is a non–rare earth metal. More than 130 such RM cI2 compounds are known to exist, and many show surprisingly high room-temperature ductility for fully ordered, stoichiometric intermetallics. YAg displays 20% elongation in polycrystalline specimens tested in ordinary air at room temperature, and it has been cold-rolled 88% without fracturing. Polycrystalline YCu and DyCu have slightly lower elongations (12 to 15%); DyCu displays a K_{IC} fracture toughness of 25 $MPa \cdot m^{0.5}$. These samples were exactly stoichiometric and contained no B or other third-element additions. High ductility has also been reported in AuZn and AgMg cI2 intermetallics, although it is uncertain whether the materials tested in those studies were fully ordered.

The RM compounds appear to be the first family of intermetallics found to deform in a ductile manner at room temperature without relying on any "contrivances" involving off-stoichiometric compositions, third-element additions, or special testing atmospheres. Single-crystal tensile tests on YCu and YAg show $\langle 010 \rangle \{100\}$ and $\langle 010 \rangle \{101\}$ slip lines, which by themselves would provide only three independent slip systems. However, TEM analysis of plastically deformed YAg shows large numbers of $\langle 111 \rangle$ Burgers vector dislocations and some $\langle 010 \rangle$ and $\langle 110 \rangle$ dislocations. It is not yet known whether the $\langle 111 \rangle$ and $\langle 110 \rangle$ dislocations result from other slip systems or from dislocation reactions of the type $[100] + [010] + [001] \rightarrow [111]$. The apparent lack of environmental embrittlement problems in these materials is surprising because the rare earth elements are even more aggressive oxide formers than Al. The unusual ductility of the RM compounds may provide useful insights into the fundamental mechanisms related to intermetallic ductility. This, in turn, could lead to development of new alloying additions for other intermetallic materials.

ADDITIONAL INFORMATION

References, Appendixes, Problem Sets, and Metal Production Figures are available at *ftp://ftp.wiley.com/public/sci_tech_med/nonferrous*

INDEX

Abrasive jet grinding, 120
Abundance of elements, 156, 157, 291, 292, Appendix C
Actinide metals, 5, 146, 220, 435. *See also* Plutonium, Thorium, Uranium
   applications, 451, 454–458, 461–464, 466–473
   bonding of 5f electrons, 451, 453, 454, 466, 469, 470
   coring in castings, 468, 469
   nuclear properties, 451, 453–458, 461, 464, 465, 467, 470–473
   physical properties, 451–454, 458–462, 465–473
   sources, 455, 456, 464, 468
   toxicity, 455, 456, 462–464, 466, 467, 471–473
Actinium, 451–454, 469–471. *See also* Actinide metals
Adhesive bonding, 126
Alloying elements, 20
Alpha-lanthanum, 1, 3
Aluminum, 5, 11, 12, 18, 29, 49, 52, 70, 75, 77, 84, 93, 96–98, 109, 112, 114, 115, 125, 126, 129, 131, 134, 137, 139, 144, 145, 162, 164, 166–168, 170–173, 175–178, 181, 188–191, 199, 213, 216, 223, 224, 233, 239, 269, 274, 275, 278, 281, 289, 301, 305, 312, 319, 322, 324, 339, 340, 343, 344, 347, 348, 354, 388, 395, 406, 408, 429, 441, 445, 448–450, 457, 474, 475, 477–480, 483–487
   alloy designations and behavior, 364–376, 378–382
   applications, 358, 360, 362, 364, 365, 367–369, 374–378, 382, 383
   Bayer process, 358, 383
   conductivity, 360, 363–366
   corrosion behavior, 360, 361,
   dispersoids in, 385, 386, 448–450
   GP zones, 371–373, 376, 378
   Hall-Héroult process, 358, 383, 384
   H solubility in, 362–364, 380, 381
   oxide layer, 360, 361, 377, 386
   precipitation hardening, 369–376, 378–382
   properties, 358–364
   pure Al, 364–367, 384
   reflectance, 360–362
   sources, 383, 384
   weldability, 364, 365, 368, 369, 375, 378, 381
Amalgams, 324, 355, 357
Americium, 451–454, 466, 469–472. *See also* Actinide metals
Amorphous metals, 90–96, 98, 99
   applications, 94, 95
   cooling rates, 90–92
   creep rates, 92
   formability, 94
   mechanical properties, 92–94
   production methods, 90–92
   thermal stability, 93, 94
Anisotropy, 33, 42, 49, 77, 104, 132, 255, 263, 264, 339, 387, 419–421, 423, 438, 444, 458–460, 479
Antimony, 8, 92, 168, 230, 299, 316, 343, 348, 382, 390, 400, 403–409, 412, 413, 415, 416, 419–422, 427, 428, 458, 477
   applications, 423–425
   properties, 420, 423, 424
   sources, 426, 427
   thermoelectric devices, 426, 427
   toxicity, 419, 425, 426
Antiphase boundary, 279, 333, 334, 378, 380, 479, 480, 481
Arc welding, 122–124
Armor/ammunition, 16, 96, 218, 219, 415, 462, 463
Arrhenius rate equation, 55
   creep effects, 80
Arsenic, 8, 305, 316, 391, 409, 423, 426, 428, 431, 458
   applications, 421, 422
   properties, 419–421
   sources, 422, 423
   toxicity, 419, 421–423
Atomic bomb, 163, 457, 458

Ball milling, 86
Bamboo defect, 115
Band gap, 9
Barium, 156, 157, 211, 325
   properties and applications, 156, 178

Structure–Property Relations in Nonferrous Metals, by Alan M. Russell and Kok Loong Lee
Copyright © 2005 John Wiley & Sons, Inc.

INDEX

Batteries, 102, 150, 154, 177, 243, 244, 249, 250, 269, 271, 285, 302, 322, 327, 337, 346, 347, 349–351, 353, 356, 388, 390, 392, 405, 410–413, 416, 421, 423, 441, 446
Bauschinger effect, 196, 197
Bayer process, 358, 383
Bearings, 326, 327, 407, 408, 414
Bend plane splitting, 160–162
Bending sheet metal, 116, 117
Berkelium, 451, 452, 469, 472. *See also* Actinide metals
Beryllium, 5, 11, 12, 49, 51, 65, 92, 93, 169, 177, 313, 458, 471, 475, 477, 486
  applications, 157, 163, 164
  cross slip in, 164, 165
  impurity effects, 162
  nucleosynthesis, 156, 157
  properties, 156–163
  slip systems, 159
  sources, 164
  toxicity, 164
Bismuth, 8, 9, 168, 177, 229, 234, 264, 291, 299, 316, 343, 382, 391, 403, 409, 415, 417, 419, 420, 426, 433, 458, 478
  alloys, 428, 429
  liquid metal embrittlement, 429, 430
  properties, 420, 427, 428
  sources, 429
  thermoelectric devices, 426, 427
  toxicity, 419, 429
Blisters and blows, 104
Body-centered cubic, 1, 3
  twinning, 26
  slip systems, 49
Bomb reduction reactions, 174
Bonding energy, 4, 14, 435, 436, 451, 453, 466, 469, 470
Brass, 308, 309
Bravais lattices, 7
Brazing, 90, 125, 126, 232, 240, 327, 328, 395, 401, 416
Bronzes (Sn, Al, and Si), 310–313
Burgers vector, 21, 22

Cadmium, 5, 25, 49, 129, 162, 168, 170, 199, 204, 285, 305, 324, 327, 337–339, 346–348, 390, 391, 405, 413, 428, 433, 439, 440, 445, 461
  applications, 349–351
  properties, 349
  slip, 352, 353
  sources, 351, 352
  toxicity, 349, 351, 352
Calcium, 102, 150, 156, 157, 173, 175, 176, 351, 409, 412, 413, 416, 417, 433, 446, 464, 468
  properties and applications, 177, 178
Californium, 451, 452, 469, 472, 473. *See also* Actinide metals

Cast iron, *see* Steel and cast iron
Casting, 102–109
  centrifugal, 108
  continuous, 104, 105
  crucibles, 102
  defects, 103, 104
  furnaces, 102
  investment (lost wax), 107
  Mg, 169–172
  permanent mold, 107, 343–346
  precision, 106
  sand mold, 103, 104, 106
  segregation, 104, 468, 469
  semipermanent mold, 107
  semi-solid, 109
  shell mold, 106
  single crystal, 108
  stir casting for composites, 132, 133
Catalysis, 295–298. *See also individual elements*
Centerburst (center cracking), 115
Cerium, *see* Rare Earths
Cesium, 17, 234, 302, 308, 331, 478, 479
  applications, 149, 150, 153, 154
  low-melting ternary eutectic, 153
  properties, 146–149
  sources, 154
Charpy test, 65
Chemiadsorption of H, 67
Chemical machining, 120
Chromium, 13, 14, 17, 126–129, 168, 190, 193, 200, 208, 209, 213, 232, 238, 263, 265, 269, 272–276, 278, 281, 287, 364, 368, 375, 376, 379, 385, 405, 449, 474, 487, 488
  applications, 223, 224
  coating/plating, 222–224
  ductility, 221, 223, 225, 226
  ferrochromium in steel, 221–225
  mechanical properties, 223, 225, 226
  properties, 221–223
Cleavage fracture, 65
Coatings
  anodizing, 129, 217
  chemical vapor deposition (CVD), 128, 129
  contaminant, 127
  dry coating, 129
  electroless plating, 129
  electroplating, 129, 222–224, 273, 281, 283, 284, 405
  galvanizing, 337, 339–343
  hard-facing, 127
  ion implantation, 128
  lubricating, 127
  painting, 129
  physical vapor deposition (PVD), 128, 129
  porcelain enameling, 129, 250
  thermal spray, 127

Cobalt, 13, 90, 97, 218, 222, 223, 238, 240, 250, 251, 281, 284–287, 313, 334, 352, 407, 441, 484, 487
  applications, 259, 263–269
  binder phase in WC, 266, 267
  coatings, 267
  crystal structures, 260
  high-temperature alloys, 263–265
  magnetism, 262–264, 267, 268
  properties, 259–263
  sources, 270, 286
  in steel, 268
  toxicity, 269, 270
Coble creep, 81–84
Cold shuts, 104
Columbium, see Niobium
Composites, metal matrix, 130–145
  Al-SiC, 133, 143–145
  applications, 132
  cavitation, 141
  creep, 142–145
  debonding, 141, 145
  interface chemical reactions, 133
  internal stresses, 137–139
  microdamage, 140
  powder processing, 133, 134
  spray deposition, 133, 134
  stir casting (vortex mixing), 132, 133
  strengthening mechanisms, 134–137
  stress relaxation, 138–141
  reinforcement, 131, 132, 140
  Ti-SiC, 142,143
  voids, 141, 145
Constitutional supercooling, 104
Cooling rates, in amorphous metals, 90–92
Cope, 103, 106
Copper, 6, 17, 19, 27, 35, 49, 68, 70, 84, 86, 88, 92, 98, 99, 107, 108, 110, 115, 119, 121, 123, 124, 126, 129, 131, 147, 150, 154, 162, 166, 168, 171, 173, 175, 177, 199, 211, 212, 218, 227, 232–234, 238, 239, 243, 251, 257, 259, 260, 262, 270, 273, 274, 286, 287, 291, 295, 321, 322, 324–328, 331, 333, 334, 337, 339, 343, 346, 348, 350, 352, 359, 360, 363–365, 369–372, 375, 378–382, 391, 401–414, 417, 421, 422, 424, 429, 430, 438, 441, 448, 450, 474, 480, 481, 486, 488
  in the biosphere, 314, 315
  brass, 308, 309
  bronzes (Sn, Al, and Si), 310–313
  color, 303, 304
  conductivity, 304–306
  corrosion, 303, 304
  Cu-Be alloys, 313
  Cu-Ni alloys, 309, 310
  DMMCs, 313, 314
  dislocation structures, 317–319
  electrical use, 306, 307, 313–315
  electron phases, 308, 309
  mechanical properties, 304, 317–320
  nanocrystalline, 318–320
  nickel silvers, 309
  plumbing/heat exchangers, 307
  properties, 302–306
  smelter pollution, 316, 317
  sources, 316, 317
  Space Shuttle engines, 306, 307
  space tethers, 314, 315
  stacking faults in, 317, 318
Coring, 104, 468, 469
Corrosion, in amorphous metals, 93. *See also individual elements*
Cracks, 27, 61–67, 369, 380. *See also* Fatigue, Fracture
  centerburst (center cracking), 115
  in creep, 78
  end effects, 63–64
  hydrogen effects, 67
  initiation, 74, 75
  opening displacement, 65
  stress corrosion, 368, 378, 382
Creep, 77–83
  in amorphous metals, 92
  Coble, 81–84
  in composites, 142–145
  deformation mechanism maps, 82, 83
  diffusion effects, 80–84
  dislocation mechanisms, 79, 80
  in irradiated metal, 99
  in Mg, 172, 175, 176
  metal-matrix composites, 78, 79
  Nabarro-Herring, 81–84
  in Re, 257, 258
  stages, 78
  steady-state, 78, 80
  in superalloys, 283
  in W, 240–242
  in Zn, 339, 344, 345
Critical resolved shear stress, 51, 59. *See also individual elements*
Cross slip, 42–47
Crucibles, 103
Crush-zone structures, 61
Crystal structures, 1, Appendix A. *See also* Single crystals and *individual elements*
  covalency effects, 5
  defects, 18–27
  ductility, 8, 244–247
  electronic structure effects, 9
  entropy effects, 2, 4
  magnetic effects, 5
  in nanocrystalline metals, 85–90
  polymorphism, 5, 402, 403, 446–448
  pressure effects, 8, 9
  in quasicrystalline metals, 96, 97

Cup and cone fracture, 68
Curium, 451, 452, 454, 469, 472. *See also*
    Actinide metals
Cutting (machining), 118–122
    speeds, 119
    tool materials, 118
    torch, 122
    workpiece-tool interactions, 118, 119
    wear, 119
Czochralski crystal pulling, 108

Deformation mechanism maps, 82, 83
Dendritic freezing, 104
Density, 15–17, 32
Density-of-states function, 11, 12
Dental alloys, 324, 355
Diamond-anvil cell, 9
Diamond structure, 6
Diffusion
    in amorphous metals, 95
    bonding, 124, 185, 239
    bulk, 89
    Coble, 81
    in composites, 139
    in creep, 80–84
    dislocation climb effect, 53
    grain boundary, 81, 89
    H, 66, 487–488
    Nabarro-Herring, 80
    in radiation-damaged metals, 98
    yield point effect, 76
Dislocation(s), 20, 38–60
    activation energy for motion, 56
    activation volume, 57
    athermal barriers, 56
    barriers to motion, 23, 26, 28, 29, 55, 56, 80
    Burgers vector, 21, 22, 40
    climb, 21, 23, 53, 80
    in composites, 136, 139
    in creep, 79, 80
    cross slip, 42–47
    curvature, 39
    density, 23, 28, 33, 55
    edge, 21
    elastic strain energy, 40
    forces on, 38
    Frank-Read source, 30, 42, 43
    Frank sessile dislocation, 47
    geometrically necessary, 28, 29
    glide, 21, 23, 53
    glissile, 46, 352, 353
    interaction with solutes, 58, 59
    intersection, 53–55
    jog, 53, 54
    kink, 53, 54
    Lomer-Cottrell barrier, 48
    multiplication, 42
    node, 22
    Orowan loop, 35
    partial, 44–48, 164, 165, 317, 318
    Peierls stress, 31, 40
    pile-up, 30, 59, 60, 88, 318–320
    in radiation-damaged metals, 98
    screw, 21, 237
    sessile, 46, 48, 352, 353
    Shockley partials, 45–48
    statistical, 28
    subsonic and supersonic, 77
    Suzuki locking, 59
    Taylor factor, 31
    thermally activated, 53–56
Dispersoids, 131, 225, 226, 368, 385, 386, 448–450
Drag, 103, 106
Drawing, 115, 116
Ductile fracture, 67–69
Ductility, 32, Appendix D. *See also individual elements*
    in Be, 160
    in Cr, 221, 223, 225, 226
    ductile-brittle transition, 225
    in engineering design, 160
    in intermetallic compounds, 474, 475, 477, 478, 480–488
    in Ir, 299–301
    liquid metal embrittlement, 429, 430
    in Mg, 169
    in Mn, 244–247
    pressure effects, hydrostatic, 235, 236
    in sheet metal forming, 116
    in Ta, 216, 218, 219
    in U, 458–460
    in Zr, 198, 201–203
Dysprosium, *see* Rare Earths

Einsteinium, 451, 473. *See also* Actinide metals
Elastic modulus, *see* Modulus of elasticity
Elastic strain energy, 40, 45, 61
Electrical conductivity. *See also individual elements*
    in Ag, 322–324
    in Al, 360, 363–366
    in Cu, 304–306
    crystal structure effect, 8–10, 20, 247
    in quasicrystalline metals, 97
    in rare earth metals, 438
    work hardening effect, 32
Electrical discharge machining, 121
Electrochemical machining, 120
Electrodeposition, in nanocrystalline metals, 86
Electron beam
    machining, 121, 122
    welding, 124, 125
Electronegativity, 15, 17
Electrical discharge machining, 121, 122

Entropy effects
  in amorphous metals, 91
  in crystal structures, 2
  in H solubility, 213
Environmental effects, see Toxicity
Erbium, see Rare Earths
Europium, see Rare Earths
Extrinsic stacking fault, 27, 47
Explosive
  shaped-charge, 218, 219
  sheet metal forming, 117
  welding, 124, 125
Extrusion, 115, 116

Face-centered cubic, 1, 3
  partial dislocations in, 44
  single crystal, 52
  slip systems, 48, 49
Failure, 61
Fasteners, 126
Fatigue, 69–75. See also Fracture, Cracks
  endurance limit, 69
  fatigue life, 74, 75
  intrusions, 71
  in Mo, 231
  Paris regime, 73
  in Pb, 417, 418
  S/N curves, 69, 70
  strain aging effect. 70
Fermi energy, 11
Fermium, 451, 473. See also Actinide metals
Firescale in sterling silver, 401
Fission reactors, 101, 152, 199–204, 350, 364, 445, 451, 453–457, 461–464, 468, 470–473
Fleischer's relation, 35
Flux (filler rod), 122, 126
Foamed metals, 95
Forging, 114
Fracture. See also Cracks
  in Be, 160, 161
  brittle, 61–67
  Charpy test, 65
  ductile, 61, 67–69
  free surface effect, 63
  H embrittlement, 66, 67
  Izod test, 65
  plate thickness effect, 63, 64
  toughness, 63–65, 88
  transition temperature, 66
Francium, 146–149
Frank sessile dislocation, 47
Frank-Read source, 30, 42, 88
Free electron gas, 11
Frenkel defect, 20
Friction welding/friction stir welding, 124, 127
Fusion reactors, 101, 151, 208

Gadolinium, see Rare Earths
Galileo spacecraft, 100
Gallium, 5–7, 170, 208–210, 358, 359, 401, 421, 422, 424, 429, 466–469, 477, 486
  applications, 387, 468, 469
  LEDs, 388
  properties, 6, 7, 387
  sources, 389
  toxicity, 389
  transport, 388
Gas atomization, 110
Gas bubbles, 27
Germanium, 7, 326, 348, 392, 421, 422, 443, 486
  applications, 398–401
  properties, 6, 7, 399, 400
Gibbs free energy, 2, 4
Gold, 5, 17, 49, 110, 129, 147, 270, 286, 290, 291, 294, 296–298, 301, 323, 329, 339, 347, 355, 356, 389, 390, 405, 406, 417, 480, 481, 488
  corrosion, 17, 302, 330, 331
  dental alloys, 334, 335
  electronic use, 334
  history, 330, 332
  jewelry, 331, 333, 334
  properties, 302, 330–332
  production, 330, 331, 335, 336
  texture in, 335, 336
GP zones, 371–373, 376, 378
Grain boundary(-ies)
  diffusion, 25, 89
  as dislocation barriers, 31
  in nanocrystalline metals, 85, 86, 318–320
  superplasticity, 25, 83, 84
  tilt and twist, 24, 25
Grain
  columnar, 160
  growth, 34, 86–90
  size, 29, 85–88, 137, 160, 318–320
  structure in castings, 104
Griffith theory, 61–65
Grinding metal, 119, 120

Hafnium, 49, 128, 152, 173, 179, 199, 201, 213, 218, 231, 239, 245, 278, 284, 289, 352, 486
  properties and uses, 204
Hall-Héroult process, 358, 383, 384
Hall-Petch relation, 29–31, 318–320
Heat-affected zone, 124
Heat shields, Mercury capsules, 162
Hexagonal close-packed, 1, 3
  basal plane, 159
  c/a ratio, 5, 48
  prism planes, 49, 159
  pyramidal planes, 49, 159
  twinning, 26, 159, 160
  slip systems, 48, 49

494   INDEX

Holmium, *see* Rare Earths
Hopkinson bar technique, 77, 218, 219
Hot tears, 103
Hydride-dehydride process, 67
Hydroforming sheet metal, 117
Hydrogen
  in Al, 103, 362–364
  in Cu, 306, 307
  embrittlement, 66, 67
  in intermetallics, 487, 488
  metallic, 10
  in Nb, 213
  in Zircaloy, 200, 201
Hydrostatic pressure effects, 235, 236

Impurities, 20
Indium, 6, 350, 358, 389–390, 412, 428, 429
Intergranular fracture, 65
Intermetallic Compounds
  applications, 477, 478
  bonding, 474, 475
  cI2 (CsCl) structure, 478–481
  cF4 (AuCu$_3$) structure, 481–483
  dislocations in, 479–488
  ductility, 474, 475, 477, 478, 480–488
  environmental (H) embrittlement, 487–488
  strength, 474–476, 481–487
  nonstoichiometric/doped, 483–487
  oxidation/pest, 474–477, 486–488
Intersection of dislocations, 53–55
Interstitial atoms, 20
  dislocation climb, 53
  in irradiated metal, 99
  strain aging effect, 76
  in Ti, 182, 183
Intrinsic stacking fault, 27, 47
Iridium, 16, 244, 245, 254, 471
  abundance and nucleosynthesis, 291, 292
  applications, 296–298
  ductility, 299–301
  properties, 290–295, 297, 298
  sources, 299
Iron, *see* Steel and Cast Iron
Izod test, 65

Jogs, 53, 54, 99
Joining, 122–126
  adhesive bonding, 126
  brazing, 125, 126
  fasteners, 126
  soldering, 125, 126
  welding, 122–125, 364, 365, 368, 369, 375, 378, 381
Jupiter, 10, 100

K_{IC}, 63–65
Kink band, 26
Kroll process, 180, 182

Lanthanides, *see* Rare Earths
Lanthanide contraction, 17
Lanthanum, *see* Rare Earths
Laser beam
  machining, 121, 122
  welding, 124, 125
Lawrencium, 451, 473. *See also* Actinide metals
Lead, 5, 90, 92, 96, 102, 103, 107, 129, 152, 166,
      168, 170, 177, 178, 234, 255, 264, 270,
      291, 306, 308, 309, 327, 328, 336, 339,
      340, 343, 346, 347, 351, 352, 357, 382,
      389–392, 401, 402, 405–408, 419,
      421–425, 427–429, 463, 478
  applications, 410–416
  batteries, 410–413
  corrosion, 410, 411, 413–415, 417, 418
  fatigue, 417, 418
  properties, 411, 412, 414
  solders, 406, 407, 414
  sources, 417
  toxicity, 416, 417
LIGA devices, 120
Lighting
  LEDs, 388
  quartz-halogen, 128
  tungsten filament, 240–242
Lithium, 2, 157, 168, 170, 177, 208, 234, 250, 269,
       306, 350, 367, 378–381, 412, 450
  applications, 149–151
  nucleosynthesis, 156, 157
  phase transformations, 154, 155
  properties, 146–149
  sources, 154
Lomer-Cottrell barrier, 29, 48
Lutetium, *see* Rare Earths

Machining, *see* Cutting
Magnesium, 6, 8, 9, 13, 49, 65, 103, 107, 114,
         116, 131, 150, 152, 163, 164, 178, 180,
         199, 201, 213, 243, 291, 339, 343, 348,
         352, 364, 365, 368–370, 375, 380, 382,
         409, 417, 433, 435, 441, 445, 450, 455, 464
  applications, 170–175
  casting, 169–172
  creep in, 172, 175, 176
  extended solubility in, 176, 177
  mechanical properties, 169
  properties, 156, 157, 166–170
  in steel and cast iron, 173
  sacrificial anodes, 174, 175
  sources, 175
  twinning, 169
Magnetic-pulse forming sheet metal, 117
Magnetism
  in amorphous metals, 93
  ferromagnetism, 13, 262–264, 267, 268, 272,
          273, 284, 438, 441–445
  in nanocrystalline metals, 93

Manganese, 5, 13, 14, 96, 167, 168, 170, 223, 256, 259, 262, 265, 270, 273, 274, 276, 279, 310, 347, 365, 368, 441, 449, 466, 486
  applications, 247–250
  crystal structures, 244–247
  ferromanganese, 247, 248, 250
  nodules, ocean floor, 251
  properties, 243–247
  sources, 250, 251
  vibration-damping, 251, 252
Mechanical alloying, 86
Melt-spinning, 91, 93
Melting temperatures. *See individual elements*
MEMS devices, 121
Mendelevium, 451, 473. *See also* Actinide metals
Mercury, 5, 153, 154, 162, 166, 178, 229, 234, 324, 325, 329, 336, 338, 343, 389, 390, 412, 429, 433
  amalgams, 324, 355, 357
  applications, 354–356
  properties, 353, 354
  sources, 356, 357
  toxicity, 337, 353–356
Metallic bond, 4, 9
Metallic hydrogen, 10
Metalloids, 7, 9
Microdamage in composites, 140
Misruns, 104
Modulus of elasticity, 20, Appendix B2. *See also individual elements*
Modulus of resilience, 94
Molybdenum, 17, 37, 49, 112–115, 118, 128, 187, 188, 190, 192, 210, 221, 225, 238, 239, 243, 244, 253–257, 264, 265, 268, 276, 278, 279, 281, 429, 475–477, 486
  applications, 229–231
  coatings on, 232
  ferromolybdenum, 229, 230
  hot work, 229
  oxide layer, 228, 229
  mechanical properties variation with temperature, 230
  mining, 228
  powder metallurgy, 229
  properties, 227–231
  sources, 232, 233

Nabarro-Herring creep, 81–84
Nanocrystalline metals, 85–90, 318–320
  deformation mechanisms in, 89
  engineering applications, 90
  magnetic properties, 88
  mechanical properties, 87, 88, 318–320
  thermal stability, 88–90
Neodymium, *see* Rare Earths
Neptunium, 451–454, 464, 469–471. *See also* Actinide metals

Nickel, 8, 13, 45, 49, 86–90, 97, 104, 106, 108, 109, 112, 114, 119, 121, 126, 129, 131, 150, 166, 168, 178, 187, 194, 195, 199, 200, 204, 205, 210–213, 218, 222, 223, 231, 233, 238–240, 249–255, 261, 262, 265, 267, 268, 299, 301, 305–307, 309, 310, 316, 333, 334, 337, 349–351, 360, 381, 382, 396, 429, 441, 455, 457, 462, 467, 478–481, 483, 485, 486
  annealing twins, 287, 288
  applications, 259, 273–285
  coatings, 273, 281, 283, 284
  corrosion, 271, 272
  superalloys, 276–284
  magnetism, 272, 273, 284
  monel, 274
  Ni-Cr(-Fe) alloys, 274–276
  properties, 259, 271–273
  sources, 285–287
  single-crystal turbine blades, 277, 281, 282, 288, 289
  in steel, 273
  topologically close-packed phases, 278–280
  toxicity, 285
Niobium, 49, 118, 128, 188, 190, 199, 200, 207–209, 218, 220, 225, 230, 239, 245, 263, 264, 267, 268, 275, 276, 278, 279, 281, 313, 314, 325, 397, 408, 460, 462, 474, 477, 478, 483
  applications, 210, 211
  ferroniobium in steels, 210
  high-temperature alloys, 212, 213
  interstitial impurities, 213–216
  mechanical properties variation with temperature, 230
  properties, 205, 206, 210
  sources, 213
  superconductivity, 211, 212
Nobelium, 451, 473. *See also* Actinide metals
Nuclear reactors, *see* Fission reactors, Fusion reactors
Nucleosynthesis, 156, 157, 291, 292

Ordnance, *see* Armor/ammunition
Orowan loop, 35, 137, 138
Osmium, 16, 119, 210, 216, 254
  abundance and nucleosynthesis, 291, 292
  applications, 297
  properties, 290–295, 297
  sources, 299
  toxicity, 298, 299

Painting, 129
Palladium, 13, 88, 92, 93, 188, 239, 240, 324, 333–335, 486
  abundance and nucleosynthesis, 291, 292
  applications, 297
  catalysis, 295–298

Palladium (*continued*)
  properties, 290–297
  sources, 299
Pauli exclusion principle, 11
Pearson symbols, 6
Peen forming sheet metal, 117
Peierls stress, 31
Periodicity of properties, 13
Phase boundaries, 27
Physisorbtion of H, 67
Planar defects, 24
Plane strain, 63, 64
  fracture toughness, 65
Plane stress, 63, 64
Plastic deformation
  crack tip, 62
  strain hardening, 33
  twinning, 26
Platinum, 13, 16, 238, 255, 267, 268, 286, 316, 324, 333–336, 419, 453, 471, 475, 486
  abundance and nucleosynthesis, 291, 292
  applications, 294–298
  catalysis, 295–298
  properties, 290–295
  sources, 299
Plutonium, 5, 152, 163, 387, 389, 435, 446, 471, 472
  applications, 451, 457–458, 466–469
  bonding of 5f electrons, 451, 453, 466, 469, 470
  coring in castings, 468, 469
  nuclear properties, 451, 458, 464, 465, 467, 468
  physical properties, 451–454, 465–470
  sources, 468
  toxicity, 467
Point defects, 18–20
Poisson's ratio, Appendix B2,
Polonium, 8, 157, 163, 419, 420, 430–432
Porosity, 104, 105
Potassium, 49, 157, 234, 239, 240–242, 285, 302, 358, 392, 456
  applications, 149, 150, 152, 153
  low-melting ternary eutectic, 153
  properties, 146–149
  sources, 154
Powder metallurgy, 109–112, 229, 240
  powder production, 109, 110
  sintering, 112
Praeseodymium, *see* Rare Earths
Precipitation hardening, 35–37, 369–376, 378–382
  aging, 35
  coherent precipitates, 36
  incoherent precipitates, 36
Preferred orientation, 25
Promethium, *see* Rare Earths
Protactinium, 451, 452, 454, 469–471. *See also* Actinide metals

Quasicrystalline metals, 96–98
  coefficient of friction, 97
  mechanical properties, 97, 98

Radiation damage, 20, 98–101
  in nuclear reactors, 101
  in Zr, 202, 203
Radium, 156, 157, 178, 470
Rare Earths, 5, 17, 92, 168, 170, 172, 173, 259, 267, 284, 285, 386, 470
  abundance, 433
  applications, 440–445
  Ce polymorphism, 446–448
  magnetic properties, 438, 441–445
  optical properties, 439, 440
  physical properties, 434–439
  Sc in Al, 448–450
  sources, 446
Recrystallization, 25, 33, 34, 367
  in composites, 139
  in nanocrystalline metals, 85, 86
  in rolling, 112
  in Zn, 340
Recovery, 79, 80, 112, 340, 367
Reinforcement particles/fibers, *see* Composites
Residual stresses, 113
Resistivity, *see* Electrical conductivity
Rhenium, 37, 49, 230, 231, 238, 239, 252, 253, 278, 279, 282, 297
  applications, 255, 256
  mechanical properties variation with temperature, 230, 254, 255, 257, 258
  physical properties, 243–245, 254–256
  sources, 257
Rheocasting, 109
Rhodium, 239, 240, 326, 471
  abundance and nucleosynthesis, 291, 292
  applications, 296
  catalysis, 296–298
  properties, 290–296
  sources, 299
Risers, 103
Roller burnishing, 127
Rolling sheet and plate, 112, 113
  roll bonding, 124
Rotating consumable electrode process, 109, 111
Rubidium, 9, 302
  applications, 149, 150, 153, 154
  properties, 146–149
  sources, 154
Ruthenium, 119, 188, 239
  abundance and nucleosynthesis, 291, 292
  applications, 296
  properties, 290–296
  sources, 299
  toxicity, 298, 299

Samarium, *see* Rare Earths
Scandium, *see* Rare Earths
Schmid's law, 28, 50–52
Segregation in castings, 104, 468, 469
Semiconductors, 393–396, 398, 399, 421, 422, 426, 427
Semisolid die casting, 109
Shape rolling, 113
Shear bands, 92
Shockley partial dislocations, 45–48
Shot-peening, 127
Shrinkage pipe, 105
Silicon, 6, 7, 9, 43, 90, 92, 103, 106, 108, 109, 119–121, 126, 131, 133, 134, 140–145, 147, 152, 157, 162, 163, 167, 169, 170, 172, 173, 175, 178, 207, 223, 224, 232, 238, 242, 248, 259, 263, 265, 267, 270, 273–276, 281, 284, 300, 305, 306, 308, 310, 318, 340, 342, 361, 363–365, 368–370, 374–376, 379, 381–383, 385, 387, 400–402, 421, 422, 426, 443, 450, 475, 476, 487
   applications, 392, 394–398
   compounds, 392, 396–398
   electronic properties, 393–396
   physical properties, 6, 7, 392–394
   sources, 398
   strained SiGe devices, 398, 399
Silver, 49, 126, 129, 166, 168, 172, 238, 270, 286, 290, 291, 295, 303, 305, 306, 333, 334, 336, 339, 347, 350, 351, 355, 356, 360, 390, 401, 404, 406–408, 415, 417, 439, 440, 486, 488
   in bearings, 326, 327
   in brazing/soldering, 327, 328
   corrosion, 302, 321, 322
   dental alloys, 324, 355
   electrical use, 322–324
   monetary, 329
   in photography, 328, 329
   properties, 302, 321–322
   sources, 329
   sterling, 324–326, 401
   in superconducting wire, 325
Single crystals, 28, 52, 277, 281, 282, 288, 289, 398
Skull melting, 110, 111, 182
Slag, 102
Slip systems, 48–50, Appendix B1,
Sodium, 2, 8, 9, 11, 12, 49, 189, 199, 208, 213, 220, 229, 234, 306, 355, 357, 358, 370, 381, 382, 384, 390, 397, 400, 417, 429
   applications, 149–152
   low-melting ternary eutectic, 153
   properties, 146–149
   sources, 154
Soldering, 125, 126, 406, 407, 414
Solid solution hardening, 34, 58, 59, 264
   chemical interaction, 59
   electrical interaction, 59

Fleischer's relation, 35
   interstitial, 35
   shear modulus effect, 59
   substitutional, 35
Spinning sheet metal, 117
Sports implements, 96, 132
Stacking
   fault, 27, 45–48, 99, 260
   fault energy, 45–47, 260, 317, 318, 359, 360, 368
   sequence, 1, 3, 27
Steel and Cast Iron, 37, 70, 84, 101–103, 106, 107, 110, 112, 114–116, 118, 127–129, 148, 152, 167, 169, 173–175, 177, 178–180, 184, 185, 192, 193, 199, 205–207, 209, 210, 212, 213, 218, 221–224, 227–233, 236–238, 243, 244, 247–250, 253, 259, 267–271, 273–276, 285, 287, 297, 298, 303, 309, 312, 326, 327, 330, 331, 337, 339, 340, 342, 343, 346–350, 357, 358, 362, 383, 384, 394, 397, 398, 402, 405, 408, 414–416, 423, 429, 435, 441, 445, 476
Strain aging, 76
Strain hardening, 32, 55
   rate, 33, 52
   recovery and recrystallization, 55
   single crystal, 52
Strain rate jumps, 57
Strain rate sensitivity, 26, 57
Strained Si-Ge devices, 398, 399
Strengthening mechanisms. *See also individual elements* and Composites, Creep, Critical resolved shear stress, Fatigue
   grain boundary, 28–32, 87
   precipitation hardening, 35–37, 239
   solid solution, 34, 35, 239
   strain hardening, 32, 33, 239
Strontium, 156, 175–178, 211
Stress intensity factor, 62–65
Superalloys, 263–265, 276–284
Superconductivity, 208, 209, 211, 212, 252, 325
Superplasticity, 83, 84, 88, 117, 348
Surface contaminants, 127

Tailpipe defect, 115
Tantalum, 112, 119, 128, 148, 149, 152, 188, 199, 211–213, 225, 230, 231, 238, 239, 255, 261, 264, 267, 269, 278, 281, 289, 477
   in capacitors, 216, 217
   in chemical processing, 217, 218
   ductility, 216, 218, 219
   mechanical properties variation with temperature, 230
   in ordnance, 218, 219
   properties, 205, 206, 216
   sources, 220
Technetium, 13
   applications, 252, 253

Technetium (*continued*)
  nuclear properties, 243, 252, 253
  physical properties, 243, 244, 252
  sources/production, 253
Terbium, *see* Rare Earths
Texture, 25, 27
  fiber, 33
  sheet, 33
Thallium, 5, 170, 358, 359, 389, 404, 412
  applications and properties, 390, 391
Thermal activation
  cross slip, 46
  dislocation motion, 53, 55
Thermal conductivity, 150, 247. *See also individual elements*
  in Ag, 322–324
  in Al, 360, 363, 364
  in Cu, 306, 307
Thermal expansion. *See also individual elements*
  strain hardening effect, 33
Thermionic generators, 148
Thermoelectric devices, 426, 427
Thixocasting, 109
Thorium, 5, 168, 172, 291, 300, 435, 446, 469
  applications, 451, 454, 455
  bonding of 5f electrons, 451, 453, 454
  nuclear properties, 451, 454, 455
  physical properties, 451–454
  sources, 455, 456
  toxicity, 455
Thulium, *see* Rare Earths
Tin, 7, 8, 107, 129, 168, 177, 188, 199, 200, 208, 209, 211, 212, 220, 230, 234, 255, 303, 305, 306, 308, 310, 311, 314, 324, 325, 327, 334, 343, 355, 365, 389, 390, 392, 394, 395, 400, 401, 412–417, 423, 428, 429, 478
  applications, 404–409
  coatings, 405
  crystal structures, 402, 403
  properties, 402–405
  solders, 406, 407
  sources, 409
  whiskers, 409, 410
Tin cry, 25
Titanium, 13, 14, 49, 70, 71, 87, 93, 102, 103, 106, 110, 111, 115, 116, 118, 120, 125, 128, 131, 143, 152, 169, 173, 176, 177, 198, 199, 201, 204–207, 210, 211, 223, 225, 231, 245, 264, 265, 267, 268, 274, 277, 278, 281, 289, 296, 309, 322, 333, 346, 355, 362, 366, 376, 379, 383, 400, 402, 441, 455, 460, 461, 474–478, 483, 484, 486
  alpha alloys, 188, 190
  alpha-beta alloys, 191–193
  Bauschinger effect, 196, 197
  beta alloys, 190, 191
  applications, 179, 180, 193
  Cambridge process, 180, 181

commercial purity Ti, 187, 188
corrosion behavior, 184, 185, 188, 189
interstitial impurities, 182, 183
Kroll process, 180, 182
mechanical properties, 185–187, 196, 197
nitinol shape memory alloy, 194, 195
phase transformations, 187
processing, 184, 185
properties, 179–187
sources, 194
TiO_2, 180–185
Torch welding, 124
Toxicity
  Ac, 471
  Am, 471, 472
  As, 270, 419, 421–423
  Ba, 178
  Be, 157, 158, 163, 164, 477
  Bi, 419, 429
  Cd, 285, 337, 349–352
  Cf, 472, 473
  Cm, 472
  Cr, 224
  Co, 269, 270
  Cu, 307, 315
  Ga, 389
  Hg, 337, 353–356
  Mg alloys, 170
  Mn, 250
  Ni, 285
  Np, 471
  Os, 297–299
  Pa, 471
  Pb, 327, 405, 407, 410, 412, 415–417
  Po, 432
  Pu, 466, 467
  Ra, 178
  Rn, 471
  Ru, 293, 298, 299
  Sb, 419, 425, 426
  Sn (organotin), 409
  Th, 455
  Tl, 391,
  U, 462–464
  V, 205, 206, 209
  WF_6, 240
  Zn, 347
Transmutation, 101
Tungsten, 13, 17, 49, 86, 93, 104, 109, 111, 128, 131, 148, 153, 168, 218, 221, 225, 227, 228, 230, 232, 252–256, 264, 266, 278, 281, 289, 300, 324, 354, 415, 429, 455, 463
  applications, 236–242
  filaments, 240–242
  hydrostatic pressure, effects on, 235, 236
  mechanical properties variation with temperature, 230, 237
  oxide, 233, 234

properties, 233–236
WC (hard metal), 240
Twin(s), Appendix B1
  annealing, 287
  in Be, 159
  in Mg, 169
  mechanical, 25, 159, 169, 198, 201–203
  mirror plane symmetry, 26
  plane, 25, 159, 169
  in Zr, 198, 201–203

Ultrasonic
  machining, 120
  welding, 127
Uranium, 5, 16, 112, 146, 152, 157, 163, 173, 174, 178, 199, 243, 253, 273, 291, 292, 430, 431, 435, 446, 469–471
  anisotropy, 458–460
  applications, 451, 456–458, 460–462
  bonding of 5f electrons, 451, 453, 454
  depleted, 16, 96, 453, 457, 458, 462–464
  nuclear properties, 451, 453, 456–458, 462
  physical properties, 451–454, 458–462
  sources, 455, 456, 464, 468
  toxicity, 464

Vacancy(-ies), 19
  clusters, 20
  diffusion in creep, 80–84
  dislocation climb, 53
  in irradiated metal, 99
van Arkel-de Boer process, 128
Vanadium, 14, 118, 128, 173, 176, 191–193, 211, 486
  ferrovanadium in steel, 205–208
  in fusion reactors, 208
  properties, 205, 206
  sources, 209
  superconductivity, 208, 209
  toxicity, 209
Viscous flow, 84, 91, 94
Vitreloy, 93
Voids, 27
  in composites, 141
  in creep, 78
  in ductile fracture, 68
  in irradiated metal, 99, 101
Volume defects, 27
von Mises criterion, 50, 159

Wash, 104
Welding, *see* Joining
Whisker, 28, 131, 409, 410
Work hardening, 26, 368
  in creep, 79
  in machining, 119

Yield point, 76
Yield stress
  work hardening effect, 32
Young's modulus, *see* Modulus of elasticity
Ytterbium, *see* Rare Earths
Yttrium, *see* Rare Earths

Zinc, 5, 25, 33, 49, 51, 65, 66, 97, 107, 108, 126, 129, 162, 166, 168–173, 175, 177, 223, 224, 229, 231, 243, 249, 250, 259, 270, 273, 303, 305, 308, 309, 316–319, 324, 327–329, 333, 349–352, 355, 357, 363–365, 367, 369, 375, 376, 378, 379, 382, 389–391, 398, 401, 404–407, 410, 412, 417, 421, 422, 429, 439, 440, 445, 450, 462, 481, 488
  as alloy addition to other metals, 343
  in the biosphere, 346, 347
  coatings on steel, 337, 339–343
  creep, 339, 344, 345
  die-casting alloys, 343–346
  properties, 337–339
  recovery/recrystallization, 340
  sources, 347, 348
  superplasticity, 348
Zirconium, 49, 92, 94, 101, 125, 128, 152, 163, 167, 168, 170–173, 176, 177, 204, 205, 210, 211, 231, 232, 238, 239, 261, 263, 264, 274, 278, 289, 291, 306, 352, 364, 378–381, 385, 386, 393, 402, 405, 441, 448, 449, 450, 455, 462, 466, 467, 475–477, 486
  alloys for nuclear use, 199–201
  hydrides in Zircaloy, 200, 201
  mechanical properties, 198, 201–203
  neutron capture, 199
  properties, 179, 197–199
  pure Zr, 199
  radiation damage, 202, 203
  sources, 201
  twinning, 198, 201–203

TA 479.3 .R84 2005
Russell, Alan M., 1950-
Structure-property relations
 in nonferrous metals